Integral methods in science and engineering

CHAPMAN & HALL/CRC
Research Notes in Mathematics Series

Submission of proposals for consideration
Suggestions for publication, in the form of outlines and representative samples, are invited by the Editorial Board for assessment. Intending authors should approach one of the main editors or another member of the Editorial Board, citing the relevant AMS subject classifications. Alternatively, outlines may be sent directly to the publisher's offices. Refereeing is by members of the board and other mathematical authorities in the topic concerned, throughout the world.

Preparation of accepted manuscripts
On acceptance of a proposal, the publisher will supply full instructions for the preparation of manuscripts in a form suitable for direct photo-lithographic reproduction. Specially printed grid sheets can be provided. Word processor output, subject to the publisher's approval, is also acceptable.

Illustrations should be prepared by the authors, ready for direct reproduction without further improvement. The use of hand-drawn symbols should be avoided wherever possible, in order to obtain maximum clarity of the text.

The publisher will be pleased to give guidance necessary during the preparation of a typescript and will be happy to answer any queries.

Important note
In order to avoid later retyping, intending authors are strongly urged not to begin final preparation of a typescript before receiving the publisher's guidelines. In this way we hope to preserve the uniform appearance of the series.

CRC Press UK
Chapman & Hall/CRC Statistics and Mathematics
Pocock House
235 Southwark Bridge Road
London SE1 6LY
Tel: 020 7450 7335

B Bertram, C Constanda,
and A Struthers
(Editors)

Integral methods in science and engineering

CHAPMAN & HALL/CRC

Boca Raton   London   New York   Washington, D.C.

Library of Congress Cataloging-in-Publication Data

Integral methods in science and engineering / B. Bertram, C. Constanda, A. Struthers, editors; corresponding editor, C. Constanda.

     p. cm. -- (Chapman & Hall/CRC research notes in mathematics series)

    Includes bibliographical references.

    ISBN 1-58488-146-1 (alk. paper)

    1. Integral equations--Numerical solutions. 2. Science--Mathematics. 3. Engineering mathematics. I. Bertram, Barbara. II. Constanda, C. (Christian) III. Struthers, A. (Allan A.) IV. Series.

QA431 .I49   2000

515′.624—dc21   99-086281

Preface

The international conferences on Integral Methods in Science and Engineering (IMSE) are a forum for the presentation of frontline research that makes use of integration in its various guises as a main investigative technique.

The first two IMSE meetings were held at the University of Texas at Arlington in 1985 and 1990. They were followed by IMSE93 (organized by Tohoku University, Sendai, Japan), IMSE96 (hosted by the University of Oulu, Finland), and IMSE98, which took place at Michigan Technological University, Houghton, Michigan. The IMSE conferences are now a well-established biennial event that offers scientists and engineers a platform for communicating their latest results and for comparing and contrasting the merits and applicability of a wide category of important mathematical methods.

IMSE98 continued the tradition of friendly atmosphere established on previous occasions, which, as before, was created through meticulous preparation by the Local Organizing Committee:

B. Bertram (Mathematical Sciences), Chairman,
A. Narain (Mechanical Engineering and Engineering Mechanics),
D. Sikarskie (Mechanical Engineering and Engineering Mechanics), *
A. Struthers (Mathematical Sciences),
M. Vable (Mechanical Engineering and Engineering Mechanics).

The participants are indebted to them all for their efforts, and to the host institution and other sponsoring agencies for their generosity. Advice and general guidance were provided by the International Steering Committee.

Plans for future conferences are well in hand. Thus, IMSE2000 will be held in Banff, Canada, IMSE2002 at the University of California-Berkeley, and IMSE2004 at the University of St. Etienne, France. Details of IMSE2000 can be found on the Internet at the address http://mece.ualberta.ca/IMSE2000.

This volume contains 3 invited papers and 57 of the contributed papers presented in Houghton, which have undergone peer review. In each category, the papers are arranged in alphabetical order by (first) author's surname.

The editors would like to acknowledge the help of the referees, the secretarial support effort, and the efficient handling of the publication process by the staff of Chapman & Hall/CRC. Special thanks are due to George Beckett and Ronnie Wallace, who, with commendable patience and good humour, employed computer wizardry to make some of the diagrams electronically printable.

Christian Constanda

---

* Sadly, David Sikarskie passed away in 1999.

Contents

Invited papers

A.G. GIBSON

Mathematical modeling of N-body quantum scattering processes

1. Introduction

Quantum scattering theory is the mathematical modeling of experiments in atomic and nuclear physics in which a target is bombarded by a projectile and the outcome is measured. The goal of the mathematical theory is to predict the probability of the various possible outcomes. It is especially difficult to model the scattering process in cases where the target breaks up into sub-particles. In this paper we describe some of the mathematical methods that have been used to model the scattering of N nonrelativistic particles for $N \geq 2$.

A starting point of the theory is the Schrödinger equation. However, since the boundary conditions are imposed at times $t \to \pm\infty$, it is useful to reformulate the problem as a system of coupled singular integral equations. For $N = 2$ there are the Lippmann-Schwinger equations (cf. p. 135 of [1]), and for $N = 3$ there are the Faddeev equations [2].

A new system of integral equations (Eq. (26) below) has recently been rigorously derived by Colston Chandler and Archie G. Gibson that is valid for all $N \geq 2$ (cf. [3–7] and references cited therein). This system of equations has many nice mathematical properties such as a compact kernel and existence and uniqueness of solution. Either product integration or B-spline numerical methods may be used for solving these equations. The nonhomogeneous and kernel terms are defined by certain high dimensional integrals, which are approximated by choosing an orthonormal basis on the surface of certain unit kinetic energy hyperspheres. Two different three-particle test problems have confirmed that these equations may be solved and yield satisfactory numerical solutions.

2. Scattering processes

Consider a nonrelativistic system of N particles moving in three dimensions and interacting via short-range pair potentials. A *channel* α is a specification of a partition of the particles into n_α bound clusters. Let 0 denote the partition into N free particles, and let N denote the set of all N particles. Denote the other possible channels by $\alpha, \beta, \gamma, \ldots$.

For example, suppose that an *incoming* two-cluster channel consists of a deuteron d (a bound state of one proton p and one neutron n) and an alpha particle α (a bound state of two protons and two neutrons). After the interaction of the deuteron with the alpha particle several possible *outgoing* channels are possible, depending on the total energy of the system. One possible outcome is $d+\alpha$, which is called the *elastic channel* if the protons and neutrons remain in the same clusters or the *exchange channel* if some protons and/or neutrons have switched clusters. A second possible outcome is the *rearrangement channel* $^3H + {}^3He$. At higher energies the *breakup channel* $n+p+\alpha$ also becomes possible.

This work was partly supported by the National Science Foundation grant PHY-9505615.

3. A mathematical model

We begin by assuming *quantum mechanics*, wherein *states* are vectors $\psi \in \mathcal{H}_N \equiv \mathcal{L}^2(\mathbf{R}^{3N})$, and *observables* are self-adjoint operators on $\mathcal{H}_N$.

The particular observable that is the *total Hamiltonian* of the system is the self-adjoint operator

$$H_N \equiv H_0 + \bar{V}_0 \equiv \sum_{i=1}^{N} \frac{-\Delta_i}{2m_i} + \sum_{i<j} V_{ij}\,, \tag{1}$$

where H_0 is the *free Hamiltonian* (a self-adjoint operator) for N noninteracting particles and V_{ij} are given *pair potentials*. The state $\psi(t) \in \mathcal{H}_N$ evolves in time as a solution

$$\psi(t) = e^{-iH_N t}\psi(0) \tag{2}$$

of *Schrödinger's equation*

$$i\frac{\partial \psi}{\partial t} = H_N \psi\,. \tag{3}$$

But this equation is solvable only in very simple cases. A serious problem is that the boundary conditions are at $t = \pm\infty$, not at $t = 0$. For scattering problems some other approach is needed.

4. Channel subspaces and P_α operators

For each channel α with n_α clusters, and the center of mass motion removed, there is a set of Jacobi momenta that divides into the *internal* (bound state) *momenta* $\vec{\mathbf{p}}_\alpha \equiv (\vec{\mathbf{p}}_\alpha^{(1)}, \ldots, \vec{\mathbf{p}}_\alpha^{(N-n_\alpha)})$, and the *external momenta* $\vec{\mathbf{q}}_\alpha \equiv (\vec{\mathbf{q}}_\alpha^{(1)}, \ldots, \vec{\mathbf{q}}_\alpha^{(n_\alpha-1)})$. The *channel subspace* $\mathcal{H}_\alpha \subset \mathcal{H}_N$ is the closed linear span of vectors of the form

$$\hat{\varphi}_\alpha(\vec{\mathbf{p}}_\alpha) f(\vec{\mathbf{q}}_\alpha)\,, \tag{4}$$

where $\hat{\varphi}_\alpha \in \mathcal{L}^2(\mathbf{R}^{3(N-n_\alpha)})$ are products of orthonormal eigenfunctions of the bound clusters, and f are arbitrary function in $\mathcal{L}^2(\mathbf{R}^{3(n_\alpha-1)})$. The *channel projection operators* P_α are the orthogonal projections of $\mathcal{H}_N$ onto $\mathcal{H}_\alpha$. Note that $P_0 = I_N$ since the free partition does not have any bound states.

Each channel α has an associated *channel Hamiltonian* H_α,

$$H_\alpha \equiv \epsilon_\alpha + \sum_{m=1}^{n_\alpha-1} \frac{|\vec{\mathbf{q}}_\alpha^{(m)}|^2}{2\mu_\alpha^{(m)}}\,, \tag{5}$$

where ϵ_α is the *threshold (bound state) energy* for channel α, and the sum is the *kinetic energy*, with $\mu_\alpha^{(m)}$ denoting the *reduced masses*.

5. The scattering operator S

The *asymptotic condition of scattering theory* says that in the remote past and distant future the particles behave like noninteracting clusters each of which is in a specific quantum mechanical bound state, i.e., given a $\phi_\alpha \in \mathcal{H}_\alpha$ there exists a $\psi \in \mathcal{H}_N$ such that

$$\|e^{-iH_N t}\psi - e^{-iH_\alpha t}\phi_\alpha\| \to 0 \quad \text{as} \quad t \to -\infty\,, \tag{6}$$

and given this $\psi \in \mathcal{H}_N$ there exists a $\phi_\beta \in \mathcal{H}_\beta$ such that

$$\|e^{-iH_\beta t}\phi_\beta - e^{-iH_N t}\psi\| \to 0 \quad \text{as} \quad t \to \infty\,. \tag{7}$$

Equations (6) and (7) define the *channel wave operators*

$$\Omega_\alpha^\pm \equiv \mathrm{s-}\lim_{t\to\pm\infty} e^{iH_N t} e^{-iH_\alpha t} , \tag{8}$$

which in turn define the *scattering operators* $S_{\beta\alpha}$ from channel α to channel β by

$$\phi_\beta = S_{\beta\alpha}\phi_\alpha \equiv \Omega_\beta^{+*}\Omega_\alpha^- \phi_\alpha . \tag{9}$$

The matrix $S \equiv [S_{\beta\alpha}]$ is the *full system scattering operator*.

The asymptotic condition of scattering theory can now be restated in terms of the scattering operators $S_{\beta\alpha}$, and this is the main problem of scattering theory.

Main problem. Given an input state $\phi_\alpha \in \mathcal{H}_\alpha$, determine all possible output states $\phi_\beta = S_{\beta\alpha}\phi_\alpha \in \mathcal{H}_\beta$, i.e., find the scattering operators $S_{\beta\alpha}$ for fixed α and for all β.

6. Transition operators

If the projectile misses the target and no scattering occurs, then S acts like the identity operator I. In practice, therefore, it is better to compute the matrix elements $T_{\beta\alpha}$ of the *transition operator* T defined implicitly by

$$S = I - 2\pi i T, \quad \text{and} \quad S_{\beta\alpha} = \delta_{\beta\alpha} - 2\pi i T_{\beta\alpha} . \tag{10}$$

The operators T and S have spectral representations (direct integral decompositions) with respect to the *total energy E* of the system. Since the total energy is conserved, it suffices to compute the *on-shell transition operator matrix elements* $\widehat{T}_{\beta\alpha}(E)$ for fixed E and α and for all open channels β.

The operators $\widehat{T}_{\beta\alpha}(E)$ are directly related to the physical scattering observables, such as the cross sections measured in accelerator experiments. The challenge then is to obtain some solvable equations whose solutions determine $\widehat{T}_{\beta\alpha}(E)$. The complexity of the problem increases with the number N of particles.

7. The Lippmann-Schwinger equation ($N = 2$)

For $N = 2$ the only open channel is the free channel 0, and

$$H_N = H_0 + \bar{V}_0 , \qquad \text{with} \qquad \bar{V}_0 = V_{12} . \tag{11}$$

The *Lippmann-Schwinger equation* in operator form is

$$T_{00}(z) = \bar{V}_0 + \bar{V}_0 R_0(z) T_{00}(z) , \tag{12}$$

where $R_0(z) \equiv (z - H_0)^{-1}$, with $z = E + iy$ [1, p. 135]. If $y \neq 0$, and $\bar{V}_0 \in \mathcal{L}^2(\mathbf{R}^3)$, then this equation has a unique solution and a compact kernel. It may be solved by Fredholm integral equation methods. The on-shell transition operator $\widehat{T}_{00}(E)$ is obtained from $T_{00}(E + iy)$ by taking the limit $y \to 0$.

Many excellent 2-body calculations have been made using the LS equation.

8. The Faddeev equations ($N = 3$)

When $N = 3$ the total Hamiltonian is of the form

$$H_N = H_0 + \bar{V}_0\,, \qquad \text{with} \qquad \bar{V}_0 \equiv V_{12} + V_{13} + V_{23}\,, \tag{13}$$

where V_{ij} denotes the pair interaction between particles i and j. The Faddeev equation solution procedure consists of the following three steps [2].

1. First, solve

$$t_\alpha(z) = V_\alpha + V_\alpha R_0(z) t_\alpha(z) \tag{14}$$

   for all *two-cluster transition operators* $t_\alpha(z)$, where α ranges over the clusterings $(12, 3), (13, 2), (23, 1)$, with $V_\alpha = V_{12}, V_{13}, V_{23}$, respectively.

2. Next, solve the 3-body system (the Lovelace form of the Faddeev equations)

$$U_{\beta\alpha}(z) = \bar{V}_\alpha + \sum_{\gamma \neq \beta} t_\gamma(z) R_0(z) U_{\gamma\alpha}(z) \tag{15}$$

   for the *transition operators*

$$U_{\beta\alpha}(z) \equiv (z - H_\beta) R_N(z) \bar{V}_\alpha \tag{16}$$

   acting on $\mathcal{H}_N$.

3. Finally,

$$T_{\beta\alpha}(z) = P_\beta U_{\beta\alpha}(z) P_\alpha\,, \tag{17}$$

   and

$$\widehat{T}_{\beta\alpha}(E) = \lim_{y \to 0} T_{\beta\alpha}(E + iy)\,. \tag{18}$$

The Faddeev equations have unique solutions and the square of the kernel is compact for $y \neq 0$, allowing use of Fredholm integral methods.

Many excellent three-body scattering calculations have been made using the Faddeev equations.

9. Yakubovskii's generalization ($N = N$)

In 1967 O.A. Yakubovskii, a student of Faddeev, generalized the Faddeev method to $N = N$ [8]. However, the "tree" structure, and the fact that the unknowns are on the full space $\mathcal{H}_N$ make these equations prohibitively complicated. For example, for $N = 8$ distinguishable particles, the final system consists of 1,587,600 integral equations containing integrals of dimension 18!

A few calculations have been made for $N = 4$, but there have been no scattering calculations for $N > 4$.

6

10. The Chandler-Gibson approach ($N = N$)

Since it is only the operators

$$T_{\beta\alpha}(z) = P_\beta U_{\beta\alpha}(z)P_\alpha = P_\beta(z - H_\beta)R_N(z)\bar{V}_\alpha P_\alpha \tag{19}$$

that are needed, why use equations for the operators $U_{\beta\alpha}(z)$? Our first step was to derive the *transition operator equations* [3]

$$T_{\beta\alpha}(z) = P_\beta \bar{V}_\alpha P_\alpha + P_\beta \bar{V}_\beta (\sum_\delta P_\delta)^{-1} \sum_\gamma P_\gamma R_\gamma(z) T_{\gamma\alpha}(z) \,, \tag{20}$$

where $R_\gamma(z) \equiv (z - H_\gamma)^{-1}$. The inverse of $\sum_\delta P_\delta$ in this equation can be avoided by using the alternative transition operator

$$M_{\beta\alpha}(z) \equiv P_\beta(z - H_\beta)(\sum_\delta P_\delta)^{-1} R_N(z)\bar{V}_\alpha P_\alpha \,. \tag{21}$$

Let $\bar{\delta}_{\beta\alpha} \equiv 1 - \delta_{\beta\alpha}$. Our *M-equations* are then [5]

$$M_{\beta\alpha}(z) = P_\beta \bar{V}_\alpha P_\alpha + \sum_\gamma P_\beta \left[\bar{V}_\gamma R_\gamma(z) - \bar{\delta}_{\beta\alpha} \right] P_\gamma M_{\gamma\alpha}(z) \,. \tag{22}$$

Solution strategy. Approximate the P_α by a sequence of orthogonal projection operators Π_α such that $\Pi_\alpha \vec{s} P_\alpha$ (strongly). This is equivalent to approximating the spaces $\mathcal{H}_\alpha$ by $\mathcal{H}_\alpha^\pi \equiv P_\pi \mathcal{H}_\alpha \subset \mathcal{H}_\alpha$, and the total Hamiltonian H_N by $H_\pi \equiv P_\pi H_N P_\pi$, where P_π is the orthogonal projection of $\mathcal{H}_N$ onto the closure of the range of $\sum_\alpha \Pi_\alpha$.

11. Approximate transition amplitudes

In our approach it is important to use *kinetic energy hyperspherical variables (KEHS)* $(\mu, \hat{\mathbf{k}}_\alpha)$, where

$$\mu \equiv \sum_{m=1}^{n_\alpha - 1} \frac{|\vec{\mathbf{q}}_\alpha^{(m)}|^2}{2\mu_\alpha^{(m)}} \,, \tag{23}$$

is the kinetic energy, and

$$\hat{\mathbf{k}}_\alpha \equiv (\hat{\mathbf{k}}_\alpha^{(1)}, \dots, \hat{\mathbf{k}}_\alpha^{(n_\alpha - 1)}), \quad \text{with} \quad \hat{\mathbf{k}}_\alpha^{(m)} \equiv \vec{\mathbf{q}}_\alpha^{(m)} / [2\mu_\alpha^{(m)}\mu]^{1/2} \,. \tag{24}$$

The quantity $\hat{\mathbf{k}}_\alpha$ ranges over the *unit kinetic energy hypersphere* Γ_α in $\mathbf{R}^{3(n_\alpha - 1)}$, and μ ranges over the half-line $\mathbf{R}^+$. The *square root of the Jacobian* of the transformation from the variables $\mathbf{q}_\alpha$ to the variables $(\hat{\mathbf{k}}_\alpha, \mu)$ is denoted by $\nu_\alpha(\mu)$.

For each channel α let $\{\chi_{\alpha i}(\hat{\mathbf{k}}_\alpha)\}$, $i = 1, 2, \dots, n_i$ be a finite subset of some infinite bounded *orthonormal basis* on the hypersphere Γ_α. Let the corresponding *approximate on-shell transition operator matrix elements* be denoted by $\widehat{T}_{\beta\alpha}^\pi(E; \hat{\mathbf{k}}_\beta, \hat{\mathbf{k}}_\alpha)$. Then

$$\widehat{T}_{\beta\alpha}^\pi(E; \hat{\mathbf{k}}_\beta, \hat{\mathbf{k}}_\alpha) = \sum_{j=1}^{n_j} \sum_{i=1}^{n_i} \chi_{\beta j}(\hat{\mathbf{k}}_\beta) \widetilde{\mathcal{M}}_{\beta j, \alpha i}^\pi(e_\beta, e_\alpha) \chi_{\alpha i}^*(\hat{\mathbf{k}}_\alpha) \,, \tag{25}$$

with $e_\alpha \equiv E - \epsilon_\alpha$, where E is the total energy, and ϵ_α is the α channel threshold energy.

12. Half-on-shell equations

The complex-valued quantities $\widetilde{\mathcal{M}}^{\pi}_{\beta j,\alpha i}(e_\beta, e_\alpha)$ in Eq. (25) are the *on-shell matrix elements* of corresponding *half-on-shell matrix elements* $\widetilde{\mathcal{M}}^{\pi}_{\beta j,\alpha i}(\lambda, e_\alpha) \in \mathcal{L}^2(\mathbf{R}^+)$, and these half-on-shell matrix elements are the unique solution of the following system of *half-on-shell $\mathcal{M}^\pi$-equations* [6,7]:

$$\widetilde{\mathcal{M}}^{\pi}_{\beta j,\alpha i}(\lambda, e_\alpha) = \tilde{\mathcal{A}}^{\pi}_{\beta j,\alpha i}(\lambda, e_\alpha) + \sum_{\gamma,k}\left[\fint_0^\infty d\eta \frac{\tilde{\mathcal{A}}^{\pi}_{\beta j,\gamma k}(\lambda, \eta)}{e_\gamma - \eta}\nu_\gamma^2(\eta)\widetilde{\mathcal{M}}^{\pi}_{\gamma k,\alpha i}(\eta, e_\alpha) \right.$$

$$\left. -i\pi\tilde{\mathcal{A}}^{\pi}_{\beta j,\gamma k}(\lambda, e_\gamma)\nu_\gamma^2(e_\gamma)\widetilde{\mathcal{M}}^{\pi}_{\gamma k,\alpha i}(e_\gamma, e_\alpha) \right], \qquad (26)$$

where $\fint$ is a Cauchy principal value integral.

13. Properties of the $\mathcal{M}^\pi$-equations

Main assumptions

- $V_{ij} \in \mathcal{L}^2(\mathbf{R}^3)$ for each pair of particles.

- $\chi_{\alpha i} \in \mathcal{L}^2(\Gamma_\alpha) \cap \mathcal{L}^\infty(\Gamma_\alpha)$ for each channel α and $i = 1, 2, \ldots, n_i$.

Properties [7,9]

- The kernel of the $\mathcal{M}^\pi$-equations are kernels of a *compact integral operator* when considered on a certain Hilbert space of Hölder continuous functions.

- A solution *exists* and is *unique*.

- Sequences of approximate on-shell transition operators $\{\widehat{T}^\pi(E)\}$ *converge* strongly to $\{\widehat{T}(E)\}$.

- The $\mathcal{M}^\pi$-equations are of the *same algebraic form* for all $N \geq 2$.

- The integral equations have *1-dimensional integrals* for all N.

- The *complexity* of the equations depends on the maximum number of clusters rather than the number N of particles.

14. Solution of the $\mathcal{M}^\pi$-equations

The system of integral equations in Eq. (26) have Cauchy singularities at the fixed points e_γ. A comparison of solution methods for integral equations with fixed Cauchy singularities was made by Bertram and Gibson and presented at IMSE 1986 [10]. Other investigations of solution methods for these equations were made in [11-16].

Some conclusions

- *Jacobians.* It is better to put in the Jacobians $\nu_\gamma^2(\eta)$ as shown in Eq. (26).

8

- *Intervals* $[0, \infty)$. It is best to use $x \in [-1, 1]$ and map to $\eta \in [0, \infty)$ with

$$\eta = e_\gamma \left(\frac{1 + x}{1 - x} \right)^2 . \qquad (27)$$

- *Knots.* Our most successful choice of knots was Chebyshev knots on $[-1, 1]$.

- *Most Successful Numerical Methods*

  1. *Product Integration* [9]. The singularity of the kernel is absorbed into the weight functions, and piecewise linear interpolating polynomials are used. The fastest convergence was obtained when product integration was combined with the Atkinson-Brakhage iteration method.

  2. *B-Spline Collocation Method* [12–14]. An approximate basis of $\hat{n} + 1$ cubic B-splines $B_{\hat{j}}(x(\lambda))$ on $[-1, 1]$ is constructed. Assuming

$$\widetilde{\mathcal{M}}^{\pi \hat{n}}_{\beta j, \alpha i}(\lambda, e_\alpha) = \sum_{\hat{j}=0}^{\hat{n}+1} c_{\beta j \hat{j}} B_{\hat{j}}(x(\lambda)) , \qquad (28)$$

the B-spline coefficients $c_{\beta j \hat{j}}$ are chosen so that the residual $r_{\beta j}(\bar{\lambda}_{\hat{j}})$ equals zero at all collocation points $\bar{\lambda}_{\hat{j}}$ for all $\hat{j} = 0, 1, \ldots, \hat{n}+1$, and for all β and j. The integrals containing singular integrals are evaluated using singularity subtraction.

15. The input terms

The input terms in the $\mathcal{M}^\pi$-equations (26) are of the form

$$\widetilde{\mathcal{A}}^\pi_{\beta j, \alpha i}(\lambda, \mu) = \widetilde{\mathcal{B}}^\pi_{\beta j, \alpha i}(\lambda, \mu) - \widetilde{\mathcal{C}}^\pi_{\beta j, \alpha i}(\lambda, \mu)(e_\alpha - \mu) , \qquad (29)$$

where $\widetilde{\mathcal{B}}^\pi_{\beta j, \alpha i}(\lambda, \mu)$ is the kernel of the operator

$$\widetilde{\mathcal{B}}^\pi_{\beta j, \alpha i} \equiv \tilde{\rho}^\pi_{\beta j} \bar{V}_\alpha \tilde{\rho}^{\pi *}_{\alpha i} \qquad (the\ Born\ terms) , \qquad (30)$$

$\widetilde{\mathcal{C}}^\pi_{\beta j, \alpha i}(\lambda, \mu)$ is the kernel of the operator

$$\widetilde{\mathcal{C}}^\pi_{\beta j, \alpha i} \equiv \bar{\delta}_{\beta \alpha} \tilde{\rho}^\pi_{\beta j} \tilde{\rho}^{\pi *}_{\alpha i} \qquad (the\ overlap\ terms) , \qquad (31)$$

and $\tilde{\rho}^\pi_{\alpha i} : \mathcal{H}_N \to \mathcal{L}^2(\mathbf{R}^+)$ is the *projection operator*

$$(\tilde{\rho}^\pi_{\alpha i} \psi_N)(\mu) \equiv \int d\vec{\mathbf{p}}_\alpha d\hat{\mathbf{k}}_\alpha \, \hat{\varphi}^*_\alpha(\vec{\mathbf{p}}_\alpha) \chi^*_{\alpha i}(\hat{\mathbf{k}}_\alpha) \psi_N(\vec{\mathbf{p}}_\alpha, \hat{\mathbf{k}}_\alpha, \mu) . \qquad (32)$$

16. On choosing the $\{\chi_{\alpha i}(\hat{\mathbf{k}})\}$ basis elements

16.1. Two-cluster channels. If the N particles are bound into two clusters, the KEHS coordinates are

$$\mu = \frac{|\vec{\mathbf{q}}_\alpha^{(1)}|^2}{2\mu_\alpha^{(1)}}, \quad \text{and} \quad \hat{\mathbf{k}}_\alpha = \frac{\vec{\mathbf{q}}_\alpha^{(1)}}{\sqrt{2\mu_\alpha^{(1)}\mu}}. \tag{33}$$

In this case we choose

$$\chi_{\alpha i}(\hat{\mathbf{k}}_\alpha) = Y_{lm}(\hat{\mathbf{k}}_\alpha), \tag{34}$$

where $Y_{lm}(\hat{\mathbf{k}}_\alpha)$ are the *spherical harmonics*.

16.2. Three-cluster channels. If the N particles are bound into three clusters, the KEHS coordinates are

$$\mu = |\vec{\mathbf{q}}_\alpha^{(1)}|^2 + \frac{3|\vec{\mathbf{q}}_\alpha^{(2)}|^2}{4}, \quad \text{and} \quad \hat{\mathbf{k}}_\alpha = \left(\frac{\vec{\mathbf{q}}_\alpha^{(1)}}{\sqrt{\mu}}, \frac{\vec{\mathbf{q}}_\alpha^{(2)}}{\sqrt{4\mu/3}}\right). \tag{35}$$

The two momenta $\vec{\mathbf{q}}_\alpha^{(1)}$ and $\vec{\mathbf{q}}_\alpha^{(2)}$ determine a plane in $\mathbf{R}^6$ with *Euler angles* $(\bar{\alpha}, \bar{\beta}, \bar{\gamma})$. We choose

$$\chi_{\beta LMNnm}(\hat{\mathbf{k}}_\alpha) = \mathcal{D}_{MN}^{(L)}(\bar{\alpha}, \bar{\beta}, \bar{\gamma})\hat{Z}_n^m(\rho, \phi), \tag{36}$$

where $\mathcal{D}_{MN}^{(L)}$ are *rotation matrices* [17], and $\hat{Z}_n^m$ are normalized *Zernicke polynomials* in the *Dalitz variables* (λ, ρ, ϕ) [18,19].

The transformation to the magnitudes $q_\alpha^{(1)}$ and $q_\alpha^{(2)}$ of $\vec{\mathbf{q}}_\alpha^{(1)}$ and $\vec{\mathbf{q}}_\alpha^{(2)}$ and the angle τ_α between $\vec{\mathbf{q}}_\alpha^{(1)}$ and $\vec{\mathbf{q}}_\alpha^{(2)}$ from the *Dalitz variables* (λ, ρ, ϕ) is given by

$$\begin{aligned}
q_\alpha^{(1)} &= (\mu/2)^{1/2}\sqrt{1 - \rho\cos(\phi - \phi_\alpha)}, \\
q_\alpha^{(2)} &= (2\mu/3)^{1/2}\sqrt{1 + \rho\cos(\phi - \phi_\alpha)}, \\
\cos(\tau_\alpha) &= \rho\sin(\phi - \phi_\alpha)/\sqrt{1 - \rho^2\cos^2(\phi - \phi_\alpha)},
\end{aligned} \tag{37}$$

where the angles ϕ_α are defined by

$$\phi_\alpha \equiv \left\{ \begin{array}{ll} 0, & \text{for } \alpha = 1 \\ 4\pi/3, & \text{for } \alpha = 2 \\ -4\pi/3, & \text{for } \alpha = 3 \end{array} \right\}. \tag{38}$$

Then

$$\hat{Z}_n^m(\rho, \phi) \equiv 4\sqrt{n+1}R_n^{|m|}(\rho)\hat{cs}_m(\phi), \tag{39}$$

where $R_n^{|m|}(\rho)$ are the *radial polynomials* of degree n defined for $n - |m|$ an even integer by

$$R_n^{|m|}(\rho) \equiv \sum_{p=0}^{(n-|m|)/2} (-1)^p \frac{(n-p)!}{p![(n+|m|)/2 - p]![(n-|m|)/2 - p]!}\rho^{n-2p}, \tag{40}$$

and the functions $\hat{cs}_m(\phi)$ are the *normalized Fourier series basis functions* defined by

$$\hat{cs}_m(\phi) \equiv \left\{ \begin{array}{ll} 1/\sqrt{2\pi}, & \text{for } m = 0 \\ \cos(m\phi)/\sqrt{\pi}, & \text{for } m > 0 \\ \sin(|m|\phi)/\sqrt{\pi}, & \text{for } m < 0 \end{array} \right\}. \tag{41}$$

17. Numerical calculations

The $\mathcal{M}^\pi$-equations (26) reduce to the integral form of the Lippmann-Schwinger equation (12) when $N = 2$, but they are very different from the Faddeev-Yakubovskii equations for $N \geq 3$. It is, therefore, necessary to test these equations. For our tests we chose two different $N = 3$ models which have been solved previously using the Faddeev equations, giving us some results with which to compare.

In our first numerical test we considered the scattering of a nonrelativistic system of three identical spin zero particles moving in one dimension and interacting through attractive delta-function potentials [14]. In this case the input Born and overlap matrices were evaluated analytically, and the $\mathcal{M}^\pi$-equations were solved used the B-spline collocation method. The computed scattering matrix elements were within 0.5% of the known exact solutions, and the corresponding scattering probabilities were within 0.001% of the exact probabilities both below and above the three-body breakup threshold.

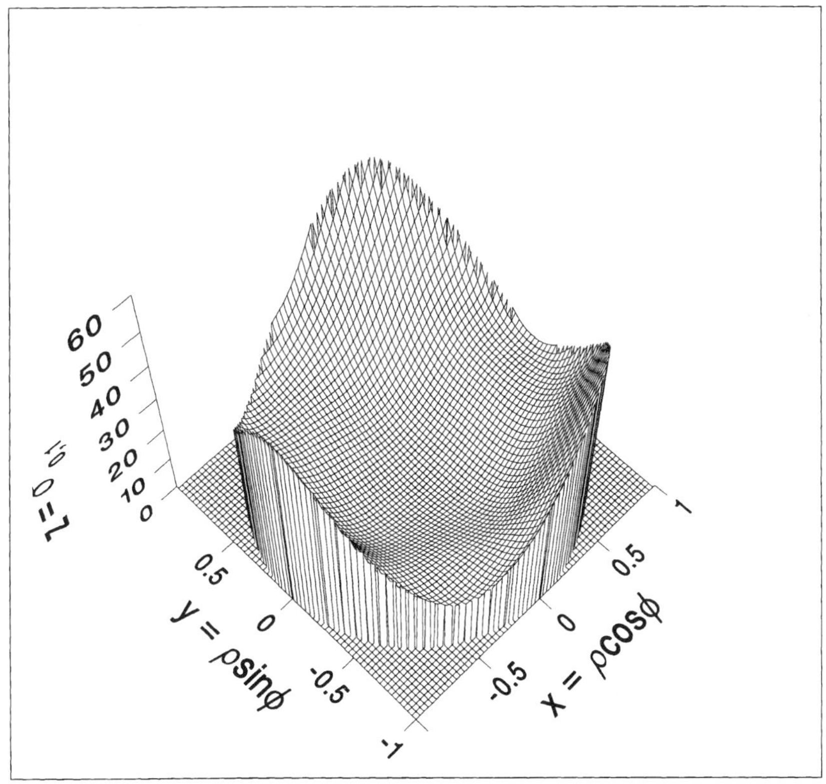

Fig. 1. $\sigma_{0,1}^{(3/2)}(E; \rho, \phi) \times 10^4$ at $E_{\mathrm{lab}} = 42.0$ MeV with $n = 6$.

In our second test we considered the spin-quartet case (i.e., total spin $= 3/2$) of the scattering of a nonrelativistic system of three identical spin-1/2 particles moving in three dimensions and interacting via Malfliet-Tjon I-III type pair potentials [15,16]. This problem has been used previously by several groups as a benchmark problem [20]. We chose the $\chi_{\alpha\iota}(\hat{\mathbf{k}})$ basis elements as shown in Sec. 16 and evaluated the input terms in Eq. (29) analytically as far as possible before numerically integrating the final one- and two-dimensional integrals. We numerically solved the $\mathcal{M}^\pi$-equations (26) using the B-spline collocation method. We then computed the approximate on-shell transition operators $\widehat{T}^\pi_{\beta\alpha}(E; \hat{\mathbf{k}}_\beta, \hat{\mathbf{k}}_\alpha)$ using Eq. (25), and the corresponding *scattering cross sections* defined by

$$\sigma^\pi_{\beta\alpha}(E; \hat{\mathbf{k}}_\beta, \hat{\mathbf{k}}_\alpha) = |\widehat{T}^\pi_{\beta\alpha}(E; \hat{\mathbf{k}}_\beta, \hat{\mathbf{k}}_\alpha)|^2 \,. \tag{42}$$

For example, the quartet scattering cross section at a lab energy of 42.0 MeV for α equal to the two-cluster channel $1 \equiv (1, 23)$ and β equal to the three-cluster breakup channel $0 \equiv (1, 2, 3)$ is shown in Fig. 1.

The interpretation of the *Dalitz plot* in Fig. 1 is the following. The z values are not scaled as probabilities, but they are proportional to the probabilities of the various outcomes. The $2\pi/3$ symmetry is a consequence of the three particles being treated as identical particles, and a rotation of $\pm 2\pi/3$ radians corresponds to a permutation of the three particles. The peak at $\rho = 1$ and $\phi = \pi$ corresponds to $q_1^{(2)} = 0$, which means that particle 1 is stationary in the center of mass frame and particles 2 and 3 are going in opposite directions with equal momenta. This is the most probable outcome at this energy. The valley at $\rho = 1$ and $\phi = 0$ corresponds to $q_1^{(1)} = 0$, which means that particles 2 and 3 are going in the same direction and particle 1 is going in the opposite direction with momentum equal to the sum of the momenta of particles 2 and 3. The valley at $\rho = 0$ corresponds to the particles separating with equal momenta at angles of $2\pi/3$, i.e., they form an equilateral triangle. This is the first time that such a complete "picture" of the breakup channel has been obtained.

Our numerical calculations [15,16] are consistent with the known benchmark results [20] and establish that the Chandler-Gibson equations are an attractive option for N-body scattering problems.

Acknowledgments. I am deeply indebted to Colston Chandler for many years of stimulating collaboration. We have benefited greatly from interaction with Gy. Bencze, G.H. Berthold, B. Bertram, G.W. Pletsch, H.J. Taijeron, A.J. Waters, and numerous other colleagues.

References

1. J.R. Taylor, *Scattering theory*, Wiley, New York, 1972.

2. L.D. Faddeev: *Mathematical aspects of the three-body problem in quantum scattering theory*, Israel Program for Scientific Translations, Jerusalem, 1965.

3. C. Chandler and A.G. Gibson, N-body quantum scattering theory in two Hilbert spaces. I. The basic equations, *J. Math. Phys.* **18** (1977), 2336–2346.

4. C. Chandler and A.G. Gibson, N-Body quantum scattering theory in two Hilbert spaces. III. Theory of approximations, *J. Functional Anal.* **52** (1983), 80–105.

5. C. Chandler and A.G. Gibson, N-body quantum scattering theory in two Hilbert spaces. IV. Approximate equations, *J. Math. Phys.* **25** (1984), 1841–1856.

6. C. Chandler and A.G. Gibson, N-body quantum scattering theory in two Hilbert spaces. V. Computation strategy, *J. Math. Phys.* **30** (1989), 1533–1544; **34** (1993), 886.

7. C. Chandler and A.G. Gibson, N-body quantum scattering theory in two Hilbert spaces. Integral equations, *Few-Body Systems* **23** (1998), 223–258.

8. O.A. Yakubovskii, On the integral equations in the theory of N particle scattering, *Soviet J. Nuclear Phys.* **5** (1967), 937–942.

9. B. Bertram, On the product integration method for solving singular integral equations in scattering theory, *J. Comput. Appl. Math.* **25** (1989), 79–92.

10. B. Bertram and A.G. Gibson, Comparison of solution methods for integral equations with Cauchy singularities, in *Integral Methods in Science and Engineering*, F.R. Payne et al., eds., Hemisphere, Washington, DC, 1986, 99–105.

11. A.G. Gibson, B. Bertram and C. Chandler, On the numerical solution of the CG-equations for a six boson problem, in *Proceedings of the 10th European Symposium on the Dynamics of Few-Body Systems*, P. Doleschall, ed., KFKI, Budapest, 1985, 47–49.

12. G.H. Berthold, C. Chandler, A..G. Gibson and H.J. Taijeron, B-spline comparison of three forms of the CG-equations, in *Few Body XII*, B.K. Jennings, Ed., TRIUMF, Vancouver, BC, 1989, F18.

13. A.G. Gibson, A.J. Waters, G.H. Berthold and C. Chandler, A new $\mathcal{K}$-matrix approach to N-body scattering, *J. Math. Phys.* **32** (1991), 3117–3124.

14. A.G. Gibson, A.J. Waters, G.H. Berthold and C. Chandler, Solution of the Chandler-Gibson equations for a three-body test problem, *Phys. Rev. C* **44** (1991), 1796–1811.

15. A.J. Waters, C. Chandler and A.G. Gibson, Solution of a benchmark problem with the Chandler-Gibson equations, in *Few Body XIV*, F. Gross, Ed., William and Mary, Williamsburg, VA, 1994, 844–847.

16. A.J. Waters, Breakup theory and computation for three-body scattering with the Chandler-Gibson equations, Ph.D. thesis, University of New Mexico, 1994.

17. A.R. Edmonds, *Angular momentum in quantum mechanics*, 2nd ed., Princeton University Press, Princeton, NJ, 1957.

18. M. Born and E. Wolfe, *Principles of optics*, 5th ed., Pergamon Press, 1975, Appendix VII.

19. A.B. Bhatia and E. Wolf, On the circle polynomials of Zernicke and related orthogonal sets, *Proc. Camb. Phil. Soc.* **50** (1954), 40–48.

20. J.L. Friar, B.F. Gibson, et al., Benchmark solutions for a model three-nucleon scattering problem, *Phys. Rev. C* **42** (1990), 1838–1840.

Department of Mathematics and Statistics, University of New Mexico, Albuquerque, NM 87131, USA

S. KIM and I. MUSTAKIS

Microhydrodynamics of sharp corners and edges

1. Introduction

Sharp corners and edges are ubiquitous on particles that appear in natural and manufacturing processes. Because of the underlying materials chemistry, the corners and edges may be viewed as "sharp" to atomistic scales. This conference presentation work provides a framework for the theoretical investigation of the dynamics of such particles in a viscous fluid, with special emphasis on hydrodynamic interaction between sharp corners or edges and another nearby surface.

In general, a better understanding of the dynamics of two surfaces in close proximity is of fundamental importance, whether for the interpretation of the performance of scientific instruments such as the atomic force microscope (AFM) [3], for mathematical models of particle aggregation [10], or for the elucidation of the role of particle stresses in the rheological behavior of suspensions [8]. The physics of hydrodynamic interactions between smooth surfaces have been well understood for over a century thanks to the pioneering work on lubrication flows by Osborne Reynolds. As summarized in standard texts [2], strong lubrication stresses in the intervening fluid resist the motion of smooth surfaces approaching each other. The result is that the hydrodynamic drag scales as ϵ^{-1} times the Stokes drag on an isolated particle; here ϵ is the surface–surface gap divided by a characteristic particle size.

In contrast, when a sharp wedge or corner approaches a smooth surface, the intervening fluid "oozes" out more readily from the gap region. We intuitively expect weaker hydrodynamic resistance to the applied motion than in the case of two smooth surfaces. The quantitative details will depend of course on the value of the dihedral angle: a broad wedge will behave more or less as a smooth surface because the escaping fluid is impeded to a considerable degree, whereas a surface with a small dihedral angle (as in a "knife") would encounter almost no lubrication resistance in approaching a target surface. Our goal is to quantify these intuitive ideas in the form of simple scaling relations for the resistance, as a function of ϵ and the dihedral angle.

2. Traction singularities

For very slow and viscous flows the linearized limit of the Navier–Stokes equations, i.e., the Stokes equations, are the appropriate mathematical model for steady flow:

$$-\nabla p + \mu \nabla^2 \mathbf{v} = \mathbf{0} \ , \qquad \nabla \cdot \mathbf{v} = 0 \ , \tag{1}$$

where $\mathbf{v}$ and p are the disturbance velocity, and pressure fields respectively and μ is the viscosity. The double layer representation of the Stokes flow allows the application of integral equation of the second kind to both the solution of the mobility problem and the traction problem, *Reizs-Rennerb* equation [7]. The resolution of the traction field on particle edges is of great importance since the tractions across the edges have a singular behavior [4]. On a two-dimensional edge there are three canonical flow patterns, the symmetric (splitting), antisymmetric, and parallel flow. The application of the continuation approach [9] on the integral equations for the limiting case very close to the particles edges allows the analytical extraction of the singular exponents directly from the Riesz-Rennerb equation. The methodology can even be extended to the three-dimensional corner. In 3-D corners the traction field $t_i(r, \theta)$ exhibits a

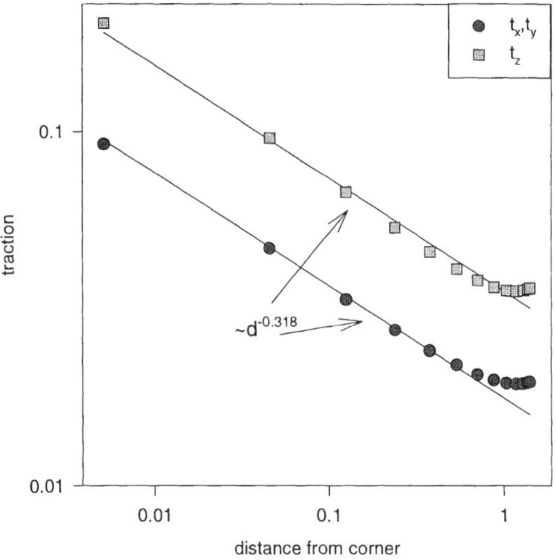

Fig. 1. Symmetric flow $\mathbf{u} = (-1, -1, -1)$ on corner $(1, 1, 1)$. Tractions across the diagonal of the top plane.

stronger singularity, and the solution can be expressed as a factor r^{-q} times an angular dependence $h_i(\theta)$. The angular function $h_i(\theta)$ also has singular behavior, since it should resolve the traction field across the edges that create the corner. The application of the continuation approach in this case needs to be augmented by numerical discretization of $h_i(\theta)$ and leads to a nonlinear eigenvalue problem for q. Only two distinguished roots for q can be found, corresponding to a flow splitting on the 3-D corner (smaller eigenvalue) and the flow normal to the splitting vector (larger eigenvalue). The latter is a double root; by symmetry any flow on the normal plane will satisfy the eigenvalue problem. Thus the complete picture of the traction field around a corner can be drawn. A detailed analysis of the traction singularity problem can be found in Mustakis and Kim [6].

3. Numerics

The main advantage of the application of an integral equation of the second kind is its well posedness and its ability to accommodate simple iterative solution techniques. These properties however are not always true for particles with sharp corners. The eigenvalue spectra of the double layer operator is analyzed. As the particle becomes sharper the gravest eigenvalue moves close to -1, leading to an almost ill-posed problem (parallelepiped of $10°$). The eigenvalue spectrum is also examined for the case of wall-particle interactions. In this case the value of the gravest eigenvalue moves to -1 less dramatically, and in proportion to the amount of surface available for close interaction. These properties lead to the surprising result that an edge-wall interaction

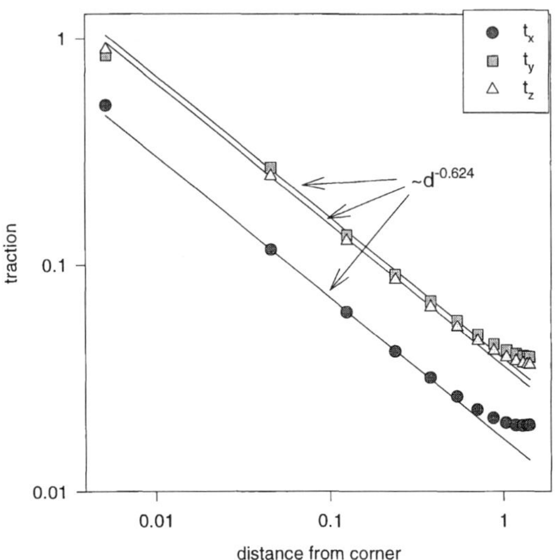

Fig. 2. Antisymmetric flow, mode II $\mathbf{u} = (-1, 2, -1)$ on corner $(1, 1, 1)$. Tractions across the diagonal of the top plane.

can be solved more accurately than the face-wall one. An iterative method based on the GMRES [11] has been applied for the solution of the discretized system.

A number of different discretization techniques have been applied to the boundary integral equation of the second kind. Here we want to distinguish between the traction solving methods and the solution of the mobility problems. As we have seen, the tractions are singular and require special attention; however, the double layer density on sharp corners and edges is just a discontinuous analytic function. The discretization methods used in this work range from low order boundary elements up to spectral and B-spline [1] formulations. In the case of the lower order (constant, linear, or quadratic elements), adaptive discretization has been applied in order to capture accurately the solution in the edge and corner region. Integration on the boundary elements is always an important issue in both the speed and the accuracy of the code. This is especially important in our case, since we are looking at very small separations. We have applied a mixture of adaptive and fixed point integration [5].

4. Particle-wall interactions

Lubrication theory predicts that flat (in small scale) particles approaching each other or a wall will experience infinite stresses and are thus restricted from touching in finite time. For particles with sharp corners and edges these lubrication stresses are no longer present and the particles will touch—but how? In this work we examined the particle-wall interaction for particles shaped as parallelepipeds.

16

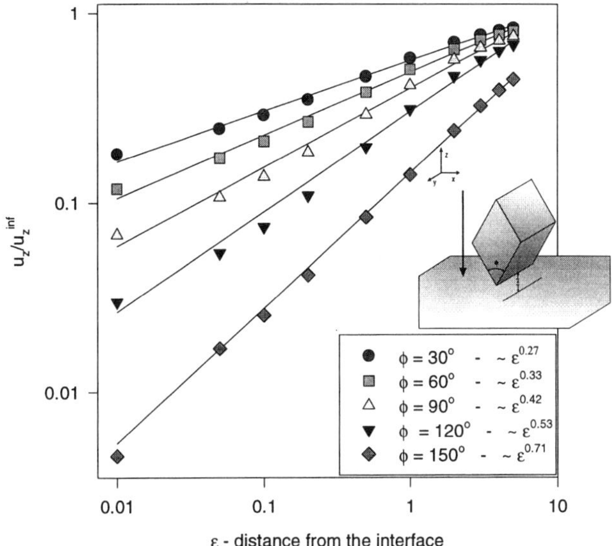

Fig. 3. Particle-wall interaction. Particle approaching with the edge: "approaching".

Due to symmetry, for this geometry we can define three separate fundamental movements of the particle with respect to the wall:

- The particle is approaching the wall.

- The particle is "skating" along the wall.

- The particle is "scraping" across the wall.

Of course there is yet another separate category where the particle is approaching with its three-dimensional corner. Numerical simulations of these scenarios have been done using the boundary integral equation. As expected the approaching velocity is heavily dependent on the dihedral angle of the particles. As the particle is moving normal to the wall the "approaching" velocity shows an algebraic decay with the particle–wall gap. The decay exponent increases as the dihedral angle increases and it covers an interval between 1 (almost flat surfaces) to 0 (needle like object). Here we should mention that in the case of the flat surface we can reproduce the lubrication results. This behavior of the exponent should be expected, since as the dihedral angle increases the particle surface that is exposed to the interaction also increases and the fluid between the particle and the wall can escape easier. However, since the decay exponent is less than 1 the particle will touch the wall within a finite time period.

Now we turn our attention to the remaining two flow modes: "scraping" and "skating". In both cases the particle motion is almost unaffected by the presence of

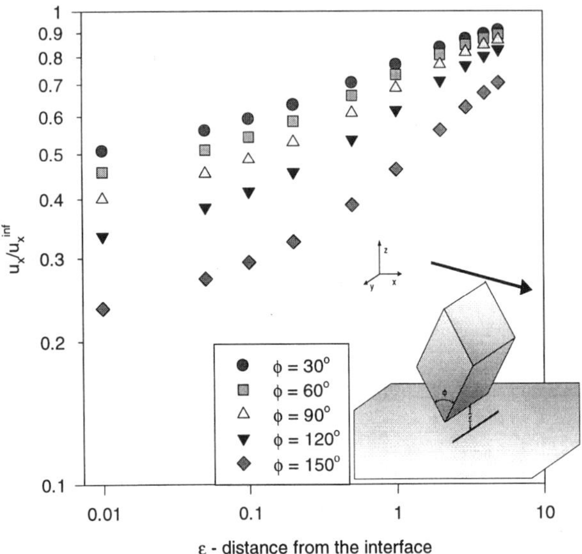

Fig. 4. Particle-wall interaction. Particle approaching with the edge: "scraping".

the wall, since it retains 20% or even 50% of its bulk velocity. For separations less that 0.1 of particle size, sedimentation velocity is insensitive to the gap. The flattening of the curve is strongly dependent on the dihedral angle. In contrast, smooth particles sliding across a plane will follow a logarithmic resistance to sliding.

In the case that the particle is approaching the wall with its 3-D corner, the sedimentation velocity remains relatively unaffected. The presence of the 3-D corner means that even less surface is available for squeezing the fluid out of the gap; as a result the particle retains up to 50% (parallel), 20% (normal) of its bulk velocity. Clearly the presence of the corner allows touching, which leads to the possibility of solid-solid friction on any of these cases. Thus any cornered particle will hit the wall with a finite velocity.

Thus we can distinguish four different interactions between edged particles in a suspension:

- Face-face interactions where the lubrication forces are dominant.

- Face-edge approaching where the squeezing action is enough to stop the particle.

- Face-edge sliding where the particle will slide with a finite velocity.

- Face-corner interactions where the particles will hit each other with a finite velocity.

18

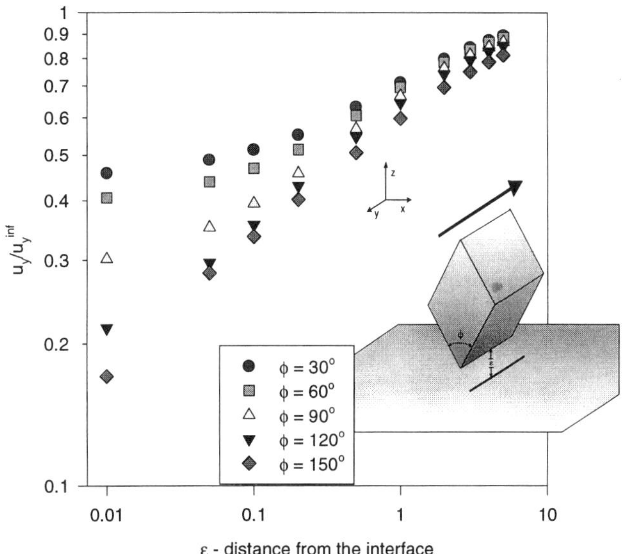

Fig. 5. Particle-wall interaction. Particle approaching with the
edge: "skating".

It is obvious that due to torques developed in a concentrated suspension, the most
frequent interaction will be the face-corner type, leading to particle collisions even in
Stokes flow. Any attempt to model the behavior of such suspensions must include
modeling of such collisions.

References

1. C. de Boor, *A practical guide to splines*, Springer-Verlag, New York, 1978.

2. S. Kim and S.J. Karrila, *Microhydrodynamics: principles and selected applications*, Butterworth-Heinemann, Boston, 1991.

3. Y. Liu, D.F. Evans, Q. Song and D.W. Grainger, Structure and frictional properties of self-assembled surfactant monolayers, *Langmuir* **12** (1996), 1235–1244.

4. H.K. Moffatt, Viscous and resistive eddies near a sharp corner, *J. Fluid Mech.* **18** (1964), 1–18.

5. G.P. Muldowney and J.J.L. Higdon, A spectral boundary element approach to three-dimensional Stokes flow, *J. Fluid Mech.* **298** (1995), 167–192.

6. I. Mustakis and S. Kim, Microhydrodynamics of sharp corners and edges: traction singularities, *AIChE J.* **44** (1998), 1469.

7. P. Pakdel, and S. Kim, Traction singularities on sharp corners and edges in Stokes flows, *Chem. Engrg. Comm.* **148-150** (1996), 257–269.

8. N. Phan-Thien, Constitutive equation for concentrated suspensions in Newtonian liquids, *J. Rheology* **39** (1995), 679–695.

9. D. Rosen and D.E. Cormack, On corner analysis in the BEM by the continuation approach, *Engrg. Anal. Boundary Elements* **16** (1995), 53–63.

10. W.B. Russel, W.R. Schowalter and D.A. Saville, *Colloidal dispersions*, Cambridge University Press, London, 1989.

11. Y. Saad, Practical use of polynomial preconditionings for the conjugate method, *SIAM J. Sci. Stat. Comput.* **6** (1985), 865–881.

Department of Chemical Engineering, University of Wisconsin, Madison, WI 53706, USA

B.D. SLEEMAN and H. CHEN

Acoustic scattering by irregular obstacles

1. Formulation of the acoustic scattering problem

We wish to discuss the problem of scattering of plane incident waves by an obstacle $\Gamma = R^n - \Omega$, where Ω is an exterior domain. This is a classic problem of acoustics with a long and distinguished history; see, for example, [1] and [2]. The problem is formulated as follows.

Let $w_0(\mathbf{p}, x)$ represent an incident plane wave propagating in the direction $\mathbf{p}$, and assume a time-harmonic dependence $\exp(-i|\mathbf{p}|t)$. Let $w_1(\mathbf{p}, x)$ be the wave scattered by the obstacle Γ. Then $w(\mathbf{p}, x) \equiv w_0(\mathbf{p}, x) + w_1(\mathbf{p}, x)$ satisfies the boundary value problem for the Helmoltz equation

$$\left(\Delta + |\mathbf{p}|^2\right) w(\mathbf{p}, x) = 0, \quad x \in \Omega, \tag{1.1}$$

$$Bw(\mathbf{p}, x) = 0, \quad x \in \partial\Gamma, \tag{1.2}$$

and $w_1(\mathbf{p}, x)$ satisfies the Sommerfeld radiation condition to ensure that the scattered wave is outgoing.

The boundary condition (1.2) may be of Dirichlet (sound-soft obstacle), Neumann (sound-hard obstacle), or impedance type.

Of particular interest in this paper is the study of solutions to (1.1), (1.2) for sound-hard obstacles whose boundary $\partial\Gamma$ is rough in the sense to be discussed below. The case of the sound-soft obstacle has been treated previously in [3].

In order to formulate the problem rigorously it is necessary first to develop a generalised notion of the Neumann boundary condition. This is motivated by the formal use of Green's formula

$$\int_\Omega \left\{(\Delta u)v + \nabla u \cdot \nabla v\right\} dx = \int_{\partial\Gamma} (\hat{\nu} \cdot \nabla u)\, v\, ds, \tag{1.3}$$

which is valid whenever $\partial\Gamma$ and the functions u and v are sufficiently regular; here $\hat{\nu}$ is the unit outward normal to $\partial\Gamma$. In particular if u satisfies the Neumann condition on $\partial\Gamma$, then the right-hand side of (1.3) is zero for all v for which (1.3) holds.

To adapt (1.3) to the case where $\partial\Gamma$ is arbitrary we follow [4] and begin by introducing the function spaces

$$L_2(\Omega) = \left\{u : u(x) \text{is Lebesque measurable on } \Omega \text{ and } \int_\Omega |u(x)|^2 dx < \infty\right\}, \tag{1.4}$$

$$L_2(\Delta, \Omega) = L_2(\Omega) \cap \{u : \Delta u \in L_2(\Omega)\}, \tag{1.5}$$

$$L_2^1(\Delta, \Omega) = L_2^1(\Omega) \cap L_2(\Delta, \Omega), \tag{1.6}$$

where

$$L_2^1(\Omega) = L_2(\Omega) \cap \{u : D^\alpha u \in L_2(\Omega) \text{ for } |\alpha| \le 1\}.$$

Research supported by grants from the Royal Society of London and the National Natural Science Foundation of China.

We now define the generalized form of the Neumann condition.

Definition 1.1. $u \in L_2^1(\Delta, \Omega)$ is said to satisfy the generalized Neumann condition for Ω if and only if

$$\int_\Omega \{(\Delta u)v + \nabla u \cdot \nabla v\}\, dx = 0 \quad \text{for all } v \in L_2^1(\Omega). \tag{1.7}$$

Note that (1.7) is meaningful for arbitrary domains Ω and defines a closed subspace

$$L_2^N(\Delta, \Omega) = L_2^1(\Delta, \Omega) \cap \{u : u \text{ satisfies } (1.7)\}$$

in the Hilbert space $L_2^1(\Delta, \Omega)$.

The condition (1.7) is equivalent to the classical Neumann condition if $\partial\Gamma$ is sufficiently smooth. In this case the functions $v \in L_2^1(\Delta, \Omega)$ will have a trace on $\partial\Gamma$ and ∇u for functions $u \in L_2^1(\Delta, \Omega)$ will have a trace on $\partial\Gamma$ by the Sobolev imbedding theorems.

Definition 1.2. Let Ω be an exterior domain. Then the outgoing scattered wave $w_1(\mathbf{p}, x)$ for Ω and $\mathbf{p} \in R^n$ is the function characterized by

$$w_0(\mathbf{p}, \cdot) + w_1(\mathbf{p}, \cdot) \in L_2^{N,\mathrm{loc}},$$

which satisfies (1.1), and $w_1(\mathbf{p}, x)$ satisfies the Sommerfeld radiation condition.

Next we need to establish existence and uniqueness of solution to our problem. In order to do this we need a local compactness result for Ω as well as a corresponding limiting absorption principle for obstacles with rough boundaries. The tool needed for these considerations is an extension of the Rellich selection theorem.

Definition 1.3. (Local compactness property, LC) A domain $\Omega \subset R^n$ is said to have the local compactness property if and only if for each set of functions $S \subset L_2^{1,\mathrm{loc}}(\bar{\Omega})$ and each $R > 0$,

$$\|u\|_{L_2^1(\Omega_R)} \leq C(R)$$

for all $u = v|_{\Omega_R}$ with $v \in S$ implies that $\{u = v|_{\Omega_R} : v \in S\}$ is pre-compact in $L_2(\Omega_R)$, where $\Omega_R = \Omega \cap \{x : |x| < R\}$.

Rellich's original selection theorem states that bounded domains with smooth boundaries are of class LC. In [5], this result is generalized to domains having a "segment property". We recall Ω has the segment property in the sense of Agmon if there exists a finite open covering of $\partial\Omega$, that is, $\partial\Omega \subset O_1 \cup O_2 \cup \ldots \cup O_N$, and corresponding non-zero vectors $x^{(1)}, x^{(2)}, \ldots, x^{(n)}$ such that

$$\{x = x_0 + tx^{(j)},\ 0 < t < 1\} \subset \Omega \quad \text{for every } x_0 \in \bar{\Omega} \cap O_j.$$

While domains having the segment property occur in many applications, there are, as pointed out in [4], some notable exceptions. For example, the disc

$$\Gamma = \left\{(x_1, x_2, x_3) : x_1^2 + x_2^2 \leq 1,\ x_3 = 0\right\} \subset R^3$$

does not have the segment property. To overcome these limitations, Wilcox [4] introduced the set of domains having the "tiling property".

To introduce this concept, suppose that there is an open set $O \subset R^n$, compact sets $K_1, K_2, \ldots, K_n \subset R^n$ and nonzero vectors $x^{(1)}, x^{(2)}, \ldots, x^{(N)}$ such that

$$\partial \Gamma \subset O, \tag{1.8}$$

$$O \cap \Omega \subset \bigcap_{j=1}^{N} K_j. \tag{1.9}$$

Each set K_j lies in a curvilinear coordinate patch O_j so that

$$\left\{ x = x_0 + t x^{(j)} : 0 < t < 1 \right\} \subset \Omega \tag{1.10}$$

for every $x_0 \in \Omega \cap K_j$ holds when $x = (x_1, \ldots, x_n)$ are the corresponding coordinates.

Definition 1.4. (Finite tiling property) An exterior domain Ω has the finite tiling property if (1.8)–(1.10) hold for a suitable set O, compact sets K_j, and coordinate patches O_j.

Remark. Domains having the finite tiling property embrace many fractal domains including the well-known Koch Island.

From this we have the following extension of the theorems of Rellich and Agmon.

Theorem 1.1. [4] *If Ω is an exterior domain having the finite tiling property, then $\Omega \in \mathrm{LC}$.*

From the above definitions we have the following result [4].

Theorem 1.2. *Let Ω be an exterior domain such that $\Omega \in \mathrm{LC}$. Then for each $\mathbf{p} \in R^n$ there exists a unique outgoing scattered wave $w_1(\mathbf{p}, x)$.*

Furthermore, we have an important corollary.

Corollary 1.1. *Under the conditions of Theorem 1.2 there is a function $\theta(\eta, P) \in C^\infty(S^{n-1} \times R^n \setminus \{0\})$ such that*

$$w_1(\mathbf{p}, x) = \frac{e^{i|\mathbf{p}||x|}}{|x|^{(n-1)/2}} \, \theta\left(\frac{x}{|x|}, \mathbf{p}\right) + q(\mathbf{p}, x),$$

where $q(\mathbf{p}, x) = O\left(|x|^{-(n+1)/2}\right)$ as $|x| \to \infty$ uniformly for $\eta = (x/|x| \in S^{n-1}$ and $\mathbf{p}$ in any compact subset of $R^n \setminus \{0\}$.

The function $\theta(\eta, \mathbf{p})$ is called the far-field amplitude, or radiation pattern.

2. The scattering phase

We now introduce the quantity which is the main subject of study, namely the scattering phase $s(\lambda)$, $\lambda = |\mathbf{p}|^2$. This can be defined in several ways. For example one can use the trace-class perturbation theory of Birman and Krein [6] and Yafaev [7] using wave operators or, more generally, as in [8], where Kato's invariance principle is employed. Alternatively, we can use a traditional approach as modified by Robert [9] and Christiansen [10]. Essentially, the scattering matrix $S(\lambda) : L^2(S^{n-1}) \to L^2(S^{n-1})$ is given by

$$S(\lambda) = \mathrm{id} + A(\lambda),$$

where $A(\lambda)$ is the operator with Schwartz kernel given by the radiation pattern $\theta(\eta, \mathbf{p})$ as defined above.

Definition 2.1. (The scattering phase) The scattering phase $s(\lambda)$ is defined as

$$s(\lambda) = \frac{1}{2\pi i} \log \det S(\lambda). \tag{2.1}$$

The first result concerning the asymptotic form of $s(\lambda)$ (as $\lambda \to \infty$) for strictly convex domains was stated without proof in [11]. Then Majda and Ralston [12] established the form of the first three terms in the expansion of $s(\lambda)$ for strictly convex, smooth domains. For non-convex domains Jensen and Kato [8] obtained the first term in the asymptotics of $s(\lambda)$ for star-like domains. Petkov and Popov [13] have established the most complete result for the Dirichlet problem by obtaining three terms in the expansion of $s(\lambda)$ for all smooth and non-trapping obstacles.

Much more recently, Robert [14] has developed a new trace perturbation formula for obstacle scattering, which considerably extends the class of obstacles for which scattering phase asymptotics can be developed. Robert's results for the Dirichlet problem are summarized in the following assertion.

Theorem 2.1. *For every $n \geq 1$, the Dirichlet scattering phase $s(\lambda)$ is given by*

$$s(\lambda) = (2\pi)^{-n} w_n |\Gamma|_n \lambda^{n/2} + R_n(\lambda) \quad \text{as } \lambda \to \infty, \tag{2.2}$$

where w_n is the volume of the unit ball in R^n, $|\cdot|_n$ is the n-dimensional Lebesgue measure, $R_n(\lambda) = o(\lambda^{n/2})$ for arbitrary compact obstacles Γ, and $R_n(\lambda) = O(\lambda^{(n-1)/2})$ for smooth compact obstacles.

In addition, if the set of billiard trajectories in $\Omega = R \setminus \Gamma$ has measure zero in the cotangent space, then

$$R_n(\lambda) = \left[4(2\pi)^{n-1}\right]^{-1} w_{n-1} |\partial\Gamma|_{n-1} \lambda^{(n-1)/2} + o(\lambda^{(n-1)/2}). \tag{2.3}$$

The asymptotic formula (2.2) also holds for the Neumann scattering phase.

The results of Robert have been further developed by Christiansen [10] to determine the asymptotics of the scattering phase for both Dirichlet and Neumann problems for obstacles with irregular, possibly fractal, boundaries. We return to these results later in Section 3.

In 1992 the authors established asymptotics of the scattering phase $s(\lambda)$ for the Dirichlet problem by different methods to Christiansen, drawing on the intimate connection between the scattering phase and the counting function $N(\lambda)$ for the associated interior Dirichlet eigenvalue problem. The methods we use to discuss the Neumann problem also differ from those of Robert and Christiansen and exploit new monotonicity properties of $s(\lambda)$.

In order to introduce our results, we first describe the concepts of Minkowski measure and Minkowski dimension.

Let Γ be a compact obstacle in R^n $(n \geq 2)$ with boundary $\partial\Gamma$. We define

$$\Gamma_\varepsilon = \{x \in R^n : d(x, \partial\Gamma) < \varepsilon\}, \quad \varepsilon > 0, \tag{2.4}$$

where $d(\cdot, \partial\Gamma)$ denotes the Euclidean distance to the boundary $\partial\Gamma$. The set Γ_ε is called the ε-neighborhood of $\partial\Gamma$. For each positive number ℓ, the numbers

$$\begin{aligned}
\mu^*(\ell, \partial\Gamma) &= \lim_{\varepsilon\to 0+} \sup \varepsilon^{-(n-\ell)} |\Gamma_\varepsilon|_n, \\
\mu_*(\ell, \partial\Gamma) &= \lim_{\varepsilon\to 0+} \inf \ \varepsilon^{-(n-\ell)} |\Gamma_\varepsilon|_n
\end{aligned} \tag{2.5}$$

are called the ℓ-upper and ℓ-lower Minkowski contents of $\partial\Gamma$, respectively. The Minkowski dimension of $\partial\Gamma$, denoted by δ, is defined as

$$\delta = \inf\left\{\ell \in R_+ : \mu^*(\ell, \partial\Gamma) = 0\right\} = \sup\left\{\ell \in R_+ : \mu^*(\ell, \partial\Gamma) = \infty\right\}. \tag{2.6}$$

Furthermore, we say that $\partial\Gamma$ is δ-Minkowski measurable if

$$0 < \mu_*(\delta, \partial\Gamma) = \mu^*(\delta, \partial\Gamma) < \infty, \tag{2.7}$$

and the common value denoted by $\mu(\delta, \partial\Gamma)$ is called the δ-Minkowski measure of $\partial\Gamma$.

In the case of the Dirichlet scattering problem the following assertion holds.

Theorem 2.2. [3] *Let Γ be an obstacle in R^n with boundary $\partial\Gamma$. If $\partial\Gamma$ is δ-Minkowski measurable with $\delta \in (n-1, n)$ and δ-Minkowski measure $\mu^*(\delta, \partial\Gamma)$, then there exists a constant $C_{n,\delta} > 0$ depending only on n and δ such that*

$$\left| s(\lambda) - C_n |\Gamma|_n \lambda^{n/2} \right| \leq C_{n,\delta} \mu^*(\delta, \partial\Gamma) \lambda^{\delta/2}$$

as $\lambda \to \infty$, and where

$$C_n = (4\pi)^{-n/2} / \Gamma\left(1 + \tfrac{1}{2}n\right).$$

In the case of the Neumann scattering problem we have recently established the following assertion.

Theorem 2.3. [15] *Let Γ be an obstacle (that is, compact with connected complement $\Omega = R^n - \Gamma$) in R^n ($n \geq 2$) with boundary $\partial\Gamma$, which satisfies the "finite tiling" property. If δ is the Minkowski dimension of $\partial\Gamma$, $\delta \in (n-1, n)$ and $\mu^*(\delta, \partial\Gamma) < \infty$, then*

(i) if $\delta = n-1$ and $s(\lambda)$ is the Neumann scattering phase corresponding to Γ, there is a positive constant d_n depending only on n such that

$$\left| s(\lambda) - C_n |\Gamma|_n \lambda^{n/2} \right| \leq d_n \mu^*(n-1, \partial\Gamma)(\log \lambda)\lambda^{n-1/2} \quad as\ \lambda \to \infty;$$

(ii) if $\delta \in (n-1, n)$ then there is a positive constant $d_{n\delta}$ depending only on n and δ such that

$$\left| s(\lambda) - C_n |\Gamma|_n \lambda^{n/2} \right| \leq d_{n\delta} \mu^*(\delta, \partial\Gamma)\lambda^{\delta/2} \quad as\ \lambda \to \infty.$$

The proof of Theorem 2.3 depends on a tessellation of domains argument together with the following monotonicity properties of the scattering phase.

Lemma 2.1. [16] *Let Γ_1, Γ_2 be two obstacles with $\Gamma_1 \subseteq \Gamma_2$, and let $s_{D,1}(\lambda)$, $s_{D,2}(\lambda)$ be the Dirichlet scattering phases corresponding to Γ_1 and Γ_2, respectively. Then*

$$s_{D,1}(\lambda) \leq s_{D,2}(\lambda).$$

Lemma 2.2. [15] *Let Γ be an obstacle with Neumann scattering phase $s_N(\lambda)$ and Dirichlet scattering phase $s_D(\lambda)$. Then*

$$s_N(\lambda) \leq s_D(\lambda) \quad for\ \lambda > 0.$$

We conclude the above discussion by pointing out that if we know the scattering phase $s(\lambda)$ as $\lambda \to \infty$, then we can determine the volume of the scattering obstacle Γ and the Minkowski dimension δ of $\partial\Gamma$. In other words, the volume and a measure of the "roughness" of the surface of an obstacle may be determined from the high-frequency asymptotics of the scattering phase.

It is also of interest to ask whether the scattering phase reveals any further geometric properties of the scattering obstacle. We consider this in the next section.

3. Further properties of $s(\lambda)$

We recall from Theorem 2.1 that if $\partial\Gamma$ is sufficiently smooth and satisfies the billiard condition, then the precise form of the second term in the asymptotics of $s(\lambda)$ can be determined.

The principal tool used by Robert [14] in obtaining this result has recently been generalized and extended by Christiansen [10] to obtain the asymptotics of $s(\lambda)$ for somewhat more general domains. We apply Christiansen's ideas to domains with fractal boundaries.

Let P be the Laplacian defined on the exterior of a compact set Ω with $R^n \setminus \Omega$ connected. Further, let $\Omega \subset B(R_0)$ (the ball of radius R_0), and let T_R, $R > R_0 + a_0$, be a flat torus. Next, let $P^\#$ be the unbounded self-adjoint operator acting on the Hilbert space

$$H^\# = H_{B(R_0)} \oplus L^2\left(T_R \setminus B(R_0)\right)$$

and given by

$$P^{\#}u = P(\chi u) + \Delta_T(1 - \chi)u, \tag{3.1}$$

where $\chi \in C_0^{\infty}\big(B(R_0 + \tfrac{1}{2}a_0)\big)$ and $\chi = 1$ on $\overline{B(R_0)}$. Then under certain conditions Christiansen proves that the scattering phase for Ω is given by

$$s(\lambda) = -N_{p\#}(\lambda) - C_n \operatorname{Vol} T_R \lambda^{n/2} + O(\lambda^{\max\{(n-1)/2, d/2\}}), \tag{3.2}$$

where $C_n = (2\pi)^{-n}\operatorname{Vol}(B(1))$, and $N_{p\#}(\lambda)$ is the counting function for $P^{\#}$ satisfying

$$N_{p\#}(\lambda) = N_{p\#}(\lambda - 1) = O(\lambda^{d/2}) \quad \text{as } \lambda \to \infty. \tag{3.3}$$

The importance of this result is that if we know enough about the counting function $N_{p\#}(\lambda)$, then we may be able to obtain more precise information concerning $s(\lambda)$. Precisely,

$$N_{p\#}(\lambda) = \#\left\{\lambda_j : \lambda_j \text{ is an eigenvalue of } P^{\#} \text{ with } \lambda_j \leq \lambda\right\}. \tag{3.4}$$

Counting function asymptotics has a long history going back to the work of Weyl, and over the past two decades has been explored intensively in relation to domains with fractal boundaries. For reference to this work we cite [17–25] and the titles therein.

To illustrate a typical result, consider the set $T_R \setminus \Omega \subset R^2$ as shown in Fig. 1. Here $T_R \setminus \Omega = \bigcup_{k=0}^{\infty}(\#A_k)\tilde{Q}_k$ is a connected set consisting of $\#A_k = k$ squares $\tilde{Q}_k$ of side a^k, $0 < a < 1$, which has small cuts or openings of width $\in_k$ on each side, where

$$a^{-k} \in_k \to 0 \quad \text{as } k \to \infty.$$

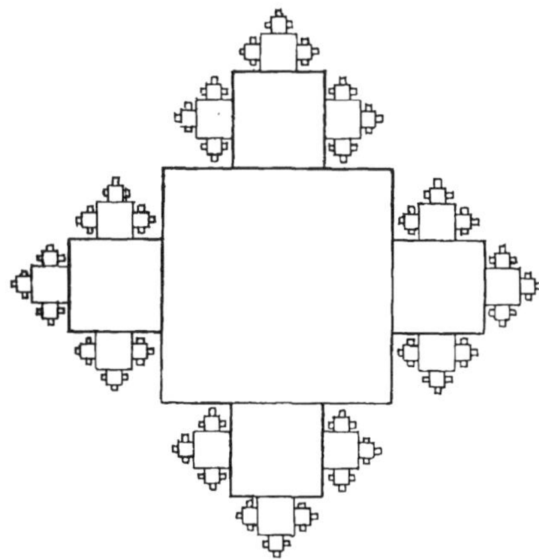

Fig. 1.

Then $\Omega = T_R \setminus (T_R \setminus \Omega)$ is compact and $R^2 \setminus \Omega$ is connected. Furthermore, the Minkowski dimension of $\Gamma = \partial\Omega$ is given by

$$\delta = (\ln 3)/\ln(1/a)$$

provided $\sqrt{3} < 1/a < 3$.

From the results of Chen and Sleeman [20] and Fleckinger and Vassiliev [22] we know that for the set $T_R \setminus \Omega$, $N_{p\#}(\lambda)$ can be estimated as

$$\pi^{-\delta} P_2\left(t + o(1)\right)\lambda^{\delta/2} + o(\lambda^{\delta/2}) \leq \phi(\lambda) - N_{\rho\#}(\lambda) \leq \pi^{-\delta} P_2(t)\lambda^{\delta/2} + o(\lambda^{\delta/2}), \quad (3.5)$$

as $\lambda \to \infty$, where $t = (\ln\lambda - 2ln\pi)/(2\ln(1/a))$ and $\phi(\lambda)$ is the Weyl term given by $\phi(\lambda) = (1/(4\pi))|T_R \setminus \Omega|_2\lambda$. Furthermore,

$$P_2(t) = \sum_{-\infty}^{\infty} a^{\delta(t-k)}\rho_2\left(a^{(k-t)}\right), \quad (3.6)$$

where $\rho_2(r) = (\pi/4)r^2 - \#\left\{(q_1, q_2) \in \mathbf{N}^2 : q_1^2 + q_2^2 < r^2\right\}$.

For this example it can be shown that (3.3) holds with $d = \delta$ and the result of Christiansen allows us to claim that

$$s(\lambda) = (4\pi)^{-1}|\Omega|_2\lambda + F_2(\lambda)\lambda^{\delta/2} + O(\lambda^{\delta/2}), \quad (3.7)$$

where $F_2(\lambda) = (4\pi)^{-1}\left[|T_1 \setminus \Omega|_2\lambda - N_{\rho\#}(\lambda)\right]/\lambda^{\delta/2}$ is a well defined, positive, bounded and oscillatory function that is left-continuous, its points of discontinuity being dense in R_+.

4. Uniqueness

It is important and natural to ask if the scattering obstacle Ω is uniquely determined by $s(\lambda)$.

The question of uniqueness of the inverse scattering problem has been studied extensively in the literature (see [1] and [26]), and the results are generally concentrated on reasonably smooth obstacles, that is, Lipschitz domains. In the case of highly irregular obstacles very little is known. Furthermore, in view of the results above and in particular the relation been $s(\lambda)$ and $N_{p\#}(\lambda)$ as $\lambda \to \infty$, it is clear that uniqueness of our inverse problem is closely related to the classic isospectral problem for $N_{\mathbf{p}\#}(\lambda)$.

Let (M, g) be a compact Riemann manifold with boundary, and let Δ be the Laplace operator defined on M. The spectrum of M is a sequence of eigenvalues of Δ. Two Riemann manifolds are isospectral if their spectra (counting multiplicities) coincide. A fundamental question concerning the relationship between spectral analysis and geometry raised by Kac [27] is whether two isospectral planar domains must necessarily be isometric. Despite many results for nonplanar domains being known for some time, it was not until the ground-breaking work of Gordon, Webb and Wolpert [28] that Kac's question received a resounding negative answer. Furthermore, these authors' work allows one to construct a whole variety of pairs of isospectral nonisometric planar domains. Following [28], Berard [29] gave a simple proof showing that the eigenvalues of two planar isospectral, nonisometric domains A, B are identical, by constructing a map which takes an eigenfunction of A and maps it onto an eigenfunction of B for the same eigenvalue. The idea of transposition maps has been developed further by Buser et al. [30] to generate new examples of isospectral domains. In [31] and [32] we have developed these ideas further to construct isospectral, nonisometric planar domains with fractal boundaries. The key to these constructions is a "paper folding" argument due to Chapman [33], which is a constructive method of realizing the transposition maps of Berard [29] and Buser et al. [30]. Examples are given in Figs. 2 and 3.

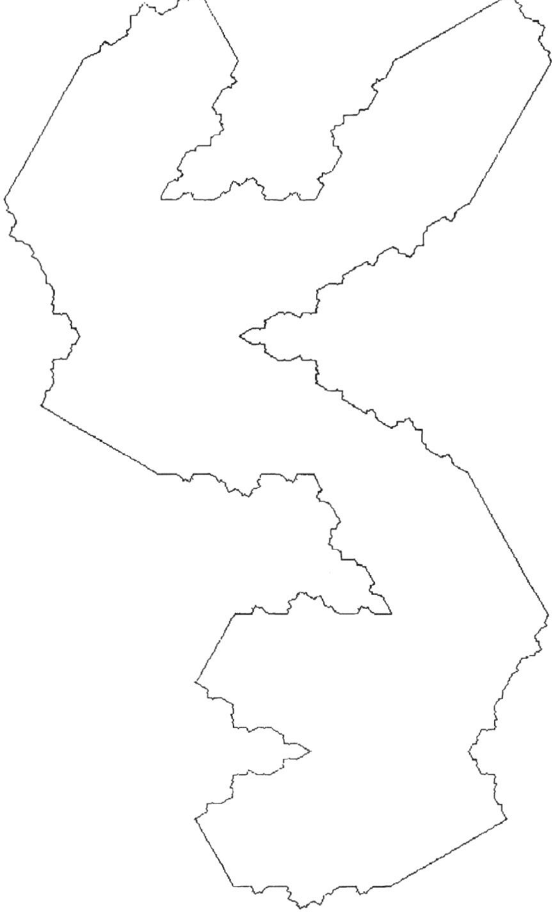

Fig. 2.

Fig. 3.

Figs. 2 and 3 show a pair of (i) isospectral nonisometric domains with piecewise continuous boundaries represented by the outer curves, and (ii) the first steps in constructing a pair of isospectral nonisometric connected fractal domains represented by the domains enclosed by the inner boundaries. The fractals are constructed following a similar procedure as for creating the Koch island curve based on the complement of the 1/3 Cantor set.

The upshot of this is that $s(\lambda)$ does *not* in general determine the scatterer uniquely, and this prompts us to ask the following questions:

1. How much information is required of the radiation pattern to determine a fractal scattering obstacle uniquely?

2. Can one derive stable numerical methods to reconstruct fractal obstacles from scattered data?

References

1. D. Colton and R. Kress, *Inverse acoustic and electromagnetic scattering*, 2nd ed., Theory in Appl. Math. Sci. **93**, Springer-Verlag, New York, 1998.

2. D.S. Jones, *Acoustic and electromagnetic waves*, Oxford Univ. Press, 1989.

3. H. Chen and B.D. Sleeman, An analogue of Berry's conjecture for the phase in fractal obstacle scattering, *IMA J. Appl. Math.* **49** (1992), 193–202.

4. C. Wilcox, *Scattering theory for the d'Alembert equation in exterior domains*, Lecture Notes Math. Ser. **442**, Springer-Verlag, Heidelberg, 1975.

5. S. Agmon, *Elliptic boundary value problems*, Van Nostrand, 1965.

6. M.S. Birman and M.G. Krein, On the theory of wave operators and scattering operators, *Dokl. Akad. Nauk SSSR* **44** (1962), 475–478.

7. D. Yafaev, *Mathematical scattering theory. General theory*, Amer. Math. Soc. **105**, Providence, RI, 1992.

8. A. Jensen and T. Kato, Asymptotics behaviour of the scattering phase for exterior domains, *Comm. Partial Diff. Equat.* **3** (1978), 1165–1195.

9. D. Robert, On the Weyl formula for obstacles, Rapport de Recherche, Université de Nantes, 95/10-1.

10. T. Christiansen, Spectral asymptotics for compactly supported perturbations of the Laplacian on R^n, *Comm. Partial Diff. Equat.* **23** (1998), 933–948.

11. V.S. Buslaev, On the asymptotic behaviour of the spectral characteristics of exterior problems of the Schrödinger operator, *Math. USSR Izvestiya* **9** (1975).

12. A. Majda and J. Ralston, An analogue of Weyl's formula for unbounded domains, *Duke Math. J.* **45** (1978), 183; **45** (1978), 513; **46** (1979), 725.

13. V. Petkov and G. Popov, Asymptotic behaviour of the scattering phase for non-trapping obstacles, *Ann. Inst. Fourier* **32** (1982), 111–149.

14. D. Robert, A trace formula for obstacles problems and applications, in *Operator theory: advances and applications* **70**, Birkhäuser Verlag, Basel, 1994, 283–292.

15. H. Chen and B.D. Sleeman, High frequency estimates for the Neumann scattering phase in non-smooth obstacle scattering (submitted for publication).

16. T. Kato, Monotonicity theorems in scattering theory, *Hadronic J.* **1** (1978), 134–154.

17. M.V. Berry, Some geometric aspects of wave motion: wave front dislocations, diffraction catastrophes, diffractals in geometry of the Laplace operator, *Proc. Symp. Pure Math.* **36**, Amer. Math. Soc., Providence, RI, 1980, 13–38.

18. J. Brossard and R. Carmona, Can one hear the dimension of a fractal, *Comm. Math. Phys.* **104** (1986), 103–122.

19. H. Chen and B.D. Sleeman, Fractal drums and the n-dimensional modified Weyl-Berry conjecture, *Comm. Math. Phys.* **168** (1995), 581–607.

20. H. Chen and B.D. Sleeman, Estimates for the remainder term in the asymptotics of the counting function for domains with irregular boundaries, *Rend. Sem. Mat. Univ. Polit. Torino* **55** (1997), 171–188.

21. H. Chen and B.D. Sleeman, Counting function asymptotics and the weak Weyl-Berry conjecture for connected domains with fractal boundaries, *Acta Math. Sinica* **14** (1998), 261–276.

22. J. Fleckinger and P.G. Vassiliev, An example of a two term asymptotics of the counting function of a fractal drum. *Trans. Amer. Math. Soc.* **337** (1993), 99–116.

23. M.L. Lapidus, Fractal drum, inverse spectral problems for elliptic operators and a partial resolution of the Weyl-Berry conjecture, *Trans. Amer. Math. Soc.* **325** (1991), 465–529.

24. M.L. Lapidus, Vibrations of fractal drums, the Riemann hypothesis, waves in fractal media and the Weyl-Berry conjecture, in *Ordinary and partial differential equations*, vol. 4, Pitman Res. Notes Math. Ser. **289**, Longman, Harlow, 1993, 126–209.

25. M. Levitan and D. Vassiliev, Spectral asymptotics, renewal theorem and the Berry conjecture for a class of fractals, *Proc. London Math. Soc.* **72** (1996), 188–214.

26. D. Colton and B.D. Sleeman, Uniqueness theorems for the inverse problem of acoustic scattering, *IMA J. Appl. Math.* **31** (1983), 253–259.

27. M. Kac, Can one hear the shape of a drum? *Amer. Math. Monthly* **73** (1996), no. 11, 1–23 (Slaught memorial paper).

28. C. Gordon, D. Webb and S. Wolpert, Isospectral plane domains and surfaces via Riemannian orbifolds, *Invent. Math.* **110** (1992), 1–22.

29. P. Berard, Domaines plans isospectraux à la Gordon-Webb-Wolpert: une preuve terra à terra (preprint). (See also Transplantation et isospectralite, I, II, *Math. Ann.* (1992).)

30. P. Buser, J. Conway, P. Doyle and K.-D. Semmler, Some planar isospectral domains (preprint).

31. B.D. Sleeman and H. Chen, On the construction of non-isometric isospectral domains and Weyl-Berry conjecture, *J. Sem. Ceremath-UT1*, University of Toulouse 1, 1996.

32. B.D. Sleeman and H. Chen, On non-isometric isospectral connected fractal domains (submitted for publication).

33. S.J. Chapman, Drums that sound the same, *Amer. Math. Monthly* (1995), 124–138.

School of Mathematics, University of Leeds, Leeds LS2 9JT, UK

Department of Mathematics, Wuhan University, Wuhan 430072, China

Contributed papers

M. AHUES and A. LARGILLIER

On generalized inverse Steklov problems

1. Introduction

Let p be a positive integer, $\Omega \subset I\!\!R^p$ a convex polygonal open set. We use the notation $\partial_i v \equiv \partial v / \partial s_i$, $\partial \equiv (\partial_1, \ldots, \partial_p)$. Let Π be a polynomial in p variables. We shall deal with the formal differential operator $\Pi(\partial)$. E denotes a suitable Banach space of (classes of) functions defined in Ω in which we expect to find the (class) of functions $\Pi(\partial)v$, for v in some subspace of E. Let $\mathcal{I}$ be a positive integer, and for each positive integer $i \leq \mathcal{I}$, let b_i be a boundary linear form with domain $\mathcal{D}(b_i) \subset E$. Similarly, for a fixed positive integer $d \leq \mathcal{I}$, and for each i such that $1 \leq i \leq d$, let f_i be a boundary linear form with domain $\mathcal{D}(f_i) \subset E$. We set $\mathcal{D}_b \equiv \bigcap_{i=1}^{\mathcal{I}} \mathcal{D}(b_i)$ and $\mathcal{D}_f \equiv \bigcap_{i=1}^{d} \mathcal{D}(f_i)$. We assume $(b_1, \ldots, b_{\mathcal{I}})$ is a linearly independent system in the algebraic dual space $\mathcal{D}_b'$. The motivation of this paper is the following eigenvalue problem:

Find all nonzero complex-valued functions $v \in E$ and all nonzero complex numbers μ such that

$$\Pi(\partial)v = \mu v \text{ in } \Omega, \ f_i v = \mu b_i v \text{ for } 1 \leq i \leq d, \ b_i v = 0 \text{ for } d < i \leq \mathcal{I},$$

which we call a *generalized Steklov problem*.

Problems of this kind are encountered in the control of the digitation effect in petrol recovery problems (cf. [3]). Certainly, if $d = \mathcal{I}$ then the last $\mathcal{I} - d$ homogeneous boundary conditions do not appear in the problem.

2. The theoretical framework

Let X be the product space $X \equiv E \times \mathbb{C}^{d \times 1}$, endowed with the norm defined by the formula $\|(u, \vec{\alpha})\| \equiv \max\{\|u\|_E, \|\vec{\alpha}\|_\infty\}$. We define $V \equiv \{v \in E : \Pi(\partial)v \in E\} \cap \mathcal{D}_b \cap \mathcal{D}_f$ and we suppose that V is not reduced to the null function. We consider first the operator $A : \mathcal{D}(A) \subset E \to E$ defined by $Av \equiv \Pi(\partial)v$, with domain $\mathcal{D}(A)$ given by $\mathcal{D}(A) \equiv \{v \in V : b_i v = 0, 1 \leq i \leq \mathcal{I}\}$.

We suppose that the operator A has a compact inverse $K : E \to E$. We say in that case that $(\Pi(\partial) ; b_1, \ldots, b_{\mathcal{I}})$ is bijective.

We define the operator $L : \mathcal{D}(L) \subset X \to X$ by

$$L(v, \vec{\beta}) \equiv (\Pi(\partial)v, fv),$$

where $f \equiv [f_1, \ldots, f_d]^\top$ with the domain $\mathcal{D}(L)$ given by

$$\mathcal{D}(L) \equiv \left\{ (v, \vec{\beta}) \in X : v \in V, \beta_i = b_i v, 1 \leq i \leq d; b_i v = 0, d < i \leq \mathcal{I} \right\}.$$

Proposition 1. $\mathcal{D}(L)$ *is a linear subspace of* X *and* L *is a linear operator.*

To prove that L has a compact inverse we consider the following problem:
Given $u \in E$ and $\vec{\beta} \in \mathbb{C}^{d \times 1}$, find $v \in V$ such that

$$\Pi(\partial)v = u \text{ in } \Omega, \ b_i v = \beta_i \text{ for } 1 \leq i \leq d, \ b_i v = 0 \text{ for } d < i \leq \mathcal{I}.$$

Let $(p_1, \ldots, p_{\mathcal{I}})$ be an ordered family of linearly independent functions in $\mathcal{D}_b$, adjoint to the ordered system of linearly independent boundary linear forms $(b_1, \ldots, b_{\mathcal{I}})$, in the sense that $b_i(p_j) = \delta_{ij}$, the Kronecker delta. The function

$$w \equiv v - \sum_{i=1}^{d} \beta_i p_i$$

solves $\Pi(\partial)w = u - \sum\limits_{j=1}^{d} \beta_j q_j$ in Ω, $\quad b_i w = 0$ for $1 \le i \le \mathcal{I}$, where
$$q_j \equiv \Pi(\partial)p_j \text{ for } 1 \le j \le d.$$

In other words, w solves
$$Aw = u - \sum_{j=1}^{d} \beta_j q_j$$

and hence
$$w = Ku - \sum_{j=1}^{d} \beta_j g_j,$$

where
$$g_j = Kq_j \text{ for } 1 \le j \le d.$$

The solution v is then uniquely defined by

$$v \equiv Ku + \sum_{j=1}^{d} \beta_j(p_j - g_j). \tag{1}$$

We define the functionals ξ_i, the matrix of functionals ξ and the matrix $B \in \mathbb{C}^{d \times d}$ as follows:
$$\xi_i \equiv f_i K, \ \xi \equiv [\xi_1, \ldots, \xi_d]^\top \text{ and } B(i,j) \equiv f_i p_j - \xi_i q_j.$$

We have proved the following assertion.

Theorem 1. *If B is invertible, then L has a compact inverse given by $T(u, \vec{\alpha}) = (v, \vec{\beta})$, where*
$$\vec{\beta} \equiv B^{-1}(\vec{\alpha} - \xi u)$$
and v is defined by (1).

Corollary. *The spectrum of L is a countable subset of $\mathbb{C}$ consisting of isolated eigenvalues of L with finite algebraic multiplicity.*

We call the eigenvalue problem for operator T a *generalized inverse Steklov problem*. A collectively compact approximation T_n to T will allow us to solve this eigenproblem numerically.

We can give a precise sense to the matrix B in terms of the following *standard direct Steklov eigenvalue problem* (see [2]):

Find all nonzero complex-valued functions $v \in V$ and all nonzero complex numbers μ such that

$$\Pi(\partial)v = 0 \text{ in } \Omega, \quad f_i v = \mu b_i v \text{ for } 1 \le i \le d, \quad b_i v = 0 \text{ for } d < i \le \mathcal{I}. \tag{2}$$

This problem is equivalent to the finite-dimensional eigenvalue problem corresponding to the operator
$$\vec{\beta} \in \mathbb{C}^{d \times 1} \mapsto fv \in \mathbb{C}^{d \times 1}$$

where v is the unique solution of
$$\Pi(\partial)v = 0 \text{ in } \Omega, \ b_i v = \beta_i \text{ for } 1 \le i \le d, \ b_i v = 0 \text{ for } d < i \le \mathcal{I},$$

that is, $v = \sum\limits_{j=1}^{d} \beta_j(p_j - g_j)$ so $f_i v = \sum\limits_{j=1}^{d} \beta_j(f_i p_j - \xi_i q_j)$. Hence, the associated eigenproblem is precisely the one of matrix B. To each eigenpair $(\mu, \vec{\beta})$ of B we can associate a function v which satisfies (2).

The preceding remarks lead to the following assertion.

Theorem 2. *If* $(\Pi(\partial)\,;\,f_1,\ldots,f_d,b_{d+1},\ldots,b_{\mathcal{I}})$ *is bijective, then* $\mu = 0$ *is not an eigenvalue of* (2) *and hence* B *is invertible.*

3. Numerical approximation

We fix the space E to be $C^0(\bar{\Omega})$. Then, the compactness of K is equivalent to the existence of a positive Radon measure m on $\bar{\Omega}$ and a kernel $\kappa : \bar{\Omega} \times \bar{\Omega} \to \mathbb{C}$ such that

(1) for all $s \in \bar{\Omega}$, the function $\kappa_s \, : \, t \in \bar{\Omega} \mapsto \kappa(s,t) \in \mathbb{C}$ belongs to $L^1(\bar{\Omega},m)$;

(2) the map $s \in \bar{\Omega} \mapsto \kappa_s \in L^1(\bar{\Omega},m)$ is continuous;

(3) the operator K is given by $Ku = \displaystyle\int_{\bar{\Omega}} \kappa(\cdot,t)u(t)\,dm(t).$

The eigenvalue problem for T consists of finding all nonzero couples $(u, \vec{\alpha}) \in X = C^0(\bar{\Omega}) \times \mathbb{C}^{d \times 1}$ and all complex numbers λ such that $T(u,\vec{\alpha}) = \lambda(u,\vec{\alpha})$. This problem leads to the following equations in the unknown $(u,\vec{\alpha})$:

$$\vec{\alpha} - \xi u = \lambda B \vec{\alpha},$$

$$\int_\Omega \Big(u(t) - \lambda \sum_{j=1}^d \alpha_j q_j(t)\Big)\kappa(\cdot,t)dm(t) + \lambda \sum_{j=1}^d \alpha_j p_j = \lambda u.$$

The last one may be rewritten as

$$\int_\Omega \kappa(\cdot,t)u(t)dm(t) = \lambda\Big(u + \sum_{j=1}^d \alpha_j(g_j - p_j)\Big).$$

We suppose that κ is a continuous kernel and that a numerical quadrature formula defined with a grid $\mathcal{G}_n$ of n points $\mathcal{G}_n \equiv \{t_{j,n} : j = 1,\ldots,n\} \subset \bar{\Omega}$, and weights $(\omega_{j,n})_{j=1}^n$ is used to approximate the integrals involved here above. Let us denote by λ_n the approximate eigenvalue and by $(u_n,\vec{\alpha}_n)$ an approximate eigenvector corresponding to λ_n. Set $\vec{u}_n \equiv [u_n(t_{1,n}),\ldots,u_n(t_{n,n})]^\top$ and define the matrices K_n, C_n and W_n by

$$K_n(i,j) \equiv \omega_{j,n}\,\kappa(t_{i,n},t_{j,n}), \ C_n(i,j) \equiv g_j(t_{i,n}) - p_j(t_{i,n}), \ W_n(i,j) \equiv \omega_{j,n}\,f_i\kappa(\cdot,t_{j,n}).$$

Then the unknowns λ_n and $(\vec{u}_n,\vec{\alpha}_n)$ satisfy

$$\vec{\alpha}_n - W_n\vec{u}_n = \lambda_n B\vec{\alpha}_n, \ K_n\vec{u}_n = \lambda_n(\vec{u}_n + C_n\vec{\alpha}_n),$$

or equivalently,

$$\begin{bmatrix} K_n & O \\ -W_n & I \end{bmatrix} \begin{bmatrix} \vec{u}_n \\ \vec{\alpha}_n \end{bmatrix} = \lambda_n \begin{bmatrix} I & C_n \\ O & B \end{bmatrix} \begin{bmatrix} \vec{u}_n \\ \vec{\alpha}_n \end{bmatrix}.$$

If B is invertible, then this generalized eigensystem may be rewritten in the standard form

$$\begin{bmatrix} K_n + C_nB^{-1}W_n & -C_nB^{-1} \\ -B^{-1}W_n & B^{-1} \end{bmatrix} \begin{bmatrix} \vec{u}_n \\ \vec{\alpha}_n \end{bmatrix} = \lambda_n \begin{bmatrix} \vec{u}_n \\ \vec{\alpha}_n \end{bmatrix}.$$

Let $\vec{\alpha}_n \equiv [\alpha_{1,n},\ldots,\alpha_{d,n}]^\top$. Then for each $\lambda_n \neq 0$ the function u_n may be recovered from its grid values $\vec{u}_n$ with the *natural interpolation* formula

$$u_n(s) = \frac{1}{\lambda_n} \sum_{j=1}^n \omega_{j,n}\,\kappa(s,t_{j,n})u_n(t_{j,n}) + \sum_{j=1}^d \alpha_{j,n}(p_j(s) - g_j(s)).$$

The eigenvector $(u_n, \vec{\alpha}_n)$ corresponds to an operator $T_n : X \to X$ defined by

$$T_n(u, \vec{\alpha}) \equiv (v_n, \vec{\beta}_n)$$

where

$$\vec{\beta}_n \equiv [\beta_{1,n}, \ldots, \beta_{d,n}]^\top = B^{-1}(\vec{\alpha} - W_n \vec{u}), \quad \vec{u} \equiv [u(t_{1,n}), \ldots, u(t_{n,n})]^\top$$

and

$$v_n \equiv \sum_{j=1}^{n} \omega_{j,n}\, \kappa(\cdot, t_{j,n}) u(t_{j,n}) + \sum_{i=1}^{d} \beta_{i,n}(p_i - g_i).$$

Theorem 3. *If B is invertible and the quadrature formula converges for continuous functions then T_n is a collectively compact approximation to T.*

Proof. Polya's theorem implies that $c_0 \equiv \sup_n \sum_{j=1}^{n} |\omega_{j,n}|$ is finite. Since the sequence $W_n \vec{u}$ is pointwise convergent in $C^0(\bar{\Omega})$, T_n is pointwise convergent to T in X. By the Banach-Steinhaus theorem, W_n is uniformly bounded with respect to n. Thus we can set $c_1 \equiv \|B^{-1}\|_\infty \left(1 + \sup_n \|W_n\|_\infty\right)$. We must prove that for some integer n_0, the set $S \equiv \{T_n(u, \vec{\alpha}) : \|u\|_\infty \leq 1, \|\vec{\alpha}\|_\infty \leq 1, n \geq n_0\}$ is relatively compact in X. Equivalently, we must prove that the set $V_S \equiv \{v \in C^0(\bar{\Omega}) : (v, \vec{\beta}) \in S\}$ is bounded and equicontinuous in $C^0(\bar{\Omega})$ and that the set $R_S \equiv \{\vec{\beta} \in \mathbb{C}^{d\times 1} : (v, \vec{\beta}) \in S\}$ is bounded in $\mathbb{C}^{d\times 1}$. If $\vec{\beta} \in R_S$ then

$$\|\vec{\beta}\|_\infty \leq \|B^{-1}\|_\infty (\|\vec{\alpha}\|_\infty + \|W_n\|_\infty \|u\|_\infty) \leq c_1,$$

so R_S is bounded. If $v \in V_S$ then

$$\|v\|_\infty \leq (c_0 + c_1 \max_{1 \leq i \leq d} \|q_i\|_\infty) \max_{s,t \in \bar{\Omega}} |\kappa(s,t)| + c_1 \max_{1 \leq i \leq d} \|p_i\|_\infty,$$

so V_S is bounded. For s and $\tilde{s}$ in $\bar{\Omega}$ we have

$$v(s) - v(\tilde{s}) = \sum_{j=1}^{n} \omega_{j,n}(\kappa(s, t_{j,n}) - \kappa(\tilde{s}, t_{j,n})) u(t_{j,n})$$

$$+ \sum_{i=1}^{d} \beta_{i,n}(p_i(s) - p_i(\tilde{s}) + g_i(\tilde{s}) - g_i(s)),$$

hence

$$|v(s) - v(\tilde{s})| \leq (c_0 + c_1 \max_{1 \leq i \leq d} \|q_i\|_\infty) \max_{t \in \bar{\Omega}} |\kappa(s,t) - \kappa(\tilde{s}, t)|$$

$$+ c_1 \max_{1 \leq i \leq d} |p_i(s) - p_i(\tilde{s})|.$$

Each function p_i is uniformly continuous in $\bar{\Omega}$ and κ is uniformly continuous in $\bar{\Omega} \times \bar{\Omega}$. Hence V_S is uniformly equicontinuous.

We recall that the collectively compact convergence of T_n to T ensures the spectral convergence in the sense of [1]. That is to say, given an eigenvalue λ of T that is an isolated point of the spectrum of T and whose spectral projection P has finite rank m, there exists a positive integer n_0 such that for each $n > n_0$, T_n has a cluster of eigenvalues whose total spectral projection P_n is a collectively compact approximation to P. This implies that P_n has finite rank m, its range converges in gap to the range of P and the sequence of complex numbers $\lambda_n \equiv \mathrm{trace}(P_n T_n)/m$ converges to λ.

38

4. Example

Consider the model problem

$$-\Delta v = \mu v \text{ in } \Omega, \qquad \frac{\partial v}{\partial \hat{n}} = \mu v \text{ on } \Gamma_1, \qquad v = 0 \text{ on } \Gamma_2 \equiv \Gamma_1^c,$$

where $\Omega \equiv \,]\,0,1\,[\,\times\,]\,0,1\,[\,$, and $\Gamma_1 \equiv \{1\} \times \,]\,0,1\,[\,$. Denoting $s_1 = s$, $s_2 = t$ and separating variables $v(s,t) = y(s)\theta(t)$ we find that for each positive integer k we must find a function y such that

$$-y''(s) + (k\pi)^2 y(s) = \mu y(s), \text{ for } s \in \,]\,0,1\,[\,, \quad y'(1) = \mu y(1), \quad y(0) = 0.$$

We identify the boundary linear forms $b_1 y = y(1)$, $b_2 y = y(0)$ and $f_1 y = y'(1)$. Multiplying the differential equation by $\bar{y}$, then integrating (the first member by parts) and taking into account the boundary conditions, we find $\|y'\|_2^2 + (k\pi)^2 \|y\|_2^2 = \mu(\|y\|_2^2 + |y(1)|^2)$, where $\|\cdot\|_2$ denotes the $L_2(\bar{\Omega})$−norm. If $\|y\|_2 = 0$ then $y = 0$ because y is continuous. Hence $\mu = (\|y'\|_2^2 + (k\pi)^2 \|y\|_2^2)/(\|y\|_2^2 + |y(1)|^2) \in I\!R_+$.

Let us fix $k = 1$. Setting $\nu^2 \equiv \pi^2 - \mu$ we find that ν must satisfy the equation

$$(\pi^2 - \nu^2)\sinh \nu - \nu \cosh \nu = 0, \tag{3}$$

which has a unique positive solution.

The eigenspace corresponding to the unique eigenvalue $\mu \in \,]\,0,\pi^2\,[\,$ is generated by the eigenfunction y_μ defined by $y_\mu(s) = \sinh(s\sqrt{\pi^2 - \mu})$.

Setting $\nu^2 = \mu - \pi^2$ we find that ν must satisfy the equation

$$\nu \cos \nu - (\nu^2 + \pi^2)\sin \nu = 0, \tag{4}$$

which has an infinite countable set of positive solutions.

The eigenspace corresponding to each eigenvalue $\mu \in \,]\,\pi^2, +\infty\,[\,$ is generated by the eigenfunction y_μ defined by $y_\mu(s) = \sin(s\sqrt{\mu - \pi^2})$.

Let us come back to the Green function approach developed in the preceding section. An easy computation gives $p_1(s) = s$, $q_1(s) = \pi^2 s$, $p_2(s) = 1 - s$, $q_2(s) = \pi^2(1-s)$, $f_1 p_1 = 1$.

The kernel κ is the Green function defined by

$$\kappa(s,t) = \frac{1}{\pi \sinh \pi} \begin{cases} \sinh \pi(1-t)\sinh \pi s & \text{if } 0 \le s \le t \le 1, \\ \sinh \pi(1-s)\sinh \pi t & \text{if } 0 \le t < s \le 1, \end{cases}$$

$$B(1,1) = \pi \frac{\cosh \pi}{\sinh \pi}, \quad f_1 \kappa(\cdot, t) = -\frac{\sinh \pi t}{\sinh \pi}, \quad g_1(s) = s - \frac{\sinh \pi s}{\sinh \pi}$$

In this example, the approximation T_n is built with the trapezoidal composite rule using 41 equally spaced points. We have used the Newton-Raphson method to approximate the unique positive root ν_0 of (3) and the first nine positive roots $(\nu_j)_{j=1}^9$ of (4). These values lead to eigenvalues

$$\lambda_0 = \frac{1}{\pi^2 - \nu_0^2} \quad \text{and, for } j = 1,\ldots,9, \quad \lambda_j = \frac{1}{\nu_j^2 - \pi^2},$$

which may be considered as being the exact ones. They are compared with the corresponding approximate eigenvalues provided by T_n, with $n = 41$, in the following table.

	APPROXIMATE EIGENVALUES	
	From (3) and (4)	Eigenvalues of T_n
j	(Newton-Raphson)	(with $n = 41$)
0	0.3700185516E-00	0.3701147779E-00
1	0.4818214643E-01	0.4823680311E-01
2	0.1963232407E-01	0.1968607326E-01
3	0.9951702595E-02	0.1000484817E-01
4	0.5894213388E-02	0.5947125240E-02
5	0.3868063892E-02	0.3920934498E-02
6	0.2723925530E-02	0.2776865444E-02
7	0.2018419648E-02	0.2071503314E-02
8	0.1554023099E-02	0.1607308060E-02
9	0.1232618216E-02	0.1286153675E-02

These results improve on those obtained by a finite difference scheme applied to the direct generalized Steklov problem. That scheme proves to be equivalent to a rectangle composite quadrature formula applied to the generalized inverse Steklov problem, as shown in [3].

The results of this paper can be easily extended to weakly singular integral operators. Also, numerical computations can be improved by iterative refinement schemes based on Newton-like procedures.

References

1. M. Ahues and F. Hocine, A note on spectral approximation of linear operators, *Appl. Math. Lett.* **2** (1994), 63–66.

2. J.H. Bramble and J.E. Osborn, Approximation of Steklov eigenvalues of non-selfadjoint second order elliptic operators, in *The mathematical foundations of the finite element method with applications to partial differential equations*, A.K. Aziz (Ed.), Academic Press, New York, 1972.

3. C. Carasso and G. Pasa, Convergence d'un schéma pour estimer la stabilité de l'interface en récupération assistée, UMR 5585 Équipe d'Analyse Numérique Lyon-St-Étienne No 279, 1998, web : http://numerix.univ-lyon1.fr/tdumont.

UMR 5585 CNRS - Université Jean Monnet de St-Étienne, 23 rue Dr. Paul Michelon, 42023 St-Étienne, France

C. AVILES-RAMOS, K.T. HARRIS and A. HAJI-SHEIKH

A hybrid root finder

1. Introduction

Standard root-finding techniques are commonly applied to boundary value problems, property estimation, etc. They are usually chosen so that the iteration never gets outside the best bracketing bounds obtained at any stage in the calculation. The secant and Newton-Raphson methods can violate this condition [1]. The bisection method does not violate this condition, but it converges more slowly toward the root. Ridders' method [2] is a variant of the false position method that is guaranteed to stay inside the bracketed interval and its convergence is superlinear. An algorithm developed in the 1960s by van Wijngaarden, Dekker and others has superlinear convergence and stays within the bracketed root [3]. An improved algorithm presented by Brent [3] guarantees convergence if there is a root inside the bracketed interval.

A hybrid root-finding method is developed that efficiently finds an accurate root. This method is an application of a high-order Newton's and the bisection methods. The information gained while executing the bisection method is preserved to compute the first and second derivatives. The hybrid root finder has a few unique features. Except for computation of $f(x)$, it contains no strenuous mathematical relations. The technique uses the bisection method until it achieves a certain criterion. Below is the description of the hybrid root finder; it is followed by relevant numerical examples.

2. The hybrid root finder

The objective is to develop an efficient root finder that can produce 10^5 to 10^6 accurate eigenvalues in a short time, and with a sufficiently high computational speed. To accomplish this purpose, the root finder must be free from computationally slow functions such as square roots and produce accurate results with a few iterations. The hybrid root finder, developed here, satisfies all these requirements. The root finder uses the bisection method to find the region where the root is located. Then, it uses a high-order Newton's method to compute the root. Meanwhile, central differencing is used to compute the first and second derivatives of the function under consideration.

Let $f(x) = 0$ be a function whose root is being sought. The Taylor series expansion of $f(x) = f(x_0 + h)$ is

$$f(x_0 + h) \approx f(x_0) + f'(x_0)h + \frac{1}{2}f''(x_0)h^2 \tag{1}$$

As short-hand notations, define $a = f'(x_0) = (f_2 - f_1)/2\Delta x$ and $b = f''(x_0) = (f_2 + f_1 - 2f_0)/\Delta x^2$ where $f_0 = f(x_0)$, $f_1 = f(x_1)$, and $f_2 = f(x_2)$. The locations of f_0, f_1, f_2, and the value of Δx are shown in Fig. 1. Based on Eq. (1), the root is located at $f(x) = f(x + h) = 0$; that is,

$$f(x_0) + ah + \frac{1}{2}bh^2 = 0. \tag{2}$$

The bisection method was used to find the approximate location of x_0 that is used to calculate f_0. Since x_1, x_2, f_1, and f_2 are readily available from earlier steps, then a central differencing scheme as in a finite difference analysis yields the values of a and b.

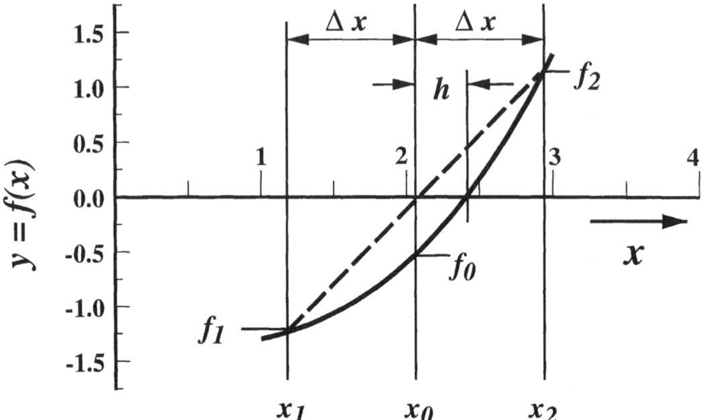

Fig. 1. A graphical description of parameters in the hybrid root finder.

Once the values of a and b are known, the parameter h points to the location of the root at $x = x_0 + h$. However, to avoid using the square root function, one can set

$$h = -f_0/a + \epsilon, \qquad (3)$$

where $\epsilon < h$ because ϵ is a slight perturbation of h. Now, it is sufficient to compute ϵ instead of h. After substituting for h in Eq. (1), one obtains

$$b(-f_0/a + \epsilon)^2 + 2a(-f_0/a + \epsilon) + 2f_0 = 0, \qquad (4)$$

which produces

$$\epsilon = \frac{1}{2}(bf_0{}^2/a)/(bf_0 - a^2) \qquad (5)$$

In the derivation of (5), it is assumed that $\epsilon < h$, hence the term containing ϵ^2 is deleted. A computer program is written to include the algebra described above. The computer program is tested using several numerical examples.

Example 1. A simple example is selected to show the methology and convergence behavior of this root finding technique. Data are processed using an Intel Pentium processor. The roots of the function $\sin(x)=0$ located between 2 and 4 is computed by the Newton's method, Table 1, and by this method, Table 2. The computations are performed using Fortran 77 with double precision accuracy. For comparison, the root is located at $x = \pi$=3.14159265358979323, computed using Mathematica.

42

Iterations	Root	Criterion	Acceleration	Percent Error
1	3.169401614135062	-0.278054E-01	0.659595E-02	8.85E-03
2	3.136345059787232	0.524757E-02	0.251164E-03	1.67E-03
3	3.142581331822654	-0.988678E-03	0.893907E-05	3.15E-04
5	3.141627744540093	-0.350910E-04	0.112620E-07	1.12E-05
10	3.141592645261746	0.832805E-08	0.634324E-15	2.65E-09
15	3.141592653591770	-0.197696E-11	0.357330E-22	6.59E-14
18	3.141592653589780	0.130010E-13	0.159189E-26	2.26E-15
19	3.141592653589796	-0.298616E-14	0.561637E-28	8.48E-16
20	3.141592653589793	0.122461E-15	0.255106E-29	2.70E-18

Table 1. Calculated root using the first-order Newton method.

Iterations	Root	Criterion	Acceleration	Percent Error
1	3.167224329702086	-0.256289E-01	0.659595E-02	8.16E-03
2	3.14595438470406	-0.278488E-05	0.164318E-03	8.86E-07
3	3.141592653589793	0.122461E-15	0.193887E-11	2.70E-18

Table 2. Calculated root using the hybrid root finder.

The first columns in Tables 1 and 2 represent the number of iterations while the second columns show the computed root. The third column is the value of the function where the computed root is located. The acceleration factor, ϕ, in the fourth column indicates the behavior of the function, hence

$$\phi = [f''(x)/f'(x)]/[f'(x)/f(x)] = [f''(x)/f(x)]/[f'(x)]^2 \tag{6}$$

This term indicates the speed that higher terms in the Taylor series become negligible. The fifth columns in Tables 1 and 2 are the magnitudes of the deviation from π.

Example 2. The next example shows the utility of this hybrid rootfinder for a demanding application. Many boundary value problems require finding eigenvalues that are often the roots of transcendental equations. The task of finding many irregularly spaced eigenvalues is demanding. The computation of Green's function in a heterogeneous medium [4] consisting an orthotropic region 1 and an isotropic region 2 is one example. The surfaces are located at $x = 0$, $x = a$, $y = 0$, and $y = c$ so that $0 < y < b$ for region 1 and $0 < y < c$ for region 2 for all x values. All boundaries are insulated. For region 1, the thermal conductivity is k_{1x} in x-direction and k_{2y} in y-direction whereas k_2 is the thermal conductivity of region 2. Also, subscripts 1 and 2 will identify the density ρ and specific heat c_p in regions 1 and 2, and therefore the respective thermal diffusivities become $\alpha_{1x} = k_{1x}/\rho_1 c_{p1}$, and $\alpha_2 = k_2/\rho_2 c_{p2}$. There is a perfect contact between regions 1 and 2.

The temperature solution [4] requires finding the roots of equation

$$\tan[\eta_n(c - b)] = -\beta(\gamma_n/\eta_n)\tan(\gamma_n k_r b), \tag{7}$$

where $\beta = (k_{1x}k_{1y})^{1/2}/k_2$ and $k_r = k_{1x}/k_{1y}$. The variables γ_n and η_n may be real or imaginary. They are related to the eigenvalue λ_n by the relations

$$\gamma_n = [\lambda_n^2/\alpha_{1x} - (n\pi/a)^2]^{1/2}, \tag{8}$$

$$\eta_n = [\lambda_n^2/\alpha_2 - (n\pi/a)^2]^{1/2}. \tag{9}$$

Equation (7) has infinite roots for a given integer n, therefore the individual roots are identified by $\lambda_{n,m}$.

In this example, the following parameters are selected: $a = 7.5$, $c = 1$, $b = 0.9$, $k_{1x} = 0.4$, $k_{1y} = 0.1$, $k_2 = 0.001$, and $\rho_1 c_{p1} = \rho_2 c_{p2} = 2.5$. The solid lines in Fig. 2 show the variation of the left-hand side and the dash lines demonstrate the variation of the right-hand side of (7). Equation (7) is satisfied where a solid line intersects a dash line. The itersections are shown by circles in Fig. 2 and they designate the location of the eigenvalues. Note that there is an eigenvalue between any two adjacent asymptotes. This is a useful property that identifies a subregion wherein a particular eigenvalue is located. Often, only 3 to 4 iterations are needed to achieve an accurate solution with as many as 16 significant figures. An alternative form of Eq. (7),

$$f(\eta_n) = \sin[\eta_n(c - b)]\cos(\gamma_n k_r b) + \beta(\gamma_n/\eta_n)\sin(\gamma_n k_r b)\cos[\eta_n(c - b)] \tag{10}$$

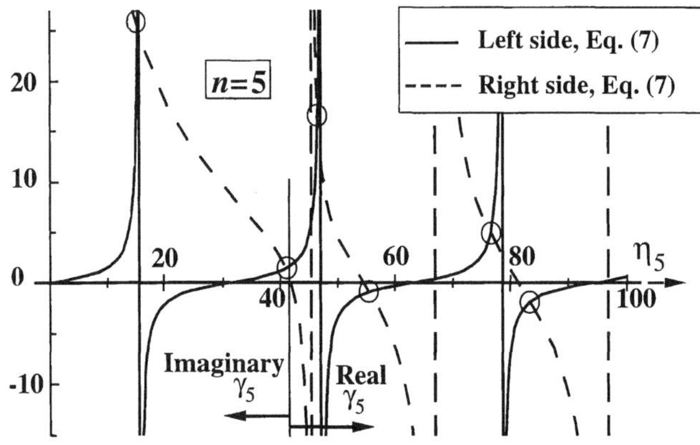

Fig. 2. Asymptotes of the left-hand side and right-hand side of
Eq. (7) within which the eigenvalues are located.

is a function whose roots are computed by the hybrid root finder by setting $f(\eta_n) = 0$. Each function has many roots and the $\lambda_{n,m}$ is the mth root of $f(\eta_n)$. As an illustration, for $n = 5$, the function $f(\eta_n)$ is plotted as a function of λ_n in Fig. 3. The calculated first 6 eigenvalues $\lambda_{n,m}$, for $n=5$, that is, $m = 1$ to 6, are given on the same figure.

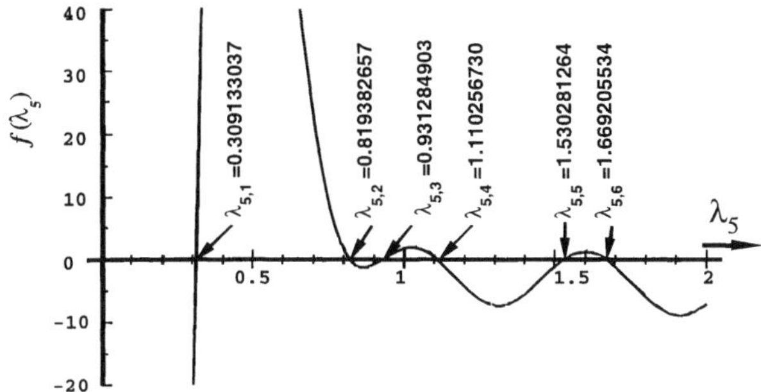

Fig. 3. Location of eigenvalues when $n = 5$.

3. Conclusion

The hybrid root finder is a useful tool for finding the roots of transcedental equations one encounters when solving the diffusion equation. An examination of data in Tables 1 and 2 shows that the hybrid root finder has excellent convergence characteristics. Figures 2 and 3 show that the computation of eigenvalues for demanding application is possible. In all cases studied, this method accommodates functions with erratic behavior by simply adjusting the convergence criterion. It was observed that any convergence criterion must agree with the computer's word size.

References

1. W.H. Press, S.A. Teukolsky, W.T. Vetterling and B.P. Flannery, *Numerical Recipes in Fortran. The Art of Scientific Computing*, Cambridge University Press, 1994.

2. C.J.F. Ridders, *IEEE Transactions on Circuits and Systems*, **CAS-26**, 1979, 979–980.

3. R.P. Brent, *Algorithms for Minimization without Derivatives*, Chapters 3 and 4, Prentice-Hall, Englewood Cliffs, NJ, 1973.

4. C. Aviles-Ramos, A. Haji-Sheikh and J.V. Beck, Exact solution of heat conduction in composites and application to inverse problems, *ASME J. Heat Transfer* **120** (1998), 592–599.

Department of Mechanical and Aerospace Engineering, The University of Texas at Arlington, Arlington, Texas 76019-0023, USA

AKBAR H. BEGMATOV

Volterra-type integral geometry problems

1. Introduction

We shall consider the following problem [1]. Let $v(x)$ be a sufficiently smooth function defined in n-dimensional space $(x_1, \ldots, x_n)$ and let $\{S(y)\}$ be a family of smooth manifolds in this space depending on a parameter $y = (y_1, \ldots, y_k)$. Suppose, further, that about the function $v(x)$ we know the integrals

$$\int_{S(y)} g(x, y) v(x)\, ds = f(y), \tag{1}$$

where $g(x, y)$ is a given weight function, ds defines the element of measure on $S(y)$. The problem of integral geometry (1) is the problem of finding the function $v(x)$ from its given integrals, i.e., from the function $f(y)$.

Problems of integral geometry arise in a natural way in the study of many mathematical models in domains having extensive applications, such as seismic prospecting, interpretation of the data of geophysical and aerospace observations, various processes described by kinetic equations, and so on [2]. The apparatus devised here is the mathematical basis of computerized tomography, a promising and intensively developed direction of modern science [3]. Wide classes of inverse problems for partial differential equations of various type reduce to the solution of problems of integral geometry (see [2] and the references given there).

V.G. Romanov obtained rather general results on uniqueness of solutions to Volterra problems of integral geometry in the case when the manifolds of integration have the form of paraboloids and the manifolds and weight functions of perturbations, i.e., additional integrals over the bodies bounded by the manifolds, are invariant under the group of all motions parallel to some $(n-1)$-dimensional hyperplane (see [4]). In a rather general formulation the problem of integral geometry was investigated by Yu.E. Anikonov and A.L. Bukhgeim in special classes of functions (see [5,6]). Some problems of integral geometry in the plane were considered in the author's article [7].

With the advance of the technical base of computerized tomography and widening of the field of application of tomography methods the problems of determination of functions from their integrals over manifolds of rather complicated shape acquire greater importance. The importance of the problem of integral geometry of rather general form is also caused by the intrinsic demands of the theory of ill-posed problems of mathematical physics and analysis as well as the fact that some inverse problems for PDE reduce to solving problems of integral geometry.

In the present article, we consider the problem of integral geometry in a three-dimensional layer on a family of paraboloids with perturbation which represents the integral over the interior of the paraboloids with a known weight function. We give a uniqueness theorem for this problem under rather general assumptions on the weight function. We show that the uniqueness question for the problem of integral geometry of rather general form reduces to the studying uniqueness of solutions of the above-mentioned operator equation (i.e., the problem of integral geometry with perturbation). Unlike the mentioned works of V.G. Romanov, we do not suppose that the manifolds over which we integrate as well as the weight functions are invariant under the group of parallel translations along some hyperplane.

This work was partly supported by a grant from the State Committee for Science and Technology of the Republic of Uzbekistan.

46

2. Integral geometry problem with perturbation

We use the notation $\xi \in \mathbf{R}^3$, $x \in \mathbf{R}^3$, $\bar{\xi} = (\xi_1, \xi_2)$, $\bar{x} = (x_1, x_2)$, $\mathbf{R}_+^3 = \{x = (x_1, x_2, x_3) : x_3 \geq 0\}$.

Let $\{P(x)\}$ be a family of paraboloids with vertices at x that are defined by the relations

$$x_3 - \xi_3 = |\bar{x} - \bar{\xi}|^2, \ \xi_3 \geq 0.$$

By $p(\bar{x})$ we denote the projection of $P(x)$ onto the plane $x_3 = 0$; by $V(x)$ we denote the part of the three-dimensional space bounded by the surface of the paraboloid $P(x)$ and the plane $x_3 = 0$.

Problem 1. Determine a function $u(x)$ given the sums of the integrals of $u(x)$ over the paraboloids $P(x)$ and the integrals of $u(x)$ with weight function $G(x, \xi)$ over the volumes $V(x)$

$$\int\limits_{p(\bar{x})} u(\xi)\, d\bar{\xi} + \int\limits_{V(x)} G(x, \xi)\, u(\xi)\, d\xi = f(x), \tag{2}$$

for all x from the layer

$$L = \{x \in \mathbf{R}_+^3 : 0 \leq x_3 \leq l, \, l < \infty\}.$$

Here $\rho = \sqrt{x_3 - \xi_3}$.

Theorem 1. *Suppose that the function $f(x)$ is known in the layer L, the function $G(x, \xi)$ has all continuous derivatives up to the order 2 inclusively and vanishes together with its derivatives on the surface of the paraboloid $P(x)$.*

Then a solution to Problem 1 is unique in the class $C_0^2(L)$.

Proof. Let us consider

$$f_0(x) = \int\limits_{p(\bar{x})} u(\xi)\, d\bar{\xi}, \quad f_1(x) = \int\limits_{V(x)} G(x, \xi)\, u(\xi)\, d\xi.$$

We introduce auxiliary functions

$$\mathcal{F}(x, t) = \int\limits_0^{x_3} \int\limits_{-\pi}^{\pi} u(x_1 + \tau \rho \cos\varphi, \ x_2 + \tau \rho \sin\varphi, \xi_3)\, d\varphi\, d\zeta$$

and

$$\mathcal{J}(x, t) = \frac{\partial}{\partial t} \mathcal{F}(x, t),$$

where $\tau = \sqrt{1 - \tau}$ and $t \in [0, 1]$.

It is easy to see that

$$\mathcal{F}(x, 0) = f_0(x),$$

$$\mathcal{F}(x, 1) = 2\pi \int\limits_0^{x_3} u(x_1, x_2, \xi_3)\, d\xi_3.$$

The function $\mathcal{J}(x, t)$ satisfies an equation:

$$\frac{\partial^2}{\partial x_3 \partial t} \mathcal{J} = -\frac{1}{4} \triangle_{x_1 x_2} \mathcal{J}.$$

Suppose that $f(x) = 0$. Then

$$\frac{\partial}{\partial x_3}\mathcal{J}(x,t) = -\frac{1}{4}\int_0^{x_3}\triangle_{x_1 x_2}\mathcal{J}(x,\tau)\,d\tau + \varphi(x), \tag{3}$$

where

$$\varphi(x) = -\frac{1}{4}\triangle_{x_1 x_2}f_0(x) = \frac{1}{4}\triangle_{x_1 x_2}f_1(x)$$

Uniqueness of the solution of an integral geometry problem with perturbation (2) is based on using the method of energy inequalities to study (3).

Let us mention that a problem of integral geometry in the plane on parabolas with a perturbation that is an integral with a weight function over the interiors of the parabolas was considered in [8].

3. Uniqueness theorem for the general problem of integral geometry

In this part we establish a uniqueness theorem for the general Volterra-type problem of integral geometry. As we show below, this problem can be reduced to the problem of integral geometry on a family of paraboloids with perturbation.

We present the formulation of the general Volterra-type problem of integral geometry in the three-dimensional space. Consider the family of surfaces $\{\Gamma(\xi)\}$ in $\mathbf{R}^3_+$ that satisfy the following conditions:

a) The surfaces $\{\Gamma(\xi)\}$ are uniquely parameterized by their vertices ξ.

b) The surfaces from the family smoothly fill the layer

$$H = \{x \in \mathbf{R}^3_+ : 0 \le x_3 \le h, \, h < \infty\}.$$

c) Any surface from the family can be smoothly continued to

$$\mathbf{R}^3_- = \{x = (x_1, x_2, x_3) : x_3 < 0\}.$$

d) The unit outer normal to an arbitrary surface $\Gamma(\xi)$ at any point $x \in \Gamma(\xi)$ belongs to the set

$$M = \{m = (m_1, m_2, m_3) : |m| = 1, -1 < k_i \le m_i \le n_i < 1,$$

$$i = 1, 2; \, 0 < k_3 \le m_3 \le 1\}.$$

A surface $\Gamma(\xi)$ from this family is specified by the equation

$$x_3 = \psi(\xi, \bar{x}).$$

Problem 2. Consider the problem of reconstructing the function of three variables $f(x)$, from its integrals over surfaces of the family of $\{\Gamma(\xi)\}$, with given weight function $a(\xi, \bar{x})$, i.e., the problem of solving the operator equation for a function $f(x)$:

$$\int_{G(\bar{\xi})} a(\xi, \bar{x})f(x)d\bar{x} = b(\xi). \tag{4}$$

Here $G(\bar{\xi})$ is the projection of $\Gamma(\bar{\xi})$ to the plane $x_3 = 0$. We denote by ∂G the boundary of $G(\bar{\xi})$.

Theorem 2. *Suppose that* $a(\xi, \bar{x}) \in C^4$, $\psi(\xi, \bar{x}) \in C^6$ *and satisfy the conditions:*

$$\psi(\xi, \bar{x}) < \xi_3, \quad \bar{x} \neq \bar{\xi}, \tag{5}$$

$$\psi(\xi, \bar{\xi}) = \xi_3, \tag{6}$$

$$\psi(\xi, \bar{x}) = 0, \qquad if \quad \bar{x} \in \partial G(\bar{\xi}), \tag{7}$$

$$\frac{\partial^2 \psi}{\partial x_1^2} n_1^2 + \frac{\partial^2 \psi}{\partial x_1 \partial x_2} n_1 n_2 + \frac{\partial^2 \psi}{\partial x_2^2} n_2^2 \leq \delta_2 < 0, \tag{8}$$

where $n = (n_1, n_2), |n| = 1,$

$$a(\xi, \bar{x}) \geq \delta_1 > 0. \tag{9}$$

Then the solution to Problem (2) is unique in the class $C_0^2(H)$.

To prove Theorem 2, we need the following lemma.

Lemma. *If conditions a)–d) and (5)–(9) are fulfilled, then an arbitrary surface from the family* $\{\Gamma(\xi)\}$ *is uniquely determined by means of 5 parameters introduced above.*

From the enumerated conditions it follows that for any point $x \in H$ and any normal $m \in M$ there exists a unique surface $\Gamma(x, m)$ which passes through the point x and has normal m at this point.

Using parameterization of surfaces $\Gamma(\xi)$ by means of x, m_1, m_2, we can reduce the problem of solving (4) to the problem of solving

$$\int\limits_{\Gamma(x, \bar{m})} a_1(x, \bar{m}, y) f_1(y) ds = C_1(x, \bar{m}(\xi, x)). \tag{10}$$

The problem of solving (10) in fact is not overdetermined, because the function $C_1(x, \bar{m}(\xi, x))$ satisfies two first-order differential equations. Equation (10) is reduced to the canonical form (see [9]) by means of linear integral operators and a differential operator.

Thus, we can reduce the general problem of integral geometry (4) to the problem of integral geometry on the family of paraboloids with perturbation which is an integral with general weight over the interiors of paraboloids. Then the uniqueness of the solution to initial problem of integral geometry in the class $C_0^2(H)$ follows from Theorem 1.

References

1. I.M. Gel'fand, M.I. Graev, and N.Ya. Vilenkin, *Generalized functions. Vol. 5. Integral geometry and representation theory*, Fizmatgiz, Moscow, 1962; English transl.: Academic Press, 1966.

2. M.M. Lavrent'ev, V.G. Romanov, and S.P. Shishatskii, *Ill-posed problems of mathematical physics and analysis*, Nauka, Moscow, 1980; English transl.: AMS, 1986.

3. F. Natterer, *The mathematics of computerized tomography*, Teubner, Stuttgart, 1986.

4. V.G. Romanov, *Some inverse problems for the equations of hyperbolic type*, Nauka, Novosibirsk, 1972; English transl.: *Integral geometry and inverse problems for hyperbolic equations*, Springer-Verlag, 1974.

5. Yu.E. Anikonov, *Some methods for the study of multidimensional inverse problems for differential equations*, Nauka, Novosibirsk, 1978 (Russian).

6. A.L. Bukhgeĭm, *Volterra equations and inverse problems*, Nauka, Novosibirsk, 1983 (Russian).

7. Akbar H. Begmatov, On a class of problems of integral geometry in the plane, *Dokl. Akad. Nauk* **331** (1993), 261–262; English transl.: *Dokl. Math.* **48** (1994), 56–58.

8. M.M. Lavrent'ev, Integral geometry problems with perturbation on the plane, *Sibirsk. Mat. Zh.* **37** (1996), 851–857; English transl.: *Siberian Math. J.* **37** (1996), 747–752.

9. Akbar H. Begmatov, Reducing problems of integral geometry in the space to canonical form, in *Mathematical analysis and differential equations*, M.M. Lavrent'ev (Ed.), Novosibirsk, 1991 (Russian).

Department of Mathematics and Physics, Cooperative Institute, Samarkand, Uzbekistan

AKBAR H. BEGMATOV and AKRAM H. BEGMATOV

Inversion of the X-ray transform and the Radon transform with incomplete data

1. Introduction

In the present article, we study two problems of reconstruction of a function from its integrals over families of linear manifolds.

In the first problem, we know the integrals of the sought function over a certain family of straight lines that are generators of cones. Such integral transformations are called X-ray transforms [1] and have various applications in study of problems of computerized tomography [2]. Inversion formulas for the ray transform connected with the cone scanning geometry used in computerized tomography are presented in [3–5] (see also the review articles [6,7] and the bibliography therein).

Uniqueness and stability theorems in the Sobolev spaces and a rather simple inversion formula for the problem of integral geometry on a family of right circular cones in an even-dimensional space are given in the second author's articles [8,9]. The problem considered in Sections 2 and 3 is connected with an auxiliary problem of analytic continuation and is strongly ill-posed, unlike those of [8,9]. We obtain a uniqueness theorem for its solution in the class of continuous compactly supported functions and a stability estimate of logarithmic type.

The second represents the problem of inversion of the Radon transform on the plane with restricted range of angles and perturbation. Here we give a uniqueness theorem for this problem. Such problems arise from consideration of some problems of integral geometry and computerized tomography (see [2,10]). Operator equations like those in the Radon problem with perturbation were considered in [11,12] under different assumptions on the weight function.

2. Uniqueness of a solution to the inversion problem for the X-ray transform with incomplete data

Introduce the following notation: $x, \xi, \lambda \in \mathbf{R}^3$; $\bar{x} = (x_1, x_2)$, $\bar{\xi} = (\xi_1, \xi_2)$, $\bar{\lambda} = (\lambda_1, \lambda_2)$; $\Theta = \{\alpha : \alpha \in [0, 2\pi]\}$, $\mu = (\cos\alpha, \sin\alpha)$; $Q = \mathbf{R}^3 \times \Theta$, $\mathcal{D} = \mathbf{R}^2 \times \Theta$, $\Omega = \{x : |\bar{x}| < 1, |x_3| < l, 0 < l < \infty\}$.

Denote by $K(x)$ the family of two-sheeted cones defined as follows:

$$K(x) = \left\{\xi \in \mathbf{R}^3 : |x_3 - \xi_3| = |\bar{x} - \bar{\xi}|\right\}.$$

Consider the operator equation in the function $u(x)$:

$$\int_{\mathbf{R}^1} u(x_1 + s\cos\alpha, x_2 + s\sin\alpha, x_3 + s)\, ds = f(x, \alpha). \tag{1}$$

The problem of solving (1) is the problem of integral geometry for a family of straight lines that are the generators of two-sheeted cones $K(x)$. The sought function $u(\cdot)$ is a function of three variables. The right-hand side of (1) depends on the four

This work was partly supported by a grant from the State Committee for Science and Technology of the Republic of Uzbekistan.

parameters x_1, x_2, x_3, and α. However, as is demonstrated in Theorem 1 below, the function $u(\cdot)$ is uniquely determined by the function $f(\cdot)$.

The right-hand side of (1) satisfies the following first-order partial differential equation:

$$\frac{\partial f}{\partial x_1} \cos \alpha + \frac{\partial f}{\partial x_2} \sin \alpha + \frac{\partial f}{\partial x_3} = 0. \tag{2}$$

We can take another parameterization for the problem (1), using three parameters z_1, z_2, and α, where $z_1 = x_1 - x_3 \cos \alpha$ and $z_2 = x_2 - x_3 \sin \alpha$. It is easy to see that the function $f_1(z_1, z_2, \alpha) = f(x, \alpha)$ satisfies (2). Thus, in fact the problem of solving (1) is not overdetermined.

Theorem 1. *Suppose that the function $f(x, \alpha)$ is known for all $(x, \alpha) \in Q$. Then the solution to (1) in the class of continuous functions compactly supported in Ω is unique.*

Proof. Applying the Fourier transform in the variables x_1 and x_2 to both sides of (1), we obtain

$$\int_{\mathbf{R}^1} e^{-is\langle \bar{\lambda}, \mu \rangle} w(\lambda_1, \lambda_2, s + x_3)\, ds = \psi(\bar{\lambda}, x_3, \alpha). \tag{3}$$

Here $w(\bar{\lambda}, x_3)$ and $\psi(\bar{\lambda}, x_3, \alpha)$ are the Fourier transforms in the variables x_1 and x_2 of the functions $u(x)$ and $f(x, \alpha)$, respectively, and $\langle \cdot, \cdot \rangle$ stands for the inner product.

We can rewrite (3) as

$$v(\lambda) = e^{ix_3 \lambda_3} \psi(\lambda, \alpha), \tag{4}$$

where $v(\lambda)$ is the Fourier transform in the variable x_3 of the function $w(\lambda_1, \lambda_2, x_3)$ and $\lambda_1 \cos \alpha + \lambda_2 \sin \alpha = -\lambda_3$.

In the space of the Fourier variables, consider the two-sheeted cone

$$\mathcal{K} = \{(\lambda_1, \lambda_2, \lambda_3) \in \mathbf{R}^3 : |\bar{\lambda}| = |\lambda_3|\}$$

with vertex at the origin, and denote $\tilde{\mathcal{K}} = \{\lambda \in \mathbf{R}^3 : |\bar{\lambda}| < |\lambda_3|\}$.

Recalling that (4) holds for all $\alpha \in [0, 2\pi]$ and $\psi(\bar{\lambda}, x_3, \alpha)$ is the Fourier transform in the variables (x_1, x_2) of the function $f(x, \alpha)$, we conclude that the value of the function $v(\lambda)$ at every $\lambda \in \mathbf{R}^3 \setminus \tilde{\mathcal{K}}$ is uniquely determined from (4). Thus, the problem of finding the function $u(x)$ in (1) reduces to the problem of extending its Fourier transform $v(\lambda)$ from $\mathbf{R}^3 \setminus \tilde{\mathcal{K}}$ into the interior of the cone $\mathcal{K}$, and the assertion of Theorem 1 follows from uniqueness of a solution to the latter problem.

3. A stability estimate for a solution to the inversion problem for the X-ray transform with incomplete data

We furnish the space of functions $f(x, \alpha)$ with the norm

$$\|f(x, \alpha)\|_1 = \left\| f_1(x_1 - x_3 \cos \alpha, x_2 - x_3 \sin \alpha, \alpha) \right\|_{C(\mathcal{D})}.$$

Theorem 2. *Suppose that the function $u(x)$ belongs to the class $C_0^5(\Omega)$ and the following inequalities are valid:*

$$\|u(\cdot)\|_{C_0^5(\Omega)} < 1,$$

$$\|f(\cdot)\|_1 < \varepsilon,$$

with $\varepsilon > 0$ sufficiently small.

Then a solution to (1) satisfies the conditional stability estimate

$$\|u(\cdot)\|_C < a_1 \left| \ln \frac{1}{\varepsilon} \right|^{-1},$$

where a_1 is some constant.

Proof. As mentioned above, the initial problem of finding the function $u(x)$ in (1) can be reduced to the problem of extending its Fourier transform $v(\lambda)$ from $\mathbf{R}^3 \setminus \tilde{\mathcal{K}}$ into the interior of the cone $\mathcal{K}$.

In the complex plane, consider the strip

$$S = \{ z = z_1 + iz_2 : z_1 \in \mathbf{R}^1, |z_2| < a\pi, a > 0 \}$$

and the rays

$$r_1 = \{ z : -\infty < z_1 \le -a, \ z_2 = 0 \},$$

$$r_2 = \{ z : a \le x < \infty, \ y = 0 \}.$$

Let $G = S \setminus \{ r_1 \cup r_2 \}$. In other words, the domain G is the strip S with cuts along the rays r_1 and r_2. Introduce the following notation: $E = r_1 \cup r_2$, $\bar{G}$ is the closure of the domain G, ∂G is the boundary of the domain G, and $\omega = \omega(z, E, G)$ is the harmonic measure of the set E with respect to the domain G.

We essentially use the following assertion in the proof of Theorem 2:

Lemma. *The harmonic measure ω satisfies the estimate*

$$\frac{2}{3} < \omega(z, E, G) \le 1.$$

Remark. We can easily generalize the above results to the case of cones of the form

$$\sum_{m=1}^{2} a_m^2 (x_m - \xi_m)^2 = (y - \eta)^2, \quad a_m \in \mathbf{R}^1.$$

The methods given above enable us to obtain uniqueness and stability results for the inversion problem for the X-ray transform in the case when the directrices of the cones are smooth closed convex planar curves.

4. Uniqueness of the solution of the Radon problem with incomplete data and perturbation

We adopt the following notation: $L = \left\{ x \in \mathbf{R}^2 : 0 < x_2 < l, \, l < \infty \right\}$ is a strip in $\mathbf{R}^2$, $P(x, m)$ is the straight line passing through the point $x \in L$ with the normal m, where $m \in M$, $M \subset S_+ = \{m = (m_1, m_2) : |m| = 1, m_2 \geq 0\}$, and $S(x, m)$ is the half-plane bounded by $P(\cdot)$ and containing the vector $-m$.

Let us consider an operator equation

$$\int\limits_{P(x,m)} g(y)\,ds + \int\limits_{S(x,m)} B(x, y, m)g(y)\,d\omega = C(x, m) \tag{5}$$

It is easy to see that the first operator on the left-hand side of the equation (5) is the well-known Radon operator with a restricted range of angles.

Put $\alpha = \cos^{-1} m_1$. Without loss of generality we can assume that $0 \leq \alpha \leq \pi/2$. Consider the integral operator

$$J(f) = \int\limits_0^{\pi/2} \frac{f(\cdot, \alpha)}{\cos \alpha + \sin \alpha}\, d\alpha \tag{6}$$

and the differential operator

$$\mathcal{L} = \frac{\partial}{\partial x_1} - \frac{\partial}{\partial x_2}. \tag{7}$$

Let $G(\cdot)$ denote the result of successively applying J and $\mathcal{L}$ to the function $B(\cdot)$:

$$G(x, y) = \mathcal{L}\Big[J(B(x, y, m_1))\Big].$$

For $e \in \mathbf{R}^3$ and $q \in \mathbf{R}^3$, define a set $\mathcal{K}(e, q) \in \mathbf{R}^2 \times \mathbf{R}^2$ as follows:

$$\mathcal{K}(e, q) = \{(x, y) : \sum_{i=1}^2 e_i x_i + e_3 > 0, \sum_{i=1}^2 q_i y_i + q_3 < 0\}.$$

Theorem 3. *Suppose that the function $B(x, y, m)$ has all continuous derivatives up to the order 2 inclusively and vanishes together with its derivatives on the $P(x, m)$; the function $C(x, m)$ is known for all $(x, m) : x \in L, m \in M$. Suppose, further, that there exist*

$$e \in \mathbf{R}^3 \quad (e_1^2 + e_2^2 \neq 0) \quad and \quad q \in \mathbf{R}^3 \, (q_1^2 + q_2^2 \neq 0)$$

such that

$$G(x, y) = 0 \quad when \quad (x, y) \in \mathcal{K}(e, q).$$

Then the solution to (5) in the class of continuously differentiable compactly supported functions is unique.

Proof. Applying the operators defined in (6) and (7) to both sides of (5) gives the equation

$$\int\limits_{\mathbf{R}^1} \frac{w(\xi_1, x_2)}{x_1 - \xi_1}\,d\xi_1 - \int\limits_{\mathbf{R}^1} \frac{w(x_1, \xi_2)}{x_2 - \xi_2}\,d\xi_2 + \int\limits_{\mathbf{R}^2} G(x, \xi)w(\xi)\,d\xi = F(x), \tag{8}$$

where $F(x) = \mathcal{L}\big[J(C(x,m))\big]$, $w(x)$ is a continuously differentiable compactly supported function.

Equation (8) is a bisingular integral equation with perturbation [13]. Note that analogous equations in the three-dimensional space are considered in [14]. The results of [13] imply that (8) has a unique solution in the class of continuously differentiable compactly supported functions. Therefore, (5) also has a unique solution in the same class of functions.

References

1. S. Helgason, *The Radon transform*, Birkhauser, Boston, 1980.

2. F. Natterer, *The mathematics of computerized tomography* , Teubner, Stuttgart, 1986.

3. H.K. Tuy, An inversion formula for cone-beam reconstruction, *SIAM J. Appl. Math.* **43** (1983), 546–552.

4. D.V. Finch, Cone beam reconstruction with sources on a curve, *SIAM J. Appl. Math.* **45** (1985), 665–673.

5. P. Grangeat, Mathematical framework of cone beam 3D reconstruction via the first derivative of the Radon transform, in *Mathematical methods in tomography*, Herman et al., Eds., Springer, 1991.

6. F. Natterer, Recent developments in X-ray tomography, in *Tomography, impedance imaging, and integral geometry*, Quinto et al., Eds., AMS, Providence, RI, 1994.

7. V.P. Palamodov, Some mathematical aspects of 3D X-ray tomography, in *Tomography, impedance imaging, and integral geometry*, Quinto et al., Eds., AMS, Providence, RI, 1994.

8. Akram H. Begmatov, The integral geometry problem for a family of cones in the n-dimensional space, *Sibirsk. Mat. Zh.* **37** (1996), 500–505; English transl.: *Siberian Math. J.* **37** (1996), 430–435.

9. Akram H. Begmatov, Two new classes of problems in integral geometry, *Dokl. Akad. Nauk* **360** (1998), 586–588; English transl.: *Dokl. Math.* **57** (1998), 427–429.

10. M.M. Lavrent'ev, V.G. Romanov, and S.P. Shishatskii, *Ill-posed problems of mathematical physics and analysis*, Nauka, Moscow, 1980; English transl., AMS, 1986.

11. M.M. Lavrent'ev and A.L. Bukhgeĭm, On a class of operator equations of the first kind, *Funktsional. Anal. i Prilozhen.* **7** (1973), No. 4, 44–53; English transl. in *Functional Anal. Appl.*, **7** (1973), No. 4.

12. M.M. Lavrent'ev, On a class of operator equations on the plane, in *Conditionally well-posed mathematical problems and problems of geophysics*, M.M. Lavrent'ev, Ed., Novosibirsk, 1979 (Russian).

13. M.M. Lavrent′ev, On a certain class of singular integral equations, *Sibirsk. Mat. Zh.* **21** (1980), 225–228 (Russian).

14. Akbar H. Begmatov, On some classes of polysingular integral equations, *Sibirsk. Mat. Zh.* **35** (1994), 515–519; English transl.: *Siberian Math. J.* **35** (1994), 459–463.

Department of Mathematics and Physics, Cooperative Institute, Samarkand, Uzbekistan

Department of Analysis, Samarkand State University, Samarkand, Uzbekistan

AKRAM H. BEGMATOV

Reconstructing a function by means of integrals over a family of conical surfaces

1. Introduction

We consider the problem of finding a function from its integrals over a family of one-sheeted cones in the n-dimensional Euclidean space of an arbitrary odd dimension. This problem is a Volterra-type integral geometry problem (see [1]). We obtain a stability estimate for a solution to the problem in spaces of finite smoothness, thereby demonstrating weak ill-posedness of the problem, and also construct a representation for a solution. Generally speaking, a uniqueness theorem for a solution can be obtained by the methods of the articles (see, for instance, [2]) by V.G. Romanov who considered strongly ill-posed problems of Volterra type. This statement is included in Theorem 1 for completeness. We also prove uniqueness theorem and stability estimates for the solution to problem with perturbation under rather general assumptions on the weight function of perturbation.

Analogous problem on surfaces of one-sheeted cones in a space of an arbitrary even dimension was considered in the author's articles [3, 4]. The uniqueness theorem was formulated, a simple representation for the solution was constructed, and the solution stability in Sobolev spaces was estimated. The problem of reconstructing a function from its integrals over an n-parameter family of conic surfaces with vertices ranging over a fixed coordinate axis was considered by S.V. Uspenskii [5]. Similar problems in operator form were studied by A.L. Bukhgeim [6].

2. The problem of integral geometry for a family of cones

Let us introduce notation: $x \in \mathbf{R}^{n-1}$, $\xi \in \mathbf{R}^{n-1}$; $n = 2m + 1$, $m \geq 2$.

Consider the following family of manifolds in the layer

$$\bar{\Omega} = \left\{ (x, y) : x \in \mathbf{R}^{n-1}, \ y \in [0, h], \ h < \infty \right\}.$$

Denote by $\{K(x, y)\}$ the family of the cones having as vertices the points (x, y) and determined by the relations

$$\sum_{m=1}^{n-1} (x_m - \xi_m)^2 = (y - \eta)^2, \quad 0 \leq \eta \leq y.$$

Consider the operator equation in the function $u(x, y)$:

$$\int_{K(x,y)} u(\xi, \eta) \, dk = f(x, y), \tag{1}$$

where dk is the area element on K. The left-hand side of (1) represents the collection of the integrals of a sought function over the family of the cones $K(\cdot)$ with vertices at the points (x, y). Solving (1) is a Volterra-type integral geometry problem.

This work was partly supported by a grant from the State Committee for Science and Technology of the Republic of Uzbekistan.

We introduce the function

$$\varphi_2(x, y) = \frac{1}{(2\pi)^{n-1}} \int\limits_{\mathbf{R}^{n-1}} e^{-\imath\langle\lambda, x\rangle} \frac{J_0(|\lambda|y)}{1 + |\lambda|^{n-1}} d\lambda, \tag{2}$$

where $J_0(\cdot)$ is the Bessel function, $\lambda = (\lambda_1, \ldots, \lambda_{n-1})$, and $\langle\lambda, x\rangle$ is the inner product.

Theorem 1. *Suppose the function* $f(x, y)$ *is known in the layer* Ω. *Then the solution to (1) in the class* $C_0^{2n+1}(\Omega)$ *is unique, the representation*

$$u(x, y) = C_1 \int\limits_0^y \int\limits_0^\zeta \int\limits_{\mathbf{R}^{n-1}} \left(\frac{\partial^2}{\partial\eta^2} - \Delta_\xi\right)^{m+1} \left(E + (-1)^m \Delta_\xi^m\right) f(\xi, \eta)$$

$$\times \varphi_2((x - \xi)(y - \eta)) \, d\xi \, d\eta \, d\zeta \tag{3}$$

is valid, and the inequality

$$\|u(x, y)\|_{W_2^{0, \ldots, 0, 1}(\Omega)} \leq C_2 \|f(x, y)\|_{W_2^{n, \ldots, n}(\Omega)}$$

holds. Here

$$C_1 = \frac{1}{2^{2m+\frac{1}{2}} \pi^{m-\frac{1}{2}} \Gamma(m + 1/2)}, \tag{4}$$

$$\Delta_\xi = \frac{\partial^2}{\partial\xi_1^2} + \ldots + \frac{\partial^2}{\partial\xi_{n-1}^2},$$

C_2 *is some constant, and* E *is the identity operator.*

Proof. Rewrite (1) as

$$\int\limits_0^y d\eta \int\limits_{S_{n-2}(y-\eta)} u(\xi, \eta) \, ds = f_1(x, y), \tag{5}$$

where $S_{n-2}(y - \eta)$ is the sphere of radius $y - \eta$ in the $(n - 1)$-dimensional space, ds is the area element on the sphere, and $f_1(x, y) = f(x, y)/\sqrt{2}$.

Applying the Fourier transform in x and the Laplace transform in y to both sides of (5), we obtain

$$v(\lambda, p) = \frac{b(p^2 + |\lambda|^2)^{n/2}}{p} \cdot \varphi(\lambda, p), \tag{6}$$

where $v(\lambda, p)$ and $\varphi(\lambda, p)$ are the images of the functions $u(x, y)$ and $f_1(x, y)$ respectively,

$$b = \frac{1}{2^{2m} \pi^{m-\frac{1}{2}} \Gamma(m + 1/2)}.$$

As is known,

$$\int\limits_0^\infty e^{-py} J_0(|\lambda|y) dy = \frac{1}{\sqrt{p^2 + |\lambda|^2}}.$$

Further, it is easy to see that the function

$$(p^2 + |\lambda|^2)^{m+1} \cdot \varphi(\lambda, p)$$

is the Laplace transform of

$$\left(\frac{\partial^2}{\partial \eta^2} + |\lambda|^2\right)^{m+1} \cdot \hat{f}_1(\lambda, \eta)$$

with respect to the second variable.

Apply the inverse Laplace transform in p to both sides of (6). Using the Inversion Theorem and the Convolution Theorem and recalling the properties of the Laplace transform, we obtain

$$\frac{\partial \hat{u}(\lambda, y)}{\partial y} = C_1 \int_0^y \left(\frac{\partial^2}{\partial \eta^2} + |\lambda|^2\right)^{m+1} \hat{f}(\lambda, \eta) J_0(|\lambda|(y - \eta))\, d\eta, \tag{7}$$

where C_1 is defined by (4).

It is known that the function $J_0(|\lambda|y))$ can be represented in the form

$$J_0(|\lambda|y)) = |\lambda|^{-1/2} y^{-1/2} \cdot \varphi_1(|\lambda|y), \tag{8}$$

where $\varphi_1(\cdot)$ is a bounded function, $\varphi_1(0) \neq 0$.

It follows from (2) and (8) that the function

$$\frac{J_0(|\lambda|(y - \eta))}{1 + |\lambda|^{2m}}$$

is the Fourier transform of $\varphi_2(x, y - \eta)$ in the first variable.

Apply the inverse Fourier transform in λ to (7). Using the Inversion Theorem, the Convolution Theorem, and the properties of the Fourier transform, we obtain

$$\frac{\partial u(x, y)}{\partial y} = C_1 \int_0^y \int_{\mathbf{R}^{n-1}} \left(\frac{\partial^2}{\partial y^2} - \Delta_\xi\right)^{m+1} \left(E + (-1)^m \Delta_\xi^m\right) f(\xi, \eta)$$

$$\times\, \varphi_2((x - \xi)(y - \eta))\, d\xi\, d\eta. \tag{9}$$

From (9) we obtain the required representation for the solution to (1):

$$u(x, y) = C_1 \int_0^y \int_0^\zeta \int_{\mathbf{R}^{n-1}} \left(\frac{\partial^2}{\partial y^2} - \Delta_\xi\right)^{m+1} \left(E + (-1)^m \Delta_\xi^m\right) f(\xi, \eta)$$

$$\times\, \varphi_2((x - \xi)(y - \eta))\, d\xi\, d\eta\, d\zeta. \tag{10}$$

The inversion formula (10) is of a local nature with respect to y. It is clear from the condition $\operatorname{supp} u \subset \Omega$ that representation (10) for a solution to (1) is also valid for $l < \infty$. Then (6) and (10) imply uniqueness of a solution to the original integral geometry problem (1) in the function class $C_0^5(\Omega)$.

Using the properties of differentiation of the Fourier and Laplace transforms, the triangle inequality for norms, (10) and the conditions on the function u, we arrive at the estimate

$$\|u(x,y)\|_{W_2^{0,\ldots,0,1}(\Omega)} \leq C_2 \|f(x,y)\|_{W_2^{n,\ldots,n}(\Omega)},$$

where C_2 is some constant.

By analyzing the proof of Theorem 1, it is not hard to see that the following proposition is valid.

Theorem 2. *Suppose the function $f(x,y)$ satisfies the following conditions:*
1) $f(x,y)$ has compact support with respect to the first variable;
2) the function $f(x,y)$ has all continuous derivatives up to the order n inclusive;

$$3) \left.\frac{\partial^k}{\partial y^k}f(x,y)\right|_{y=0} = \left.\frac{\partial^k}{\partial y^k}f(x,y)\right|_{y=l} = 0, \quad (k=0,1,\ldots,n+1).$$

Then there exists a solution to (1) in the class of continuous functions finite in the variable x, and this solution is determined by (3).

3. The problem of integral geometry with perturbation

We denote by $Q(x,y)$ the part of the n-dimensional space that is bounded by the surface of the cone $K(x,y)$ and the hyperplane $y=0$.

Consider an operator equation in the function $u(x,y)$

$$\int_{K(x,y)} u(\xi,\eta)\,dk + \int_{Q(x,y)} g(x,y,\xi,\eta)u(\xi,\eta)\,dq = F(x,y), \tag{11}$$

where dq is the volume element of Q.

Equation (11) corresponds to the integral geometry problem with perturbation.

Theorem 3. *Suppose the function $F(x,y)$ is known in the layer Ω. The function $g(x,y,\xi,\eta)$ has all continuous derivatives up to the order n inclusively and vanishes together with its derivatives on the surface of the cone $K(x,y)$.*

Then the solution to (11) is unique and the following estimate holds:

$$\|u(x,y)\|_{W_2^{0,\ldots,0,1}(\Omega)} \leq C_3 \|F(x,y)\|_{W_2^{n,\ldots,n}(\Omega)}, \tag{12}$$

where C_3 is some constant.

Proof. Consider the second summand on the left-hand side of (11):

$$\int_{Q(x,y)} g(x,y,\xi,\eta)u(\xi,\eta)\,dq = f_0(x,y). \tag{13}$$

Recalling the constraints imposed on the weight function $g(\cdot)$ and using expressions for the corresponding derivatives of the function $f_0(\cdot)$, we obtain the following estimate for $y < y_0$, wherein y_0 is sufficiently small:

$$\|f_0(x,y)\|_{W_2^{n,\ldots,n}(\Omega)} \leq \varepsilon \|u(x,y)\|_{W_2^{0,\ldots,0,1}(\Omega)}, \quad 0 < \varepsilon < 1. \tag{14}$$

We denote the integral operators on the left-hand side of (1) and (13) by A and A_1. Then (1) and (11) take the form

$$Au = f, \tag{15}$$

$$Au + A_1 u = F. \tag{16}$$

From the above we infer that the operator A in (15) has a left inverse A^{-1}. Acting by the operator A^{-1} from the left on both sides of (16), we arrive at the equality

$$u + A^{-1} A_1 u = A^{-1} F. \tag{17}$$

The estimates obtained in Theorem 1 and the fact established above imply that the operator A_1 defined on the functions $u(\cdot)$ is continuous. Hence, the operator $A^{-1} A_1$ in (17) is continuous on the functions $u(\cdot)$.

Thus, the operator $A^{-1} A_1$ satisfies the inequality

$$\|A^{-1} A_1\| \leq \varepsilon < 1 \tag{18}$$

for $y < y_0$ sufficiently small.

The contraction mapping principle applied to the operator on the right-hand side of (11) yields uniqueness for a solution to (11) for y sufficiently small. Since (17) is a Volterra-type equation in the sense of the definition of [1], uniqueness takes place not only at small y but also in the whole layer Ω. Thus, inequalities (14) and (18) and Theorem 1 imply the estimate (12).

Remark. It is obvious that the results of the article can be easily translated to the case of the more general family of conical surfaces of the form

$$\sum_{m=1}^{n-1} a_m^2 (x_m - \xi_m)^2 = (y - \eta)^2, \quad 0 \leq \eta \leq y, \quad a_m \in \mathbf{R}^1.$$

References

1. M.M. Lavrent'ev, V.G. Romanov and S.P. Shishatskii, *Ill-posed problems of mathematical physics and analysis*, Nauka, Moscow, 1980; English transl.: AMS, 1986.

2. V.G. Romanov, *Some inverse problems for the equations of hyperbolic type*, Nauka, Novosibirsk, 1972; English transl.: *Integral geometry and inverse problems for hyperbolic equations*, Springer-Verlag, 1974.

3. Akram H. Begmatov, The integral geometry problem for a family of cones in the n-dimensional space, *Sibirsk. Mat. Zh.* **37** (1996), 500–505; English transl.: *Siberian Math. J.* **37** (1996), 430–435.

4. Akram H. Begmatov, Two new classes of problems in integral geometry, *Dokl. Akad. Nauk* **360** (1998), 586–588; English transl.: *Dokl. Math.* **57** (1998), 427–429.

5. A.L. Bukhgeĭm, *Volterra equations and inverse problems*, Nauka, Novosibirsk, 1983 (Russian).

6. S.V. Uspenskii, On reconstruction of a function given by integrals over a family of conical surfaces, *Sibirsk. Mat. Zh.* **18** (1977), 675–684; English transl.: *Siberian Math. J.* **18** (1977), 675–684.

Department of Analysis, Samarkand State University, Samarkand, Uzbekistan

B. BERTRAM

On the use of wavelet expansions and the conjugate gradient method for solving first kind integral equations

1. Introduction

In this paper we investigate the use of the conjugate gradient method in conjunction with wavelet expansions for the solution of first kind integral equations. Fredholm integral equations of the first kind with non-degenerate kernels represent some of the most frequent examples of *ill-posed* problems [1,2]. Such examples occur in many imaging problems [3,4,7–9]. We consider

$$\int K(x-y)f(y)dy = g(x), \tag{1}$$

where (without loss of generality) we may consider the interval of integration to be $[0,1]$. If K is compact and non-degenerate, there will typically not be a continuous dependence of the solution on the input data, that is, $f(x)$ does not depend continuously on $g(x)$. The problem becomes variational in nature and an approximation to $f(x) = K^\dagger g(x)$ is sought ($K^\dagger$ is the generalized inverse of K), where typically, K has a nontrivial null space. Therefore, a best approximation is computed. Sophisticated algorithms for such approximate solutions date back to Tikhonov's concept of *regularization* [3]. In this paper, we use the conjugate gradient method to solve the discretized system, and the number of iterations can be considered to be the regularization parameter [3]. The wavelet expansions used were Daubechies (orthonormal, of compact support, once differentiable), quadratic B-splines, and Shannon.

2. The conjugate gradient method

The Conjugate Gradient Method is known as a Krylov subspace method and and can be used directly to solve $K^T K f = K g$ for f. The iterative scheme expresses $z = x_k - x_0$ in terms of the span of $\{z, Kz, \cdots, K^{k-1}z\}$ [1]. This method uses K-Orthogonality: i.e., x is K-orthogonal to y if $\langle x, Ky \rangle = 0$, and at each step, the search direction p_k is chosen to minimize $\|p - r_{k-1}\|$ over all vectors $p \in \text{span}\{Kp_1, Kp_2, \cdots, Kp_{k-1}\}^\perp$ (where $r_{k-1} = g - Kx_k$) and

$$x_k = x_{k-1} + \alpha_k p_k \tag{2}$$

where $\alpha_k = p_k^T r_{k-1} / p_k^T A p_k$. The iteration is then applied to $K^T K$ (<u>C</u>onjugate <u>G</u>radient <u>N</u>ormal <u>E</u>quations). If we define $\|w\|_K = \sqrt{w^T K w}$ in K norm, then if the condition number of K in Euclidean norm is nearly equal to one, the method will converge very rapidly [1] and [5, p. 530]

In contrast, in this work, we expand f in terms of a wavelet basis defined by a set of scaling functions, (mother wavelets), $\{\phi_i\}$, with

$$f_n(x) = \sum_j a_j \phi(2^n x - j), \tag{3}$$

The author wishes to thank John W. Hilgers, Otto G. Ruehr, and Monica M. Alger for their time and expertise in guiding and encouraging pursuance of this application of first kind integral equations.

where $f_n(x)$ is the projection of f onto the subspace V_n of the multi-resolution analysis defined by ϕ (see [6]). The scaling functions satisfy the two-scale equation

$$\phi(x) = \sum c_i \phi(2x - i) \tag{4}$$

for an appropriate choice of wavelet coefficients c_i, the limits of summation being defined by the type of wavelet and the interval of compact support. To be specific, $\phi(x)$ is determined from the two scale equation recursively by evaluation at *dyadic* points of the form $i/2^j$. For each choice of wavelet, we use the length of the interval of compact support and the length of the interval of integration in following way: Let $[a, b]$ be the interval of integration and $[c, d]$ be the interval of compact support for a given scaling function. Then, letting $R = d - c$ we find the limits for the summation index j using

$$\lfloor 2^n a - R \rfloor \leq j \leq \lceil 2^n b \rceil.$$

This procedure will yield the number of indices, m, of the nonzero coefficients a_i in (3). Since $m < n + 1$ (where n is the number of intervals of length $2^i/n$ characterizing $[a, b]$), the resulting system is underdetermined and, hence, to obtain a square system, we choose $n + 1 - m$ points from the next smallest dyadic level (points of the form $2^i/(n+1)$).

In this paper, we explore the use of three types of scaling functions: Daubechies (D6) orthonormal (differentiable), quadratic B-spline, and truncated, shifted Shannon; three types of image functions: twin peaks, three triangles, and parabola; and, in addition, two kernels: sinc and stove pipe. Details are available from the author.

3. Computational results

In this section we compare the respective errors of the different methods investigated (see Table 1). All experiments were conducted using $n = 6$ and using a repeatable noise vector (no statistical sampling was employed). From our numerical evidence, we can make several observations.

For the twin peaks object function (a discontinuous case), we find, for the sinc kernel, that the three types of wavelet expansions produce virtually indistinguishable errors. On the other hand, for the stove pipe kernel, with a noise level of 0.01 or below, the Daubechies wavelets are superior. This is to be expected, since the level of Daubechies wavelets guarantees exact reconstruction in the presence of no noise. However, this trend does not continue, and at the noise level of 0.1, we see that either of the other two wavelets is superior. Considering the parabolic object function (infinite differentiability), and either type of kernel, both the quadratic B-spline wavelets and the Shannon wavelet expansions give better results than the Daubechies wavelets. Lastly, for the three triangles object function (continuous but not differentiable), considering both kernels, we see that, for the noise levels investigated, the quadratic B-spline expansions worked best.

It should be noted that there are two obvious deviations of patterns contained in the table. The first occurs for the stove pipe kernel, the quadratic B-spline wavelet and a noise level of 0.01. This appears to be an anomaly. However, with an increase in the level of dyadic points to $n = 7$, we find a decrease in the error to 0.0566. Secondly, it should be noted that the errors for the stove pipe kernel, D6 wavelet, and three triangles object function grow in magnitude at the same rate as the amplitude of the noise.

Kernel	Wavelet	Noise Level	Twin Peaks Error	Parabola Error	Three Triangles Error
sinc	B-spline	0.0	0.1899	0.0019	0.0165
sinc	B-spline	0.001	0.2282	0.0036	0.0226
sinc	B-spline	0.01	0.2310	0.0043	0.0269
sinc	B-spline	0.1	0.3060	0.0052	0.0319
sinc	Shannon	0.0	0.1898	0.0012	0.1349
sinc	Shannon	0.001	0.2278	0.0024	0.1823
sinc	Shannon	0.01	0.2294	0.0032	0.2136
sinc	Shannon	0.1	0.2426	0.0035	0.3016
sinc	D6	0.0	0.1943	0.1521	0.1315
sinc	D6	0.001	0.2372	0.2646	0.1623
sinc	D6	0.01	0.2384	0.3077	0.1968
sinc	D6	0.1	0.2472	0.3189	0.2005
stove pipe	B-spline	0.0	0.0115	0.0008	0.0014
stove pipe	B-spline	0.001	0.02486	0.0008	0.0308
stove pipe	B-spline	0.01	0.2233	0.0009	0.0317
stove pipe	B-spline	0.1	2.0155	0.0007	0.0367
stove pipe	shannon	0.0	0.0935	0.0003	0.0013
stove pipe	shannon	0.001	0.1859	0.0003	0.0178
stove pipe	shannon	0.01	2.4660	0.0003	0.2646
stove pipe	shannon	0.1	0.6363	0.0003	0.3769
stove pipe	D6	0.0	0.00195	0.2262	0.0016
stove pipe	D6	0.001	0.00183	0.2262	0.8998
stove pipe	D6	0.01	0.3130	0.2262	4.4738
stove pipe	D6	0.1	1.4505	0.2262	50.875

Table 1. Errors for the three wavelets.

We make the further observation that this measurement of error may not always communicate all the information we would wish. With that in mind, we introduce Figs. 1–3.

We see from Fig. 1a that the conjugate gradient method (CGNE) with no wavelet expansion, using the sinc kernel with twin peaks object function, produces a viable reconstruction with noise levels from 0.00 to 0.01. On the other hand, from Fig. 1b we see that with the stove pipe kernel and the three triangle object function, we achieve "exact" reconstruction with no noise present. Conversely, with a noise level of 0.01, none of the wavelets tested produced a satisfactory reconstruction (see Fig. 2a). Fig. 2b suggests that when the smooth parabola object function is coupled with the sinc kernel, we have very good results for a noise level of 0.1. Finally, Fig. 3 demonstrates that for the sinc kernel and the twin peaks object function, the D6 wavelet produces a viable reconstruction with a definite economy of points used.

The computation was done on an Ultra1 Sun workstation using Fortran77 and double precision.

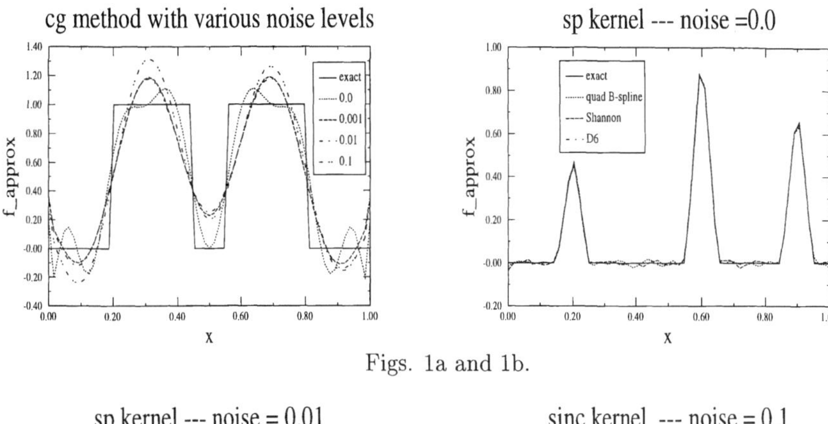

Figs. 1a and 1b.

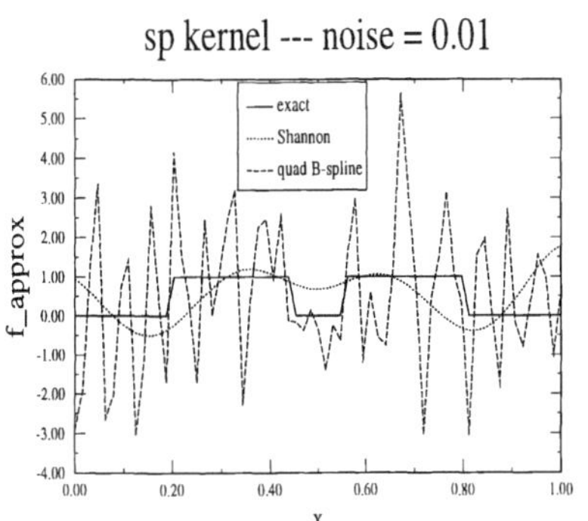

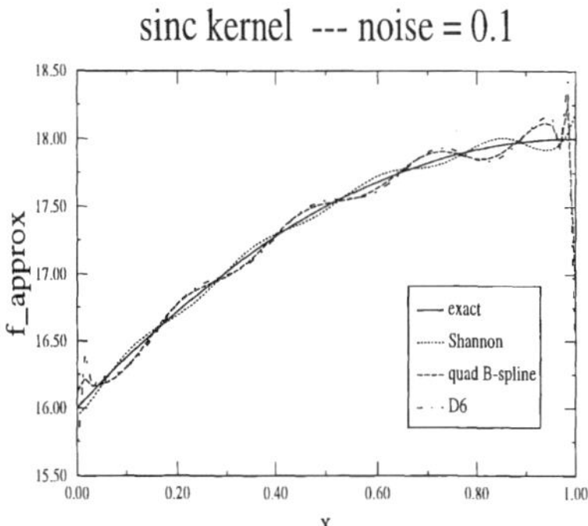

Figs. 2a and 2b.

References

1. M. Hanke, *Conjugate gradient type methods for ill-posed problems*, Longman Scientific and Technical, **327**, New York, 1995.

2. A. Neumaier, Solving ill-conditioned and singular linear systems: a tutorial on regularization, *SIAM Rev.* **40** (1998), 636–666.

3. C. W. Groetsch, *The theory of Tikhonov regularization for Fredholm equations of the first kind*, Pitman Advanced Publishing Program, **105**, London, 1995.

4. J. Hilgers and W. Reynolds, Existence of an optimal regularization operator, SGR TR998 (Technical Report, Signature Research, Inc.).

5. G.H. Golub and C. Van Loan, *Matrix computations*, 3rd ed., Johns Hopkins, Baltimore, 1996.

6. B. Bertram, Numerical solution of Fredholm integral equations using simple wavelet expansions, in *Integral methods in science and engineering*, Pitman Res. Notes Math. Series **375**, Constanda et al., Eds., Longman, Harlow, 1997, 50–53.

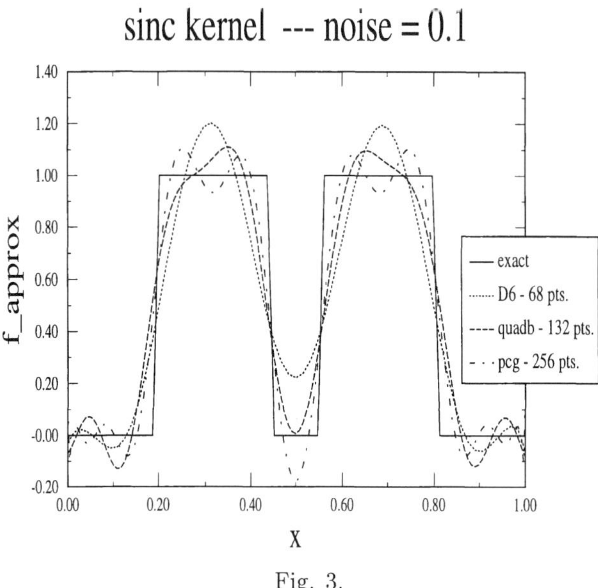

Fig. 3.

7. J.W. Hilgers, Non-iterative methods for solving operator equations of the first kind. Ph.D. thesis, University of Wisconsin, 1973 (Math Research Center TSR # 1413, Jan. 1974).

8. J.W. Hilgers, B.S. Bertram, M.M. Alger and W.R. Reynolds, Extensions of the cross referencing method for choosing good regularized solutions to image recovery problems, *Proc. SPIE*, Computational, experimental, and numerical methods for solving ill-posed inverse problems: medical and non-medical applications, Barbour et al., Eds., **3171**, 1997, 234–237.

9. M.M. Alger, J.W. Hilgers, B.S. Bertram and W.R. Reynolds, An extension of the Tikhonov regularization based on varying the singular values of the regularization operator, *Proc. SPIE*, Computational, experimental, and numerical methods for solving ill-posed inverse problems: medical and non-medical applications, Barbour et al., Eds., **3171**, 1997, 225–231.

Department of Mathematical Sciences, Michigan Technological University, 1400 Townsend Drive, Houghton, MI 49931-1295, USA; e-mail: bertram@mtu.edu

I.V. BOIKOV and A.N. ANDREEV

Optimal algorithms for the calculation of singular integrals

1. Introduction

In solving numerous problems in mathematics, mechanics, and physics, one is faced with the necessity of calculating different singular integrals.

Many hundreds of works have been devoted to approximate methods for the calculation of singular integrals since the 1950s and the flow of publications increases. In this connection, it is necessary to develop criteria for which these methods can be compared. One of such tests is optimality of algorithms. The optimal methods of calculating the singular and hypersingular integrals are investigated in [1–5].

This paper is devoted to algorithms which are optimal with respect to accuracy for the calculation of singular integrals with fixed singularity, Cauchy and Hilbert kernels, polysingular, many-dimensional singular and hypersingular integrals. We will use the definitions of optimal with respect to accuracy and complexity algorithms from [2, 3].

Several authors have studied the calculation of singular integrals on various classes of analytical functions [6,7]. We shall give the quadrature rules for calculation of singular integrals on one class of analytical functions .

We shall need the following classes of functions. Let $1 \le p \le \infty$. Let D be a unit disk in the complex plane C, $\gamma = \{z \in C : |z| = 1\}$ and define the class of functions $F = H_p^r(D)$ as $F = H_p^r(D) = \{f : D \to C, f \text{ is analytic in } D \text{ and } \|f^{(r)}\|_p < \infty\}$, where

$$\|f^{(r)}\|_p = \sup_{0<\rho<1} \left(\frac{1}{2\pi} \int_0^{2\pi} |f^{(r)}(\rho e^{i\theta})|^p d\theta \right)^{1/p}, \qquad \text{if} \qquad 1 \le p < \infty,$$

$$\|f^{(r)}\|_p = \sup_{0<\rho<1} \sup_{0\le\theta\le 2\pi} |f^{(r)}(\rho e^{i\theta})|, \qquad \text{if} \qquad p = \infty.$$

If $r = 0$, the class $F = H_p^r(D)$ is the well-known Hardy class of functions.

Let D_i be a unit disk in the complex plane $C_i, i = 1, 2, \ldots, l$. Let $D^* = \cup_{i=1}^l D_i$. Denote $\gamma_i = \{z_i \in C_i : |z_i| = 1\}, i = 1, 2, \ldots, l$. Define the class of functions $F_l = H_p^{r,\ldots,r}(D^*)$ by $F_l = H_p^{r,\ldots,r}(D^*) = \{f(z_1,\ldots,z_l) : D^* \to C, f(z_1,\ldots,z_l) \text{ is analytic in } D_i \text{ with respect to variable } z_i, i = 1,\ldots,z_l, \text{ and if } 1 \le p < \infty,$

$$\sup_{0<\rho<1} \sup_{1\le i\le l} \sup_{(z_1,\ldots,z_{i-1},z_{i+1},\ldots,z_l)} \left| \int_0^{2\pi} |f_i^{(r)}(z_1,\ldots,z_{i-1},\rho e^{i\theta_i}, z_{i+1},\ldots,z_l)|^p d\theta_i \right|^{1/p} < \infty,$$

$$\sup_{0<\rho<1} \sup_{1\le i\le l} \sup_{(z_1,\ldots,z_{i-1},z_{i+1},\ldots,z_l)} \sup_{0\le\theta\le 2\pi} |f_i^{(r)}(z_1,\ldots,z_{i-1},\rho e^{i\theta_i}, z_{i+1},\ldots,z_l)| < \infty,$$

if $p = \infty$, where $f_i^{(r)}(z_1,\ldots,z_l) = \frac{\partial^r f(z_1,\ldots,z_l)}{\partial z_i^r} \}$.

The work was supported by Russian Foundation of Fundamental Investigation (grant 97-01-00621) and by a grant from Novosibirsk State University.

2. Optimal algorithms of calculation of singular integrals

In this section we start by investigating the calculation methods for singular integrals of the following kind:

$$J\varphi = \int\limits_{-1}^{1} \frac{\varphi(\tau)}{\tau} d\tau. \tag{1}$$

As the method of evaluation we use the quadrature rule (q.r.)

$$J\varphi = \sum_{k=-N}^{N} \sum_{l=0}^{\rho} p_{kl} \varphi^{(l)}(t_k), \tag{2}$$

where $-1 \le t_{-N} < \cdots < t_{-1} < t_0 < t_1 < \cdots < t_N \le 1$.

Calculation of the integral (1) with quadrature rules of the type (2) was constructed to be optimal with respect to the accuracy of quadrature rules on $W^1(1)$ and $H_1(1)$ classes of functions, and asymptotically optimal with respect to accuracy and complexity of quadrature rules on Hölder class $H_\alpha(1), 0 < \alpha \le 1$, and Sobolev $W^r(1), r = 1, 2, \ldots$, classes of functions [1-4]. The rule was constructed optimal with respect to the order of accuracy of algorithms of calculation of integrals of type (1) on Hölder and Sobolev classes of functions and its stability was investigated in [1].

Here we shall discuss calculation of integral (1) on the class $H_p^r(1), r = 1, 2, \ldots, 1 \le p \le \infty$. To this end, we construct the spline for approximation of the integrand of (1). Let us consider a segment $[a, b]$. Let $\zeta_k, k = 1, \ldots, r$, be the roots of Legendre polynomial of degree r. We map the segment $[-1, 1]$ onto segment $[a, b]$. As a result of this mapping the points $\zeta_1, \ldots, \zeta_r$ are mapped onto the points $\zeta_1', \ldots, \zeta_r'$. Using the points $\zeta_1', \ldots, \zeta_r'$, we construct the interpolation polynomial $P_r(f, [a, b])$ for approximation of the function $f(x)$ on the segment $[a, b]$. For approximation of the function $f(x_1, \ldots, x_l)$ in the cube $[a_1, b_1; \ldots; a_l, b_l]$ we use the interpolation polynomial

$$P_{r,\ldots,r}(f(x_1, \ldots, x_l), [a_1, b_1; \ldots; a_l, b_l])$$

$$= P_r^{x_1}(P_r^{x_2}(\ldots\ldots(P_r^{x_l}(f(x_1, \ldots, x_l); [a_1, b_1]), \ldots, [a_l, b_l]).$$

The upper index $x_i, i = 1, 2, \ldots, l$, defines a variable of interpolation.

Let $t_0^1 = -1, t_k^1 = -1 + e^{(k-N)v}, k = 1, \ldots, N_0, N_0 = [N - \frac{1}{v} \ln 2], t_{N_0+1}^1 = -\frac{1}{2}$, $t_0^2 = 0, t_k^2 = -e^{(k-N)v}, k = 1, \ldots, N_0, t_{N_0+1}^2 = -\frac{1}{2}, t_0^3 = 0, t_k^3 = e^{(k-N)v}, k = 1, \ldots, N_0$, $t_{N_0+1}^3 = \frac{1}{2}, t_0^4 = 1, t_k^4 = 1 - e^{(k-N)v}, k = 1, \ldots, N_0, t_{N_0+1}^4 = \frac{1}{2}$, where $[a]$ is greatest integer in a. The value v will be defined below. Let $\Delta_k^1 = [t_k^1, t_{k+1}^1], \Delta_k^2 = [t_{k+1}^2, t_k^2]$, $\Delta_k^3 = [t_k^3, t_{k+1}^3], \Delta_k^4 = [t_{k+1}^4, t_k^4], k = 0, 1, \ldots, N_0$.

We introduce the following quadrature rule:

$$Jf = \sum_{k=0}^{N_0} \int\limits_{t_k^1}^{t_{k+1}^1} \frac{P_{s_k}(f, \Delta_k^1)}{\tau} d\tau + \sum_{k=1}^{N_0-1} \int\limits_{t_{k+1}^2}^{t_k^2} \frac{P_{s_k}(f, \Delta_k^2)}{\tau} d\tau$$

$$+ \sum_{k=1}^{N_0} \int\limits_{t_k^3}^{t_{k+1}^3} \frac{P_{s_k}(f, \Delta_k^3)}{\tau} d\tau + \sum_{k=0}^{N_0-1} \int\limits_{t_{k+1}^4}^{t_k^4} \frac{P_{s_k}(f, \Delta_k^4)}{\tau} d\tau + \int\limits_{t_1^2}^{t_1^3} \frac{P_{s^*}(f, [t_1^2, t_1^3])}{\tau} d\tau + R_N(f), \tag{3}$$

where $s^* = r + 1, s_0 = r, s_k = [kv(r + 1 - 1/p)] + 1, k = 1, 2, \ldots, N_0, v = ln(1 + 1/e)$.

Theorem 2.1. *Let $f(t) \in H_p^r, r = 1, 2, \ldots, 1 \leq p \leq \infty, \ v = ln(1 + 1/e)$, and let n be the number of functionals $f(t_k^i), i = 1, 2, 3, 4, k = 0, 1, \ldots, N_0 + 1$ which are used for construction of the quadrature rule (3). Then $R_N[H_p^r(1)] \leq Ae^{-\sqrt{nv(r+1-1/p)/2}}$.*

Now let $L(f, [a, b], n)$ be a Gauss quadrature rule with n knots for calculation of the integral $\int_a^b f(t)dt$.

We introduce the following quadrature rule:

$$Jf = \sum_{k=0}^{N_0} L(f, \Delta_k^1, s_k) + \sum_{k=1}^{N_0-1} L(f, \Delta_k^2, s_k) + \sum_{k=1}^{N_0} L(f, \Delta_k^3, s_k) + \sum_{k=0}^{N_0-1} L(f, \Delta_k^4, s_k)$$
$$+ L((f(t) - f(0))/t, [t_1^2, t_1^3], r + 2) + R_N(f), \tag{4}$$

where $s^* = r + 1, s_0 = r, s_k = [kv(r + 1)] + 1, k = 1, 2, \ldots, N_0, v = ln(1 + 1/(2e))$.

Theorem 2.2. *Let $f(t) \in H_p^r, r = 1, 2, \ldots, 1 \leq p \leq \infty, \ v = ln(1 + 1/(2e))$, and n be the number of functionals $f(t_k^i), i = 1, 2, 3, 4, k = 1, 2, \ldots, N_0 + 1$ that are used for constructing quadrature rule (4). Then $R_N[H_p^r(1)] \leq Ae^{-\sqrt{nv(r+1)/2}}$.*

Next we consider singular integrals of the following kind: $Kf = \int_{-1}^{1} \frac{f(\tau)}{\tau - t}d\tau, \ -1 < t < 1$. Let $\Delta_0^1 = [-1, t_1^1], \Delta_0^2 = [t_1^2, 1], \Delta_k^1 = [t_k^1, t_{k+1}^1], \Delta_k^2 = [t_{k+1}^2, t_k^2], k = 1, 2, \ldots, N - 1$, where $t_k^1 = -1 + e^{(k-N)v}, t_k^2 = 1 - e^{(k-N)v}, k = 1, 2, \ldots, N$, and constant v will be defined below.

Let $t \in \Delta_j^1, 0 \leq j \leq N - 1$. (The case when $t \in \Delta_j^2$ can be considered in a similar way.) We calculate the integral Kf by the next quadrature rule,

$$Kf = \begin{cases} \int_{-1}^{t_1^1} \frac{T_{r-1}(f, \Delta_0^1, -1)}{\tau - t}d\tau & + \sum_{k=1, k \neq j-1, j, j+1}^{N-1} \int_{t_k^1}^{t_{k+1}^1} \frac{P_{n_k}(f, \Delta_k^1)}{\tau - t}d\tau \\ + \int_{t_{j-1}^1}^{t_{j+2}^1} \frac{P_{n_k}(f, [t_{j-1}^1, t_{j+2}^1])}{\tau - t}d\tau & + \sum_{k=1}^{N-1} \int_{t_{k+1}^2}^{t_k^2} \frac{P_{n_k}(f, \Delta_k^2)}{\tau - t}d\tau \\ + \int_{t_2^2}^{1} \frac{T_{r-1}(f, \Delta_0^2, 1)}{\tau - t}d\tau & + R_N, \ \ 2 \leq j \leq N - 1, \end{cases} \tag{5a}$$

$$Kf = \begin{cases} \int_{-1}^{t_2^1} \frac{T_{r-1}(f, [-1, t_2^1], -1)}{\tau - t}d\tau & + \sum_{k=2}^{N-1} \int_{t_k^1}^{t_{k+1}^1} \frac{P_{n_k}(f, \Delta_k^1)}{\tau - t}d\tau \\ + \sum_{k=1}^{N-1} \int_{t_{k+1}^2}^{t_k^2} \frac{P_{n_k}(f, \Delta_k^2)}{\tau - t}d\tau & + \int_{t_1^2}^{1} \frac{T_{r-1}(f, \Delta_0^2, 1)}{\tau - t}d\tau + R_N, \ \ 0 \leq j < 2, \end{cases} \tag{5b}$$

where $n_k = [vk(r - 1/p)/(ln2 - ln(e^v - 1))] + 1$, and $T_r(f, [a, b], c) = f(c) + \frac{1}{1!}f^{(1)}(c)(t - c) + \frac{1}{2!}f^{(2)}(c)(t - c)^2 + \cdots + \frac{1}{r!}f^{(r)}(c)(t - c)^r$.

The error of the quadrature rule (5) on the function class $H_p^r(1)$ is estimated by the inequality $R_N(H_p^r(1)) \leq A\sqrt{n}e^{-\sqrt{n(r-1/p)vln(2/(e^v-1))}}$, where n is the number of functionals $f(t_k^i), i = 1, 2, k = 1, 2, \ldots, N$, that are used for constructing q.r. (5). It is impossible to calculate the exact maximum of $\psi(v) = v \ln\left(2/(e^v - 1)\right)$. The approximate value of $\max_{0 < v < \ln 3} \psi(v)$ is equal to 0.57 with exactness 10^{-6}, and $v = 0.45$.

Theorem 2.3. *Let $f(t) \in H_p^r, r = 1, 2, \ldots, 1 \le p \le \infty, v = 0.45$, and n be the number of functionals $f(t_k^i), i = 1, 2, k = 1, 2, \ldots, N$, that are used for constructing quadrature rule (5). Then $R_N[H_p^r(1)] \le An^{1/2}e^{-0.752492\sqrt{n(r-1/p)}}$.*

The optimal algorithm (with respect to accuracy and complexity) for the calculating the singular integrals $Ff = \int_0^1 \cdots \int_0^1 \frac{f(\tau_1, \cdots, \tau_l)}{(\tau_1 - t_1) \ldots (\tau_l - t_l)} d\tau_1 \ldots d\tau_l$ and $Mf = \int\limits_G \frac{g(\theta)f(u)}{r^l(u,v)} du$, where $G = [-1, 1]^l, \theta = (u - v)/r(u, v), u = (u_1, \ldots, u_l), v = (v_1, \ldots, v_l), r(u, v) = \left[(u_1 - v_1)^2 + \cdots + (u_l - v_l)^2\right]^{1/2}$ were considered in [1,2,4].

As an example of the integral Mf, we describe the construction of the quadrature rule on the class of functions F_l.

Let Λ_0 be the set of points $t \in G$ satisfying $0 \le \rho(t, \Gamma) \le e^{(1-N)}$. Let $\Lambda_k, k = 1, 2, \ldots, N - 1$, be the points $t \in G$ satisfying $e^{(k-N)} \le \rho(t, \Gamma) \le e^{(k+1-N)}$.

The distance $\rho(t, \Gamma)$ between the point t and the boundary Γ of the domain G is calculated by formula $\rho(t, \Gamma) = \min_{1 \le i \le l}(\min(|-1 - t_i|, |1 - t_i|))$.

We cover the domain Λ^k by cubes $\Lambda_{i_1, \ldots, i_l}^k$, edges of which are equal to $h_0 = e^{(1-N)}$, $h_k = e^{(k+1-N)} - e^{(k-N)}, k = 1, \ldots, N - 1$. Let n be the number of cubes $\Lambda_{i_1, \ldots, i_l}^k$ by which the domain G was covered. This number is equal to $n \asymp N^{(l-1)}$.

Let $v \in \Delta_{q_1, q_2, \ldots, q_l}^k$. To calculate Mf we use the quadrature rule

$$
Mf = \sum_{k=0}^{N-1} \sum_{i_1, \ldots, i_l} {}^* \int\int_{\Delta_{i_1 \cdots i_l}^k} \frac{g(\theta)P_{s_k, \ldots, s_k}(f, \Delta_{q_1, \ldots, q_l}^k)du}{r^2(u, v)}
$$

$$
+ \sum_{k=0}^{N-1} \sum_{i_1, \ldots, i_l} {}^{**} \frac{g(\theta)P_{s_k, \ldots, s_k}(f, \Lambda_{i_1, \ldots, i_l}^k)du}{r^2(u, v)} + R_N, \tag{6}
$$

where $s_0 = r, s_k = [k(r - 1/p)/(\ln 2 - \ln(e - 1))] + 1, k = 1, 2, \ldots, N; \sum^*$ denotes the summing with respect to squares, which are those with square $\Delta_{q_1, \ldots, q_l}^k$, and $\sum^{**}$ denotes the summing with respect to other squares.

Theorem 2.4. *Let $f(t) \in H_p^{r, \ldots, r}(D^*), r = 1, 2, \ldots, 1 \le p \le \infty$, and m be the number of functionals used to construct quadrature rule (6). Then $R_N[H_p^{r, \ldots, r}(D^*)] \le Am^{-(r-1/p)/(l-1)} \ln m \ln(\ln m)$.*

3. Optimal algorithms for calculation of Hadamard finite-part integrals

In this section we investigate calculation methods for Hadamard finite-part integrals of the following kind

$$
Hf = \int_{-1}^1 \frac{f(\tau)}{\tau^m} d\tau \quad m = 2, 3, \ldots \tag{7}
$$

As the method of evaluation we use the quadrature rule

$$
Hf = \sum_{k=-N}^N \sum_{l=0}^{\rho} p_{kl} f^{(l)}(t_k), \tag{8}
$$

where $-1 \leq t_{-N} < \cdots < t_{-1} < t_0 < t_1 < \cdots < t_N \leq 1$.

For approximation of the function $f(\tau)$ on the segment $[v_k, v_{k+1}]$ we introduce a function $\tilde{f}(\tau, [v_k, v_{k+1}])$ with corresponding formula

$$\tilde{f}(\tau, [v_k, v_{k+1}]) = \sum_{l=0}^{r-1} \left(\frac{f^{(l)}(v_k)}{l!} (\tau - v_k)^l + B_l \delta^{(l)}(v_{k+1}) \right),$$

$$\delta(\tau) = f(\tau) - \sum_{l=0}^{r-1} \frac{f^{(l)}(v_k)}{l!} (\tau - v_k)^l,$$

where the coefficients B_l are determined from the equality

$$(v_{k+1} - \tau)^r - \sum_{j=0}^{r-1} \frac{B_j r! (v_{k+1} - v_k)}{(r-j-1)!} (v_{k+1} - \tau)^{r-j-1} = (-1)^r R_{rq}(v_k', h_k, \tau),$$

in which $R_{rq}(v_k', h_k, \tau) = \tau^r + \sum_{k=0}^{r-1} a_k \tau^k$ is the polynomial of degree r of least deviation from zero in the space $L_q(1/p + 1/q = 1)$ on the segment $[v_k, v_{k+1}], v_k' = (v_k + v_{k+1})/2, h_k = v_{k+1} - v_k$.

Theorem 3.1. *We set $\Psi = W_p^r(1), r \geq m, 1 \leq p \leq \infty$. Among all possible quadrature rules of type (8), where $\rho = r - 1$, the formula*

$$Hf = \sum_{k=0}^{r-1} \frac{f^{(k)}(0)}{k!(k+1-m)} t_1^{k+1-m}(1 - (-1)^{k+1-m})$$

$$+ \sum_{k=-N, k \neq -1, 0}^{N-1} \int_{t_k}^{t_{k+1}} \frac{\tilde{f}(t, [t_k, t_{k+1}]) dt}{t^m} + R_N, \tag{9}$$

where $t_k = \pm(k/N)^{(r+1/q)(r+1/q-m)}$, $1/p + 1/q = 1$, $k = 0, 1, \ldots, N$, is asymptotically optimal. The error is

$$R_N[\Psi] = \frac{(1 + o(1))R_{rq}(1)}{2^{r-1/q}(rq+1)^{1/q}r!} \left(\frac{r+1/q}{r+1/q-m} \right)^{r+1/q} \frac{1}{N^r}.$$

Let $t_0^1 = -1, t_k^1 = -1 + e^{(k-N)v}, k = 1, \ldots, N_0, N_0 = [N - \frac{1}{v} ln\, 2], t_{N_0+1}^1 = -\frac{1}{2}, t_0^2 = 0, t_k^2 = -e^{(k-N)v}, k = 1, \ldots, N_0, t_{N_0+1}^2 = -\frac{1}{2}, t_0^3 = 0, t_k^3 = e^{(k-N)v}, k = 1, \ldots, N_0, t_{N_0+1}^3 = \frac{1}{2}, t_0^4 = 1, t_k^4 = 1 - e^{(k-N)v}, k = 1, \ldots, N_0, t_{N_0+1}^4 = \frac{1}{2}$. Let $v = ln(1 + 1/(2e))$. Let $\Delta_k^1 = [t_k^1, t_{k+1}^1], \Delta_k^2 = [t_{k+1}^2, t_k^2], \Delta_k^3 = [t_k^3, t_{k+1}^3], \Delta_k^4 = [t_{k+1}^4, t_k^4], k = 0, 1, \ldots, N_0$.

We introduce the following quadrature rule:

$$Hf = \sum_{k=0}^{N_0} L(\psi, \Delta_k^1, s_k) + \sum_{k=1}^{N_0-1} L(\psi, \Delta_k^2, s_k) + \sum_{k=1}^{N_0} L(\psi, \Delta_k^3, s_k) + \sum_{k=0}^{N_0-1} L(\psi, \Delta_k^4, s_k)$$

$$+ \int_{t_1^2}^{t_1^3} \frac{P_{s^*}(f, [t_1^2, t_1^3])}{t^m} dt + R_N(f), \tag{10}$$

where $s^* = r + q + 1, s_0 = r, s_k = [Nv(r + 1 - 1/p)] + 1, k = 1, 2, \ldots, N_0$, and $v = ln(1 + 1/(2e))$.

Theorem 3.2. *Let $f(t) \in H_p^r, r = m, m+1, \ldots, 1 \leq p \leq \infty$, $v = ln(1 + 1/(2e))$, and n be the number of functionals $f(t_k^i), i = 1, 2, 3, 4, k = 1, 2, \ldots, N_0 + 1$ which are used for construction the quadrature rule (10). Then*

$$R_N[H_p^r(1)] \leq An^{m/2}e^{-\sqrt{nv(r+1-1/p)/(r+m+1-1/p)}}.$$

References

1. I.V. Boikov, *Optimal with respect to accuracy algorithms of approximate calculation of singular integrals*, Saratov State University Press, 1983 (Russian).

2. I.V. Boikov, *Passive and adaptive algorithms for the approximate calculation of singular integrals*, Ch. 1, Penza Technical State Univ. Press, 1995 (Russian).

3. I.V. Boikov, *Passive and adaptive algorithms for the approximate calculation of singular integrals*, Ch. 2, Penza Technical State Univ. Press, 1995 (Russian).

4. I.V. Boikov, N.F. Dobrunina and L.N. Domnin, *Approximate methods of calculation of Hadamard integrals and solution of hypersingular integral equations*, Penza Technical State University Press, 1996 (Russian).

5. I.V. Boikov, The optimal algorithms of calculation of singular integrals, decision of singular integral equations and its applications, in 15th *IMACS World Congress Sci. Comp., Mod. and Appl. Math.* **1**. *Comp. Math.*, A. Sydow et al., Eds., Wissenschaft, Technik Verlag, Berlin, 1997.

6. F. Stenger, Approximation via Whittaker's cardinal functions, *J. Approximation Theory* **17** (1976), 222–240.

7. B. Bialecki, A sinc quadrature rule for Hadamard finite-part integrals, *Numer. Math.* **57** (1990), 263–269.

IVB: Department of Higher Mathematics; ANA: Department of Radio Design and Production, Penza State University, 40 Krasnaya St., Penza 440017, Russia

A.I. BOIKOVA

An order one approximate method of solution of differential and integral equations

1. Introduction

Many problems of elasticity theory (theory of thin hulls), electrodynamics (diffraction on the surface with boundary), aerodynamics, and building mechanics are reduced to different types of singular integral equations, singular integro-differential equations and differential equations.

Solving these equations by classical numerical methods (collocation, Galerkin, etc.) we obtain systems of N algebraic equations with N unknown variables. The coefficient matrices, A, of these systems have $N \times N$ elements, most which of are not equal to zero. Thus, solution of these systems by classical methods (such as Gauss' method, etc.) requires $O(N^3)$ arithmetical operations. In this paper, we present a method using only $O(N)$ arithmetical operations.

This paper is devoted to approximate methods of solutions of differential and integral equations using the characteristic functions of the differential and integral operators as the interpolation polynomials.

Using these interpolating polynomials we investigate approximate solutions of singular integral equations with constant coefficients of the form

$$Kx \equiv ax(t) + \frac{b}{\pi} \int_{-1}^{1} \frac{x(\tau)}{\tau - t} d\tau = f(t); \tag{1}$$

an approximate solution of a singular integro-differential equation from the diffraction problem

$$\frac{d}{dt} \int_{-1}^{1} \sqrt{1 - \tau^2} \frac{x(\tau)}{\tau - t} d\tau = f(t); \tag{2}$$

and an approximate particular solution of the equation

$$\frac{d}{dt} \{(1 - t^2) \frac{dx}{dt}\} + ax(t) = f(t), \tag{3}$$

where $f(t) \in W^r(1)$, and a is constant, with $a \neq l(l + 1), l = 1, 2, \ldots$

The rate of convergence and error of numerical methods for the solution of (1)–(3) were investigated.

2. Interpolation

Let $\{P_n(x)\}$ be a set of polynomials of degree n, orthonormal on the segment $[a, b]$. Let $\mu_k, k = 0, 1, \ldots, n$, be the roots of the polynomial $P_{n+1}(x)$.

In [1], it is shown that the polynomial

$$L_n f = \sum_{k=0}^{n} (\frac{1}{\gamma_k} \sum_{i=0}^{n} P_i(\mu_k) P_i(x)) f(\mu_k) \tag{4}$$

where $\gamma_k = \sum_{k=0}^{n} P_i^2(\mu_k)$, is the interpolation polynomial with the knots $\mu_k, k = 0, 1, \ldots, n$, for the function $f(x)$.

Theorem 2.1. [1] *The following estimate holds*

$$\| f(x) - L_n(f) \|_C \leq AE_n(f)\lambda_n,$$

where $E_n(f)$ is the best approximation of the function $f(x)$ by algebraic polynomials of degree n, λ_n is the Lebesgue constant of the interpolation by roots of orthonormal polynomials $P_{n+1}(x)$.

3. Approximate solution of ordinary differential equations

As an example, we apply (4) to Legendre's equation. We seek a particular solution of

$$\frac{d}{dt}\{(1 - t^2)\frac{dx}{dt}\} + ax(t) = f(t), \tag{5}$$

where $f(t) \in W^r(1), a$ is constant, and $a \neq l(l + 1), l = 1, 2 \ldots$ It is known, [2], that the eigenfunctions of the operator

$$\frac{d}{dt}\{(1 - t^2)\frac{dx}{dt}\} \tag{6}$$

are the Legendre polynomials. Therefore, we shall approximate the right-hand side of (5) by interpolating polynomials built on the knots of the Legendre polynomials of order $n + 1$ and seek an approximate solution in the form of a linear combination of Legendre polynomials of order up to and including n.

At first we turn our attention to interpolation of the function $f(t)$.

Let us denote by $\mu_k (k = 0, 1, \ldots, n)$ the knots of Legendre polynomial $P_{n+1}(t)$ of degree $n + 1$. As it was shown above, the polynomial

$$L_n f = \sum_{k=0}^{n} \frac{1}{\gamma_k} \sum_{i=0}^{n} \frac{2i + 1}{2} P_i(t) P_i(\mu_k) f(\mu_k) \tag{7}$$

where $P_i(t)$ is the Legendre polynomial of order i, and

$$\gamma_k = \sum_{i=0}^{n} \frac{2i + 1}{2} P_i^2(\mu_k)$$

is the interpolating polynomial of degree n with respect to t interpolating the function $f(t)$ at the nodes μ_k.

We seek an approximate solution of (5) in the form of the interpolating polynomial

$$x_N(t) = \sum_{k=0}^{N} \frac{1}{\gamma_k} \left(\sum_{l=0}^{N} \frac{2l + 1}{2} \frac{1}{a - l(l + 1)} P_l(\mu_k) P_l(t) \right) x_k \tag{8}$$

with unknown values $x_k, k = 0, 1, \ldots, N$.

Having replaced $f(t)$ on the right-hand side of (5) with the interpolating polynomial $f_N(t)$ and finding a solution in the form of the polynomial $x_N(t)$, we have $x_k = f(t_k), k = 0, 1, \ldots, N$.

Therefore a particular solution of (5) has the form

$$x_N(t) = \sum_{k=0}^{N} \frac{1}{\gamma_k} \left(\sum_{l=0}^{N} \frac{2l + 1}{2} \frac{1}{a - l(l + 1)} P_l(t_k) P_l(t) \right) f(t_k). \tag{9}$$

4. Singular integral equations with constant coefficients

Let us consider the singular integral equation

$$Kx \equiv ax(t) + \frac{b}{\pi} \int_{-1}^{1} \frac{x(\tau)}{\tau - t} d\tau = f(t) \tag{10}$$

with constant coefficients.

We seek solutions $x(t)$ of (10) where $x(t)$ is contained in the Banach space $X = H_\delta$ of functions satisfying the Hölder condition with exponent δ and norm $\|x\| = \max_{t \in [-1,1]} |x(t)| + \sup_{t_1 \neq t_2} \frac{|x(t_1) - x(t_2)|}{|t_1 - t_2|^\delta}$.

The index ξ of this equation takes the values $-1, 0, 1$ and its solution has the form $x(t) = \omega_\xi(t) z(t)$, where

$$\omega_\xi(t) = (1 - t)^\alpha (1 + t)^\beta, \ 0 < |\alpha|, |\beta| < 1, \ \xi = -(\alpha + \beta),$$

and $z(t)$ is a smooth function.

The value α is defined by the formula

$$a + b\, ctg(\pi\alpha) = 0. \tag{11}$$

Let us denote by $P_n^{\alpha,\beta}$ the Jacobi polynomials of degree n orthogonal with weight $(1 - t)^\alpha (1 + t)^\beta$ on the segment $[-1, 1]$.

We need the following formula [3,4]:

$$a\omega_\xi(t) P_n^{\alpha,\beta}(t) + \frac{b}{\pi} \int_{-1}^{1} \frac{\omega_\xi(\tau) P_n^{\alpha,\beta}(\tau) d\tau}{\tau - t} = -\frac{b}{2^\xi \sin \alpha\pi} P_{n-\xi}^{-\alpha,-\beta}(t). \tag{12}$$

For (10), below we propose and justify calculating schemes for the indices $\xi = 0, 1$ separately.

4.1. The index $\xi = 0$. First of all let us build an interpolating polynomial for approximating the function $f(t)$.

We denote by t_k $(k = 0, 1, \ldots, n)$ the knots of the polynomial $P_{n+1}^{-\alpha,-\beta}(t)$, input designations

$$\gamma_k^{-\alpha,-\beta} = \sum_{l=0}^{n} (P_l^{-\alpha,-\beta}(t_k))^2, \ \ k = 0, 1, \ldots, n.$$

Then the polynomial

$$f_n(t) = \sum_{k=0}^{n} \left(\frac{1}{\gamma_k^{-\alpha,-\beta}} \sum_{l=0}^{n} P_l^{-\alpha,-\beta}(t_k) P_l^{-\alpha,-\beta}(t) f(t_k) \right) \tag{13}$$

interpolates the function $f(t)$ on knots $t_k, k = 0, 1, \ldots, n$.

It follows from arguments carried out in the second section that

$$\|f(t) - f_n(t)\| \leq A E_n(f) \lambda_n^{(-\alpha,-\beta)}, \tag{14}$$

where $\lambda_n^{(-\alpha,-\beta)}$ is the Lebesgue constant for interpolation on knots of the Jacobi polynomial $P_n^{-\alpha,-\beta}(t)$.

We shall look for an approximate solution of (10) in the form

$$x_n(t) = \omega_0(t)z_n(t),$$

where

$$z_n(t) = -\frac{\sin \alpha\pi}{b} \sum_{k=0}^{n} \left(\frac{x_k}{\gamma_k^{-\alpha,-\beta}} \sum_{l=0}^{n} P_l^{-\alpha,-\beta}(t_k)P_l^{\alpha,\beta}(t) \right)$$

with coefficients x_k to be defined.

Substituting the function $x_n(t)$ in (10) and interchanging the function $f(t)$ with the polynomial $f_n(t)$, we have $x_k = f_k, k = 0, 1, \ldots, n$.

Theorem 4.1. *Let the operator K have a continuous inverse in the space X. Then the estimate $\|x^*(t) - x_n^*(t)\| \leq AE_n(f)n^\delta \lambda_{n+1}^{(-\alpha,-\beta)}$ is valid. Here x^* is the solution of (10) and $x_n^*(t) = -\frac{\sin \alpha\pi}{b} \sum_{k=0}^{n} \left(\frac{f_k}{\gamma_k^{-\alpha,-\beta}} \sum_{l=0}^{n} P_l^{-\alpha,-\beta}(t_k)P_l^{\alpha,\beta}(t) \right).$*

Proof. It is not difficult to see that in the metric of the space X, $\|x^*(t) - x_n^*(t)\| \leq An^\delta E_n(f)\lambda_n^{(-\alpha,-\beta)}$ and $x_n^*(t)$ is the solution of equation $Kx = f_n(t)$. The assertion follows from these remarks.

4.2. The index $\xi = 1$. We denote by $\mu_k, k = 0, 1, \ldots, n-1$, the knots of the Jacobi polynomial $P_n^{-\alpha,-\beta}(t)$ of degree n.

We shall approximate the function $f(t)$ by the interpolating polynomial $f_n(t)$ built on the knots $\mu_k, k = 0, 1, \ldots, n-1$. The method of building the polynomial $f_n(t)$ is described in the previous section. We seek an approximate solution of (10) in the form

$$x_n(t) = \omega_1(t)z_n(t),$$

where

$$z_n(t) = -\frac{2\sin \alpha\pi}{b} \sum_{k=0}^{n-1} (\frac{x_k}{\gamma_k^{-\alpha,-\beta}} \sum_{l=0}^{n-1} P_l^{(-\alpha,-\beta)}(\mu_k)P_{l+1}^{(\alpha,\beta)}(t)) + C,$$

where C is a constant.

Substituting the polynomial $x_n(t)$ into equation (10) and replacing $f(t)$ with the polynomial $f_n(t)$, we have $x_k = f_k, k = 0, 1, \ldots, n-1$.

To find the coefficients x_i it is sufficient to take advantage of the additional condition imposed on the solution of (10): $\int_{-1}^{1} x_n(t)dt = A = $ const. We denote the resulting solution by $x_n^*(t)$. In the case where $\xi = 1$ a statement similar to that formulated in Theorem 4.1 is valid.

5. Approximate solution of the diffraction problem

Let us consider a singular integro-differential equation

$$\frac{d}{dt}\frac{1}{\pi} \int_{-1}^{1} \sqrt{1 - \tau^2}\frac{x(\tau)}{\tau - t}d\tau = f(t), \tag{15}$$

$-1 < t < 1$, which describes many diffraction problems.

In this section we propose a method for writing an approximate solution of (15) in analytical form. We look for such a solution in the form

$$x_N(t) = \sum_{k=0}^{N} x_k \frac{1}{\gamma_k} \sum_{i=0}^{N} \frac{1}{i+1} U_i(\mu_k)\, U_i(t), \tag{16}$$

where $U_i(t) = [2/(\pi(1-t^2))]^{1/2} \sin[(i+1)\arccos(t)]$ are the Chebyshev polynomials of the second kind of order i, x_k are unknown numbers, μ_k are the roots of $U_{N+1}(t)$, and $\gamma_k = \sum_{i=0}^{N} U_i^2(\mu_k)$.

It is well known that $\frac{1}{\pi} \int_{-1}^{1} \sqrt{1-\tau^2}\frac{U_k(\tau)}{\tau-t}\,d\tau = -T_{k+1}$ $(k \geq 0)$, where $T_k(t) = (2/\pi)^{1/2}\cos(k\arccos t)$ are the Chebyshev polynomials of the first kind of order k.

Using the formula [5]

$$\frac{d}{dt} T_{k+1}(t) = (k+1)U_k(t),$$

we obtain

$$\frac{d}{dt}\frac{1}{\pi}\int_{-1}^{1} \sqrt{1-\tau^2}\frac{U_k(\tau)}{\tau-t}d\tau = -(k+1)U_k(t).$$

Substituting (16) into (15) and replacing $f(t)$ in (15) by the interpolating polynomial

$$f_N(t) = \sum_{k=0}^{N} f(\mu_k)\left(\frac{1}{\gamma_k}\sum_{k=0}^{N} U_i(\mu_k)U_i(t)\right),$$

we arrive at

$$-\sum_{k=0}^{N} x_k \frac{1}{\gamma_k}\sum_{i=0}^{N} U_i(\mu_k)U_i(t) = \sum_{k=0}^{N} f(\mu_k)\frac{1}{\gamma_k}\sum_{i=0}^{N} U_i(\mu_k)U_i(t). \tag{17}$$

So, $x_k = -f(\mu_k)$ and (15) has the approximate solution

$$x_N(t) = -\sum_{k=0}^{N} f(\mu_k)\frac{1}{\gamma_k}\sum_{i=0}^{N} U_i(\mu_k)U_i(t).$$

It easy to see that

$$\|x^* - x_N\| \leq A E_N(f)\lambda_{N+1}.$$

References

1. A.I. Boikova, Methods for the approximate calculation of the Hadamard integral and solution of integral equations with Hadamard integrals, in *Integral methods in science and engineering*, C. Constanda, J. Saranen and S. Seikkala, Eds., Pitman Res. Notes Math. Ser. **375**, Addison Wesley Longman, Harlow-New York, 1997, 59–63.

2. E.W. Hobson, *The theory of spherical and ellipsoidal harmonics*, Cambridge University Press, 1931.

3. A.V. Jeshkareanee, The solution of singular integral equations by approximate projective methods, *Zh. Vychisl. Mat. i Mat. Fiz.* **15** (1979), 1149–1151.

4. F. Erdogan, G.D. Gupta and T.S. Cook, The numerical solution of singular integral equations, in *Mechanics of fracture. Vol. 1. Methods of analysis and solution of crack problems*, Noordhoff, Leyden, 1973, 368–425.

5. P.K. Suetin, *The classical orthogonal polynomials*, Nauka, Moscow, 1976 (Russian).

Department of Higher Mathematics, Penza State University, 40 Krasnaya St., Penza 440017, Russia; e-mail: boikov@diamond.stup.ac.ru

C. CHIU

A multigrid method for solving reaction-diffusion systems

1. Introduction

Reaction-diffusion equations have been used successfully to model chemical and biological processes that involve pattern formations. Numerical solutions of such equations are important for computer simulation of patterns and data analysis. In this paper, general two-dimensional reaction-diffusion systems are considered. A multigrid method is proposed for solving the system. Analysis and applications will be discussed.

It has been known for a long time that pattern formation problems are closely related to reaction-diffusion equations. A.M. Turing first suggested in 1952 that some patterns that occur in chemistry are resulted from interaction between chemical reaction and diffusion. Since then, a substantial amount of research has been done on this subject. See [10] for a survey of related developments. According to the reaction-diffusion theory, patterns are formed by the linear instability of the system and this instability eventually will be controlled by nonlinearity. So ultimate stable patterns can be obtained [12]. Because of this nature of the problem, highly stable numerical methods are necessary for computer simulations of these patterns to ensure that patterns obtained by computer simulations are formed by the instability of the original system not by numerical instability. One approach is to use ADI (alternating direction implicit) type of schemes [6], [7]. Another intuitive approach is using a fully implicit scheme. However, with the fully implicit scheme, we need to invert a sparse matrix at each step. Then finding a fast solver for the related linear systems becomes an important issue. In this paper, a multigrid method is designed for solving this problem.

Let us consider the following two-dimensional reaction-diffusion system:

$$\begin{cases} u_t = D_u \Delta u + f(u,v) \\ v_t = D_v \Delta v + g(u,v) \end{cases} \tag{1}$$

in the minimally smooth bounded spatial domain $\Omega \in R^2$ with Neumann boundary conditions $\left.\frac{\partial u}{\partial n}\right|_{\partial\Omega} = \left.\frac{\partial v}{\partial n}\right|_{\partial\Omega} = 0$. Here u and v are considered as two chemical concentrations. $\partial\Omega$ is the boundary of Ω. Assume that the initial distributions of u and v are known as $u(x,0) = u_0(x)$ and $v(x,0) = v_0(x)$. In system (1), $D_u > 0$ and $D_v > 0$ are two positive diffusion parameters; $f(u,v)$ and $g(u,v)$ are two parameter functions which simulate chemical reactions. For a general introduction of such systems, see [2].

Reaction-diffusion systems as described above are useful for modeling patterns in chemistry and biology. A few examples can be found in [5], [8], and [9]. In these examples, the concentration of nutrient and the concentration of buffer satisfy a reaction-diffusion system as given in (1).

The main purpose of this paper is to propose an efficient numerical scheme for solving system (1). The scheme is a fully implicit finite difference scheme together with a multigrid solver. In Section 2, the finite difference scheme is constructed and its stability analysis is given. The idea of using multigrid is given in Section 3. In Section 4, some applications to pattern formation problems are discussed.

2. Fully implicit discretization

For simplicity, we consider a uniform rectangular mesh on the square $[-L, L]^2$ with mesh size $h = 2L/N$. Other geometries can be accommodated; however, approximation of the boundary conditions is less trivial [7]. Let $x_{ij} = (\alpha_{1i}, \alpha_{2j}) \in \Omega$ be a grid point then $\alpha_{1i+1} = \alpha_{1i} + h$ and $\alpha_{2j+1} = \alpha_{2j} + h$, etc. A time step size $t^{n+1} - t^n$ will be denoted by Δt, and we adopt the standard notation $u_{ij}^n \approx u(x_{ij}, t_n)$, etc. Letting

$$\gamma_i = \begin{cases} 1, & 1 \leq i \leq N - 1, \\ 2, & i = 0 \text{ or } N. \end{cases}$$

The fully implicit finite difference scheme is given by:

$$\begin{cases} \dfrac{u_{ij}^{n+1} - u_{ij}^n}{\Delta t} = D_u\left(\dfrac{\Delta_{xh} u_{ij}^{n+1}}{\gamma_i h^2} + \dfrac{\Delta_{yh} u_{ij}^{n+1}}{\gamma_j h^2}\right) + f(u_{ij}^n, v_{ij}^n), \\ \dfrac{v_{ij}^{n+1} - v_{ij}^n}{\Delta t} = D_v\left(\dfrac{\Delta_{xh} v_{ij}^{n+1}}{\gamma_i h^2} + \dfrac{\Delta_{yh} v_{ij}^{n+1}}{\gamma_j h^2}\right) + g(u_{ij}^n, v_{ij}^n), \end{cases} \tag{2}$$

where Δ_{xh} and Δ_{yh} are the centered second-order difference operators such that

$$(\Delta_{xh} + \Delta_{yh}) u_{ij}^n = (u_{i+1j}^n - 2u_{ij}^n + u_{i-1j}^n) + (u_{ij+1}^n - 2u_{ij}^n + u_{ij-1}^n).$$

The zero flux boundary condition is approximated by

$$\begin{cases} u_{iN+1}^n = u_{iN-1}^n, & u_{i(-1)}^n = u_{i1}^n, & \text{for all } i, n, \quad \text{and} \\ u_{N+1j}^n = u_{N-1j}^n, & u_{(-1)j}^n = u_{1j}^n, & \text{for all } j, n. \end{cases}$$

The boundary treatment for v can be defined in the same way. Let $\underline{u}^n = \{u_{ij}^n\}$ and $\underline{v}^n = \{v_{ij}^n\}$ be the vectors obtained by the usual ordering. Then (2) can be written in the following matrix forms:

$$(I + \tau A)\underline{u}^{n+1} = \underline{u}^n + \Delta t \underline{f}^n \tag{3}$$

$$(I + \tau' A)\underline{v}^{n+1} = \underline{v}^n + \Delta t \underline{g}^n \tag{4}$$

where $\tau = D_u \Delta t/h^2$, $\tau' = D_v \Delta t/h^2$, A is the standard matrix resulting from the discretization operator, $-\Delta_{xh} - \Delta_{yh}$, $\underline{f}^n = \{f(u_{ij}^n, v_{ij}^n)\}$ and $\underline{g}^n = \{g(u_{ij}^n, v_{ij}^n)\}$. The matrix A has the property that it is symmetric and nonnegative definite [6]. All the eigenvalues of A are real and nonnegative.

The following theorem shows that the above scheme is unconditionally stable. Then any instability showed in computation should come from the original reaction-diffusion system.

Theorem 1. *Let $\underline{u}^n$ and $\underline{v}^n$ be the numerical solution by (3) and (4) with the initial values: $\{u_{ij}^0 = u_0(x_{ij})\}$ and $\{v_{ij}^0 = v_0(x_{ij})\}$. Let $\underline{\tilde{u}}^n$ and $\underline{\tilde{v}}^n$ be the numerical solution by*

(3) and (4) with the perturbed initial values: $\{\tilde{u}_{ij}^0\}$ and $\{\tilde{v}_{ij}^0\}$. Define $\underline{w} = \begin{pmatrix} u \\ v \end{pmatrix}$ and

$\underline{\tilde{w}} = \begin{pmatrix} \tilde{u} \\ \tilde{v} \end{pmatrix}$. *If the partial derivatives, f_u, f_v, g_u and g_v, are continuous and uniformly bounded, then*

$$\|\underline{w}^n - \underline{\tilde{w}}^n\| \;\leq\; (1 + C\Delta t)^n \|\underline{w}^0 - \underline{\tilde{w}}^0\|, \tag{5}$$

where $C > 0$ is a constant independent of h and Δt; $\|\cdot\|$ is the Euclidean norm (2-norm). (Then the general definition of stability for finite difference schemes can be found in [11].)

Proof. Because of (3), we have

$$\underline{u}^n - \underline{\tilde{u}}^n = (I + \tau A)^{-1}(\underline{u}^{n-1} - \underline{\tilde{u}}^{n-1}) + \Delta t(I + \tau A)^{-1}(\underline{f}^{n-1} - \underline{\tilde{f}}^{n-1}). \tag{6}$$

Since

$$f(u_{ij}^n, v_{ij}^n) - f(\tilde{u}_{ij}^n, \tilde{v}_{ij}^n) = f_u(\zeta_{ij}^n, \eta_{ij}^n)(u_{ij}^n - \tilde{u}_{ij}^n) + f_v(\zeta_{ij}^n, \eta_{ij}^n)(v_{ij}^n - \tilde{v}_{ij}^n),$$

where $(\zeta_{ij}^n, \eta_{ij}^n)$ is a midpoint between (u_{ij}^n, v_{ij}^n) and $(\tilde{u}_{ij}^n, \tilde{v}_{ij}^n)$, (6) can be expressed as

$$\begin{aligned} \underline{u}^n - \underline{\tilde{u}}^n \;=\; & (I + \tau A)^{-1}(\underline{u}^{n-1} - \underline{\tilde{u}}^{n-1}) \\ + \; & \Delta t(I + \tau A)^{-1}B_{11}^n(\underline{u}^{n-1} - \underline{\tilde{u}}^{n-1}) + \Delta t(I + \tau A)^{-1}B_{12}^n(\underline{v}^{n-1} - \underline{\tilde{v}}^{n-1}), \end{aligned} \tag{7}$$

where B_{11}^n and B_{12}^n are diagonal matrices depending on the values of f_u and f_v. Similarly

$$\begin{aligned} \underline{v}^n - \underline{\tilde{v}}^n \;=\; & (I + \tau' A)^{-1}(\underline{v}^{n-1} - \underline{\tilde{v}}^{n-1}) \\ + \; & \Delta t(I + \tau' A)^{-1}B_{21}^n(\underline{u}^{n-1} - \underline{\tilde{u}}^{n-1}) + \Delta t(I + \tau' A)^{-1}B_{22}^n(\underline{v}^{n-1} - \underline{\tilde{v}}^{n-1}), \end{aligned} \tag{8}$$

where B_{21}^n and B_{22}^n are diagonal matrices depending on the values of g_u and g_v. Combine (7) and (8), we have

$$\begin{aligned} \underline{w}^n - \underline{\tilde{w}}^n \;=\; & \begin{pmatrix} (I + \tau A)^{-1} & 0 \\ 0 & (I + \tau' A)^{-1} \end{pmatrix} (\underline{w}^{n-1} - \underline{\tilde{w}}^{n-1}) \\[2mm] + \; & \Delta t \begin{pmatrix} (I + \tau A)^{-1}B_{11}^n & (I + \tau A)^{-1}B_{12}^n \\ (I + \tau' A)^{-1}B_{21}^n & (I + \tau' A)^{-1}B_{22}^n \end{pmatrix} (\underline{w}^{n-1} - \underline{\tilde{w}}^{n-1}) \end{aligned} \tag{9}$$

Because of the properties of A, f_u, f_v, g_u, and g_v, it is trivial to show that

$$\left\| \begin{pmatrix} (I + \tau A)^{-1} & 0 \\ 0 & (I + \tau' A)^{-1} \end{pmatrix} \right\| \leq \max\left\{\|(I + \tau A)^{-1}\|, \|(I + \tau' A)^{-1}\|\right\} \leq 1 \tag{10}$$

and

$$\left\| \begin{pmatrix} (I + \tau A)^{-1}B_{11}^n & (I + \tau A)^{-1}B_{12}^n \\ (I + \tau' A)^{-1}B_{21}^n & (I + \tau' A)^{-1}B_{22}^n \end{pmatrix} \right\| \leq C, \tag{11}$$

where $C >$ is a constant independent of h and Δt.

3. Multigrid method

Since the matrix A in (3) and (4) is banded and sparse, the best way for solving (3) and (4) is using an iterative scheme. Then a multigrid technique becomes a natural way to accelerate the convergence. There are many iterative methods that can be used for this purpose. In the following, we will use the Gauss-Jacobi iteration as an example to introduce the idea of multigrid.

Let $I + \tau A = D - B$, where D is a diagonal matrix which consists of the diagonal elements of $I + \tau A$ and $B = D - I - \tau A$, which consists of the off-diagonal elements of $-(I + \tau A)$. Then the Gauss-Jacobi iteration for (3) is given as follows:

$$
\begin{aligned}
\underline{u}^{(0)} &= D^{-1}(B\underline{u}^n + \underline{u}^n + \Delta t \underline{f}^n) \\
\underline{u}^{(j)} &= D^{-1}(B\underline{u}^{(j-1)} + \underline{u}^n + \Delta t \underline{f}^n) \\
\underline{u}^{n+1} &= \underline{u}^{(J)}.
\end{aligned}
\tag{12}
$$

Such an iteration is known to be efficient for smoothing the high frequency error modes but not efficient for the low frequency modes. The multigrid technique can be used for solving this problem. Let $\{\Omega_j\}_{j=0}^{\infty}$ be a sequence of vector spaces corresponding to different levels of discretization. Ω_j corresponds to the spatial discretization size, $h = h_j = h_{j-1}/2$. Then the low frequency modes in a fine grid become high frequency modes in a coarse grid. Let P_J^{J-1} be a projection operator from Ω_J to Ω_{J-1} and Q_{J-1}^J be an interpolation operator from Ω_{J-1} to Ω_J. Then a two-grid scheme is constructed as follows:

$$
\begin{aligned}
\underline{u}^{(0)} &= D_J^{-1}(B_J\underline{u}^n + \underline{u}^n + \Delta t \underline{f}^n), \tag{13}\\
\underline{u}^{(1)} &= \underline{u}^{(0)} - Q_{J-1}^J \underline{q}, \tag{14}\\
(I + \tau A)_{J-1}\underline{q} &= P_J^{J-1}((I + \tau A)_J\underline{u}^{(0)} - \underline{u}^n - \Delta t \underline{f}^n), \tag{15}\\
\underline{u}^{n+1} &= D_J^{-1}(B_J\underline{u}^{(1)} + \underline{u}^n + \Delta t \underline{f}^n), \tag{16}
\end{aligned}
$$

where $(I + \tau A)_J$ refers to the discretization matrix at J^{th} level grid. Other similar notations are defined by the same way. This scheme consists of three steps: two iterations on the fine grid ((13) and (16) at J–level) and one correction on the coarse grid ((14) at (J-1)–level). Note that equation (15) for finding the correction vector $\underline{q}$ is of the same form as (3) at the coarse level. So we can solve (15) by the same manner as (13)–(16) and going to the next level $((J-2)$–level). Continuation in this direction results in a full multigrid scheme. The convergence and other analysis of this technique are given in [1].

Consider a multigrid V-cycle scheme with one iteration sweep on each level. Let WU be the cost of one iteration sweep on the finest grid. Suppose there are L levels, then

$$
\text{Computational cost} = 2WU(1 + 2^{-2} + \cdots + 2^{-2L}) < \frac{8}{3}WU.
$$

We know that for the systems (3) and (4), $WU = O(N^2)$. So the computational cost for each time step is of order $O(N^2)$. Therefore, fully implicit scheme combined with multigrid technique is a competitive method for solving reaction-diffusion equations. Here, details such as choices of P_J^{J-1} and Q_{J-1}^J as well as implementation and comparison with other methods remain to be studied.

82

4. Discussion of applications

Pattern formation in chemistry and biology has been an important research area for a long time, because it is related to many unanswered but fundamental questions in these fields. Not long ago, mathematical modeling of pattern formation was just a tool for qualitative analysis. Rapid development in computer science during the last two decades opens a new direction for mathematical modeling of pattern formation and other biological events. Fast speed and large storage of advanced computers make it possible to simulate real biological experiments and to do real time computing. These computational results can be used as guide to laboratory experiments and other practical purposes such as medical treatment. Many reaction-diffusion equations arising in cell growth are of the form given in this paper. Some recent research results in this area can be found in [3] and [4].

References

1. J.H. Bramble, *Multigrid methods*, Longman Scientific & Technical, 1993.

2. N.E. Britton, Reaction-diffusion equations and their applications to biology, Academic Press, 1986.

3. E.O. Budrene and H.C. Berg, Complex patterns formed by motile cells of Escherichia coli, *Nature* **349** (1991), 630–633.

4. E.O. Budrieneé, A.A. Polezhaev and M.O. Ptitsyn, Mathematical modeling of intercellular regulation causing the formation of spatial structures in bacterial colonies, *J. Theor. Biol.* **135** (1988), 323–341.

5. C. Chiu, F.C. Hoppensteadt and W. Jäger, Analysis and computer simulation of accretion patterns in bacterial cultures, *J. of Math. Biology* **32**) (1994), 841–855.

6. C. Chiu and N. Walkington, An ADI method for reaction-diffusion systems and hysteretic reaction-diffusion torpor-systems, *SIAM J. Numer. Anal.* **34** (1997), 1185–1206.

7. C. Chiu and N. Walkington, Analysis of hysteretic reaction-diffusion systems, *Quart. Appl. Math.* **56** (1998), 89–106.

8. F.C. Hoppensteadt and W. Jäger, Pattern formation by bacteria, in *Lect. Notes in Biomath.* **38** (1980), 68–81.

9. F.C. Hoppensteadt, *Mathematical methods of population biology*, Cambridge University Press, 1982.

10. J.D. Murray, *Mathematical biology*, Biomathematics Texts, Springer, 1989.

11. R.D. Richtmyer and K.W. Morton, *Difference methods for initial value problems*, Interscience, New York, 1967.

12. J. Smoller, *Shock waves and reaction-diffusion equations*, Springer, 1983.

Department of Mathematics, Michigan State University, E. Lansing, MI 48824, USA

I. CHUDINOVICH and C. CONSTANDA

Time-dependent bending of plates with transverse shear deformation

1. Preliminaries

We consider a homogeneous and isotropic elastic plate of density ρ and Lamé constants λ and μ, which occupies a region $\bar{S} \times [-h_0/2, h_0/2] \subset \mathbb{R}^3$, $h_0 = \text{const}$, where S is a domain in $\mathbb{R}^2$ bounded by a simple, closed $C^{0,1}$-curve that divides $\mathbb{R}^2$ into an interior domain S^+ and an exterior one S^-. We write $G = S \times (0, \infty)$, $G^{\pm} = S^{\pm} \times (0, \infty)$, and $\Sigma^+ = \partial S \times (0, \infty)$. In the transverse shear deformation model, the displacement field is of the form $\big(x_3 u_1(x_1, x_2, t), x_3 u_2(x_1, x_2, t), u_3(x_1, x_2, t)\big)^{\mathrm{T}}$, where the superscript T denotes matrix transposition. Using the classical averaging procedure [1] and the notation $h^2 = h_0^2/2$, $u = (u_1, u_2, u_3)^{\mathrm{T}}$, we can write the equations of motion in the form

$$M\partial_t^2 u(x,t) - Au(x,t) = q(x,t), \quad (x,t) \in G, \tag{1}$$

where A is the matrix differential operator

$$\begin{pmatrix} h^2[\mu\Delta + (\lambda+\mu)\partial_1^2] - \mu & h^2(\lambda+\mu)\partial_1\partial_2 & -\mu\partial_1 \\ h^2(\lambda+\mu)\partial_1\partial_2 & h^2[\mu\Delta + (\lambda+\mu)\partial_2^2] - \mu & -\mu\partial_2 \\ \mu\partial_1 & \mu\partial_2 & \mu\Delta \end{pmatrix},$$

$M = \text{diag}\{\rho h^2, \rho h^2, \rho\}$ and $q(x,t)$ is a combination of the body forces and moments, and of the forces and moments on the faces $x_3 = \pm h_0/2$.

2. Formulation of the problem

Let $\mathcal{L}$ and $\mathcal{L}^{-1}$ be the Laplace transformation with respect to t and its inverse, and let $\mathbb{C}_\kappa = \{p = \sigma + i\tau \in \mathbb{C} : \sigma > \kappa\}$. For $\kappa > 0$ fixed and $m, k \in \mathbb{R}$, we consider the following function spaces.

$H_{m,p}(\mathbb{R}^2)$ coincides as a set with $H_m(\mathbb{R}^2)$ but is equipped with the norm

$$\|u\|_{m,p}^2 = \int_{\mathbb{R}^2} \left(1 + |p|^2 + |\xi|^2\right)^m |\tilde{u}(\xi)|^2 \, d\xi,$$

where $\tilde{u}$ is the distributional Fourier transform of $u \in \mathcal{S}'(\mathbb{R}^2)$.

$\mathring{H}_{m,p}(S) = \big\{u \in H_{m,p}(\mathbb{R}^2) : \text{supp}\, u \subset \bar{S}\big\}$.

$H_{m,p}(S) = \big\{u : \exists v \in H_{m,p}(\mathbb{R}^2) \text{ such that } v|_S = u\big\}$, equipped with the norm

$$\|u\|_{m,p;S} = \inf_{v \in H_{m,p}(\mathbb{R}^2):\, v|_S = u} \|v\|_{m,p}.$$

$H_{1/2,p}(\partial S) = \{u : \exists v \in H_{1,p}(S) \text{ such that } \gamma v = u\}$, where γ is the (continuous) trace operator, equipped with the norm

$$\|u\|_{1/2,p;\partial S} = \inf_{v \in H_{1,p}(S):\, \gamma v = u} \|v\|_{1,p;S}.$$

$H_{-1/2,p}(\partial S)$ is the dual of $H_{1/2,p}(\partial S)$ with respect to the duality generated by the inner product $(\cdot,\cdot)_{0;\partial S}$ on $L^2(\partial S)$.

$H_{\mathcal{L};m,k,\kappa}(S)$ consists of all $u(x,p)$, $x \in S$, $p \in \mathbb{C}_\kappa$, such that $U(p) = u(\cdot,p)$ is holomorphic from $\mathbb{C}_\kappa$ to $H_m(S)$ and

$$\|u\|^2_{m,k,\kappa;S} = \sup_{\sigma > \kappa} \int_{\mathbb{R}} \left(1 + |p|^2\right)^k \|U(p)\|^2_{m,p;S}\, d\tau < \infty.$$

$H_{\mathcal{L};\pm 1/2,k,\kappa}(\partial S)$ consists of all $f(x,p)$, $x \in \partial S$, $p \in \mathbb{C}_\kappa$, such that $F(p) = f(\cdot,p)$ is holomorphic from $\mathbb{C}_\kappa$ to $H_{\pm 1/2}(\partial S)$ and

$$\|f\|^2_{\pm 1/2,k,\kappa;\partial S} = \sup_{\sigma > \kappa} \int_{\mathbb{R}} \left(1 + |p|^2\right)^k \|F(p)\|^2_{\pm 1/2,p;\partial S}\, d\tau < \infty.$$

$\mathring{H}_{\mathcal{L};-1,k,\kappa}(S^\pm)$ consists of all $q(x,p)$, $x \in S^\pm$, $p \in \mathbb{C}_\kappa$, such that $Q(p) = q(\cdot,p)$ is holomorphic from $\mathbb{C}_\kappa$ to $\mathring{H}_{-1}(S^\pm)$ and

$$\|q\|^2_{-1,k,\kappa} = \sup_{\sigma > \kappa} \int_{\mathbb{R}} \left(1 + |p|^2\right)^k \|Q(p)\|^2_{-1,p}\, d\tau < \infty.$$

$H_{r;m,k,\kappa}(G) = \mathcal{L}^{-1}\left[H_{\mathcal{L};m,k,\kappa}(S)\right]$, equipped with the norm

$$\|u(x,t)\|_{m,k,\kappa;G} = \|u(x,p)\|_{m,k,\kappa;S}.$$

$H_{r;\pm 1/2,k,\kappa}(\Sigma^+) = \mathcal{L}^{-1}\left[H_{\mathcal{L};\pm 1/2,k,\kappa}(\partial S)\right]$, equipped with the norm

$$\|f(x,t)\|_{\pm 1/2,k,\kappa;\Sigma^+} = \|f(x,p)\|_{\pm 1/2,k,\kappa;\partial S}.$$

$\mathring{H}_{r;-1,k,\kappa}(G^\pm) = \mathcal{L}^{-1}\left[\mathring{H}_{\mathcal{L};-1,k,\kappa}(S^\pm)\right]$, equipped with the norm

$$\|q(x,t)\|_{-1,k,\kappa} = \|q(x,p)\|_{-1,k,\kappa}.$$

Also, $\gamma^\pm : H_{r;1,k,\kappa}(G^\pm) \to H_{r;1/2,k,\kappa}(\Sigma^+)$ are the obvious (continuous) trace operators and $(\cdot,\cdot)_{0;S^\pm}$ is the inner product on $L^2(S^\pm)$.

We consider the energy bilinear forms [1]

$$a_\pm(u, v) = 2 \int\limits_{S^\pm} E(u, v)\, dx,$$

where

$$2E(u, v) = h^2 E_0(u, v) + h^2 \mu(\partial_2 u_1 + \partial_1 u_2)(\partial_2 \bar{v}_1 + \partial_1 \bar{v}_2)$$
$$+ \mu\big[(u_1 + \partial_1 u_3)(\bar{v}_1 + \partial_1 \bar{v}_3) + (u_2 + \partial_2 u_3)(\bar{v}_2 + \partial_2 \bar{v}_3)\big],$$
$$E_0(u, v) = (\lambda + 2\mu)\big[(\partial_1 u_1)(\partial_1 \bar{v}_1) + (\partial_2 u_2)(\partial_2 \bar{v}_2)\big]$$
$$+ \lambda\big[(\partial_1 u_1)(\partial_2 \bar{v}_2) + (\partial_2 u_2)(\partial_1 \bar{v}_1)\big].$$

We assume that $\lambda + \mu > 0$ and $\mu > 0$, so that both $a_\pm(u, u)$ are positive quadratic forms.

The classical dynamic problems with Dirichlet and Neumann boundary conditions for a smooth contour ∂S are formulated as follows.

$(DD^\pm)$ Find $u \in C^2(G^\pm) \cap C^1(\bar{G}^\pm)$ satisfying (1) in $G^\pm$ and

$$u(x, t) = f(x, t) \quad \text{on } \Sigma^+, \qquad u(x, 0) = \partial_t u(x, 0) = 0 \quad \text{in } S^\pm.$$

$(DN^\pm)$ Find $u \in C^2(G^\pm) \cap C^1(\bar{G}^\pm)$ satisfying (1) in $G^\pm$ and

$$Tu(x, t) = g(x, t) \quad \text{on } \Sigma^+, \qquad u(x, 0) = \partial_t u(x, 0) = 0 \quad \text{in } S^\pm,$$

where the boundary moment and stress matrix operator T is defined by

$$\begin{pmatrix} h^2[(\lambda + 2\mu)\nu_1\partial_1 + \mu\nu_2\partial_2] & h^2(\mu\nu_2\partial_1 + \lambda\nu_1\partial_2) & 0 \\ h^2(\lambda\nu_2\partial_1 + \mu\nu_1\partial_2) & h^2[\mu\nu_1\partial_1 + (\lambda + 2\mu)\nu_2\partial_2] & 0 \\ \mu\nu_1 & \mu\nu_2 & \mu\nu_a\partial_\alpha \end{pmatrix}$$

and $\nu = (\nu_1, \nu_2)^{\mathrm{T}}$ is unit outward normal to ∂S.

It is not difficult to construct the variational formulation of these problems.

$(DD^\pm)$ Find $u \in H_{r;1,0,\kappa}(G^\pm)$ satisfying $\gamma^\pm u = f$ and

$$\int\limits_0^\infty \big[a_\pm(u, v) - (M^{1/2}\partial_t u, M^{1/2}\partial_t v)_{0, S^\pm}\big]\, dt = \int\limits_0^\infty (q, v)_{0, S^\pm}\, dt$$

for all $v \in C_0^\infty(\bar{G}^\pm)$ such that $\gamma^\pm v = 0$.

$(DN^\pm)$ Find $u \in H_{r;1,0,\kappa}(G^\pm)$ such that, for all $v \in C_0^\infty(\bar{G}^\pm)$,

$$\int\limits_0^\infty \big[a_\pm(u, v) - (M^{1/2}\partial_t u, M^{1/2}\partial_t v)_{0, S^\pm}\big]\, dt = \int\limits_0^\infty (q, v)_{0, S^\pm}\, dt \pm \int\limits_0^\infty (g, v)_{0, \partial S}\, dt.$$

86

3. Solvability of the variational problems

Applying $\mathcal{L}$ in (1), we arrive at the elliptic system

$$p^2 M u(x,p) - A u(x,p) = q(x,p), \quad x \in S, \ p \in \mathbb{C}_\kappa, \ \kappa > 0.$$

The corresponding variational boundary value problems for this system are as follows.

$(D_{\mathcal{L}}^{\pm})$ Find $u \in H_{1,p}(S^{\pm})$ such that, $\forall v \in \overset{\circ}{H}_{1,p}(S^{\pm})$,

$$p^2 (M^{1/2}u, M^{1/2}v)_{0,S^{\pm}} + a_{\pm}(u,v) = (q,v)_{0,S^{\pm}},$$

and satisfying $\gamma^{\pm} u = f$.

$(N_{\mathcal{L}}^{\pm})$ Find $u \in H_{1,p}(S^{\pm})$ such that, $\forall v \in H_{1,p}(S^{\pm})$,

$$p^2 (M^{1/2}u, M^{1/2}v)_{0,S^{\pm}} + a_{\pm}(u,v) = (q,v)_{0,S^{\pm}} \pm (g,v)_{0,\partial S}.$$

Below we denote by c constants that are independent of $p \in \mathbb{C}_\kappa$ but may depend on κ.

Theorem 1. *For any* $\kappa > 0$, $p \in \bar{\mathbb{C}}_\kappa$, $f \in H_{1/2,p}(\partial S)$ *and* $q \in H_{-1,p}(S^{\pm})$, $(D_{\mathcal{L}}^{\pm})$ *have unique weak solutions* $u \in H_{1,p}(S^{\pm})$, *which satisfy*

$$\|u\|_{1,p;S^{\pm}} \leq c|p| \big(\|q\|_{-1,p;S^{\pm}} + \|f\|_{1/2,p;\partial S} \big).$$

Theorem 2. *For any* $\kappa > 0$, $p \in \bar{\mathbb{C}}_\kappa$, $g \in H_{-1/2,p}(\partial S)$ *and* $q \in H_{-1,p}(S^{\pm})$, $(N_{\mathcal{L}}^{\pm})$ *have unique weak solutions* $u \in H_{1,p}(S^{\pm})$, *which satisfy*

$$\|u\|_{1,p;S^{\pm}} \leq c|p| \big(\|q\|_{-1,p;S^{\pm}} + \|g\|_{-1/2,p;\partial S} \big).$$

The proofs for the interior problems are based on the coerciveness and continuity of $a_+(u,v)$ in $\overset{\circ}{H}_{1,p}(S^+)$. The problems are then reduced to equations whose operators, by Rellich's theorem, are compact, and Fredholm theory is applied.

Since Rellich's theorem is not valid in S^-, the proofs for the exterior problems use a different approach; they are based on the coerciveness and continuity in $\overset{\circ}{H}_1(S^-)$ of the form

$$a_{-,\kappa}(u,v) = \tfrac{1}{2}\kappa^2 (M^{1/2}u, M^{1/2}v)_{0,S^-} + a_-(u,v), \quad \kappa > 0.$$

The problems are reduced to equations with selfadjoint nonnegative operators, on which spectral arguments are used.

Estimates derived for the solutions of the transformed problems [2], which show that $U(p) = u(\cdot,p) : \mathbb{C}_\kappa \to H_1(S^{\pm})$ is holomorphic, enable us to apply $\mathcal{L}^{-1}$ in the appropriate spaces and establish existence results for the originals $u(x,t)$ of the solutions $u(x,p)$ supplied by Theorems 1 and 2.

Theorem 3. *For any $\kappa > 0$, $q \in H_{r;-1,1,\kappa}(G^{\pm})$, and $f \in H_{r;1/2,1,\kappa}(\Sigma^{+})$, $(\mathrm{DD}^{\pm})$ have unique weak solutions $u \in H_{r;1,0,\kappa}(G^{\pm})$. If, furthermore, $q \in H_{r;-1,k,\kappa}(G^{\pm})$, $f \in H_{r;1/2,k,\kappa}(\Sigma^{+})$ and $k \in \mathbb{R}$, then $u \in H_{r;1,k-1,\kappa}(G^{\pm})$ and*

$$\|u\|_{1,k-1,\kappa;G\pm} \leq c\big(\|q\|_{-1,k,\kappa;G\pm} + \|f\|_{1/2,k,\kappa;\Sigma+}\big).$$

Theorem 4. *For any $\kappa > 0$, $q \in \mathring{H}_{r;-1,1,\kappa}(G^{\pm})$, and $g \in H_{r;-1/2,1,\kappa}(\Sigma^{+})$, $(\mathrm{DN}^{\pm})$ have unique weak solutions $u \in H_{r;1,0,\kappa}(G^{\pm})$. If, furthermore, $q \in \mathring{H}_{r;-1,k,\kappa}(G^{\pm})$, $g \in H_{r;-1/2,k,\kappa}(\Sigma^{+})$ and $k \in \mathbb{R}$, then $u \in H_{r;1,k-1,\kappa}(G^{\pm})$ and*

$$\|u\|_{1,k-1,\kappa;G\pm} \leq c\big(\|q\|_{-1,k,\kappa} + \|g\|_{-1/2,k,\kappa;\Sigma+}\big).$$

4. Integral representation of solutions

Let ∂S be a C^2-curve. We define the single layer and double layer transformed potentials by

$$(V_p \alpha)(x,p) = \int\limits_{\partial S} D(x-y,p)\alpha(y,p)\,ds_y,$$

$$(W_p \beta)(x,p) = \int\limits_{\partial S} P(x-y,p)\beta(y,p)\,ds_y,$$

where $P(x-y,p) = \left[T_y D(y-x,p)\right]^{\mathrm{T}}$, D is a matrix of fundamental solutions for the elliptic operator of the transformed system and α and β are density functions. At the same time, we define the retarded single layer and double layer potentials by

$$(V\alpha)(x,t) = (\mathcal{L}^{-1}V_p\alpha)(x,p) = \int\limits_{0}^{t}\int\limits_{\partial S} D(x-y,t-\tau)\alpha(y,\tau)\,ds_y\,d\tau,$$

$$(W\beta)(x,t) = (\mathcal{L}^{-1}W_p\beta)(x,p) = \int\limits_{0}^{t}\int\limits_{\partial S} P(x-y,t-\tau)\beta(y,\tau)\,ds_y\,d\tau.$$

If we seek the solution of $(\mathrm{DD}^{\pm})$ as $u = V\alpha$, then α satisfies the weakly singular boundary integral equation

$$V\alpha = f \quad \text{on } \Sigma^{+}. \tag{2}$$

On the other hand, if we seek the solution of $(\mathrm{DD}^{\pm})$ as $= W\beta$, then β satisfies the singular boundary integral equation

$$(W\beta)^{\pm} = f \quad \text{on } \Sigma^{+}. \tag{3}$$

Theorem 5. *For any $\kappa > 0$, $k \in \mathbb{R}$ and $f \in H_{r;1/2,k,\kappa}(\Sigma^+)$, (2) and (3) have unique solutions $\alpha \in H_{r;-1/2,k-1,\kappa}(\Sigma^+)$ and $\beta \in H_{r;1/2,k-2,\kappa}(\Sigma^+)$, in which case $u \in H_{r;1,k-1,\kappa}(G^\pm)$. If $k \geq 1$, then u is the weak solution of* $(DD^\pm)$.

If we now seek the solution of $(DN^\pm)$ as $u = V\alpha$, then α satisfies the singular boundary integral equation

$$(TV\alpha)^\pm = g \quad \text{on } \Sigma^+. \tag{4}$$

If we seek the solution of $(DN^\pm)$ as $u = W\beta$, then β satisfies the hypersingular boundary integral equation

$$TW\beta = g \quad \text{on } \Sigma^+. \tag{5}$$

Theorem 6. *For any $\kappa > 0$, $k \in \mathbb{R}$ and $g \in H_{r;-1/2,k,\kappa}(\Sigma^+)$, (4) and (5) have unique solutions $\alpha \in H_{r;-1/2,k-2,\kappa}(\Sigma^+)$ and $\beta \in H_{r;1/2,k-1,\kappa}(\Sigma^+)$, in which case $u \in H_{r;1,k-1,\kappa}(G^\pm)$. If $k \geq 1$, then u is the weak solution of* $(DN^\pm)$.

The ranges of the boundary integral operators generated by V and W are not closed in the appropriate (natural) spaces, so Fredholm theory cannot be applied. Instead, the proofs of Theorems 5 and 6 make use of the properties of the dynamic analogues of the Poincaré-Steklov operators [3], the continuity and injectivity of the boundary integral operators, and density arguments.

The case of nonhomogeneous initial conditions in $(DD^\pm)$ and $(DN^\pm)$ can be reduced to the homogeneous one by means of some "initial" potentials [4].

References

1. C. Constanda, *A mathematical study of elastic plates with transverse shear deformation*, Pitman Res. Notes Math. Ser. **219**, Longman, Harlow-New York, 1990.

2. I. Chudinovich and C. Constanda, Existence theorems in the dynamic theory of plates, *Strathclyde Math. Res. Report* **32** (1997).

3. I. Chudinovich and C. Constanda, Non-stationary integral equations for elastic plates, *C.R. Acad. Sci. Paris Sér. I* **329** (1999), 1115–1120.

4. I. Chudinovich and C. Constanda, The Cauchy problem in the theory of plates with transverse shear deformation, *Math. Models Methods Appl. Sci.* (to appear).

Department of Mathematics and Mechanics, Kharkov National University, Kharkov, Ukraine

Department of Mathematics, University of Strathclyde, Glasgow, UK

S.G. CIUMASU and D. VIERU

Polarization gradient in piezoelectric micropolar elasticity

1. Introduction

We discuss the linear theory of piezoelectric micropolar elasticity with polarization gradient in the case of quasi-electrostatic approximation. Mindlin [1] generalizes the classical Voight model including among the constitutive variables the spatial polarization gradient as an effect of some results from the ionic crystaline networks. Nowacki [2] has derived the linear theory of the piezoelectric micropolar thermoelasticity in the case of a quasi-static electric field. In this paper, based on Mindlin's and Nowacki's work, we derive the linear theory of micropolar piezoelectricity with polarization gradient. We deduce the basic equations of the theory and formulate mixed problems. We also derive the equations for an isotropic material. We establish reciprocity relations and new uniqueness results by using a method suggested by Ieşan [3], avoiding the use of the Laplace transform and the incorporation of the initial conditions into the equations of motion. All considerations are presented in tensor form.

The motion of the body is referred to a fixed system of rectangular Cartesian axes Ox_i, and $\mathbf{e}_i$, $i = 1, 2, 3$, are the unit vectors of the axes. Let $\mathcal{D}$ be the regular region of the Euclidean three-dimensional space occupied by the body at time $t = 0$, S the boundary of $\mathcal{D}$ and $\mathbf{n}$ the outward unit normal at a point of S; also, $(\mathbf{x}, t)$ is a point in $\overline{\mathcal{D}} \times I$, where $I = [0, t_1)$, $t_1 > 0$, is a time interval, $\mathcal{R}$ is the set of real numbers, $\mathcal{R}^p$ is the Euclidean p-dimensional vector space, $\mathcal{T}_p = \mathcal{R}^3 \otimes \mathcal{R}^3 \cdots \otimes \mathcal{R}^3$ is the space of p-th order tensors (tensors of order $p \geq 1$ are in bold type), $\mathbf{A} \cdot \mathbf{B} = \mathrm{tr}(\mathbf{A}\mathbf{B}^{\mathrm{T}})$ is the inner product of any two p-th order tensors $\mathbf{A}$ and $\mathbf{B}$, $\mathbf{B} = \mathbf{L}[\mathbf{A}]$ is the linear tensor function of Cartesian components $B_{km} = L_{kmpq}A_{pq}$, the superscript T denotes matrix transposition defined by $\mathbf{u} \cdot \mathbf{A}\mathbf{v} = \mathbf{A}^{\mathrm{T}}\mathbf{u} \cdot \mathbf{v}$, $\mathbf{u}, \mathbf{v} \in \mathcal{T}_1$, $\partial\Psi/\partial\mathbf{L} = \Psi_L(\mathbf{L})$, where $\Psi(\mathbf{L})$ is a scalar function and Ψ_L is defined by $\Psi_L(\mathbf{L}) \cdot \mathbf{C} = (d/ds)(\Psi(\mathbf{L} + s\mathbf{C}))|_{s=0}$, summation over repeated indices is understood, subscripts preceded by a comma denote partial differentiation with respect to the corresponding Cartesian coordinate, a superposed dot denotes the time derivative and the convolution $\mathbf{u} * \mathbf{v}$ of two integrable functions $\mathbf{u}$ and $\mathbf{v}$ in $\mathcal{T}_p$ is defined by

$$(\mathbf{u} * \mathbf{v})(\mathbf{x}, t) = \int_0^t \mathbf{u}(\mathbf{x}, t - \tau) \cdot \mathbf{v}(\mathbf{x}, \tau) d\tau$$

$(\mathbf{u}, \mathbf{v}$ may be scalar functions).

2. Basic equations

We consider a micropolar piezoelectric anisotropic body. The quasi-electrostatic approximation is assumed in the sequel of this paper. The local field equation and the associated boundary conditions that govern motion are obtained by means of the generalized formulation of Hamilton's principle, where the polarization gradient is added to the set of the independent constitutive variables.

90

We introduce the notation

$$\mathcal{U} = (\mathbf{u}, \mathbf{w}, \mathbf{P}, \varphi), \ \mathcal{U} : \overline{\mathcal{D}} \times I \to \mathcal{R}^{10}, \ \mathcal{F} = (\mathbf{F}, \mathbf{G}, \mathbf{E}^0, -f), \ \mathcal{F} : \overline{\mathcal{D}} \times I \to \mathcal{R}^{10},$$
$$\mathcal{T} = (\mathbf{t}, \mathbf{m}, \mathbf{s}, -f_1), \ \mathcal{T} : S \times I \to \mathcal{R}^{10}, \tag{2.1}$$

where $\mathbf{u}, \mathbf{w}, \mathbf{P} : \overline{\mathcal{D}} \times I \to \mathcal{T}_1$ are the displacement, microrotation and polarization vectors, respectively, $\varphi : \overline{\mathcal{D}} \times I \to \mathcal{R}_1$ is the electric potential, $\mathbf{F}, \mathbf{G}, \mathbf{E}^0 : \overline{\mathcal{D}} \times I \to \mathcal{T}_1$ are the body force and body couple vectors per unit volume and the given electric field, $f : \overline{\mathcal{D}} \times I \to \mathcal{R}$ is the volume density of free charge, $\mathbf{t}, \mathbf{m}, \mathbf{s} : S \times I \to \mathcal{T}_1$ are the stress, couple stress, and surface electric force vectors, and $f_1 : S \times I \to \mathcal{R}$ is the surface density of free charge. The inner products of $\mathcal{F}, \mathcal{U}$ and $\mathcal{T}, \mathcal{U}$ are

$$\mathcal{F} \cdot \mathcal{U} = \mathbf{F} \cdot \mathbf{u} + \mathbf{G} \cdot \mathbf{w} + \mathbf{E}^0 \cdot \mathbf{P} - f\varphi, \quad \mathcal{T} \cdot \mathcal{U} = \mathbf{t} \cdot \mathbf{u} + \mathbf{m} \cdot \mathbf{w} + \mathbf{s} \cdot \mathbf{P} - f_1\varphi. \tag{2.2}$$

We define the kinetic energy density by $\mathcal{K} = \frac{1}{2}\rho\dot{\mathbf{u}} \cdot \dot{\mathbf{u}} + \dot{\mathbf{w}} \cdot \mathbf{I}\dot{\mathbf{w}}$, where ρ is the material density, $\mathbf{I} = \rho\mathbf{J}$, $\mathbf{J} = J_{ij}\mathbf{e}_i \otimes \mathbf{e}_j$, and J_{ij} are the microinertia coefficients. The inertia tensor is symmetric and positive definite; that is,

$$\mathbf{J}\mathbf{a} \cdot \mathbf{b} = \mathbf{a} \cdot \mathbf{J}\mathbf{b}, \quad \mathbf{a} \cdot \mathbf{J}\mathbf{a} > 0 \quad \forall \mathbf{a}, \mathbf{b} \in \mathcal{T}_1. \tag{2.3}$$

The electric enthalpy density is defined by

$$H = W(\mathbf{e}, \kappa, \mathbf{P}(\nabla\mathbf{P})^{\mathrm{T}}) - \tfrac{1}{2}\varepsilon_0 \mathbf{E} \cdot \mathbf{E} - \mathbf{E} \cdot \mathbf{P}, \tag{2.4}$$

where

$$\mathbf{e} = (\nabla\mathbf{u})^{\mathrm{T}} + \mathbf{R} \cdot \mathbf{w}, \quad \kappa = (\nabla\mathbf{w})^{\mathrm{T}}, \tag{2.5}$$

$\mathbf{e}, \kappa : \overline{\mathcal{D}} \times I \to \mathcal{T}_2$ are the strain displacement tensors, ε_0 is the permitivity of vacuum, $\mathbf{E} : \overline{\mathcal{D}} \times I \to \mathcal{T}_1$ is the Maxwell self-field given by

$$\mathbf{E} = -\nabla\varphi, \tag{2.6}$$

and the components of $\mathbf{R} \cdot \mathbf{w}$ are $(\mathbf{R} \cdot \mathbf{w})_{ji} = \varepsilon_{jik}\varphi_k$, where $\mathbf{R}$ is the third-order tensor having the alternating symbols ε_{ijk} as components and $\mathbf{w}$ is the energy density of deformation and polarization.

The generalized Hamilton principle states that for all admissible variations of motion of the body we have

$$\delta \int_{t_0}^{t_1} dt \int_{\mathcal{D}^*} (\mathcal{K} - H)dv + \int_{t_0}^{t_1} dt \int_{\mathcal{D}} \mathcal{F} \cdot \delta\mathcal{U} \, dv + \int_{t_0}^{t_1} dt \int_S \mathcal{T} \cdot \delta\mathcal{U} \, da = 0, \tag{2.7}$$

where t_0 and t_1 are two arbitrary moments, $\mathcal{D}^* = \mathcal{D} \cup \mathcal{D}'$ and $\mathcal{D}'$ is the exterior of the body, which is vacuum.

From (2.7) we conclude that the equations of motion and quasistatic electric field are

$$\nabla \cdot \tau^{\mathrm{T}} + \mathbf{F} = \rho\ddot{\mathbf{u}}, \quad \nabla \cdot \mu^{\mathrm{T}} + \mathbf{R}[\tau] + \mathbf{G} = \mathbf{I}\ddot{\mathbf{w}}, \tag{2.8}$$

$$\nabla \cdot \pi^{1\mathrm{T}} - \nabla\varphi + \pi + \mathbf{E}^0 = 0, \quad \nabla \cdot \mathbf{P} - \varepsilon_0\nabla\varphi = f \ \text{ in } \mathcal{D}, \tag{2.9}$$

$$\Delta\varphi = 0 \ \text{ in } \mathcal{D}', \tag{2.10}$$

and that the boundary conditions are

$$\mathbf{t} = \tau^{\mathrm{T}}\mathbf{n}, \quad \mathbf{m} = \mu^{\mathrm{T}}\mathbf{n}, \quad \mathbf{s} = \pi^{1\mathrm{T}}\mathbf{n}, \quad (\varepsilon_0 <\nabla\varphi> -\mathbf{P})\cdot\mathbf{n} = f_1. \tag{2.11}$$

In (2.8)–(2.11), $\tau = W_e(\mathbf{e})$, $\mu = W_\kappa(\kappa)$ and $\pi^1 = W_{(\nabla P)\mathrm{T}}((\nabla\mathbf{P}))^{\mathrm{T}} : \overline{\mathcal{D}} \times I \to \mathcal{T}_2$ are the stress, couple stress and dipolar electric tensors, and $\pi : \overline{\mathcal{D}} \times I \to \mathcal{T}_1$ is the effective local electric force vector.

In the linear theory, under the assumption that the body is initially free from stress and hyperstress, the energy density of deformation and polarization is

$$\begin{aligned}
W = &\tfrac{1}{2}\mathbf{e}\cdot\mathbf{A}[\mathbf{e}] + \mathbf{e}\cdot\mathbf{A}^1[\kappa] - \mathbf{e}\cdot\mathbf{A}^2\mathbf{P} - \mathbf{e}\cdot\mathbf{G}[(\nabla\mathbf{P})^{\mathrm{T}}] + \tfrac{1}{2}\kappa\cdot\mathbf{B}[\kappa] - \kappa\cdot\mathbf{B}^1\mathbf{P} - \kappa\cdot\mathbf{G}^1[(\nabla\mathbf{P})^{\mathrm{T}}] \\
&+ \tfrac{1}{2}\mathbf{P}\cdot\mathbf{A}^3\mathbf{P} + \mathbf{P}\cdot\mathbf{A}^4[(\nabla\mathbf{P})^{\mathrm{T}}] + \tfrac{1}{2}(\nabla\mathbf{P})^{\mathrm{T}}\cdot\mathbf{D}[(\nabla\mathbf{P})^{\mathrm{T}}] + \mathbf{C}[(\nabla\mathbf{P})^{\mathrm{T}}],
\end{aligned}$$

where $\mathbf{A}, \mathbf{A}^1, \mathbf{B}, \mathbf{D}, \mathbf{G}, \mathbf{G}^1 : \overline{\mathcal{D}} \to \mathcal{T}_4$, $\mathbf{A}^2, \mathbf{A}^4, \mathbf{B}^1 : \overline{\mathcal{D}} \to \mathcal{T}_3$ and $\mathbf{A}^3, \mathbf{C} : \overline{\mathcal{D}} \to \mathcal{T}_2$ are tensors characterizing the material and electrical properties of the body.

We obtain the constitutive relations

$$\begin{aligned}
\tau = &\,\mathbf{A}[\mathbf{e}] + \mathbf{A}^1[\kappa] - \mathbf{A}^2\mathbf{P} - \mathbf{G}[(\nabla\mathbf{P})^{\mathrm{T}}], \quad \mu = \mathbf{e}\mathbf{A}^1 + \mathbf{B}[\kappa] - \mathbf{B}^1\mathbf{P} - \mathbf{G}^1[(\nabla\mathbf{P})^{\mathrm{T}}], \\
&\pi = \mathbf{e}\mathbf{A}^2 + \kappa\mathbf{B}^1 - \mathbf{A}^3\mathbf{P} - \mathbf{A}^4[(\nabla\mathbf{P})^{\mathrm{T}}], \\
&\pi^1 = -\mathbf{e}\mathbf{G} - \kappa\mathbf{G}^1 + \mathbf{P}\mathbf{A}^4 + \mathbf{D}[(\nabla\mathbf{P})^{\mathrm{T}}] + \mathbf{C}.
\end{aligned} \tag{2.12}$$

The coefficients in (2.12) have the symmetries

$$\begin{aligned}
\mathbf{A}^3\mathbf{a}\cdot\mathbf{b} = \mathbf{a}\cdot\mathbf{A}^3\mathbf{b} \quad &\forall \mathbf{a}, \mathbf{b} \in \mathcal{T}_1, \\
\mathbf{a}\cdot\mathbf{A}[\mathbf{b}] = \mathbf{b}\cdot\mathbf{A}[\mathbf{a}], \quad \mathbf{a}\cdot\mathbf{B}[\mathbf{b}] = \mathbf{b}\cdot\mathbf{B}[\mathbf{a}], \quad \mathbf{a}\cdot\mathbf{D}[\mathbf{b}] = \mathbf{b}\cdot\mathbf{D}[\mathbf{a}] \quad &\forall \mathbf{a}, \mathbf{b} \in \mathcal{T}_2.
\end{aligned} \tag{2.13}$$

To the system of field equation (2.5),(2.8),(2.9),(2.12) we adjoin the initial conditions

$$\mathbf{u}(\mathbf{x}, 0) = \mathbf{a}'(\mathbf{x}), \quad \dot{\mathbf{u}}(\mathbf{x}, 0) = \mathbf{b}'(\mathbf{x}), \quad \mathbf{w}(\mathbf{x}, 0) = \mathbf{c}'(\mathbf{x}), \quad \dot{\mathbf{w}}(\mathbf{x}, 0) = \mathbf{d}'(\mathbf{x}), \tag{2.14}$$

and, for a mixed problem, the boundary conditions

$$\begin{aligned}
\mathbf{u} = \mathbf{u}' \text{ on } \overline{S}_1 \times I, \quad &\mathbf{w} = \mathbf{w}' \text{ on } \overline{S}_3 \times I, \quad \varphi = \varphi' \text{ on } \overline{S}_5 \times I, \\
\mathbf{P} = \mathbf{P}' \text{ on } \overline{S}_7 \times I, \quad &\tau^{\mathrm{T}}\mathbf{n} = \mathbf{t}' \text{ on } S_2 \times I, \quad \mu^{\mathrm{T}}\mathbf{n} = \mathbf{m}' \text{ on } S_4 \times I, \\
(\varepsilon_0 \langle\nabla\varphi\rangle - \mathbf{P})\mathbf{n} = f_1 \text{ on } S_6 \times I, \quad &\pi^{1\mathrm{T}} = \mathbf{s}' \text{ on } S_8 \times I,
\end{aligned} \tag{2.15}$$

where S_r, $r = 1, 2, \ldots, 8$, denotes subsets of S such that $\overline{S}_1 \cup S_2 = \overline{S}_3 \cup S_4 = \overline{S}_5 \cup S_6 = \overline{S}_7 \cup S_8 = S$ and $S_1 \cap S_2 = S_3 \cap S_4 = S_5 \cap S_6 = S_7 \cap S_8 = \emptyset$, and $\mathbf{a}', \mathbf{b}', \mathbf{c}', \mathbf{d}', \mathbf{u}', \mathbf{t}', \mathbf{w}', \mathbf{m}', \varphi', f_1, \mathbf{P}'$ and $\mathbf{s}'$ are prescribed functions.

Suppose that

(i) the characteristic constants (2.3) and (2.13) of the material are continuously differentiable on $\overline{\mathcal{D}}$;

(ii) the external data system $\{\mathbf{a}', \mathbf{b}', \mathbf{c}', \mathbf{d}', \mathbf{u}', \mathbf{t}', \mathbf{w}', \mathbf{m}', \varphi', f_1, \mathbf{P}', \mathbf{s}', \mathbf{F}, \mathbf{G}, \mathbf{E}^0,$ $f\}$ is such that $\mathbf{F}, \mathbf{G}, \mathbf{E}^0, f \in C^{0,0}(\overline{\mathcal{D}} \times I)$, $\mathbf{u}' \in C^{0,0}(S_1 \times I)$, $\mathbf{w}' \in C^{0,0}(S_3 \times I)$, $\varphi \in C^{0,0}(S_5 \times I)$, $\mathbf{P}' \in C^{0,0}(S_7 \times I)$, $\mathbf{t}', \mathbf{m}', f_1, \mathbf{s}'$ are continuous in time and piecewise regular, respectively, on $S_2 \times I$, $S_4 \times I$, $S_6 \times I$, $S_8 \times I$, and $\mathbf{a}', \mathbf{b}', \mathbf{c}', \mathbf{d}'$ are continuous on $\overline{\mathcal{D}}$.

An admissible process in $\overline{\mathcal{D}}$ is an ordered array of functions $\{u_i, \varphi_i, P_i, \varphi, e_{ij},$ $\kappa_{ij}, \tau_{ij}, \mu_{ij}, \pi_i, \pi_{ij}^1\}$ that satisfy certain smoothness conditions in $\overline{\mathcal{D}} \times I$, namely, $u_i \in C^{2,2}$, $\varphi_i \in C^{1,2}$, $P_i \in C^{1,0}$, $\varphi \in C^{2,0}$, $e_{ij} \in C^{0,0}$, $\kappa_{ij} \in C^{0,0}$, $\tau_{ij} \in C^{1,0}$, $\mu_{ij} \in C^{1,0}$, $\pi_i \in C^{0,0}$, and $\pi_{ij}^1 \in C^{1,0}$.

A solution of the mixed problem is an admissible state that satisfies the field equations, the geometric relations (2.5), the equations of motion (2.8), the equations of the quasistatic electric field (2.9), the constitutive equations (2.12), the initial conditions (2.14) and the boundary conditions (2.15).

By means of the method given in [4] and [5], in the linear theory for a homogeneous and isotropic elastic solid we obtain the energy density of deformation and polarization in the form

$$W = \tfrac{1}{2}\lambda_1 e_{ii} e_{jj} + \tfrac{1}{2}(\lambda_2 + \lambda_3) e_{ij} e_{ij} + \tfrac{1}{2}\lambda_3 e_{ij} e_{ji} - \lambda_4 e_{ii} P_{j,j} - \lambda_5 e_{ij} P_{j,i}$$
$$- \lambda_6 e_{ij} P_{i,j} + \tfrac{1}{2}\lambda_7 \kappa_{ii} \kappa_{jj} + \tfrac{1}{2}\lambda_8 \kappa_{ij} \kappa_{ij} + \tfrac{1}{2}\lambda_9 \kappa_{ij} \kappa_{ji} - \lambda_{10} \varepsilon_{ijk} \kappa_{ij} P_k$$
$$+ \tfrac{1}{2}\lambda_{11} P_i P_i + \tfrac{1}{2}\lambda_{12} P_{i,i} P_{j,j} + \tfrac{1}{2}\lambda_{13} P_{j,i} P_{j,i} + \tfrac{1}{2}\lambda_{14} P_{j,i} P_{i,j} + C^0 P_{i,i}. \quad (2.16)$$

The field equations of the linear theory for isotropic solids are

$$\Diamond_1 \mathbf{u} + (\lambda_1 + \lambda_3)\operatorname{grad}\operatorname{div}\mathbf{u} + \lambda_2\operatorname{curl}\mathbf{w} - \lambda_5\Delta\mathbf{P} - (\lambda_4 + \lambda_6)\operatorname{grad}\operatorname{div}\mathbf{P} + \mathbf{F} = \mathbf{0},$$
$$\Diamond_2 \mathbf{w} + \lambda_2\operatorname{curl}\mathbf{u} + (\lambda_7 + \lambda_9)\operatorname{grad}\operatorname{div}\mathbf{w} + (\lambda_{10} + \lambda_6 - \lambda_5)\operatorname{curl}\mathbf{P} - 2\lambda_2\mathbf{w} + \mathbf{G} = \mathbf{0},$$
$$\lambda_5\Delta\mathbf{u} + (\lambda_4 + \lambda_6)\operatorname{grad}\operatorname{div}\mathbf{u} - (\lambda_{11} + \lambda_6 - \lambda_5)\operatorname{curl}\mathbf{w} \quad (2.17)$$
$$-\lambda_{13}\Delta\mathbf{P} + \lambda_{11}\mathbf{P} + \operatorname{grad}\varphi - \mathbf{E}^0 = \mathbf{0},$$
$$\operatorname{div}\mathbf{P} - \varepsilon_0\Delta\varphi = f,$$

where $\Diamond_1 = (\lambda_2 + \lambda_3)\Delta - \rho\partial^2/\partial t^2$ and $\Diamond_2 = \lambda_8\Delta - \mathbf{I}\partial^2/\partial t^2$.

3. Reciprocity theorem and uniqueness theorem

We consider the body subjected to two different external data systems $\mathcal{L}^{(\alpha)} = \{\mathbf{F}^{(\alpha)}, \mathbf{G}^{(\alpha)}, \mathbf{E}^{0(\alpha)}, f^{(\alpha)}, \mathbf{u}'^{(\alpha)}, \mathbf{t}'^{(\alpha)}, \mathbf{w}'^{(\alpha)}, \mathbf{m}'^{(\alpha)}, \varphi'^{(\alpha)}, f_1^{(\alpha)}, \mathbf{s}'^{(\alpha)}, \mathbf{H}^{(\alpha)}, \mathbf{a}'^{(\alpha)}, \mathbf{b}'^{(\alpha)},$ $\mathbf{c}'^{(\alpha)}, \mathbf{d}'^{(\alpha)}\}$, $\alpha = 1, 2$, where $\mathbf{H} = \mathbf{P} \otimes \mathbf{n}$ with components $H_{ij} = P_i n_j$.

Let $\mathcal{S}^{(\alpha)} = \{\mathbf{u}^{(\alpha)}, \mathbf{w}^{(\alpha)}, \mathbf{e}^{(\alpha)}, \kappa^{(\alpha)}, \tau^{(\alpha)}, \mu^{(\alpha)}, \varphi^{(\alpha)}, \mathbf{P}^{(\alpha)}, \pi^{(\alpha)}, \pi^{1(\alpha)}\}$ be a solution corresponding to $\mathcal{L}^{(\alpha)}$.

Theorem 3.1. *Let $r, s \in I = [0, t_1)$ and*

$$N_{\alpha\beta}(r, s) = \tau^{(\alpha)}(r)\cdot\mathbf{e}^{(\beta)}(s) + \mu^{(\alpha)}(r)\cdot\kappa^{(\alpha)}(s) - \pi^{(\alpha)}(r)\cdot\mathbf{P}^{(\beta)}(s)$$
$$+ \varepsilon_0\mathbf{E}^{(\alpha)}(r)\cdot\mathbf{E}^{(\beta)}(s) + (\pi^{1(\alpha)}(r) - \xi^{(\alpha)}(r))\cdot(\nabla\mathbf{P}^{(\beta)}(s))^{\mathrm{T}}, \quad (3.1)$$

where

$$\xi^{(\alpha)} = \mathbf{C} \cdot \delta^{(\alpha)}, \qquad \xi_{ij}^{(\alpha)} = C_{ik}\delta_{kj}^{(\alpha)}. \tag{3.2}$$

If the coefficients $\mathbf{A}, \mathbf{A}^3, \mathbf{B}, \mathbf{D}$ *satisfy the symmetry relation* (2.13), *then*

$$N_{\alpha\beta}(r,s) = N_{\beta\alpha}(s,r) \quad \forall r,s \in I, \ \alpha,\beta = 1,2. \tag{3.3}$$

Proof. Using the constitutive relations (2.12) and taking into consideration the symmetries of the coefficients, we conclude that (3.3) holds.

Theorem 3.2. *Let* $r,s \in I$ *and*

$$M_{\alpha\beta}(r,s) = \int_{\mathcal{D}} \mathcal{F}^{(\alpha)}(r) \cdot \mathcal{U}^{(\beta)}(s)\,dv$$

$$- \int_{\mathcal{D}} (\rho\ddot{\mathbf{u}}^{(\alpha)}(r) \cdot \mathbf{u}^{(\beta)}(s) + \mathbf{I}\ddot{\mathbf{w}}^{(\alpha)}(r) \cdot \mathbf{w}^{(\beta)}(s))\,dv$$

$$+ \int_{S} (\mathcal{T}^{(\alpha)}(r) \cdot \mathcal{U}^{(\beta)}(s) - \xi^{\mathrm{T}(\alpha)}(r) \cdot \mathbf{H}^{(\beta)}(s))\,da. \tag{3.4}$$

If the coefficients $\mathbf{I}, \mathbf{A}, \mathbf{A}^3, \mathbf{B}, \mathbf{D}$ *satisfy the symmetry relations* (2.3) *and* (2.13), *then*

$$M_{\alpha\beta}(r,s) = M_{\beta\alpha}(s,r) \quad \forall r,s \in I, \ \alpha,\beta = 1,2. \tag{3.5}$$

Proof. Formula (3.5) is obtained by inserting the geometric relations (2.5) in (3.1), using (2.8), (2.9), (2.15) and the divergence theorem, integrating (3.1) over $\mathcal{D}$, and applying Theorem 3.1.

Theorem 3.3. (Reciprocity theorem) *Suppose that the characteristic constants of the material satisfy the symmetric relations* (2.3) *and* (2.13). *If a piezoelectric micropolar material is subjected to two external data systems* $\mathcal{L}^{(\alpha)}$, $\alpha = 1,2$, *then the corresponding solutions* $\mathcal{S}^{(\alpha)}$, $\alpha = 1,2$, *satisfy the reciprocity relation*

$$\int_{\mathcal{D}} [\mathbf{f}^{(1)} * \mathbf{u}^{(2)} + \mathbf{g}^{(1)} * \mathbf{w}^{(2)} + \gamma * (\mathbf{E}^{0(1)} * \mathbf{P}^{(2)} - f^{(1)} * \varphi^{(2)})]\,dv$$

$$+ \int_{S} \gamma * (\mathcal{T}^{(1)} * \mathcal{U}^{(2)} - \xi^{\mathrm{T}(1)} * \mathbf{H}^{(2)})\,da$$

$$= \int_{\mathcal{D}} [\mathbf{f}^{(2)} * \mathbf{u}^{(1)} + \mathbf{g}^{(2)} * \mathbf{w}^{(1)} + \gamma * (\mathbf{E}^{0(2)} * \mathbf{P}^{(1)} - f^{(2)} * \varphi^{(1)})]\,dv$$

$$+ \int_{S} \gamma * (\mathcal{T}^{(2)} * \mathcal{U}^{(1)} - \xi^{\mathrm{T}(2)} * \mathbf{H}^{(1)})\,da, \tag{3.6}$$

where

$$\gamma(t) = t, \quad \mu(t) = 1,$$

$$\mathbf{f}^{(\alpha)} = \gamma * \mathbf{F}^{(\alpha)} + \rho(t\mathbf{b}'^{(\alpha)} + \mathbf{a}'^{(\alpha)}), \quad \mathbf{g}^{(\alpha)} = \gamma * \mathbf{G}^{(\alpha)} + \mathbf{I}(t\mathbf{d}'^{(\alpha)} + \mathbf{c}'^{(\alpha)}).$$

94

Proof. If we take $\alpha = 1$, $\beta = 2$, $r = \tau$, $s = t - \tau$ in (3.5), integrate with respect to τ from 0 to t and use the definition and commutativity of convolution, we arrive at

$$
\int_{\mathcal{D}} \mathcal{F}^{(1)} * \mathcal{U}^{(2)} dv - \int_{\mathcal{D}} (\rho \ddot{\mathbf{u}}^{(1)} * \mathbf{u}^{(2)} + \mathbf{I} \ddot{\mathbf{w}}^{(1)} * \mathbf{w}^{(2)}) dv
$$
$$
+ \int_{S} (\mathcal{T}^{(1)} * \mathcal{U}^{(2)} - \xi^{\mathrm{T}(1)} * \mathbf{H}^{(2)}) da
$$
$$
= \int_{\mathcal{D}} \mathcal{F}^{(2)} * \mathcal{U}^{(1)} dv - \int_{\mathcal{D}} (\rho \ddot{\mathbf{u}}^{(2)} * \mathbf{u}^{(1)} + \mathbf{I} \ddot{\mathbf{w}}^{(2)} * \mathbf{w}^{(1)}) dv
$$
$$
+ \int_{S} (\mathcal{T}^{(2)} * \mathcal{U}^{(1)} - \xi^{\mathrm{T}(2)} * \mathbf{H}^{(1)}) da. \tag{3.7}
$$

Taking the convolution of (3.7) with γ and using the equality $(\gamma * \ddot{\mathbf{u}})(\mathbf{x}, t) = \mathbf{u}(\mathbf{x}, t) - t\dot{\mathbf{u}}(\mathbf{x}, 0) - \mathbf{u}(\mathbf{x}, 0)$, the initial conditions (2.14) and the symmetry of $\mathbf{I}$, we obtain (3.6).

Theorem 3.4. (Uniqueness theorem) *Suppose that*
 (i) *the characteristic constants of the material satisfy (2.3) and (2.13);*
 (ii) $\mathbf{A}^3$, $\mathbf{A}^4$ *and* $\mathbf{D}$ *are positive definite.*
Then the mixed problem formulated in §2 has at most one solution if S_5 is non-empty. If S_5 is empty, then any two solutions of the mixed problem differ only by their electric potentials, up to a constant on $\overline{\mathcal{D}}$.

The proof of this assertion is similar to that of Brun's theorem in linear elastodynamics [6].

References

1. R.D. Mindlin, Polarization gradient in elastic dielectrics, *Internat. J. Solids Structures* **4** (1968), 637–642.

2. W. Nowacki, *Theory of asymmetric elasticity*, P.W.N., Warszawa, and Pergamon Press, Oxford, 1986.

3. D. Ieşan, Reciprocity, uniqueness and minimum principles in the linear theory of piezoelectricity, *Int. J. Engrg. Sci.* **28** (1990), 1139–1149.

4. I. Beju, E. Soós and P.P. Teodorescu, *Euclidean tensor calculus with applications*, Abacus Press, Tunbridge Wells, UK, 1983.

5. A.C. Eringen, *Continuum physics. Vol. I. Mathematics*, Academic Press, New York, 1971.

6. L. Brun, *C.R. Acad. Sci. Paris* **261** (1965), 2584–2595.

Department of Theoretical Mechanics, Technical University "Gh.Asachi", 6600 Iaşi, Romania

T. COMLEKCI, R. HAMILTON and J.T. BOYLE

Thermoelastic stress separation via Poisson equation solution by means of the boundary element method

1. Introduction

Thermoelastic stress analysis (TSA) is a full-field, noncontacting experimental stress analysis technique. The basis of this technique is the thermoelastic effect that describes a temperature change in a body due to straining [1]. Kelvin's formula describes the thermoelastic effect:

$$\Delta T = -\kappa T \Delta \sigma_m,$$

where $\kappa = \alpha/\rho c_p$ is the thermoelastic constant and $\sigma_m = \sigma_1 + \sigma_2 + \sigma_3$ is the first stress invariant (sum of principal stresses). TSA is noncontacting since infrared detectors are used to measure small temperature changes correlated to a stress quantity. TSA is also full-field, since either scanning mechanisms or sophisticated focal plane array infrared detectors are used to measure stresses on the full visible surface of a structure. SPATE (stress pattern analysis by thermal emission) is such a TSA system based on a single IR detector and scanning mechanism. A review of the theory, practice and literature of TSA and the SPATE system can be found in [1]–[3]. Although TSA is a very convenient experimental system due to its full-field non-contacting nature, one of its main criticisms is that the thermoelastic signal can only be interpreted as the sum of the principal stresses for isotropic homogeneous materials [4]. However, engineers designing and testing a structural component ideally require individual stress components to obtain the Von Mises stress; therefore, a method of separating the individual stress components is required.

A hybrid stress separation method based on the second-order form of the equilibrium equations (which are expressed as two Poisson differential equations) has been demonstrated previously over rectangular regions be means of the finite element thermal analogy [3] and boundary element Monte Carlo integration techniques [5]. Experimental TSA data and the results of an independent stress analysis method (finite element analysis) were used together in order to separate the stresses. The current work is on the extension of the hybrid stress separation algorithm to arbitrarily shaped two-dimensional domains using the boundary element method (BEM). Solutions of the Poisson equations over arbitrarily shaped structures are attempted via the BEM cell integration method (CIM), the dual reciprocity method (DRM) and the Monte Carlo method (MCM). A benchmark problem of a plate with a circular perforation under bending and tension load is used to test the stress separation algorithm. Since experimental TSA scan data were not available, the TSA scans are simulated by means of boundary element elastostatic analysis results and computer generated random noise.

This work was funded through an Extra Mural Research Agreement with the National Engineering Laboratory, East Kilbride, UK.

96

2. Basic stress separation theory

A deforming two-dimensional component as shown in Fig. 1 is considered in order to define the basic stress separation theory. The differential equations of equilibrium are written for a small rectangular element inside the solid in the form

$$\frac{\partial \sigma_x}{\partial x} + \frac{\partial \tau_{xy}}{\partial y} + X = 0, \quad \frac{\partial \sigma_y}{\partial y} + \frac{\partial \tau_{xy}}{\partial x} + Y = 0, \tag{1}$$

where X and Y are the body forces per unit volume. Then it is possible to formulate the stress separation problem as two Poisson equations by differentiating the equations of equilibrium, neglecting the body force terms and using the relation for the first stress invariant $\sigma_m = \sigma_x + \sigma_y$:

$$\nabla^2 \sigma_d = \left(\frac{\partial^2 \sigma_m}{\partial x^2}\right) - \left(\frac{\partial^2 \sigma_m}{\partial y^2}\right), \quad \nabla^2 \tau_{xy} = -\left(\frac{\partial^2 \sigma_m}{\partial x \partial y}\right), \tag{2}$$

where $\sigma_d = \sigma_x - \sigma_y$ is the stress difference. The derivatives of the stress sum σ_m give the right-hand side in the equations (2), for the stress difference and the shear stress problems.

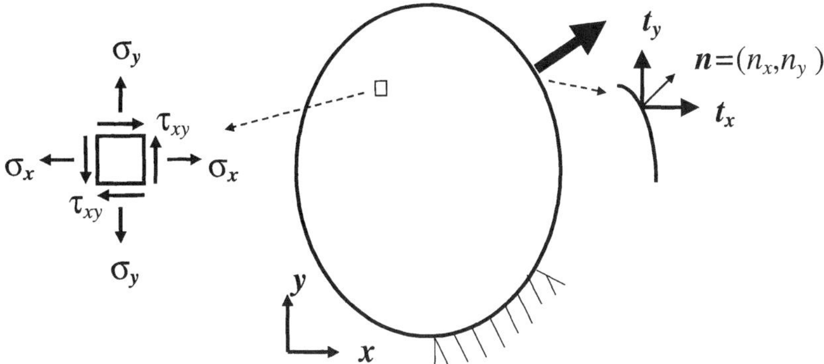

Fig. 1. A solid with applied boundary tractions.

The boundary conditions for a general curved boundary are given in terms of combinations of normal and shear stresses (Fig. 1):

$$\sigma_x n_x + \tau_{xy} n_y = t_x, \quad \tau_{xy} n_x + \sigma_y n_y = t_y, \tag{3}$$

where n_x and n_y are the Cartesian components of the normal to the boundary and t_x and t_y are the Cartesian components of the specified boundary tractions. The boundary conditions for the Poisson equations (2) are the stress difference and the shear stress values. Free surfaces and experimentally measured stress sum data over the boundary may be used to separate the stresses on those boundary segments. The boundary conditions on the remaining segments of the boundary may be obtained from an independent elastostatic analysis in a hybrid method.

The Poisson equation is $\nabla^2 u = b$, where $u = \sigma_d$ or $u = \tau_{xy}$ and b is formed by the right-hand side in (2). The boundary conditions for the stress separation problem are prescribed as $u = \bar{u}$ on the complete boundary. Natural boundary conditions $q = \partial u / \partial n = \bar{q}$, where n is the unit outward normal to the boundary are not used in this algorithm. The boundary integral equation required by the boundary element method can be written as [6]

$$c_i u_i + \int_S u q^* dS + \int_R b u^* dR = \int_S q u^* dS, \tag{4}$$

where c_i are constants and u^* and q^* are the *kernel functions*. The techniques that are used here to obtain a boundary element formulation to this boundary integral equation are CIM, DRM, and MCM. These techniques differ in their approach in dealing with the domain integral term. CIM requires that the integration domain is subdivided into cells and on each cell a numerical integration scheme applied. DRM only requires internal nodes with known values of the function b instead [6].

If the analytic form of b was known, then it would be possible to use particular solutions where the function u is split into a particular solution and the solution of the associated homogenous (Laplace) equation, that is, $u = \tilde{u} + \hat{u}$ (here $\tilde{u}$ is the solution of the Laplace equation and $\hat{u}$ is a particular solution of the Poisson equation). Then $\nabla^2 \hat{u} = b$, and the domain integral term in (4) can be written as

$$\int_R b u^* dR = \int_R (\nabla^2 \hat{u}) u^* dR. \tag{5}$$

This equation is integrated by parts and yields the boundary-only expression

$$c_i u_i + \int_S u q^* dS - \int_S q u^* dS = c_i \hat{u}_i + \int_S \hat{u} q^* dS - \int_S \hat{q} u^* dS. \tag{6}$$

DRM is a generalization of the use of particular solutions and may be adopted here since the analytic form of b is not known for the general stress separation problem. The basis of DRM as described in [6] is the summation of localized particular solutions where the coefficients are to be determined. A series of particular solutions $\hat{u}_j$ is used, where the total number of nodes n gives the number of particular solutions in the series. A set of approximating functions f_j with initially unknown coefficients α_j are used to approximate b:

$$b \cong \sum_{j=1}^{n} \alpha_j f_j. \tag{7}$$

The first few terms of the series $f = 1 + r + r^2 + \cdots + r^m$ are used, where r is the distance between the nodes. The unknown α_j coefficients are calculated from the known values of b at all nodes. In matrix this is written as $\mathbf{b} = \mathbf{F}\alpha$, where $\mathbf{F}$ is a symmetric matrix formed by vectors of the approximating function f_j and α is a vector to be calculated by solving this system of equations. The particular solutions

98

are linked to the approximating functions by the formula $\nabla^2 \hat{u}_j = f_j$; substituting this in (7) and rewriting the Poisson equation leads to

$$\nabla^2 u = \sum_{j=1}^{n} \alpha_j (\nabla^2 \hat{u}_j). \tag{8}$$

The next step is to express both the left- and right-hand sides of (8) as boundary integrals:

$$c_i u_i + \int_S u q^* dS - \int_S q u^* dS = \sum_{j=1}^{n} \alpha_j \left(c_i \hat{u}_{ij} + \int_S \hat{u}_j q^* dS - \int_S \hat{q}_j u^* dS \right). \tag{9}$$

An alternative method that does not require cells for the computation of the domain integral is MCM [7]. The technique is based on the idea that the average value of a function over a domain multiplied by the area of the domain is equal to the integral of the function over that domain. In practice, a large number (N) of coordinates inside the domain are randomly generated, and the function f which involves b and the kernel function is calculated at each point. The true average of the function can be approximated if the sample is large enough and the coordinates are distributed in a uniformly random manner. The average value is then multiplied by the area A_R of the domain to give the integral value

$$c_i u_i + \int_S u q^* dS + \frac{A_R}{N} \sum_{k}^{N} f_k = \int_S q u^* dS. \tag{10}$$

Numerical algorithms for the solution of the Poisson equation are developed starting from (4), (9) and (10). The next section describes the application of the stress separation algorithm to a practical example.

3. Numerical tests on a benchmark problem

A rectangular steel plate with a circular perforation is subjected to combined tension and bending loads as shown in Fig. 2(left). The geometry is modeled using the ANSYS finite element program with eight node finite elements. The geometry model is translated into a boundary element model which has 64 quadratic boundary elements, 128 boundary nodes, 832 internal nodes, and 224 cells (represented as nine-node Lagrangian-type elements). The loads are applied and the problem is analyzed using the developed boundary element elastostatics program. The result of the elastostatic analysis is used to simulate noisy experimental TSA scan data from the complete surface (resolution $210 * 250$ scan points) and zoomed-in region of the circular perforation (resolution $256 * 256$ scan points). Computer generated random noise with a Gaussian distribution is added in the process of TSA data simulation. The maximum value of the noise added to TSA data is 5 N/mm^2, which can be compared with the maximum equivalent stress at the stress concentration calculated by the elastostatic analysis as 368 N/mm^2. A finite element smoothing algorithm (see [8] and [9]) is used to combine the two noisy TSA scans and smooth the TSA data over the cells of the structure. The results of the smoothing include first- and second-order Cartesian derivatives of the TSA data, which form the b functions in the stress difference and shear stress problems.

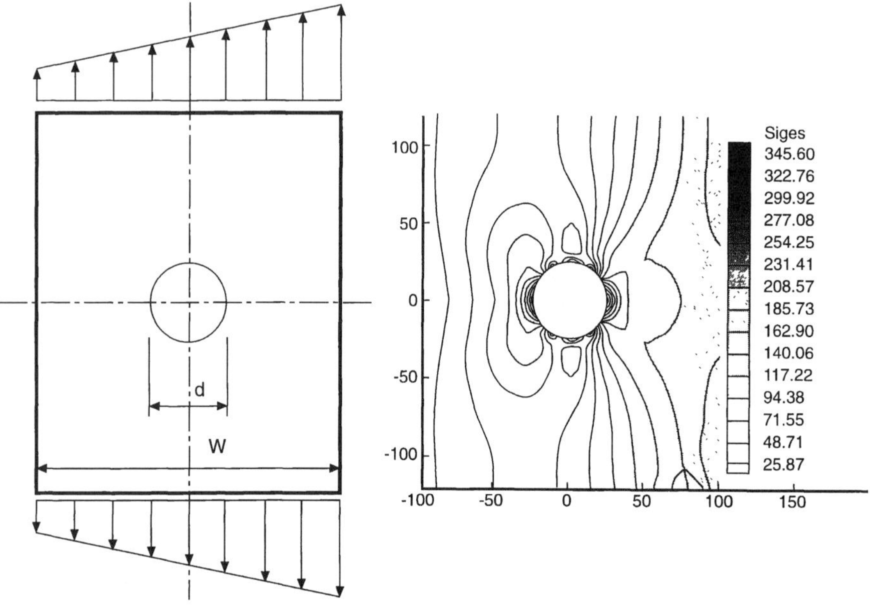

Fig. 2. (left) Benchmark problem: a rectangular steel plate with
a circular perforation subjected to tension and bending;
(right) Von Mises equivalent stresses obtained from CIM
separated stresses.

In order to solve the Poisson equations for the stress difference and shear stress problems, the boundary conditions need to be evaluated. It is possible to use the results of the BE elastostatic analysis to specify the boundary conditions on the complete boundary of the domain. Otherwise, the boundary conditions can be evaluated partly from the BE elastostatic analysis and partly using the smoothed TSA data on the boundary with the prescribed traction values. The stresses can then be separated by means of one of the Poisson equation solution algorithms. The results of the separated stresses are combined to give the equivalent stresses. A contour plot of the Von Mises equivalent stresses from a CIM stress separation analysis is shown in Fig. 2(right). It has been observed that the equivalent stress distributions obtained from the stress separation algorithm compare well with that of the boundary element elastostatic analysis result.

4. Discussion

The boundary element method for Poisson equation solution has been applied to separate the individual stress components from TSA data over an arbitrarily shaped region. CIM with numerical integration schemes proved to be robust. Since internal cells are already created automatically with the finite element mesh generation algorithms of ANSYS, geometry modeling is not too time-consuming. Where stress separation results are required only at a few internal points, it may be possible to employ the DRM or MCM techniques more efficiently. However, more research effort is required to find the optimum number of internal nodes, since that may determine the accuracy of the stress separation analysis.

References

1. N. Harwood and W.M. Cummings, Eds., *Thermoelastic stress analysis*, Adam Hilger, Bristol, 1991.

2. T. Comlekci, *Development of hybrid experimental-numerical methods for thermoelastic stress analysis*, Ph.D. thesis, University of Strathclyde, Glasgow, 1995.

3. R. Hamilton, *Development of a method of thermographic stress separation*, Ph.D. thesis, University of Strathclyde, Glasgow, 1992.

4. J.T. Boyle, Post processing SPATE data, in *Thermoelastic stress analysis*, Ch. 9, N. Harwood and W.M. Cummings, Eds., Adam Hilger, 1991.

5. R. Hamilton, J.T. Boyle and D. Mackenzie, The boundary element method in thermoelastic stress separation. I. Formulation and basic solution using a Monte Carlo method, in *Proc. Internat. Assoc. Boundary Element Methods*, Hawaii, 1995.

6. P.W. Partridge, C.A. Brebbia and L.C. Wrobel, *The dual reciprocity boundary element method*, Computational Mechanics Publications, Southampton, 1992.

7. G.S. Gipson, *Boundary element fundamentals: basic concepts and recent developments in the Poisson equation*, Computational Mechanics Publications, Southampton, 1987.

8. T. Comlekci and J.T. Boyle, Finite element smoothing of experimental thermographic data on arbitrary two-dimensional regions, *Proc. Computational Mechanics in UK*, Manchester, 1994, pp. 53–56.

9. Z. Feng and R.E. Rowlands, Continuous full-field representation and differentiation of three-dimensional experimental vector data, *Computers and Structures*, **26** (1987), 979–990.

10. T. Comlekci and J.T. Boyle, The boundary element method in thermoelastic stress separation. II. A general hybrid numerical/experimental technique, in *Proc. Internat. Assoc. Boundary Element Methods*, Hawaii, 1995.

Department of Mechanical Engineering, University of Strathclyde, Glasgow, UK

C. CORDUNEANU and M. MAHDAVI

On neutral functional differential equations with causal operators. II

1. Introduction

This paper is a continuation of our preceding paper [2], and is concerned with existence results in various function spaces for neutral functional differential equations of the form

$$\frac{d}{dt}[x(t) + (Vx)(t)] = (Wx)(t), \tag{1}$$

with V and W standing for certain causal (abstract Volterra) operators acting on the function spaces under consideration. In our preceding paper [2], we investigated the existence (both local and global) of solutions to the equation (1), choosing the space of continuous maps from $[0, T]$, $T > 0$, into R^n as underlying space. In this paper we shall concentrate on the case of spaces of measurable maps. More precisely, we shall deal with the Lebesgue spaces L^p, $1 \leq p \leq \infty$. Of course, the solution will be meant in Carathéodory's sense, i.e., almost everywhere on $[0, T]$, or on a smaller interval (on which the solution does exist). We shall also consider the case of linear equations of the form (1), when the existence results are globally valid. To (1) we will adjoin the usual initial condition

$$x(0) = x^0 \in R^n. \tag{2}$$

Further discussion of the global existence, in the nonlinear cases, will be conducted in the last part of the paper.

2. Local existence in L^p-spaces

The following conditions will be imposed on the operators V and W in (1), in order to secure the existence of a (local) solution satisfying the initial condition (2).

(H_1) V and W are causal operators on the space $L^p([0, T], R^n)$, $1 \leq p \leq \infty$.

(H_2) V is a continuous compact operator, such that for each $x \in L^p$ one has

$$\lim_{t \to 0^+} (Vx)(t) = \theta, \tag{3}$$

where θ stands for the null element of R^n.

(H_3) W is a continuous operator that takes bounded sets of L^p into bounded sets.

Before we state the existence result, let us point out that condition (3) is a slight generalization of the so-called "fixed initial value property" for causal operators. It was formulated for the first time by L. Neustadt (see [1], [4], for instance). Obviously, instead of (3), one may impose the apparently more general condition

$$\lim_{t \to 0^+} (Vx)(t) = c_0 \in R^n.$$

But (1) shows that an additive constant is immaterial for V.

An example of a classical operator satisfying the condition (3) is the Volterra operator

$$(Vx)(t) = \int_0^t K(t, s, x(s))ds,$$

under appropriate conditions on K (so that $K(t, s, x(s))$ is integrable in s for almost all $t \in [0, T]$, and for any $x \in L^p$). The following result can be proven for (1), with initial condition (3). It is appropriate mentioning that condition (2) has to be understood in the sense

$$\lim_{t \to 0^+} x(t) = x^0. \tag{4}$$

Theorem 1. *Consider equation (1) under initial condition (3). Assume that the operators V and W in (1) are satisfying the hypotheses $(H_1) - (H_3)$. Then there exists a solution of (1,2) on some interval $[0, a]$, $a \leq T$, say $x(t)$, such that $x \in L^p([0, a], R^n)$, while $x(t) + (Vx)(t)$ is an absolutely continuous function.*

Proof. First, let us note that the problem (1,2), or (1,4), is equivalent in L^p with the functional equation

$$x(t) + (Vx)(t) = x^0 + \int_0^t (Wx)(s)ds. \tag{5}$$

Indeed, if $x(t)$ is a solution of (1,3) in L^p, then $(Wx)(t) \in L^p \in L^1$, and consequently both sides of (1) can be integrated on any interval $[0, t]$, $t < T$ (or $t \leq a$). Taking (H_2) into account, the result can be written in the form (5). Since the right-hand side of (5) is absolutely continuous, one can differentiate (5), almost everywhere, and obtain (1). Also, letting $t \to 0^+$ in (5), one obtains (4). Therefore, we need to prove the existence of a solution to (5). Obviously, (5) is equivalent on the interval of definition of x with

$$x(t) = \int_0^t (Wx)ds - (Vx)(t) + x^0, \tag{6}$$

which means that it has the typical form

$$x(t) = (\mathcal{V}x)(t). \tag{7}$$

Equation (7) constitutes a functional equation of Volterra-Tonelli type (see, for instance, [1,4]), with causal operator, $\mathcal{V}$. From our hypotheses $(H_1) - (H_3)$, one obtains the existence of a solution in $L^p = L^p([0, a], R^n)$, $a \leq T$, $1 \leq p \leq \infty$. Indeed, the operator $\mathcal{V}$ in the right-hand side is acting on the space $L^p([0, t], R^n)$ according to our hypotheses. It is obviously continuous and compact (taking bounded sets into relatively compact sets). While this last property is part of the hypotheses in regard to the operator V, we have to notice that the (causal) operator

$$x(t) \to \int_0^t (Wx)(s)ds \tag{8}$$

is compact on L^p. Since the right-hand side in (8) belongs to the space of continuous functions on $[0, T]$, it is an elementary exercise to prove that the operator defined by (8) is compact from L^p into C. Hence, this operator is compact from L^p into itself (the convergence in C is stronger than the convergence in L^p). At this point in the proof we can directly apply Theorem 3.4.1 in [1], which leads to the result of Theorem 1.

Remark 1. As shown in [1], the singularly perturbed equation

$$\epsilon \dot{x}_\epsilon(t) = -x_\epsilon(t) + (\mathcal{V}x_\epsilon)(t) \tag{9}$$

is solvable in L^p for each positive ϵ, sufficiently small. Its solution can be regarded as a regularized approximate solution to (7), as $\epsilon \to 0^+$. Of course, each $x_\epsilon(t)$ is absolutely continuous on its interval of existence (which turns out to be common to all $x_\epsilon(t)$ with sufficiently small ϵ).

3. The linear case

If we admit that both operators V and W in (1) are linear operators on L^p, $1 \leq p \leq \infty$, then the existence result stated in Theorem 1 becomes a global result (on the whole $[0,T]$, or even on $[0,T)$, with $0 < T \leq \infty$). In order to obtain this result, we shall rely on the results of the paper [3] by M. Mahdavi and Y. Li.

We shall return now to (6), which can be rewritten as

$$x(t) = (Lx)(t) + f(t), t \in [0,T), \tag{10}$$

where

$$(Lx)(t) = \int_0^t [(Wx)(s) - (W\theta)(s)]ds - [(Vx)(t) - (V\theta)(t)],$$

and $f(t) = \int_0^t (W\theta)(s)ds - (V\theta)(t) + x^0$. The hypotheses $(H_1) - (H_3)$ must be slightly reformulated. More exactly, the following conditions will be admitted for the operators V and W, in order to secure the continuity and compactness of the operator L in (10).

$C_1)$ V and W are causal affine continuous operators on $L^p_{loc}([0,T), R^n)$, $1 \leq p \leq \infty$, $0 < T \leq \infty$.

$C_2)$ V is compact on $L^p_{loc}([0,T), R^n)$ and satisfies (3).

Theorem 2. *Consider the linear neutral functional differential equation (1), under initial condition (2) or (4). Assume that V and W verify the conditions C_1 and C_2 above. Then, there exists a unique solution of (1,2), defined on the interval $[0,T)$, belonging to the space $L^p_{loc}([0,T), R^n)$, and such that $x(t) + (Vx)(t)$ is locally absolutely continuous on $[0,T)$.*

Proof. The linearity and continuity of L are obvious from its definition. The compactness of V is assumed by the condition C_2 (let us point out that compactness on L^p_{loc} means compactness on each $L^p([0,a], R^n)$, for each $a < T$). The compactness of the operator $x(t) \to \int_0^t [(Wx)(s) - (W\theta)(s)]ds$, which appears in the definition of $(Lx)(t)$, on the space $L^p_{loc}([0,T], R^n)$, can be obtained by the same argument as in the nonlinear case (Proof of Theorem 1). Indeed, the continuity in the linear case assumes that bounded sets in L^p are taken into bounded sets. Now by applying the Theorem 1 in the paper [3], one obtains the result stated in Theorem 2 above. This ends the proof of Theorem 2.

Remark 2. Instead of assuming the linearity of V and W, one can obtain the result of Theorem 2 by admitting a Lipschitz type condition on these operators. For instance, in case of space L^∞, a condition of the form

$$||(Vx)(t) - (Vy)(t)|| \leq g(t) \, ess \sup_{0 \leq s \leq t} ||x(s) - y(s)||,$$

with $g(t)$ a nonnegative nondecreasing function on $[0,T)$, can lead to the global existence and uniqueness of the solution to the problem (1), (2). Similar conditions can be formulated in spaces L^p, $1 \leq p < \infty$.

104

Remark 3. In their paper [3], Mahdavi and Li considered the quasilinear case, i.e., $\dot{x}(t) = (Lx)(t) + (Nx)(t)$, where N stands for a nonlinear operator. Their result could also be used if we assume, for instance, $(Vx)(t) = (V_0x)(t) + (V_1x)(t)$, with V_0 linear and V_1 a nonlinear component, which has to be "small" in an appropriate sense (see [3] for details). A similar representation should be assumed for the operator W. Of course, the terms in the representation must also be causal operators.

4. An application

We shall apply Theorem 2 to the functional differential equation

$$\frac{d}{dt}[x(t) + \int_0^t K(t,s)x(s)ds] = (Wx)(t), \tag{11}$$

in which W is still a "general" causal operator. Of course, we attach to (11) the initial condition (2).

Assume we choose the space $L^2_{loc}([0,T], R^n)$ as underlying space (i.e., $p = 2$). Hence, we must provide conditions such that the integral operator

$$x(t) \rightarrow \int_0^t K(t,s)x(s)ds \tag{12}$$

be continuous on $L^2_{loc}([0,T], R^n)$. It is generally known (see, for instance [1]) that the operator (12) is continuous and compact on $L^2_{loc}([0,T], R^n)$ if the matrix valued kernel $K(t,s)$ is measurable in its domain of definition, $0 \leq s \leq t < T$, and for each $\bar{t} < T$ one has

$$\int_0^{\bar{t}} dt \int_0^t |K(t,s)|^2 ds < \infty, \tag{13}$$

where $|K(t,s)|$ denotes any $n \times n$ matrix norm. If one compares (11) with (1), then one sees that we have chosen

$$(Vx)(t) = \int_0^t K(t,s)x(s)ds. \tag{14}$$

As noticed, V is continuous and compact on $L^2_{loc}([0,T], R^n)$, provided (13) is verified. Condition (3) is obviously satisfied by V. Hence, the conditions C_1 and C_2 are satisfied for (11), which leads to the following statement on behalf of Theorem 2:

Theorem 3. *Consider equation (11) with initial condition (2). Assume the kernel $K(t,s)$ is measurable for $0 \leq s \leq t < T$ and verifies (13). Moreover, let W be a causal linear or affine continuous operator on $L^2_{loc}([0,T], R^n)$. Then there exists a unique solution $x(t)$ of (11), satisfying the initial condition (2) and such that*

$$x(t) + \int_0^t K(t,s)x(s)ds \tag{15}$$

is locally absolutely continuous on $[0,T)$.

Remark 4. One can particularize the operator W in different ways. For instance, one can choose $(Wx)(t) = \int_0^t K_1(t-s)x(s)ds + f(t)$, with $K_1(t)$ satisfying the condition $|K_1| \in L^1_{loc}([0,T], R^n)$, and $f \in L^2_{loc}([0,T], R^n)$. Another possible choice is $(Wx)(t) = \sum_{k=1}^m A_k x(t-t_k) + f(t)$, where $0 < t_1 < t_2 < \cdots < t_m < T$ are some fixed numbers,

A_k are $n \times n$ given matrices and $f \in L^2_{loc}([0,T], R^n)$. In this case, one must assign the solution not only at $t = 0$, but on the whole interval $[-t_m, 0)$. In order to assure W is continuous on $L^2_{loc}([0,T], R^n)$, we can assign $x(t) = x_0(t) \in L^2([-t_m, 0), R^n)$, and $x(0) = x^0 \in R^n$.

Remark 5. Since (11), with initial condition (2), is equivalent to

$$x(t) + \int_0^t K(t,s)x(s)ds = x^0 + \int_0^t (Wx)(s)ds, \tag{16}$$

another approach can be easily described. Indeed, the operator (15) which appears in the left-hand side of (16) has an inverse on $L^2_{loc}([0,T], R^n)$ that is also linear and causal (the resolvent operator!). Using this inverse one can reduce (16) to an equation of the form (10). See the paper [3] by Mahdavi and Li for details, where quasilinear equations are also investigated. In other words, the operator W in (16) could be a nonlinear operator on $L^2_{loc}([0,T], R^n)$, satisfying a generalized Lipschitz condition.

References

1. C. Corduneanu, *Integral equations and applications*, Cambridge University Press, 1991.

2. C. Corduneanu and M. Mahdavi, On neutral functional differential equations with causal operators, in *Systems science and its applications*, Proc. Third Workshop of the International Institute of General Systems Science, Tianjin, China, 1998, 43–48.

3. M. Mahdavi and Yizeng Li, Linear and quasilinear equations with abstract Volterra operators, in *Volterra equations and applications*, Gordon and Breach, 1999.

4. L. Neustadt, *Optimization (a theory of necessary conditions)*, Princeton University Press, 1976.

Department of Mathematics, The University of Texas at Arlington, Arlington, Texas, USA

Department of Mathematics, Bowie State University, Bowie, Maryland, USA

N. CRETU, G. NITA, I. STURZU and C. ROSCA

A semi-analytic method for the study of acoustic pulse propagation in inhomogeneous elastic 1-D media

1. Introduction

In recent years, particular attention has been devoted to the analysis of propagation of pulses in media with varying Young modulus. For instance, [2] studies Gaussian pulse propagation when the Young's modulus varies linearly with the propagation distance employing the approximations in [1].

In this paper, we present a general analysis of acoustic pulse propagation in a 1-D inhomogeneous elastic medium using the equation of motion mentioned in [1]. The inhomogeneity is specified by functions (from a specific class) of the spatial coordinate for the Young modulus, cross sectional area, and bulk density.

This semi-analytic method is based on the Fourier transform of the wave equation in conjunction with Taylor expansions of both the functions describing the inhomogeneities and the solution. Numerical computations were performed using a suitable iterative scheme. The truncation error and the convergence of the iterative scheme are analyzed.

2. General considerations

We start from the 1-D equation for a longitudinal wave in an elastic bar:

$$\frac{\partial}{\partial x}\left(p\left(x\right)\frac{\partial \Psi\left(x,t\right)}{\partial x}\right) - \frac{1}{c^2}g\left(x\right)\frac{\partial^2 \Psi\left(x,t\right)}{\partial t^2} = 0, \tag{1}$$

where $p(x)$ and $q(x)$ are two functions which describes the relative inhomogeneities relative to an homogeneous medium in which the phase velocity of the longitudinal wave is c. The Fourier transform of (1) is

$$\frac{\partial}{\partial x}\left(p\left(x\right)\frac{\partial u\left(x,\omega\right)}{\partial x}\right) + q\left(x,\omega\right)u\left(x,\omega\right) = 0, \tag{2}$$

where

$$\Psi\left(x,t\right) = \frac{1}{2\pi}\int\limits_{-\infty}^{\infty} u\left(x,\omega\right)\exp\left(-i\omega t\right)dt, \tag{3}$$

and

$$q\left(x,\omega\right) = \left(\frac{\omega}{c}\right)^2 g\left(x\right). \tag{4}$$

Considering the polar decomposition

$$u\left(x,\omega\right) = A\left(x,\omega\right)\exp\left[i\varphi\left(x,\omega\right)\right], \tag{5}$$

This presentation was partly supported by EC Program Copernicus No. CIPACT940132; Ioan Neculae, General Manager S.C. PRESCON S.A. Brasov, Romania; S.C. TUNELE S.A. Brasov, Romania; S.C. MALVA S.R.L. Brasov, Romania; Sorin Stegaru, Anton Tone, Ciprian Suciu, friends.

the general solution of wave-equation, $\Psi(x,t)$, become a linear superposition of ω-monochromatic waves expressed by

$$\psi_\omega(x,t) = A(x,\omega)\exp\{i[\varphi(x,\omega) - \omega t]\}.$$ (6)

According to (6), one can define a local wave-vector, local phase-velocity, and local group-velocity:

$$k(x,\omega) = \frac{\partial\varphi(x,\omega)}{\partial x}; \qquad v_\varphi(x,\omega) = \frac{\omega}{k(x,\omega)}; \qquad v_g(x,\omega) = \left[\frac{\partial k(x,\omega)}{\partial\omega}\right]^{-1}.$$ (7)

These quantities characterize the dispersion of the medium, and can be calculated after solving (2).

3. A method for controlling and limiting errors

Equation (2) admits, in general, only numerical solutions, which have to be completed with methods for controlling and limiting the errors. We give in this paper a general method, based on a property of the solution $u(x,\omega)$ originally developed in [3].

Introducing the polar decomposition (5) in (2), and integrating the imaginary part of the equation, one has

$$F(x,\omega) = p(x)A^2(x,\omega)k(x,\omega) = \text{const},$$ (8)

which is nontrivial if the solution is not purely stationary. Equation (8) is useful for controlling the truncation error.

Given the solution $u(x)$, one may calculate the implied magnitudes using

$$A(x,\omega) = \sqrt{[Re(u)]^2 + [Im(u)]^2}$$

$$k(x,\omega) = Im\left[\frac{u'(x,\omega)}{u(x,\omega)}\right].$$ (9)

The relative deviation of F

$$Err(x,\omega) = \frac{F(x,\omega) - F(x_0,\omega)}{F(x_0,\omega)},$$ (10)

(from its value at the starting point x_0) provides useful error control through the condition

$$Err(x,\omega) < \varepsilon,$$ (11)

which can be maintained by adjusting the numerical integration parameters.

4. Generalization of a numerical method

We present a generalization to analytical $p(x)$ and $q(x)$ of a numerical method originally developed in [2] for solving (2) with linear $p(x)$. This method was extended [4] to harmonic dependence of $p(x)$ and $q(x)$. The technique is based upon the Taylor expansions of both the solution and inhomogeneities:

$$u(x) = \sum_{n=0}^{\infty} c_n(x - x_0)^n \;\; ; \;\; p(x) = \sum_{n=0}^{\infty} a_n(x - x_0)^n \;\; ; \;\; q(x) = \sum_{n=0}^{\infty} b_n(x - x_0)^n.$$ (12)

108

We have shown [4] that the coefficients c_n are determined by the recurrence relation

$$c_n = \frac{1}{(n+1)(n+2)a_0} \left\{ b_0 c_n + \sum_{m=1}^{n+1} [b_m c_{n-m} + a_m (n+1)(n-m+2) c_{n-m+2}] \right\} \quad (13)$$

with initial conditions

$$c_{-1} = 0; \quad c_0 = u(x_0); \quad c_1 = u'(x_0). \quad (14)$$

Truncating the expansion at N gives the expressions

$$u(x_0 + h) = \sum_{n=0}^{N} c_n h^n, \quad u'(x_0 + h) = \sum_{n=0}^{N} n c_n h^{n-1} \quad (15)$$

for the numerical solution and its derivative in a neighborhood of x_0. The algorithm can be summarized as follows:

1. Calculate the quantities (8) and (9) at the initial points x_0, from $u(x_0)$ and $u'(x_0)$.

2. Compute the solution (and its derivative) at $x_0 + h$ using (13), (14), and (15).

3. Check the quantities (8) and (9) at $x_0 + h$ using condition (11) for a selected tolerance ε, adjusting the stepsize h and truncation N if necessary.

4. Use the new values as initial values for the next iteration and advance the solution through to the end of the desired interval.

5. Numerical results

To illustrate the method, we show some results for a constant Young modulus and exponential variation of the bulk density. In this case, the medium is described by the following functions:

$$p(x) = 1, \qquad g(x) = \exp[-\eta (x - x_0)] . \quad (16)$$

Define the relative magnitude:

$$\xi = \frac{\omega}{2\pi c \eta} = \frac{1}{\lambda_0 \eta}, \quad (17)$$

where λ_0 is wavelength of the ω-monochromatic solution for the reference homogeneous medium.

The values of ξ distinguishes three distinct cases:

1. $\xi \gg 1$, the high frequency case;

2. $\xi \cong 1$, the intermediate frequency case;

3. $\xi \ll 1$, the low frequency case.

5.1. Error control. We solved (2) for different values of N and different steps h, throughout a large frequency domain. Better results were obtained by decreasing the stepsize h rather than increasing the truncation order N. The general conclusion from this analysis is that $N = 5$ and $h = 0.001$, is a good choice for monochromatic solutions given by $\xi \in [0, 15]$. This choice gave a relative error of less than $10^{-4}\%$ after 5000 iterations.

5.2. Wave-groups. The temporal evolution of wave packets can be studied by summing several monochromatic components in the Fourier domain of the initial pulse.

For simplicity we chose $x_0 = 0$ and $c = \sqrt{E(0)/\rho(0)} = 1$, and we are assuming that, at $t = 0$, all monochromatic components have the same phase at the starting point, $\varphi(x_0, \omega) = 0$. Considering these components as pure monochromatic waves injected at $t = 0$ in the starting point x_0, the initial conditions for the studied case are:

$$c_0 = A(x_0, \omega); \qquad c_1 = i\frac{\omega}{c}A(x_0, \omega). \tag{18}$$

Fig. 1 shows the normalized Fourier sum

$$\Psi(x, t) = \frac{\sum_{\omega} u(x, \omega)\exp(-i\omega t)}{\sum_{\omega} u(x_0, \omega)}, \tag{19}$$

at 3 different times, for 64 uniformly spaced high-frequency components corresponding to $\xi \in [5, 15]$. The amplitudes of these components are a Gaussian shape, centered at $\xi_0 = 10$.

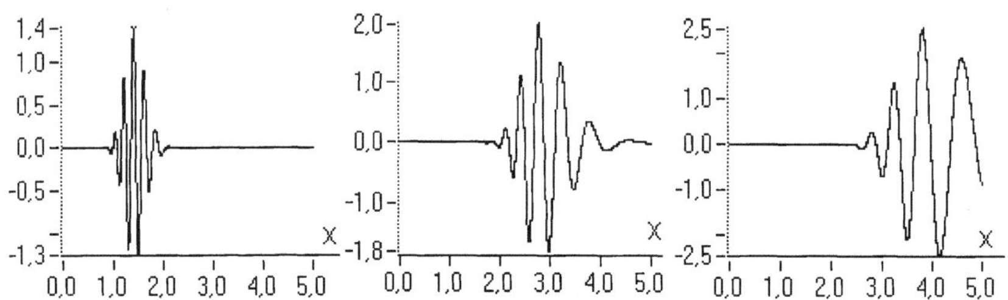

Fig. 1. Evolution of high-frequencies wave-packet, $\xi \in [5, 15]$, exponential bulk density, $\eta = 1$.

Fig. 2 shows the law of movement and the spatial behavior of the amplitude and group velocity of this wave-packet.

The wave-packet's evolution leads to the conclusion that the studied medium displays a local dispersion due to the dependence of the monochromatic phase velocities on both space and frequency. But, in the case of high frequencies, $\xi \gg 1$, the numerical results show that the phase velocities of the monochromatic waves are independent of the frequency and can be approximate by the asymptotic expression:

$$v_\varphi(x, \omega) \cong \sqrt{E(x)/\rho(x)}. \tag{20}$$

110

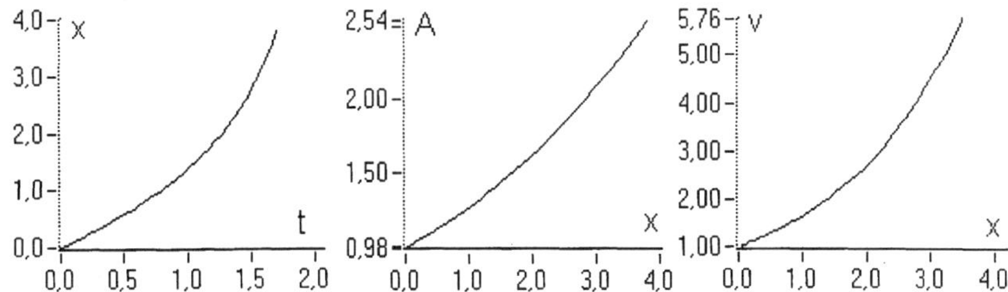

Fig. 2. Law of movement, group-amplitude, and group-velocity.

From (8) we have found a general approximation for the monochromatic solutions which may be used in practical application:

$$u\left(x,\omega\right) \cong A\left(x_0,\omega\right) \left[\frac{p\left(x_0\right)q\left(x_0,\omega\right)}{p\left(x\right)q\left(x,\omega\right)}\right]^{\frac{1}{4}} \exp\left[i\int_{x_0}^{x}\sqrt{\frac{q\left(x,\omega\right)}{p\left(x,\omega\right)}}\,dx\right]. \tag{21}$$

In the case presented in this paper, the asymptotic equation (21) becomes

$$u\left(x,\omega\right) \cong A\left(x_0,\omega\right)\exp\left\{\frac{\eta}{4}\left(x-x_0\right)+i\frac{\omega}{c}\frac{2}{\eta}\left[1-\exp\left(-\frac{\eta}{2}\left(x-x_0\right)\right)\right]\right\}. \tag{22}$$

Relation (22) has been verified by the numerical results obtained for $\xi = 10$ with a relative error less then 1%.

6. Conclusions

In this paper we have presented a semi-analytical approach for solving the wave-equation of an inhomogeneous medium described by two analytical functions.

After Fourier transforming the wave-equations, the problem is reduced to solving a second-order linear differential equation with variable coefficients.

We have presented a generalization of an iterative method which solves such equations by matching Taylor expansions of the solutions and of the functions describing the medium.

We have found an interesting correlation between the amplitude and the phase of the solutions that enables us to develop a general method for controlling and estimating the numerical errors.

Based on the same correlation, we also found, in the case of high frequencies, an approximate analytical solution which may be used in practical applications.

This results are not specific to the studied case and may be applied to any physical phenomena described by a linear differential equation of second order with variable coefficients.

References

1. G.B. Whitam, A general approach to linear and non-linear dispersive waves using a Lagrangian, *J. Fluid Mech.* **22** (1965), 273.

2. R. Oberle and R.C. Cammarata, Acoustic pulse propagation in elastically inhomogeneous media, *J. Acoust. Soc. Amer.* **94** (1993), 2947.

3. G. Nita, About a conservation law specific to the homogeneous linear differential equation of 2nd order with variable coefficients, *Bull. Transylvania Univ. Brasov* **2** (1997), 41–44.

4. N. Cretu, P.P. Delsanto, G. Nita, M. Scalerandi, I. Sturzu and C. Rosca, Ultrasonic pulse propagation in inhomogeneous one-dimensional media, *J. Acoust. Soc. Amer.* **104** (1998), 57–64.

Department of Physics, "Transylvania" University, Brasov, Romania
e-mail: cretu.c@unitbv.ro, gelu@lim.deltanet.roknet.ro

L. DELLA CROCE and T. SCAPOLLA

Efficient finite elements for the numerical approximation of cylindrical shells

1. Introduction

The numerical solution of shell problems involves two main questions: the physical model assumed to represent the deformation of the structure under load and the numerical method selected for the approximation. In this note, after choosing a reliable physical model, we deal with the construction of numerical methods based upon efficient and robust finite elements.

Shell models can be derived according to different physical assumptions. When the fibers are supposed to remain normal to the middle surface after deformation, i.e., the Kirchhoff hypotheses are assumed, Koiter's model is obtained. When the normals to the undeformed middle plane remain straight but not necessarily normal to the deformed middle surface, i.e., the Reissner–Mindlin hypotheses are assumed, another family of models is derived [1]. In these models the internal energy of the shell is the sum of bending, membrane and shear energy components.

Finite element schemes for shell problems suffer from so-called *membrane locking*, i.e., the finite element approximation of the membrane component of the energy is unstable with respect to the shell thickness. In the Naghdi model the inclusion of transverse shear strain introduces an undesirable numerical phenomenon known as *shear locking*.

The most common approaches proposed to overcome the locking effect are: the use of standard displacement formulation with high-order finite elements; modified variational forms (such as mixed or hybrid); techniques of reduced or selectively reduced integration. Numerical results show that high-order elements are able to counteract locking for shell problem in the displacement formulation. However, for very small thickness and when the degree of the element is not sufficiently high, the numerical solution exhibits a loss in the rate of convergence that can compromise the quality of the results.

Our basic idea is to combine the Naghdi model with different formulations. First, we consider a displacement formulation solved with Serendipity finite elements of hierarchic type, with both full and reduced integration. Then we deal with a mixed formulation introducing nonstandard finite elements, suitable for numerical solution with both uniform and distorted decompositions.

To analyze the behavior of the finite elements with respect to membrane and shear locking, we consider three benchmark problems often used to assess the performance of numerical approximations. The first is the Scordelis-Lo roof, a membrane dominated shell problem. The second is the classical pinched shell. The third is an hemicylindrical shell subject to a peculiar load that makes the shell problem bending dominated.

2. The shell problem and finite elements

In the model proposed by Naghdi the (five) unknowns are the covariant components $\{u_i\}_{i=1,2,3}$ of the displacement $\vec{u}$ of the middle surface of the shell and the two components $\{\theta_\alpha\}_{\alpha=1,2}$ of the rotation $\vec{\theta}$ of the unit normal vector to the middle surface. We

This work was partly supported by grants of CNR, Italy (Progetto Strategico "Modelli e metodi per la matematica e l'ingegneria"), MURST, Italy (Progetto Nazionale "Analisi numerica: metodi e software matematico") and FAR, Pavia, Italy ("Metodi e modelli numerici per problemi differenziali").

define $V = \{v \in H^1(\Omega) : v|_{\partial\Omega_0} = 0\}$ where $\partial\Omega_0$ is a nonempty subset of the boundary $\partial\Omega$ of the middle surface and $V^5 = \{(\vec{v}, \tilde{\psi}) : v_i, \psi_\alpha \in V\}$. Denoting by Υ the change of curvature tensor, Σ the transverse shear strain tensor and Λ the membrane strain tensor (see [1] for definitions), a pair $(\vec{u}, \tilde{\theta}) \in V^5$ solves the Naghdi model if

$$\int_\Omega \frac{a^{\alpha\beta\gamma\delta}}{12} \Upsilon_{\alpha\beta}(\vec{u}, \tilde{\theta}) \Upsilon_{\gamma\delta}(\vec{v}, \tilde{\psi})\, d\xi_1 d\xi_2 + \frac{1}{t^2} \int_\Omega a^{\alpha\beta} \frac{E}{2(1+\nu)} \Sigma_\alpha(\vec{u}, \tilde{\theta}) \Sigma_\beta(\vec{v}, \tilde{\psi})\, d\xi_1 d\xi_2$$

$$+ \frac{1}{t^2} \int_\Omega a^{\alpha\beta\gamma\delta} \Lambda_{\alpha\beta}(\vec{u}) \Lambda_{\gamma\delta}(\vec{v})\, d\xi_1 d\xi_2 = \int_\Omega \vec{f}\vec{v}\, d\xi_1 d\xi_2 \qquad \forall(\vec{v}, \tilde{\psi}) \in V^5, \qquad (1)$$

where $\vec{f} = \vec{p}/t^3$ is the scaled vector of the external forces $\vec{p}$, E the Young's modulus, ν the Poisson ratio and

$$a^{\alpha\beta\gamma\delta} = \frac{E}{2(1+\nu)} \left[a^{\alpha\gamma} a^{\beta\delta} + a^{\alpha\delta} a^{\beta\gamma} + \frac{2\nu}{1-\nu} a^{\alpha\beta} a^{\gamma\delta} \right], \qquad (2)$$

where $a^{\alpha\gamma}$ denotes the components of the inverse matrix of $(a_{\alpha\gamma})$, giving the first fundamental form of the midsurface of the shell.

We now introduce the finite elements approximation space

$$V_h = \left\{ (\vec{v}_h, \tilde{\psi}_h) : (v_h)_i, (\psi_h)_\alpha \in W_p(\mathcal{T}_h) \right\} \subset V^5, \qquad (3)$$

where

$$W_p(\mathcal{T}_h) = \left\{ w \in H^1(\Omega) : w|_{\mathcal{Q}_h} \in \mathcal{S}_p(\mathcal{Q}_h) \quad \forall \mathcal{Q}_h \in \mathcal{T}_h \right\} \qquad (4)$$

and $\mathcal{S}_p$ denotes the standard Serendipity space of degree p (see [2–4]).

The stiffness matrix is $K = K_b + K_s + K_m$ where K_b is the bending component, K_s the shear component, K_m the membrane component. The reduced element is obtained with underintegration of shear and membrane components of the stiffness matrix [5].

To avoid locking we use a mixed formulation modifying the bilinear forms that describe shear and membrane energy. We introduce a linear operator that weakens both shear and membrane constraints. We introduce a linear operator $\mathcal{P} = (\mathcal{P}_1, \mathcal{P}_2)$, which takes values in a suitable subspace $\mathcal{B}_h$ of $V_h \times V_h$, $V_h \subset V$, acting on the strain tensors Σ and Λ. We define modified strains: $\Sigma_\alpha^* = \mathcal{P}_\alpha(\Sigma_\alpha)$, $\Lambda_{\alpha\alpha}^* = \mathcal{P}_\alpha(\Lambda_{\alpha\alpha})$, $\Lambda_{12}^* = \frac{1}{2}[\mathcal{P}_1(u_{1,2}) + \mathcal{P}_2(u_{2,1})]$ and solve the approximate formulation of problem (1) replacing shear and membrane strain tensors Σ and Λ by Σ^* and Λ^* (see [6] for details).

3. Numerical results

We consider three classical cylindrical shell problems and for each test we present some numerical results.

1) *Scordelis–Lo shell* (see Fig. 1): a cylindrical shell known in the literature as a *barrel vault*. The shell is loaded by its own weight and is an example of a membrane dominated problem. In Fig. 2 (displacement vs. d.o.f.) we consider a very thin shell and we compare the Serendipity element S3 (of degree 3) with the corresponding mixed element MS3. The elements have been tested on both uniform and distorted decompositions. The results show that the mixed elements gives better performance and exhibits more robustness.

2) *Pinched shell* (see Fig. 3): the shell is simply supported at each end by rigid diaphragms and singularly loaded by two opposed forces acting at midpoint of the shell.

114

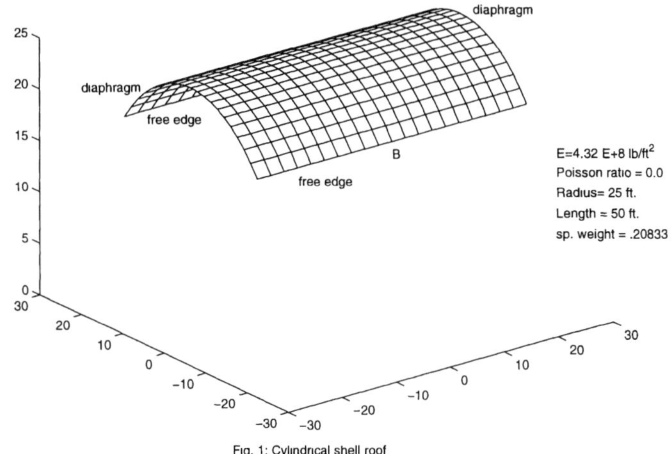

Fig. 1: Cylindrical shell roof

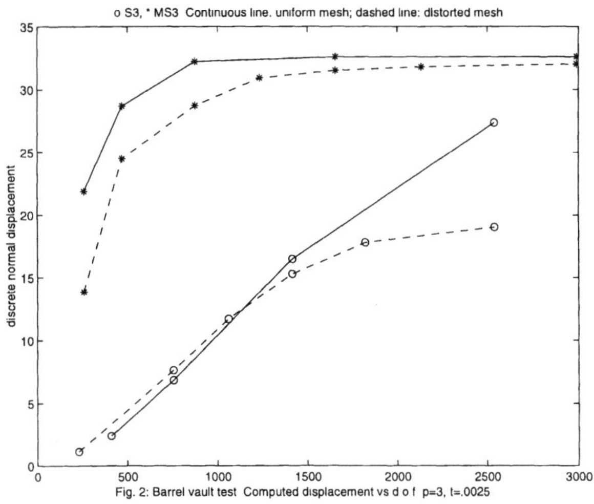

Fig. 2: Barrel vault test Computed displacement vs d o f p=3, t=.0025

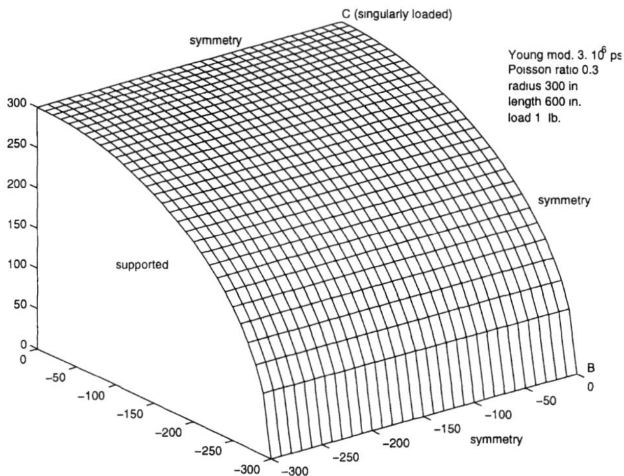

Fig 3. An octant of pinched cylindrical shell

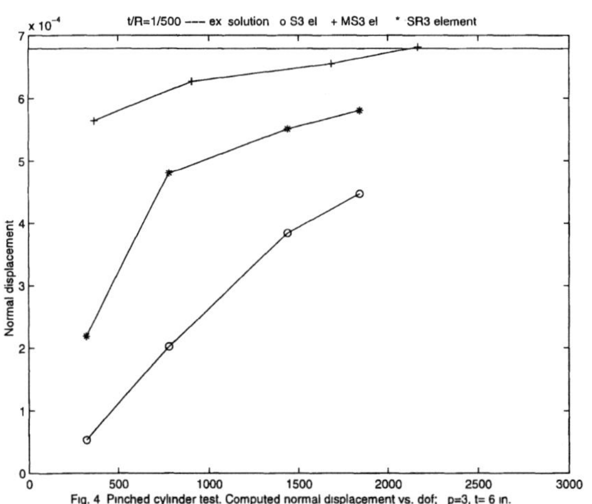

Fig. 4 Pinched cylinder test. Computed normal displacement vs. dof; p=3, t= 6 in.

116

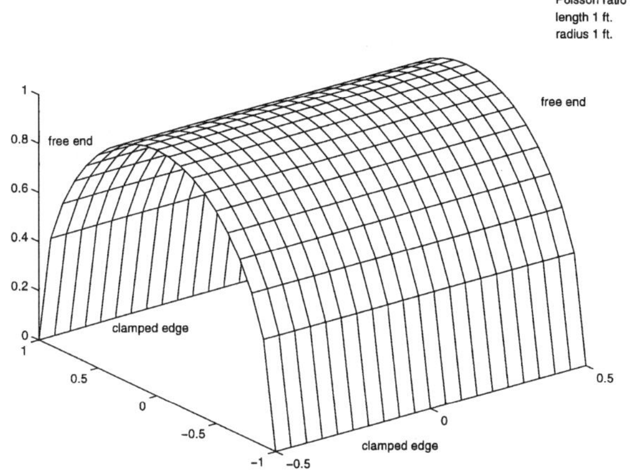

Fig.5: The hemicylindrical shell

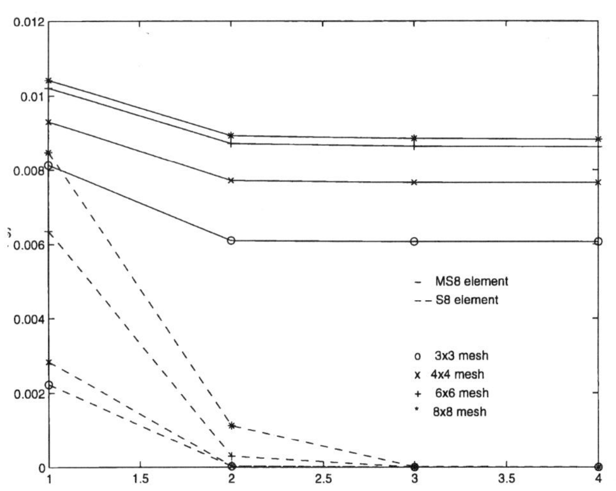

Fig. 6: Scaled energy vs.(–Log t) for p=2 and different meshes

117

In Fig. 4 (displacement vs. d.o.f.) we compare the performances of Serendipity S3, mixed MS3 and Serendipity reduced SR3 elements, all of degree 3. The mixed element outperforms the other elements. The reduced integration improves the performance of the Serendipity element.

3) *Hemicylindrical shell* (see Fig. 5): a structure presented in [7] as an example of a bending dominated problem. The following load is applied: $f_1 = f_2 = 0$, $f_3 = (1 + \xi_1/L)\cos(2\xi_2/R)$, where R is the radius and L the length of the shell. In Fig. 6 we show the scaled energy as a function of the logarithm of the thickness t for values $t/R = 10^{-1}$ to 10^{-4}. S2 denotes the Serendipity element of degree 2 and MS2 the corresponding mixed element. The comparison between the performances of S2 and MS2 is shown for several $N \times N$ decompositions of the domain. As N grows the Serendipity element is unable to avoid locking, while the mixed element shows the correct behavior, decreasing its error as N grows.

References

1. P.M. Naghdi, The theory of shells and plates, in *Handbuch der Physik*, VI, Springer-Verlag, Berlin, 1972, 425–640.

2. I. Babuška, The p and h-p versions of the finite element method. The state of the art, in *Finite elements: theory and application*, Springer-Verlag, New York, 1988, 199–239.

3. C. Chinosi, L. Della Croce and T. Scapolla, Hierarchic finite elements for thin Naghdi shell model, *Int. J. Solids Structures* **35** (1988), 1863–1880.

4. C. Chinosi, L. Della Croce and T. Scapolla, Numerical results on the locking for cylindrical shells, *Comp. Assisted Mech. Engrg. Sci.* **5** (1988), 31–44.

5. C. Chinosi, L. Della Croce and T. Scapolla, Solving thin Naghdi shells with special finite elements, *Math. Model. Scient. Computing* **8** (1997).

6. C. Chinosi and L. Della Croce, Mixed-interpolated elements for thin shells, *Comm. Numer. Methods Engrg.* **14** (1998).

7. Y. Leino and J. Pitkäranta: On the membrane locking of h-p finite elements in a cylindrical shell problem. *Internat. J. Numer. Methods Engrg.* **37** (1994), 1053–1070.

Dipartimento di Matematica, Università di Pavia, 27100 Pavia, Italy

J. DING and H. QIU

Error estimates for computing fixed densities of Markov integral operators

1. Introduction

In this paper we study the convergence rate problem for numerically computing a fixed density of Markov operators defined by stochastic kernels. Markov operators are widely used in the study of density evolutions under dynamical systems. For their general properties and applications, see the monograph [5]. Our purpose here will be toward obtaining an upper bound of the error estimate for some numerical methods for computing a fixed density of the Markov operator. Such an estimate is of practical importance since many problems in applied probability, stochastic analysis, and ergodic theory and dynamical systems are related to the computation of a fixed density of Markov operators [5].

A commonly used method for solving an integral equation is the so-called collocation method, which may not preserve some useful properties of the original operator. Because the linear operator of our integral equation is a Markov operator, it is natural and beneficial to approximate it with finite dimensional Markov operators for its numerical analysis such as in [1].

Let $L^1(0,1)$ be the space of Lebesgue integrable functions f on $[0,1]$ with L^1-norm $\|f\|_{0,1} \equiv \int_0^1 |f(x)|dx$, and let $W^{1,1}(0,1)$ be the Sobolev space of all absolutely continuous functions f on $[0,1]$ with $W^{1,1}$-norm $\|f\|_{1,1} \equiv \|f\|_{0,1} + \|f'\|_{0,1}$. The Lebesgue measure of a set A will be written as $m(A)$. Let D denote the set of all densities, that is, all nonnegative functions $f \in L^1(0,1)$ such that $\|f\|_{0,1} = 1$. Denote by M the vector subspace of $W^{1,1}(0,1)$ such that $\int_0^1 f(x)dx = 0$. A linear operator $P : L^1(0,1) \to L^1(0,1)$ is called a *Markov operator* if $PD \subset D$. Here we are interested in a class of Markov operators that are defined by stochastic kernels. In other words,

$$Pf(x) = \int_0^1 K(x,y)f(y)dy, \ f \in L^1(0,1), \tag{1}$$

where the kernel $K : [0,1] \times [0,1] \to R$ is nonnegative and satisfies

$$\int_0^1 K(x,y)dx = 1, \ y \in [0,1] \ m - \text{a.e.}$$

Since P is a positive integral operator of L^1-norm 1, there is a fixed density of P by a standard application of the Leray-Schauder fixed point theorem. A direct proof is referred to [1] in which the convergence of a class of Markov finite approximations for computing the fixed density has also been shown. Here we want to explore the convergence rate for such numerical methods.

We give a general error estimate result under some conditions in Section 2 and use it in Section 3 to get an explicit error bound for numerically computing a fixed density of (1). Some computational issues will be presented in Section 4.

2. Error bounds for Markov operators

We first present a general result on error estimates. Let $P : L^1(0,1) \to L^1(0,1)$ be a Markov operator (not necessarily defined by a stochastic kernel) that satisfies the condition that $PW^{1,1}(0,1) \subset W^{1,1}(0,1)$ and there are two nonnegative constants $\alpha < 1$ and β such that

$$\|Pf\|_{1,1} \leq \alpha\|f\|_{1,1} + \beta\|f\|_{0,1}, \ \forall \ f \in W^{1,1}(0,1). \tag{2}$$

By the Kakutani-Yosida theorem (see [5]), P has a fixed density f^* of bounded variation. Throughout the paper we assume that $f^* \in W^{1,1}(0,1)$. Now suppose that a class of numerical methods for computing f^* satisfy the assumption that if $P_n \equiv Q_n P$ is a sequence of finite approximations of P corresponding to partitions $0 = x_0 < x_1 < \cdots < x_n = 1$ and corresponding finite element subspaces $\Delta_n \subset W^{1,1}(0,1)$, then $\|Q_n f\|_{1,1} \leq \|f\|_{1,1}$ for all $f \in W^{1,1}(0,1)$. Thus from (2), we have

$$\|P_n f\|_{1,1} \leq \alpha\|f\|_{1,1} + \beta\|f\|_{0,1}, \ \forall \ f \in W^{1,1}(0,1). \tag{3}$$

The following lemma is basically due to [4].

Lemma 1. *For all $f \in M$,*

$$\|f\|_{0,1} \leq \frac{1}{3}\|f\|_{1,1}. \tag{4}$$

Proof. It is enough to show that $2\|f\|_{0,1} \leq \|f'\|_{0,1}$. Let A and B be the sets on which f is positive and negative respectively. Then

$$\int_A f(x)dx = -\int_B f(x)dx = \frac{1}{2}\|f\|_{0,1}$$

since $\int_0^1 f(x)dx = 0$. Since $m(A) + m(B) \leq 1$,

$$\frac{1}{m(A)} + \frac{1}{m(B)} \geq 4.$$

Hence

$$\begin{aligned}
\|f'\|_{0,1} &= \bigvee_0^1 f \geq \max_A f(x) - \min_B f(x) \geq \frac{\int_A f(x)dx}{m(A)} - \frac{\int_B f(x)dx}{m(B)} \\
&= \frac{\|f\|_{0,1}}{2m(A)} + \frac{\|f\|_{0,1}}{2m(B)} \geq 2\|f\|_{0,1},
\end{aligned}$$

where $\bigvee_0^1 f$ is the variation of f over $[0,1]$.

Theorem 1. *Suppose that P satisfies (2) and $f^* \in W^{1,1}(0,1)$ is a fixed density of P, and the sequence P_n satisfies (3). Let $f_n \in W^{1,1}(0,1)$ be a sequence of fixed points of P_n such that $\int_0^1 f_n(x)dx = 1$. If $\alpha + \beta/3 < 1$, then f^* is the unique fixed density of P in $W^{1,1}(0,1)$ and*

$$\|f^* - f_n\|_{1,1} \leq \frac{1}{1 - \alpha - \beta/3}\|f^* - Q_n f^*\|_{1,1}. \tag{5}$$

Proof. Let $g^* \in W^{1,1}(0,1)$ be another fixed density of P. Then, since $f^* - g^* \in M$, (2) and Lemma 1 imply that

$$\|f^* - g^*\|_{1,1} = \|P(f^* - g^*)\|_{1,1} \le \alpha \|f^* - g^*\|_{1,1} + \beta \|f^* - g^*\|_{0,1}$$

$$\le (\alpha + \frac{\beta}{3})\|f^* - g^*\|_{1,1},$$

which is impossible since $\alpha + \beta/3 < 1$. Hence f^* is the unique fixed density of P in $W^{1,1}(0,1)$. Now from (3) and

$$f^* - f_n = f^* - P_n f^* + P_n f^* - P_n f_n = P_n(f^* - f_n) + f^* - Q_n f^*,$$

$$\|f^* - f_n\|_{1,1} \le \|P_n(f^* - f_n)\|_{1,1} + \|f^* - Q_n f^*\|_{1,1}$$
$$\le \alpha \|f^* - f_n\|_{1,1} + \beta \|f^* - f_n\|_{0,1} + \|f^* - Q_n f^*\|_{1,1}.$$

Since $\int_0^1 (f^*(x) - f_n(x))dx = 0$ by assumption, from Lemma 1,

$$\|f^* - f_n\|_{1,1} \le (\alpha + \frac{\beta}{3})\|f^* - f_n\|_{1,1} + \|f^* - Q_n f^*\|_{1,1}.$$

Therefore (5) follows.

Remark 1. Using a quasi-compactness argument which implies that $I - P$ has closed range in $W^{1,1}(0,1)$ (see [3] for more details), one can show that without the condition $\alpha + \beta/3 < 1$, if f^* is the unique fixed density of P in $W^{1,1}(0,1)$, there is a constant C such that

$$\|f^* - f_n\|_{1,1} \le C\|f^* - Q_n f^*\|_{1,1}, \ \forall \, n. \tag{6}$$

This means that under the $W^{1,1}$-norm, the error of the numerical solutions f_n is about the same order as the "local error" of the numerical method applied to the exact solution f^*. However, an explicit expression for C in (6) may not be easy to get in general.

3. Application to Markov integral equations

Now we apply the general result of the previous section to our Markov operators with stochastic kernels. For this purpose assume that the kernel K in (1) is continuous on $[0,1] \times [0,1]$ and satisfies the condition:

(I) $K'_x(\cdot, y) \in L^1(0,1)$ for $y \in [0,1]$ and there is a constant β' such that

$$\int_0^1 |K'_x(x,y)|dx \le \beta', \ \forall \, y \in [0,1]. \tag{7}$$

Lemma 2. *Under the assumption (I), there is a density $f^* \in W^{1,1}(0,1)$ such that $Pf^* = f^*$. Moreover, with $\beta = 1 + \beta'$,*

$$\|Pf\|_{1,1} \le \beta \|f\|_{0,1}, \ \forall \, f \in W^{1,1}(0,1). \tag{8}$$

Proof. The existence of f^* is a direct consequence of the Leray-Schauder fixed point theorem since $P : W^{1,1}(0,1) \to W^{1,1}(0,1)$ is compact. Now let $f \in D \cap W^{1,1}(0,1)$ be given. Then $Pf \in D \cap W^{1,1}(0,1)$ and

$$
\begin{aligned}
\|(Pf)'\|_{0,1} &= \int_0^1 |(Pf)'(x)|dx = \int_0^1 |\int_0^1 K'_x(x,y)f(y)dy|dx \\
&\leq \int_0^1 f(y)dy \int_0^1 |K'_x(x,y)|dx \leq \beta' \int_0^1 f(y)dy = \beta'.
\end{aligned}
$$

Hence (8) is true from the definition of the $W^{1,1}$-norm.

Remark 2. Another proof for the existence of f^* using the Kakutani-Yosida theorem [5] was given in [1].

Combining Lemma 2 and Theorem 1, we have

Theorem 2. *Suppose that $f^* \in W^{1,1}(0,1)$ is a fixed density of (1) in which K satisfies (7). If $\beta < 3$, then f^* is the unique fixed density of P and for any sequence f_n of fixed points of P_n with $\int_0^1 f_n(x)dx = 1$,*

$$
\|f^* - f_n\|_{1,1} \leq \frac{3}{3-\beta}\|f^* - Q_n f^*\|_{1,1}. \tag{9}
$$

Corollary 1. *If $\|f^* - Q_n f^*\|_{1,1} = O(h^k)$, then so is $\|f^* - f_n\|_{1,1}$.*

Before ending this section, we present a structure-preserving result for computing the fixed density f^*.

Proposition 1. *Suppose that $K'_x(x,y) \geq 0$ on $[0,1] \times [0,1]$. Then every fixed density f^* of (1) is nondecreasing. If in addition the sequence P_n not only satisfies (3), but also maps nondecreasing densities to nondecreasing densities, then P_n has a fixed density f_n which is also nondecreasing.*

Proof. Differentiating

$$
f^*(x) = \int_0^1 K(x,y)f^*(y)dy
$$

both sides, we have $(f^*)'(x) \geq 0$ by the assumption. Now let D_1 be the set of all nondecreasing densities. Then D_1 is a closed convex subset of D. By the Brouwer fixed point theorem, P_n has a fixed density f_n in D_1.

Remark 3. A similar result can be obtained for nonincreasing fixed densities if $K'_x(x,y) \leq 0$.

4. Numerical considerations

In this last section we use one example to illustrate some numerical considerations. Here the Markov operator P in (1) is defined by the stochastic kernel

$$
K(x,y) = \frac{ye^{xy}}{e^y - 1}. \tag{10}
$$

If we define $K(x,0) = 1$, K is continuous on $[0,1] \times [0,1]$. It is easy to see that $\int_0^1 |K'_x(x,y)|dx \leq y$. So if we let $\beta' = 1$, the condition of Theorem 2 is satisfied. Moreover, since $K(x,y) \geq 1/(e-1)$, by Proposition 5.7.1 in [5], $\{P^n\}$ is *asymptotically stable.* That is, P has a unique fixed density f^* such that $\lim_{n\to\infty} P^n f = f^*$ for any $f \in D$. Since

$$K'_x(x, y) = \frac{y^2 e^{xy}}{e^y - 1} > 0, \ \forall \, y > 0,$$

f^* is strictly increasing on $[0, 1]$.

We employ Ulam's piecewise constant approximations method [6] and the method of piecewise linear Markov approximations [2] to compute f^* and to compare their performances. Divide $[0, 1]$ into n equal subintervals $I_i = [x_{i-1}, x_i]$ with length $h = 1/n$. Denote by Δ_n^0 and Δ_n^1 respectively the corresponding subspaces of (discontinuous) piecewise constant and continuous piecewise linear functions of $L^1(0, 1)$. Bases for Δ_n^0 and Δ_n^1 respectively are

$$e_i^0(x) = \chi_{I_i}(x), \ i = 1, \ldots, n,$$

$$e_i^1(x) = w(\frac{x - x_i}{h}), \ i = 0, 1, \ldots, n,$$

where χ_A is the characteristic function of A and $w(x) = (1 - |x|)\chi_{[-1,1]}(x)$.

Let $f_i = \frac{1}{h} \int_{x_{i-1}}^{x_i} f(x)dx$ be the average value of f over I_i. Define $Q_n^0 : L^1(0, 1) \to \Delta_n^0$ and $Q_n^1 : L^1(0, 1) \to \Delta_n^1$ by

$$Q_n^0 f(x) = \sum_{i=1}^{n} f_i e_i^0(x), \tag{11}$$

$$Q_n^1 f(x) = f_1 e_0^1(x) + \sum_{i=1}^{n-1} \frac{f_i + f_{i+1}}{2} e_i^1(x) + f_n e_n^1(x), \tag{12}$$

respectively. See [1] [2] [6] for some useful properties of Q_n^j. Now for $j = 0, 1$ define $P_n^j = Q_n^j P$ and solve

$$P_n^j f = f, \ f \in \Delta_n^j$$

to get a numerical solution f_n^j to the original fixed point problem $Pf = f$. It follows that the numerical methods are well-posed and the piecewise linear method satisfies all of the requirements in the previous section, although Ulam's piecewise constant method is not convergent in general under the $W^{1,1}$-norm.

The numerical methods were implemented in C. In the algorithms, the interval $[0, 1]$ was divided into $n = 2^r$ equal subintervals with $r = 2, 3, \ldots, L$ for some given L. The integration technique of the trapezoid rule was employed for the evaluation of the matrix representation of P_n^0 and P_n^1 with respect to the density basis $\{e_i^0/h\}$ of Δ_n^0 and the density basis $\{e_i^1/\|e_i^1\|_{0,1}\}$ of Δ_n^1, respectively. Because of the integration error, each column of the matrix was normalized so that the resulting matrix $\tilde{P}_n^j$ is a stochastic one. Then the direct iteration was used to find a normalized fixed nonnegative vector v_n^j of $\tilde{P}_n^j$, starting with the unit positive vector of the same components. The convergence was obtained after a couple of iterations (less than 10 for all dimensions in the computation).

The computational results show that for the piecewise linear method, the $W^{1,1}$-norm error reduces about the same order as h, which is consistent with our theoretical result (Theorem 2). Furthermore, the L^1-norm error reduces at the order of h^2, which can be explained with the fact that $\|f - Q_n^1 f\|_{0,1} = O(h^2)$. On the other hand, although the piecewise constant method does converge in the L^1-norm, it is not so under the $W^{1,1}$-norm since $\bigvee_0^1(f - Q_n^0 f) \geq \bigvee_0^1 f$ in general ([3]; Proposition 3.3). Moreover, the approximate function values of the piecewise constant and piecewise linear approximate fixed densities are increasing since Q_n^j preserve the monotone property of monotonic functions, which is guaranteed from our theoretical result (Proposition 1).

References

1. J. Ding, Computing fixed densities of Markov operators defined by stochastic kernels, *Appl. Math. Lett.* **10**, 85–88.

2. J. Ding and T.-Y. Li, Markov finite approximations of Frobenius-Perron operator equations, *Nonlinear Anal. TMA* **17**, 759–772.

3. J. Ding and T.-Y. Li, A convergence rate analysis for Markov finite approximations to a class of Frobenius-Perron operators, *Nonlinear Anal. TMA* **31**, 765–776.

4. B. Hunt, Estimating invariant measures and Lyapunov exponents, *Ergodic Theory Dynam. Sys.* **16**, 735–749.

5. A. Lasota and M. Mackey, *Chaos, fractals, and noise: stochastic aspects of dynamics*, Springer-Verlag, New York, 1994.

6. T.-Y. Li, Finite approximation for the Frobenius-Perron operator, a solution to Ulam's conjecture, *J. Approx. Theory* **17**, 177–186.

Program in Scientific Computing, University of Southern Mississippi, Hattiesburg, MS 39406-10057, USA

J. DING and Z. WANG

A modified Monte Carlo approach to the approximation of invariant measures

1. Introduction

In this paper we will propose an efficient numerical algorithm for computing absolutely continuous invariant measures of chaotic discrete dynamical systems. The algorithm applies the idea of the Monte Carlo method to the implementation of Ulam's original scheme to solve the difficult inverse image problem, especially when the mapping of the dynamical system has complicated expression or the expression is hard to obtain from physical experiments.

The concept of absolutely continuous invariant probability measures is related to the problem of studying various statistical properties of orbits of the chaotic dynamics [2]. Specifically, let us consider a discrete dynamical system

$$x_n = S^n(x), \quad n = 0, 1, \ldots$$

on a measurable space (X, Σ). We are interested in the *time mean* of an integrable function $f \in L^1(\mu) \equiv L^1(X, \Sigma, \mu)$ at x, where μ is an ergodic invariant probability measure. The Birkhoff individual ergodic theorem says that for μ-a.e. $x \in X$, this time mean equals the *space mean* of f. That is,

$$\lim_{n \to \infty} \frac{1}{n} \sum_{k=0}^{n-1} f(S^k(x)) = \int_X f d\mu, \quad x \in X \ \mu - a.e.$$

Here the invariance of μ with respect to S means that $\mu(S^{-1}(A)) = \mu(A)$ for all $A \in \Sigma$. The ergodicity means that $S^{-1}(A) = A$ implies that $\mu(A) = 0$ or 1.

In physical sciences such as statistical physics and neural networks, many problems are closely related to the problem of the existence of an absolutely continuous invariant probability measure μ of a nonsingular transformation of $X \subset R^N$ [2]. That is, μ can be viewed as a *physical measure* in the sense that there is an $f^* \in L^1(m)$ such that

$$\mu(A) = \int_A f^* dm, \quad \forall A \in \Sigma.$$

The nonnegative function f^* is called the density of μ. Here m is the Lebesgue measure on X, Σ is the Borel σ-algebra of subsets of X, and the nonsingularity of S means that $m(A) = 0$ implies $m(S^{-1}(A)) = 0$.

It is well-known [2] that the density f^* of an absolutely continuous invariant probability measure is a fixed density of the so-called *Frobenius-Perron operator* $P : L^1(m) \to L^1(m)$ associated with S, which is defined by

$$\int_A Pf dm = \int_{S^{-1}(A)} f dm, \quad \forall f \in L^1(m), \ A \in \Sigma. \tag{1}$$

In this paper we are more interested in the numerical computation of the fixed density f^* of the Frobenius-Perron operator P. For $S : [0, 1] \to [0, 1]$ Ulam [4] proposed a piecewise constant approximation method to calculate f^*, and he conjectured that the sequence of piecewise constant approximate densities f_n computed from his numerical scheme should converge in $L^1(0,1)$ to f^*. In 1976, Li [3] proved this conjecture for a class of piecewise C^2 and stretching mappings $S:[0,1]\to[0,1]$.

Numerical experiments have shown that Ulam's method converges not only for the class of piecewise monotonic mappings, but also for many other classes of mappings. But this method has one shortage. That is, the numerical evaluation in implementing Ulam's method becomes difficult or even impossible if the mapping S has a complicated expression or the expression of S is hard to obtain, which is often the case when S results from some physical experiment. In this paper we propose a new algorithm to overcome this difficulty. Here we employ the Monte Carlo approach which was first proposed by Hunt [1] of the National Institute of Standards and Technology. But unlike the random choice of points in [1], we use a uniform distribution of points, so our method can be classified as a modified Monte Carlo method, or quasi Monte Carlo method. In the next section, we describe the Ulam method and show how it leads to the problem of evaluating a matrix and solving the resulting linear system of equations. In Section 3 we introduce the Monte Carlo idea. The numerical comparison is presented in Section 4.

2. Ulam's piecewise constant approximations

Now we introduce the idea behind Ulam's piecewise constant approximations for computing the fixed density of the Frobenius-Perron operator associated with one dimensional mappings of the interval. Let us assume that $S:[0,1]\to[0,1]$ is a nonsingular transformation such that the corresponding Frobenius-Perron operator $P:L^1(0,1)\to L^1(0,1)$ has a fixed density f^*.

Let $[0,1]$ be divided into n sub-intervals $I_1,I_2,\ldots,I_n$ with equal length $h=1/n$. Let $1_i=\frac{1}{h}1_{I_i}$ for each i, where 1_{I_i} is the characteristic function of I_i. Then each $1_i\in L^1(0,1)$ is a density with its *support* $\mathrm{supp}\,1_i=I_i$. Let Δ_n be the n-dimensional subspace of $L^1(0,1)$ spanned by the density basis $1_1,\ldots,1_n$, i.e., Δ_n is the space of all piecewise constant functions associated with the partition.

For any function $f\in L^1(0,1)$, we define its approximation $Q_nf\in\Delta_n$ in the following way:

$$Q_nf = \sum_{i=1}^n c_i 1_{I_i},$$

$$s = h\sum_{i=1}^n c_i 1_i, \tag{2}$$

where for $i=1,2,\ldots,n$,

$$c_i = \frac{1}{h}\int_{I_i} f\,dm \tag{3}$$

is the average value of f over I_i. Now we define the finite dimensional approximation $P_n:\Delta_n\to\Delta_n$ of P as

$$P_nf = Q_nPf$$
$$= \sum_{i=1}^n \int_{I_i} Pf\,dm\cdot 1_i \tag{4}$$
$$= \sum_{i=1}^n \int_{S^{-1}(I_i)} f\,dm\cdot 1_i.$$

126

In particular, if we write

$$P_n 1_i = \sum_{j=1}^{n} p_{ij} 1_j, \tag{5}$$

then from (1), (4), and (5) it follows that

$$\sum_{j=1}^{n} p_{ij} 1_j = \sum_{j=1}^{n} \int_{S^{-1}(I_j)} 1_i dm \cdot 1_i.$$

We can now obtain

$$\begin{aligned} p_{ij} &= \int_{S^{-1}(I_j)} 1_i dm \\ &= \frac{m(I_i \cap S^{-1}(I_j))}{m(I_i)} \\ &= \frac{m(I_i \cap S^{-1}(I_j))}{h}. \end{aligned} \tag{6}$$

Therefore, the finite dimensional operator P_n is represented by the $n \times n$ matrix $[p_{ij}]$. Note that $[p_{ij}]$ is a *stochastic matrix*, i.e., $p_{ij} \geq 0$ and $\sum_{j=1}^{n} p_{ij} = 1$ for all i. This means that the column vector $(1, 1, \ldots, 1)^T$ is a right eigenvector of the matrix corresponding to the eigenvalue 1. Without causing any confusion, this matrix will still be denoted by P_n.

It is easy to see that if $f = \sum_{i=1}^{n} c_i 1_i \in \Delta_n$, then f is a fixed point of P_n if and only if the row vector $(c_1, c_2, \ldots, c_n)$ is a left eigenvector of the matrix P_n associated with the eigenvalue 1, i.e.,

$$(c_1, c_2, \ldots, c_n) P_n = (c_1, c_2, \ldots, c_n). \tag{7}$$

From the theory of nonnegative matrices it follows that a normalized nonnegative left eigenvector must exist. Therefore Ulam's method is well-posed. In other words, for any n, there is a piecewise constant density f_n such that $P_n f_n = f_n$.

3. Monte Carlo algorithms

Now we introduce the Monte Carlo approach to implementing the basic Ulam method. In Ulam's scheme we need to evaluate the entries of the matrix given by (6). But generally speaking, this is a difficult problem since the inverse image of a subset under the mapping S is hard to get in many cases, especially for multidimensional mappings. The Monte Carlo implementation has the advantage of not requiring explicit evaluation of such entries of the approximate Frobenius-Perron operator P_n. In fact since P_n is only an approximation to P, it is by no means necessary to evaluate the entries of P_n *exactly* if it is time-consuming or difficult to do so. Thus the Monte Carlo method, which is a probability method, is an ideal means for approximating P_n.

The basic idea of the Monte Carlo approach is that within each subinterval I_i of the partition of $[0, 1]$, K points are selected which are called $\{z_{i,k}\}_{k=1}^{K}$. For any pair (i, j) with $i, j = 1, \ldots, n$, let q_{ij} be the number of points $S(z_{i,k})$ in I_j for $k = 1, \ldots, K$. Then we have

$$\frac{q_{ij}}{K} \approx \frac{m(I_i \cap S^{-1}(I_j))}{h}. \tag{8}$$

The original Monte Carlo method first proposed in [1] selected the K numbers randomly. That is, the points $z_{i,k}$ are obtained from a random number generator. Numerical experiments of [1] indicate that the resulting error may be relatively large compared with the exact method. To get a better approximation in (8), we use the idea of the *quasi Monte Carlo method*. Here the test points $z_{i,k}$ are chosen *deterministically*. In other words, if $I_i = [x_{i-1}, x_{i-1} + h]$, then

$$z_{i,k} = x_{i-1} + \frac{k}{K}h, \ k = 1, 2, \ldots, K.$$

Our numerical results in the next section show that the new Monte Carlo approach, which may be called a modified Monte Carlo method, is much better than the original Monte Carlo method.

After the evaluation of the matrix, the main work of Ulam's method is the computation of a fixed density of the finite dimensional operator P_n. In numerical linear algebra, there are two approaches to solving the linear system (7). One is the direct method such as the Gaussian elimination method, and the other is the iteration method. Since iterating P_n several times usually gives a fixed density of P_n from the theory of Markov chains if S is ergodic, the simple iteration method seems to be a good choice of the solver to (7) since it takes only $O(n^2)$ float point operations. As an alternative, a Gaussian elimination method is also proposed here for the comparison purpose. Thus two algorithms using Ulam's method and the Monte Carlo approach for a chosen n are as follows.

The Iteration Algorithm (IA):

1. Using the modified Monte Carlo method to evaluate the matrix P_n.

2. Select a starting vector $c = (c_1, c_2, \ldots, c_n)^T$; a usual choice is $c = (1, 1, \ldots, 1)^T$.

3. Calculate $d = P_n^T c$ and Error $= \|d - c\|$.

4. Let $c = d$ and repeat the above step until Error $< \epsilon$, where ϵ is a desired tolerance.

The Gaussian Algorithm (*GA*):

Suppose P_n has been calculated by using the Monte Carlo method. To solve (7), we just need to solve the homogeneous system of linear equations

$$(I - P_n^T)c = 0. \tag{9}$$

In general, the rank of $I - P_n$ is $n - 1$ for large n if S is ergodic. We can use the Gaussian elimination to find the unique normalized solution.

4. Numerical results

In this section, we present some numerical results. The test mappings are

$$S_1(x) = \begin{cases} 0 & \text{if } x = 0 \\ \{\frac{1}{x}\} & \text{if } x \neq 0, \end{cases}$$

$$S_2(x) = \left(\frac{1}{8} - 2|x - \frac{1}{2}|^3\right)^{1/3} + \frac{1}{2},$$

$$S_3(x) = \begin{cases} \frac{1}{\sqrt{2}} - \sqrt{2}|\frac{1}{2} - x| & \text{if } x \in [0, \frac{1}{\sqrt{8}}] \cup [1 - \frac{1}{\sqrt{8}}, 1] \\ 1 - \frac{1}{\sqrt{2}}\sqrt{1 - (1 - |1 - 2x|)^2} & \text{if } x \in [\frac{1}{\sqrt{8}}, 1 - \frac{1}{\sqrt{8}}]. \end{cases}$$

S_1 is called the Gaussian transformation in which $\{a\}$ is the fractional part of a, and S_2 and S_3 are the test examples τ_1 and τ_2, respectively, in [1]. The unique fixed densities of S_i are given by

$$
\begin{aligned}
f_1^*(x) &= \frac{1}{\ln 2}\frac{1}{1+x}, \\
f_2^*(x) &= 12\left(x - \frac{1}{2}\right)^2, \\
f_3^*(x) &= 2(1 - |1 - 2x|).
\end{aligned}
$$

The experiments were made on an SAG Pentium Pro computer with C codes. In the following two tables, n denotes the number of the equal sub-intervals of the partition of $[0,1]$. The L^1-error of the computed fixed density f_n to the exact fixed density f^* from the original Monte Carlo method (MC) and our modified Monte Carlo method (MMC) with $K = 1000$ points chosen in each sub-interval is presented in Table 1. The results of the MC for S_2 and S_3 are about the same as in [1]. However, the MMC gives much smaller errors than the MC with the increasing number of the sub-intervals. In fact we have observed that the errors are almost as small as those from the *exact* Ulam method in which the entries of the matrix P_n are evaluated exactly (see page 112 of [1] on the error of the exact method for S_2).

	S_1		S_2		S_3	
n	MC	MMC	MC	MMC	MC	MMC
16	2.3132E-2	1.3538E-2	1.1644E-1	1.0254E-1	7.7969E-2	7.2897E-2
32	2.4921E-2	8.4698E-3	5.7815E-2	5.2642E-2	5.0783E-2	3.8255E-2
64	2.2484E-2	6.7886E-3	3.8867E-2	2.5817E-2	3.0469E-2	1.7884E-2
128	2.6148E-2	7.1641E-4	3.7367E-2	1.3436E-2	3.1693E-2	8.6400E-3
256	2.6702E-2	6.7251E-4	3.3161E-2	6.6441E-3	2.6904E-2	4.4750E-3

Table 1. L^1-Error comparison of MC and MMC.

Table 2 shows the CPU time (in seconds) comparison for the Gaussian algorithm (GA) and the iteration algorithm (IA), which shows clearly that the IA outperforms the GA since the former only requires $O(n^2)$ operations while the latter needs $O(n^3)$ ones in solving the linear system of equations.

	S_1		S_2		S_3	
n	GA	IA	GA	IA	GA	IA
16	0.00	0.00	0.00	0.00	0.00	0.00
32	0.01	0.00	0.01	0.00	0.01	0.01
64	0.09	0.00	0.09	0.01	0.09	0.02
128	0.70	0.02	0.68	0.06	0.76	0.09
256	5.64	0.14	5.53	0.30	6.28	0.47

Table 2. Time(s) comparisons of GA and IA.

References

1. F. Hunt, A Monte Carlo approach to the approximation of invariant measures, *Random Comput. Dynamics* **2**, 111–133.

2. A. Lasota and M. Mackey, *Chaos, fractals, and noise: stochastic aspects of dynamics*, Springer-Verlag, New York, 1994.

3. T.-Y. Li, Finite approximation for the Frobenius-Perron operator, a solution to Ulam's conjecture, *J. Approx. Theory* **17**, 177–186.

4. S. Ulam, *A collection of mathematical problems*, Interscience, New York, 1960.

Program in Scientific Computing, University of Southern Mississippi, Hattiesburg, MS 39406-10057, USA

P. DJONDJOROV and V. VASSILEV

Acceleration waves in von Kármán plate theory

1. Introduction

The von Kármán plate theory is governed by two coupled nonlinear fourth-order partial differential equations in three independent variables (Cartesian coordinates on the plate middle-plane x^1, x^2, and the time x^3) and two dependent variables (the transversal displacement function w and Airy's stress function Φ), namely

$$
\begin{aligned}
D\Delta^2 w - \varepsilon^{\alpha\mu}\varepsilon^{\beta\nu} w_{,\alpha\beta}\,\Phi_{,\mu\nu} + \rho w_{,33} &= 0, \\
(1/Eh)\,\Delta^2\Phi + (1/2)\,\varepsilon^{\alpha\mu}\varepsilon^{\beta\nu} w_{,\alpha\beta} w_{,\mu\nu} &= 0,
\end{aligned}
\tag{1}
$$

where Δ is the Laplace operator with respect to x^1 and x^2, $D = Eh^3/12(1 - \nu^2)$ is the bending rigidity, E is Young's modulus, ν is Poisson's ratio, h is the thickness of the plate, ρ is the mass per unit area of the plate middle-plane, $\delta^{\alpha\beta}$ is the Kronecker delta symbol, and $\varepsilon^{\alpha\beta}$ is the alternating symbol. Here and throughout the work: Greek (Latin) indices range over 1, 2 (1, 2, 3), unless explicitly stated otherwise; the usual summation convention over a repeated index is used and subscripts after a comma denote partial derivatives, that is $f_{,i} = \partial f/\partial x^i$, $f_{,ij} = \partial f/\partial x^i \partial x^j$, etc.

The von Kármán equations (1) completely describe the motion of a plate, the membrane stress tensor $N^{\alpha\beta}$, moment tensor $M^{\alpha\beta}$, shear-force vector Q^α, strain tensor $E^{\alpha\beta}$ and bending tensor $K_{\alpha\beta}$ are given in terms of w and Φ through:

$$
N^{\alpha\beta} = \varepsilon^{\alpha\mu}\varepsilon^{\beta\nu}\,\Phi_{,\mu\nu}, \quad M^{\alpha\beta} = -D\left\{(1 - \nu)\delta^{\alpha\mu}\delta^{\beta\nu} + \nu\delta^{\alpha\beta}\delta^{\mu\nu}\right\} w_{,\mu\nu},
$$

$$
Q^\alpha = M^{\alpha\mu}_{,\mu} + N^{\alpha\mu} w_{,\mu}, \quad E^{\alpha\beta} = (1/Eh)\left\{(1 + \nu)\varepsilon^{\alpha\mu}\varepsilon^{\beta\nu} - \nu\delta^{\alpha\beta}\delta^{\mu\nu}\right\}\Phi_{,\mu\nu}, \quad K_{\alpha\beta} = w_{,\alpha\beta}.
$$

The theory under consideration has an exact variational formulation, the von Kármán equations being the Euler-Lagrange equations [1] for the action functional

$$
I\left[w, \Phi\right] = \int\int\int L dx^1 dx^2 dx^3, \quad L = T - \Pi, \qquad \text{where} \tag{2}
$$

$$
\begin{aligned}
\Pi = {}& (D/2)\left\{(\Delta w)^2 - (1 - \nu)\,\varepsilon^{\alpha\mu}\varepsilon^{\beta\nu} w_{,\alpha\beta} w_{,\mu\nu}\right\} \\
& - (1/2Eh)\left\{(\Delta\Phi)^2 - (1 + \nu)\,\varepsilon^{\alpha\mu}\varepsilon^{\beta\nu}\,\Phi_{,\alpha\beta}\,\Phi_{,\mu\nu}\right\} + (1/2)\varepsilon^{\alpha\mu}\varepsilon^{\beta\nu}\,\Phi_{,\alpha\beta} w_{,\mu} w_{,\nu},
\end{aligned}
$$

is the strain energy per unit area of the plate middle-plane and

$$
T = (\rho/2)\left(w_{,3}\right)^2,
$$

is the kinetic energy per unit area of the plate middle-plane.

This work was supported by Contract No. MM 517/1995 with NSF, Bulgaria.

2. Conservation laws

In the recent paper [2], all Lie point symmetries of system (1) are shown to be variational symmetries of the functional (2), and all corresponding (via Noether's theorem) conservation laws admitted by the smooth solutions of the von Kármán equations are established. Each such conservation law is a linear combination of the basic linearly independent conservation laws

$$\frac{\partial \Psi_{(j)}}{\partial x^3} + \frac{\partial P_{(j)}^\mu}{\partial x^\mu} = 0 \quad (j = 1, 2, \ldots, 14),$$

whose densities $\Psi_{(j)}$ and fluxes $P_{(j)}^\mu$ are presented (together with the generators of the respective symmetries) in Table 1 in terms of Q^α, $M^{\alpha\beta}$, $G^{\alpha\beta}$ and F^α,

$$G^{\alpha\beta} = (1/Eh)\left\{(1+\nu)\delta^{\alpha\mu}\delta^{\beta\nu} - \nu\delta^{\alpha\beta}\delta^{\mu\nu}\right\}\Phi_{,\mu\nu} - (1/2)\,\varepsilon^{\alpha\mu}\varepsilon^{\beta\nu}w_{,\mu}w_{,\nu}, \quad F^\alpha = G_{,\nu}^{\alpha\nu}.$$

3. Balance laws

Given a region Ω in the plate middle-plane with sufficiently smooth boundary Σ of outward unit normal n_α, a balance law

$$\frac{d}{dt}\int_\Omega \Psi_{(j)}dx^1 dx^2 + \int_\Sigma P_{(j)}^\alpha n_\alpha d\Sigma = 0, \tag{3}$$

corresponds to each of the conservation laws listed in Table 1. It holds, just as the respective conservation law, for every smooth solution of the von Kármán equations.

The balance laws are applicable even if Ω is intersected by a discontinuity (singular) manifold (on which densities $\Psi_{(j)}$ and fluxes $P_{(j)}^\alpha$ may suffer jump discontinuities) provided the integrals exist. We extend the "continuous" von Kármán plate theory to situations when physical quantities may suffer jump discontinuities at a curve.

4. Acceleration waves

Definition 1. A discontinuity solution of the von Kármán equations is a couple of functions (w, Φ), defined in a certain region Ω, such that the two balance laws corresponding to the von Kármán equations themselves, namely

$$\frac{d}{dt}\int_{\tilde\Omega} \rho w_{,3}dx^1 dx^2 - \int_{\tilde\Sigma} Q^\alpha \tilde n_\alpha d\Sigma = 0, \quad \int_{\tilde\Sigma} F^\alpha \tilde n_\alpha d\Sigma = 0, \tag{4}$$

hold $\forall\, \tilde\Omega \subset \Omega$ with boundary $\tilde\Sigma$ of outward unit normal $\tilde n_\alpha$, and (w, Φ) is a solution of the (local) von Kármán equations (1) almost everywhere in Ω except for a moving curve Γ at which some of the derivatives of w or Φ have jumps.

Definition 2. A discontinuity solution of the von Kármán equations is an acceleration wave if at the wave front—a smoothly propagating connected singular curve Γ,

$$\Gamma : \gamma(x^1, x^2, x^3) = 0, \quad (x^1, x^2) \in \Omega \subset \mathbf{R}^2, \quad x^3 \in \mathbf{R}^+, \quad \gamma \in C^1(\Omega \times \mathbf{R}^+),$$

w - translations $X_1 = \frac{\partial}{\partial w}$	**transversal linear momentum** (first von Kármán eqn) $P^\alpha_{(1)} = -Q^\alpha, \ \Psi_{(1)} = \rho\, w_{,3}$
Φ - translations $X_{14} = \frac{\partial}{\partial \Phi}$	**compatibility condition** (second von Kármán eqn) $P^\alpha_{(14)} = F^\alpha, \ \Psi_{(14)} = 0$
time - translations $X_4 = \frac{\partial}{\partial x^3}$	**energy** $P^\alpha_{(4)} = -w_{,3}Q^\alpha - \Phi_{,3}F^\alpha + w_{,3\beta}M^{\alpha\beta} + \Phi_{,3\beta}G^{\alpha\beta}$ $\Psi_{(4)} = T + \Pi$
x^1 & x^2- translations $X_2 = \frac{\partial}{\partial x^1}$ $X_3 = \frac{\partial}{\partial x^2}$	**wave momentum** $P^\alpha_{(2)} = \delta^{\alpha 1}L + w_{,1}Q^\alpha + \Phi_{,1}F^\alpha - w_{,1\beta}M^{\alpha\beta} - \Phi_{,1\beta}G^{\alpha\beta}$ $\Psi_{(2)} = -\rho\, w_{,1}w_{,3}$ $P^\alpha_{(3)} = \delta^{\alpha 2}L + w_{,2}Q^\alpha + \Phi_{,2}F^\alpha - w_{,2\beta}M^{\alpha\beta} - \Phi_{,2\beta}G^{\alpha\beta}$ $\Psi_{(3)} = -\rho\, w_{,2}w_{,3}$
rotations $X_6 = x^2\frac{\partial}{\partial x^1} - x^1\frac{\partial}{\partial x^2}$	**moment of the wave momentum** $P^\alpha_{(6)} = x^2 P^\alpha_{(2)} - x^1 P^\alpha_{(3)} + \varepsilon^\mu_\nu w_{,\mu}M^{\alpha\nu} + \varepsilon^\mu_\nu \Phi_{,\mu}G^{\alpha\nu}$ $\Psi_{(6)} = x^2 \Psi_{(2)} - x^1 \Psi_{(3)}$
rigid body rotations $X_7 = x^1\frac{\partial}{\partial w}$ $X_8 = x^2\frac{\partial}{\partial w}$	**angular momentum** $P^\alpha_{(7)} = M^{\alpha 1} - x^1 Q^\alpha + w\varepsilon^{\alpha\nu}\Phi_{,\nu 2}, \ \Psi_{(7)} = \rho x^1 w_{,3}$ $P^\alpha_{(8)} = M^{\alpha 2} - x^2 Q^\alpha + w\varepsilon^{\nu\alpha}\Phi_{,\nu 1}, \ \Psi_{(8)} = \rho x^2 w_{,3}$
scaling $X_5 = x^\mu\frac{\partial}{\partial x^\mu} + 2x^3\frac{\partial}{\partial x^3}$	$P^\alpha_{(5)} = x^1 P^\alpha_{(2)} + x^2 P^\alpha_{(3)} - 2x^3 P^\alpha_{(4)} - w_{,\beta}M^{\alpha\beta} - \Phi_{,\beta}G^{\alpha\beta}$ $\Psi_{(5)} = x^1 \Psi_{(2)} + x^2 \Psi_{(3)} - 2x^3 \Psi_{(4)}$
Galilean boost $X_9 = x^3\frac{\partial}{\partial w}$	**center-of-mass theorem** $P^\alpha_{(9)} = -x^3 Q^\alpha, \ \Psi_{(9)} = \rho\left(x^3 w_{,3} - w\right)$
$X_{10} = x^1 x^3\frac{\partial}{\partial w}$ $X_{11} = x^2 x^3\frac{\partial}{\partial w}$ $X_{12} = x^1\frac{\partial}{\partial \Phi}$ $X_{13} = x^2\frac{\partial}{\partial \Phi}$	$P^\alpha_{(10)} = x^3 P^\alpha_{(7)}, \ \Psi_{(10)} = x^1 \Psi_{(9)}$ $P^\alpha_{(11)} = x^3 P^\alpha_{(8)}, \ \Psi_{(11)} = x^2 \Psi_{(9)}$ $P^\alpha_{(12)} = x^1 F^\alpha - G^{\alpha 1}, \ \Psi_{(12)} = 0$ $P^\alpha_{(13)} = x^2 F^\alpha - G^{\alpha 2}, \ \Psi_{(13)} = 0$

Table 1. Conservation laws.

we have

$$[w] = [\Phi] = [w_{,i}] = [\Phi_{,i}] = 0, \ [w_{,33}] \neq 0. \tag{5}$$

(Here and in what follows, the square brackets are used to denote the jump of any field f across the curve Γ, i.e., $[f] = f_2 - f_1$, where f_2 and f_1 are the limit values of f behind Γ and ahead of Γ.)

The moving curve Γ divides the region Ω into two parts Ω^+ and Ω^- and forms the common border between them. It is assumed that ahead the wave front (in the region Ω^+) we have the known unperturbed fields $w^+(x^1, x^2, x^3)$, $\Phi^+(x^1, x^2, x^3)$ and behind it (in the region Ω^-)—the unknown perturbed fields $w^-(x^1, x^2, x^3)$, $\Phi^-(x^1, x^2, x^3)$. At the wave front Γ, we have the jump conditions (5).

The jumps of the derivatives of w and Φ across Γ are permissible if they obey the compatibility conditions following by Hadamard's lemma [3]. Thus, the following assertion holds.

133

Proposition 1. *If $\lfloor w_{,33}\rfloor \neq 0$, then*

$$[w_{,\alpha\beta}] = \lambda n_\alpha n_\beta, \quad [w_{,\alpha 3}] = -\lambda C n_\alpha, \quad [w_{,33}] = \lambda C^2,$$

where λ is an arbitrary factor, C and n_α,

$$C = -\,|\nabla\gamma|^{-1}\,\partial\gamma/\partial x^3, \quad n_\alpha = |\nabla\gamma|^{-1}\,\partial\gamma/\partial x^\alpha, \quad |\nabla\gamma| = \sqrt{(\partial\gamma/\partial x^1)^2 + (\partial\gamma/\partial x^2)^2},$$

are the speed of displacement and the direction of propagation of the wave front Γ.

Proposition 2. *If at least one of the third derivatives of w suffers a jump at Γ, then the compatibility conditions for the jumps of the third derivatives of the displacement field across Γ are*

$$
\begin{aligned}
[w_{,\alpha\beta\gamma}] \;=\;& \lambda^* n_\alpha n_\beta n_\gamma + \partial\lambda/\partial s \,(n_\alpha n_\beta t_\gamma + n_\alpha t_\beta n_\gamma + t_\alpha n_\beta n_\gamma) \\
& + \lambda a \,(t_\alpha t_\beta n_\gamma + t_\alpha n_\beta t_\gamma + n_\alpha t_\beta t_\gamma)\,,
\end{aligned}
$$

where t_α is the unit tangent vector to Γ, λ^* is an arbitrary factor, while $\lambda = [w_{,\alpha\beta}]\, n^\alpha n^\beta$ and $a = t_\alpha \partial n^\alpha/\partial s$, s being the natural parameter (arc-length) of the curve Γ.

Proposition 3. *If at least one of the second derivatives of Φ suffers a jump at Γ, then*

$$[\Phi_{,\alpha\beta}] = \mu n_\alpha n_\beta, \quad [\Phi_{,\alpha 3}] = -\mu C n_\alpha, \quad [\Phi_{,33}] = \mu C^2,$$

where μ is an arbitrary factor, are the compatibility conditions for the jumps of the second derivatives of the stress field across Γ.

Proposition 4. *If at least one of the third derivatives of Φ suffers a jump at Γ, then the compatibility conditions for the jumps of the third derivatives of the stress field across Γ are*

$$
\begin{aligned}
[\Phi_{,\alpha\beta\gamma}] \;=\;& \mu^* n_\alpha n_\beta n_\gamma + \partial\mu/\partial s \,(n_\alpha n_\beta t_\gamma + n_\alpha t_\beta n_\gamma + t_\alpha n_\beta n_\gamma) \\
& + \mu a \,(t_\alpha t_\beta n_\gamma + t_\alpha n_\beta t_\gamma + n_\alpha t_\beta t_\gamma)\,,
\end{aligned}
$$

where $\mu = [\Phi_{,\alpha\beta}]n^\alpha n^\beta$ and μ^* is an arbitrary factor.

According to the divergence theorem (see, e.g., [4]), a couple of functions $(w,\ \Phi)$ suffering jump discontinuities at a singular curve Γ is a discontinuity solution of the von Kármán equations in the sense of Definition 1 iff the following jump conditions

$$C\,[\rho w_{,3}] + [Q^\alpha]\,n_\alpha = 0, \quad [F^\alpha]\,n_\alpha = 0, \tag{6}$$

hold at Γ, and a balance law of form (3) holds on this solution iff at Γ:

$$C\left[\Psi_{(j)}\right] - \left[P^\alpha_{(j)}\right] n_\alpha = 0. \tag{7}$$

Definition 2, Propositions 1, 2, 3, 4 and jump conditions (6) imply the following assertion.

Proposition 5. *If an acceleration wave in the von Kármán plate theory is such that $[w_{,\alpha\beta\gamma}] \neq 0$ $([\Phi_{,\alpha\beta\gamma}] \neq 0)$ at the curve of discontinuity Γ, then $\lambda^* = -\lambda a$ $(\mu^* = -\mu a)$.*

134

time - translations	energy
$X_4 = \frac{\partial}{\partial x^3}$	$\frac{C}{2}\left(D\lambda^2 - \frac{\mu^2}{Eh}\right) = \left(D\lambda X_4^+(w_{,\beta}) - \frac{\mu}{Eh}X_4^+(\Phi_{,\beta})\right)n^\beta$
x^1 & x^2- translations	wave momentum
$X_2 = \frac{\partial}{\partial x^1}$	$\frac{1}{2}\left(D\lambda^2 - \frac{\mu^2}{Eh}\right)n^1 = -\left(D\lambda X_2^+(w_{,\beta}) - \frac{\mu}{Eh}X_2^+(\Phi_{,\beta})\right)n^\beta$
$X_3 = \frac{\partial}{\partial x^2}$	$\frac{1}{2}\left(D\lambda^2 - \frac{\mu^2}{Eh}\right)n^2 = -\left(D\lambda X_3^+(w_{,\beta}) - \frac{\mu}{Eh}X_3^+(\Phi_{,\beta})\right)n^\beta$
rotations	moment of the wave momentum
$X_6 = x^2\frac{\partial}{\partial x^1} - x^1\frac{\partial}{\partial x^2}$	$\frac{1}{2}\varepsilon^\alpha_{\ \beta}n_\alpha x^\beta\left(D\lambda^2 - \frac{\mu^2}{Eh}\right) =$ $-\left\{D\lambda\left(\varepsilon^\alpha_{\ \beta}w_{,\alpha} + X_6^+(w_{,\beta})\right) - \frac{\mu}{Eh}\left(\varepsilon^\alpha_{\ \beta}\Phi_{,\alpha} + X_6^+(\Phi_{,\beta})\right)\right\}n^\beta$
scaling	
$X_5 = x^\mu\frac{\partial}{\partial x^\mu} + 2x^3\frac{\partial}{\partial x^3}$	$\frac{1}{2}\left(x^\alpha n_\alpha - 2Cx^3\right)\left(D\lambda^2 - \frac{\mu^2}{Eh}\right) =$ $-\left\{D\lambda\left(w_{,\beta} + X_5^+(w_{,\beta})\right) - \frac{\mu}{Eh}\left(\Phi_{,\beta} + X_5^+(\Phi_{,\beta})\right)\right\}n^\beta$

Table 2. Jump conditions.

Given a discontinuity solution of the von Kármán equations, the two corresponding balance laws (4) being satisfied, the other balance laws do not necessarily hold for this solution. The jump conditions associated with the most important conservation laws from Table 1 are derived using (7) and presented on the Table 2, where $X_j^+(w_{,\beta})$ and $X_j^+(\Phi_{,\beta})$ denote the limit values of $X_j(w_{,\beta})$ and $X_j(\Phi_{,\beta})$ ahead of Γ.

The center-of-mass theorem holds for any discontinuity solutions of the von Kármán equations. The balance laws associated with the infinitesimal symmetries X_7, X_8, X_{10} and X_{11} hold iff $\lambda = 0$, while those associated with X_{12} and X_{13} – iff $\mu = 0$. For this reason, there do not exist acceleration waves in the von Kármán plate theory satisfying all balance laws.

Obviously, when dealing with discontinuity solutions, from a physical point of view it seems reasonable that at least the balance of energy should hold in addition to the balance laws corresponding to the fundamental equations considered. Observing Table 2 it is evident that for acceleration waves propagating into an undisturbed plate the balance of energy implies also the balances of wave momentum, moment of wave momentum as well as the balance related to the scaling symmetry.

5. Examples

As an example, we consider acceleration waves such that behind and ahead of the wave front the plate motion is described by solutions of the von Kármán equations invariant under the group generated by X_3 and $X_2 + (1/c)X_4$, where c is an arbitrary constant. The most general form of such group-invariant solutions is

$$w = u(\xi) = u_0 + u_1\xi + u_2\sin\omega\xi + u_3\cos\omega\xi,$$
$$\Phi = \varphi(\xi) = \varphi_0 + \varphi_1\xi + \varphi_2\xi^2 + \varphi_3\xi^3,$$

where $\xi = x^1 - cx^3$, u_j, φ_j are arbitrary constants, and $\omega = c\sqrt{\rho/D}$. Let (u^+,φ^+)—an arbitrary solution of that kind—describe the plate motion ahead of the wave front. Then, Definition 2 and Propositions 1 to 4 imply that each acceleration wave of the type considered reads

$$u = \begin{cases} u^+ + c_1(1 - \cos\omega\xi), & \xi < 0, \\ u^+, & \xi > 0, \end{cases} \qquad \varphi = \begin{cases} \varphi^+ + c_2\xi^2, & \xi < 0, \\ \varphi^+, & \xi > 0, \end{cases} \tag{8}$$

135

where c_1 and c_2 are arbitrary constants, but $c_1 \neq 0$; the wave front in this case is the moving straight line $\Gamma : \xi = 0$. In general, however, an acceleration wave of form (8) does not satisfy the balance laws other than (4). Indeed, after a little manipulation, the jump conditions from Table 2 simplify to

$$DEh\omega^4 c_1(c_1 - 2u_3^+) = 4c_2(c_2 + 2\varphi_2^+), \tag{9}$$

$$DEh\omega^2 c_1(u_1^+ + \omega u_2^+) = 2c_2\varphi_1^+, \tag{10}$$

where u_j^+, φ_j^+ are the constants in u^+ and φ^+, respectively. The jump condition (9) is necessary and sufficient for the balances of energy, wave momentum and moment of wave momentum to hold, while the balance related to X_5 requires both (9) and (10). The second relation is treated in a different manner according to the acceleration wave under consideration. If the wave is such that $c_2 = 0$, then (10) holds for this wave only if $u_1^+ = -\omega u_2^+$. On the other hand, if we consider waves with $c_2 \neq 0$, then choosing the coefficient φ_1^+ in a suitable manner we could satisfy (10) identically (note that adding a linear function of the independent variables to Airy's stress function does not change the membrane stress tensor $N^{\alpha\beta}$). Hence, the balance associated with the scaling symmetry (X_5) holds only for acceleration waves of form (8) satisfying (9) and which are such that either $c_2 \neq 0$ or $u_1^+ = -\omega u_2^+$.

Another example, discussed in details in [2], is an axisymmetrically expanding acceleration wave composed by solutions of the von Kármán equations that are joined invariants of the rotation (X_6) and scaling (X_5) symmetries.

References

1. P.J. Olver, *Applications of Lie groups to differential equations*, 2nd ed., Graduate Texts in Mathematics **107**, Springer-Verlag, New York, 1993.

2. P. Djondjorov and V. Vassilev, Conservation laws and group-invariant solutions of the von Kármán equations, *Internat. J. Non-Linear Mech.* **31** (1996), 73–87.

3. C. Truesdell, *A first course in rational continuum mechanics*, The Johns Hopkins University, Baltimore, Maryland, 1972.

4. D. Edelin, *Nonlocal variations and invariance of fields*, Elsevier, New York, 1969.

Institute of Mechanics, Bulgarian Academy of Sciences, Acad. G. Bontchev St., Bl. 4, 1113 Sofia, Bulgaria; e-mail: padjon@bgcict.acad.bg, vassil@bgcict.acad.bg

M. EISSA

Dynamics and resonance of a nonlinear mechanical oscillator subjected to parametric and external excitation

1. Introduction

We study the dynamics and resonance of a nonlinear mechanical oscillator subject to combined parametric and external excitation. The governing equation is

$$\ddot{X} + \zeta\dot{X} + (\omega_0^2 + \beta\cos(vt))X + \alpha X^3 = \gamma\cos(\omega t)$$

where X is the displacement and the dot represents differentiation with respect to time t; here ζ is the damping coefficient, α the nonlinear spring coefficient, ω_0 the natural frequency, β the parametric amplitude, v the parametric frequency, ω the external force frequency, and γ the external force amplitude.

This and similar systems were considered in 1971 by Ness [1], and Troger and Hsu [2] in 1977, who used averaging methods to obtain steady state or periodic solutions. Plaut and Hsieh [3] in 1987 and Haquang et al. [4] in 1987 observed chaotic motion in the system. In 1990, Yagasaki et al. [5] applied averaging techniques to study system stability and showed chaotic dynamics using Melnikov techniques.

In the present work, a multiple-time scale perturbation technique [MTSPT] is used to determine the system response, examine the behavior in different resonant cases. The first section develops a solution of the governing equation through fourth order. The second section illustrates numerical results for different resonance cases. Results are compared to previously published work.

2. Mathematical analysis

The single-degree-of-freedom mechanical oscillator is governed by

$$\ddot{X} + \varepsilon\delta\dot{X} + (\omega_0^2 + \varepsilon^2\beta\cos(vT))X + \varepsilon\alpha X^3 = \varepsilon\gamma\cos(\omega T) \tag{1}$$

where ε is a small perturbation parameter. We assume a solution of (1) of the form

$$X(t,\varepsilon) = \sum_{n=0}^{\infty}\varepsilon^n X_n(T_0, T_1), \qquad T_1 = \varepsilon T_0, \tag{2}$$

$$\dot{X}(t,\varepsilon) = \sum_{n=0}^{\infty}\varepsilon^n(D_0 X_n + \varepsilon D_1 X_n + \cdots) \tag{3}$$

$$\ddot{X}(t,\varepsilon) = \sum_{n=0}^{\infty}\varepsilon^n(D_0^2 X_n + 2\varepsilon D_0 D_1 X_n + \varepsilon^2 D_1^2 X_n), \tag{4}$$

$$\text{where}\qquad D_0 = \partial/\partial T_0, D_1 = \partial/\partial T_1. \tag{5}$$

Substituting in (2)–(4) in (1) and equating powers of ε gives

$$\varepsilon^0: \quad (D_0^2 + \omega_0^2)X_0 = 0 \tag{6}$$

$$\varepsilon^1: \quad (D_0^2 + \omega_0^2)X_1 = \gamma\sin\omega T - 2D_0 D_1 X_0 - \delta(D_0 X_0) - \alpha X_0 \tag{7}$$

$$\varepsilon^2: \quad (D_0^2 + \omega_0^2)X_2 = -D_1^2 X_0 - 2D_0 D_1 X_1 - \delta(D_1 X_0 + D_0 X_2) - X_0\cos(vt) \tag{8}$$

$$\varepsilon^3: \quad (D_0^2 + \omega_0^2)X_3 = -D_1^2 X_0 - 2D_0 D_1 X_2 - \delta(D_1 X_1 + D_0 X_2) - X_1\cos(vt)$$

$$-\alpha(3X_0 X_1^2 + 3x_0^2 X_2). \tag{9}$$

The general solution of (6) (the overbar denotes a complex conjugate (c.c.)) is

$$X_0 = A_0(T_1)\, e^{j\omega_0 T_0} + \bar{A}_0(T_1)\, e^{-j\omega_0 T_0}. \tag{10}$$

Substituting (10) into (7) yields (where $f = \gamma/2j$)

$$\begin{aligned}
(D_0^2 + \omega_0^2)X_1 &= \gamma\left[e^{j\omega T_0} - e^{-j\omega T_0} - 2j\omega_0\left(D_1 A_0 e^{j\omega_0 T_0} - D_1 \bar{A}_0 e^{-j\omega_0 T_0}\right)\right] \\
&\quad - \delta\left[j\omega_0 A_0 e^{j\omega_0 T_0} - j\omega_0 A_0 e^{-j\omega_0 T_0}\right] \\
&\quad - \alpha\left[A_0^3 e^{3j\omega_0 T_0} + \bar{A}_0^3 e^{-3j\omega_0 T_0} + 3a_0^2 \bar{A}_0 e^{j\omega_0 T_0} + 3a_0 A_0^2 e^{-j\omega_0 T_0}\right]. \tag{11}
\end{aligned}$$

For a bounded solution the secular terms must cancel in (11). This gives

$$2j\omega_0 D_1 A_0 = (J\delta\omega_0 + 3\alpha C)a_0, \quad C = A_0 \bar{A}_0, \tag{12}$$

with $A_0(T_1) = \lambda_1 e^{j\rho_1 T_1}$, $\lambda_1 = \text{const}$ and $\rho_1 = J\delta/2 + 3\alpha C/2\omega_0$. Similarly,

$$\bar{A}_O = \lambda_2 e^{-j\rho_2 T_1}, \tag{13}$$

$\lambda_2 = \text{const}$, $\rho_2 = j\delta/2 + 3\alpha C_1/2\omega_0$, and $C_1 = \lambda_1 \lambda_2$ giving the first-order approximation

$$X_0 = \lambda_1 e^{j\omega_0 T_1} e^{j\rho_1 T_1} + \lambda_2 e^{-j\omega_0 T_0} e^{-j\rho_2 T_1}. \tag{14}$$

Substituting (12)–(14) into (11) and solving gives

$$X_1 = A_1(T_1)e^{j\omega_0 T_0} + (\alpha/8\omega_0^2)A_0^3 e^{3j\omega_0 T_0} + \left[F/(\omega_0^2 - \omega^2)\right]e^{j\omega T_0} + \text{c.c.} \tag{15}$$

Substituting (14)–(15) into (8) and eliminating secular terms gives

$$\begin{aligned}
A_1(T_1) &= \left[\lambda_1 \rho_1^2/2j\omega_0(2j\rho_1 - 1)\right]e^{j\rho_1 T_1} - \left[\delta\lambda_1 J\rho_1/j\omega_0(2j\rho_1 - 1)\right]e^{j\rho_1 T_1} \tag{16} \\
&= \left[\lambda_2/2j\omega_0(2j\rho_2 + 1)\right]e^{-j\rho_2 T_1} - \left[\delta\lambda_2 j\rho_2/j\omega_0(2j\rho_2 + 1)\right]e^{-j\rho_2 T_1}. \tag{17}
\end{aligned}$$

Substituting into (8) and solving gives the third approximation:

$$\begin{aligned}
X_2 &= A_2(T_1)e^{j\omega_0 T_0} + (9j\alpha/32\omega_0^3)\left[A_0^2 A_0' e^{3j\omega_0 T_0}\right] - (3j\alpha/64\omega_0^3)\left[A_0^3 e^{3j\omega_0 T_0}\right] \\
&\quad + \frac{A_0}{2j}\left[(1/(\omega_0^2 - (\omega_0 + v)^2)e^{j(\omega_0 + v)T_0} + (1/(\omega_0^2 - (\omega_0 - v)^2)e^{j(\omega_0 - v)T_0}\right] \\
&\quad + \left[jf\omega/(\omega_0^2 - \omega^2)\right]e^{j\omega_0 T_0} + \text{c.c.} + \text{n.s.t.} \tag{18}
\end{aligned}$$

where A_2 and $\bar{A}_2$ are functions of T_1 only. Similarly, the fourth approximation is

$$\begin{aligned}
X_3 &= A_3(T_1)e^{j\omega_0 T_0} + A_0^3/2j\left[e^{j(3\omega_0 + v)T_0} + e^{j(3\omega_0 - v)T_0}\right] \\
&\quad + \frac{A_0 + A_0'}{2j}\left[\frac{(j(\omega_0 + v)e^{j(\omega_0 + v)T_0}}{(\omega_0^2 - (\omega_0 + v)^2)^2} + \frac{J(\omega_0 - v)e^{j(\omega_0 - v)T_0}}{(\omega_0^2 - (\omega_0 - v)^2)^2}\right]
\end{aligned}$$

138

$$+ \quad \frac{\gamma}{\omega_0^2 - \omega^2} \left[\frac{e^{j(\omega+v)T_0}}{\omega_0^2 - (\omega + v)^2} + \frac{e^{j(\omega-v)T_0}}{\omega_0^2 - (\omega - v)^2} \right]$$

$$+ \quad \frac{\gamma}{\omega_0^2 - \omega^2} \left[A_0 \frac{e^{j(2\omega+\omega_0)T_0}}{\omega_0^2 - (2\omega + \omega_0)^2} + A_0 \frac{e^{-j(2\omega-\omega_0)T_0}}{\omega_0^2 - (2\omega - \omega_0)^2} \right]$$

$$+ \quad \frac{2f}{(\omega_0^2 - \omega^2)} \left[A_1 A_0 \frac{e^{j(\omega+2\omega_0)T_0}}{\omega_0^2 - (\omega + 2\omega_0)^2} - A_1 A_0 \frac{e^{j(2\omega_0-\omega)T_0}}{\omega_0^2 - (2\omega_0 - \omega)^2} \right]$$

$$+ \quad \frac{(2\alpha/8\omega_0^2)(\gamma}{\omega_0^2 - \omega^2} \left[A_0^4 \frac{e^{j(4\omega_0+\omega)T_0}}{\omega_0^2 - (4\omega_0 + \omega)^2} - \frac{e^{j(4\omega_0-\omega)T_0}}{\omega_0^2 - (4\omega_0 - \omega)^2} \right]$$

$$- \quad \frac{2\alpha\gamma}{8\omega_0^2(\omega_0^2 - \omega^2)} \left[\frac{e^{-j(2\omega_0-\omega)T_0}}{\omega_0^2 - (2\omega_0 - \omega)^2} - \frac{e^{-j(2\omega_0+\omega)T_0}}{\omega_0^2 - (2\omega_o + \omega)^2} \right] + \text{c.c.} + \text{n.s.t.} \quad (19)$$

where $A_3 = A_3(T_1)$ and n.s.t. contains nonsecular terms containing exponentials of $(\pm 2j\omega_0 T_0)$, $(\pm 3j\omega_0 T_0), \cdots, (\pm 7\omega_0 T_0)$. The solution gives the following resonances:

Trivial resonance: a. $\omega_0 \cong v \cong 0$; b. $\omega \cong \omega_0 \cong 0$; c. $\omega \cong v \cong 0$; d. $\omega \cong \omega_0 \cong v$.

Primary resonance: a. $\omega \cong \omega_0$; b. $\omega_0 \cong v$; c. $\omega \cong v$; d. $v \cong \omega \cong \omega_0$.

Subharmonic resonance: a. $\omega \cong 2\omega_0$; b. $\omega \cong 3\omega_0$; c. $\omega \cong 4\omega_0$; d. $\omega \cong 5\omega_0$; e. $v \cong 2\omega_0$; f. $v \cong 4\omega_0$; g. $v \cong 5\omega_0$; h. $v \cong 6\omega_0$.

Combined resonance: a. $v \cong \omega_0 - \omega$; b. $v \cong \omega_0 + \omega$; c. $v \cong \omega - \omega_0$.

3. Numerical analysis

The equations were approximated using a fourth order Runge-Kutta method. Initial conditions for all cases are $x(0) = x'(0) = 0$: the output was insensitive to the initial conditions. The figures plot amplitude x vs. time t or for phase-planes amplitude x vs. velocity x'. Fig. 1 shows typical non-resonant output.

	$\omega = 2\omega_0$	$\omega = 3\omega_0$	$\omega = 4\omega_0$	$\omega = 5\omega_0$	$v = 2\omega_0$	$v = 3\omega_0$	$v = 4\omega_0$	$v = 5\omega_0$	$v = 6\omega_0$	$v = \omega_0 - \omega$	$v = \omega - \omega_0$	$v = \omega_0 + \omega$
Amp Ratio	.8	.75	.6	.2	1.	5.2	10.2	13.2	2.5	1.2	1.2	13.2

Table 1. Subharmonic and combined resonances.

Primary resonance, external force $\omega \cong \omega_0$: Fig. 2a-c show double limit cycles or chaotic behavior.

Primary resonance, parametric freq. $v \cong \omega_0$: Fig. 3 shows slightly increased amplitude and noise.

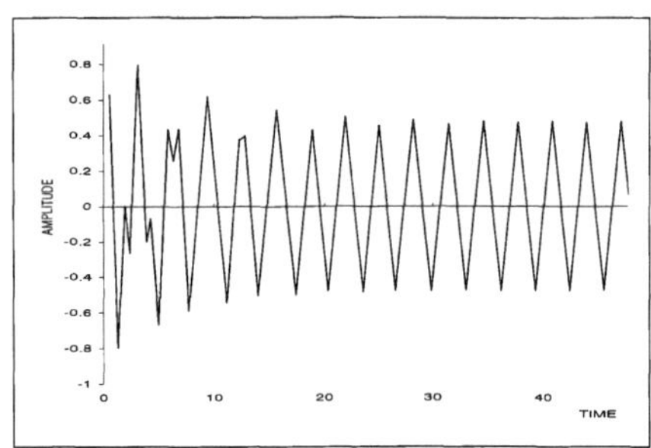

Fig. 1. Nonresonant case: $\zeta = 0.25$, $\beta = 0.05, \alpha = 0.2, \omega = 2, \omega_0 = 5, v = 3.5$, $\gamma = 10$.

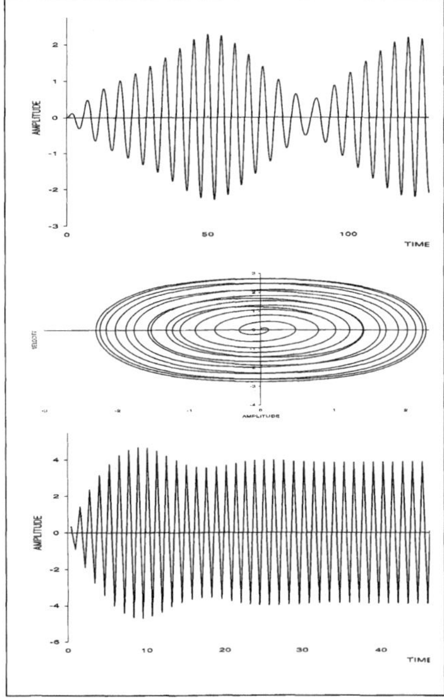

Fig. 2. Primary resonance $\omega = \omega_0$: (a) $\zeta = 0.005, \beta = 0.01, \alpha = 0.2, \omega = 1.14$, $\omega_0 = 1, v = 0.2, \gamma = 0.15$; (b) Phase plane; (c) $\zeta = 0.25, \beta = 0.05$, $\alpha = 0.2, \omega = \omega_0 = 5, v = 3.5, \gamma = 10$.

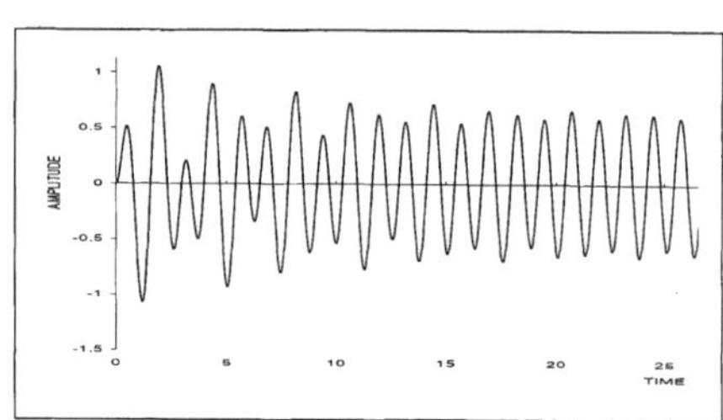

Fig. 3. Primary resonance $\omega_0 = v$: $\zeta = 0.25, \beta = 0.05, \alpha = 0.2, \omega = 2$, $\omega_0 = v = 3, \gamma = 10$.

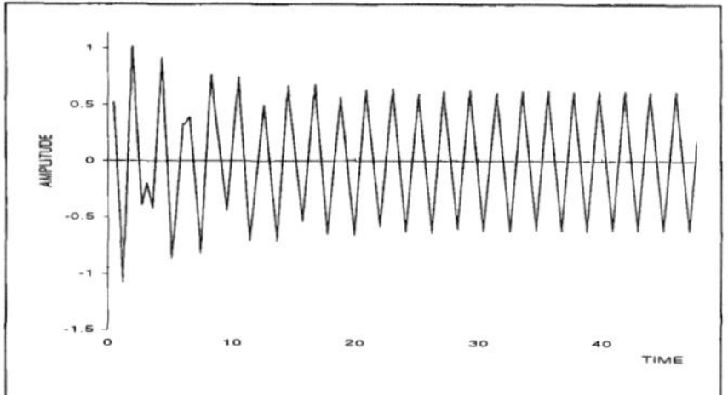

Fig. 4. Primary resonance $\omega = v$: $\zeta = 0.25, \beta = 0.05, \alpha = 0.2$, $\omega_0 = 5, \omega = v = 3, \gamma = 10$.

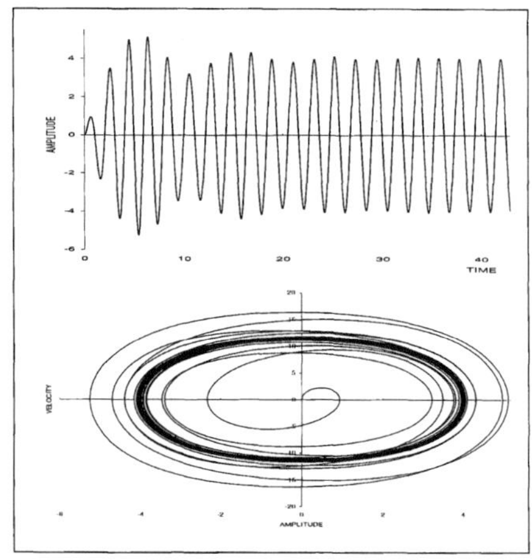

Fig. 5. Primary resonance $\omega_0 = \omega = v$: $\zeta = 0.25, \beta = 0.05, \alpha = 0.2$, $\omega = \omega_0 = v = 3, \gamma = 10$.

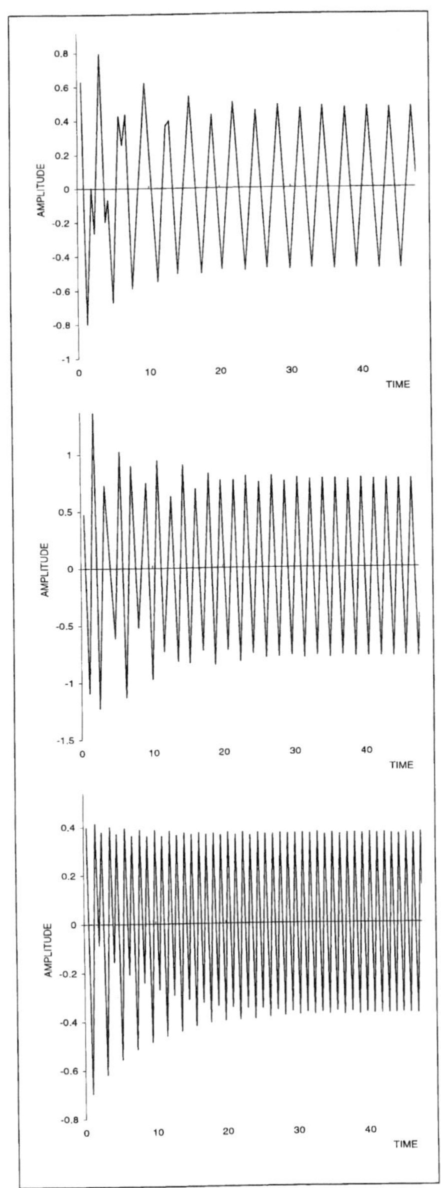

Fig. 6. Subharmonic resonance:
(a) $v = 2\omega_0, v = 5\omega_0$; (b) $v = 2\omega_0$;
(c) $2\omega_0 = \omega$.

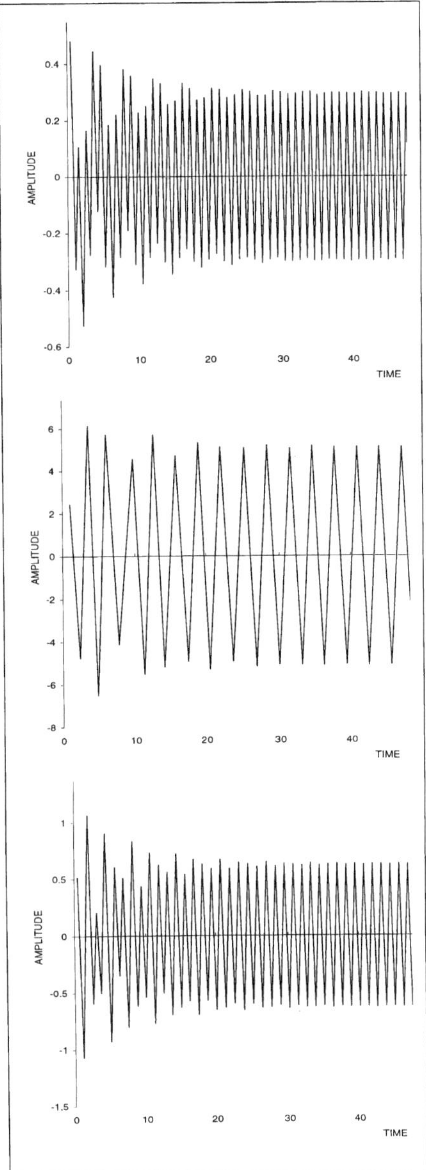

Fig. 6. Subharmonic resonance:
(d) $4\omega_0 = \omega$; (e) $4\omega_0 = v, 3\omega = v$;
(f) $v = 5\omega_0 = v = 3\omega$.

141

Primary resonance, incidence resonance $v \cong \omega$: Fig. 4 shows slightly increased amplitude, noise, and a secondary wave.

Primary resonance, incidence resonance $v \cong \omega \cong \omega_0$: Fig. 5 shows amplification of ~ 7 over the nonresonant case.

Secondary subharmonic, and combined resonances: Table 1, Fig. 6. Most interesting: $v \cong 2\omega_0$ amp ~ 0.4; $v \cong 3\omega_0$ amp ~ 2.6; $v \cong 4\omega_0$ amp ~ 5.1; $v \cong 5\omega_0$ amp ~ 6.6; $v \cong 6\omega_0$ amp ~ 1.24; $v \cong (\omega_0 + \omega)$ amp ~ 6.6.

As the external force excitation frequency varies from $(\omega/\omega_0) = 0.1$ to 6.0 the maximum amplitude occurs at the primary resonance $\omega \cong \omega_0$. The behavior is different for the parametric frequency with the peak amplitude at the sub-harmonic $v = 5\omega_0$. The only combination resonance showng significant enhancement is $v \cong (\omega + \omega_0)$.

4. Conclusions

Nonresonant results agree with [5]. No prior work was found for resonance. Analysis suggests avoiding the following resonances: $\omega \cong \omega_0$; $4\omega_0 \leq v \leq 5\omega_0$; $v = \omega + \omega_0$.

References

1. D.J. Ness, Resonance classification in a cubic system, *ASME J. Appl. Mech.* **38** (1971), 585–590.

2. H. Troger and C.S. Hsu, Response of a non-linear system under combined parametric and forcing excitation, *ASME J. Appl. Mech.* **44** (1977), 179–181.

3. R. Plaut and J.C. Hsieh, Chaos in mechanism with delays under parametric and external excitation, *J. Sound Vibr.* **114** (1987), 73–90.

4. N. Haquang, D.T. Mook and R.H. Plaut, A non-linear analysis of the interactions between parametric and external excitations, *J. Sound Vibr.* **118** (1987), 425–439.

5. K. Yagasaki, M. Sakata and K. Kimura, Dynamics of a weakly non-linear spring subjected to combined parametric and external excitation, *Trans. ASME* **57** (1990), 209–217.

Department of Engineering Mathematics and Physics, Faculty of Electronic Engineering, Menoufia University, Menouf 32952, Egypt

M. EISSA and M.M. KAMEL

On the vibration of helical springs

1. Introduction

Springs are widely used in most engineering applications and biomechanics. In fact, the spring is a nonlinear mechanical element, and an extension or compression is always accompanied by torsion and vice versa [1]. In addition, stress-strain analysis of wires, ropes, and rods can also be based on spring theory [2].

Michell [3] in 1890 showed the existence of independent radial and axial waves in spring oscillations. Yoshimura and Murata [4], Wittrick [5], and Wittrick and Pearson [6] contributed complicated equations describing spring vibrations. Castello [7,8] derived some simple spring vibration theories. Jiang et al. [9] used Laplace transforms to obtain a closed form solution for the free vibration of a helical spring and (by modifying the equations) a solid rod.

The present paper solves the coupled partial differential equations describing vibrations of a helical spring [10,11] using separation of variables. The effect of different parameters on the spring and rod response is investigated using well-known numerical techniques [12].

2. Mathematical procedure

The stress-strain relations [9] for the axial force F and twisting moment M are

$$F = k_1\varepsilon + k_2\varphi, \qquad M = K_3\varepsilon + k_4\varphi,$$

where ε and φ are the axial and rotational strain with stiffness constants k_i, $i = 1, 2, 3, 4$:

$$
\begin{aligned}
k_1 &= \pi E R^4/[r^2\Delta\{1 + \tau^2(1 + \nu)\}], & k_3 &= -\pi E R^4\tau[4\nu + (R/r)^2]/(4r\Delta), \\
k_2 &= -\pi\nu E R^4\tau/(r\Delta), & k_4 &= \pi E R^4[4(1 + \nu + \tau^2) + (\tau R/r)^2]/(4\Delta), \\
\end{aligned}
$$
$$\text{and } \Delta = [4(1 + \nu)(1 + \tau^2)^2 + (\tau R/r)^2\{(1 - \nu) + \tau^2(1 + \nu)\}]/\tau(1 + \tau^2)^{1/2},$$

where R and r are, respectively, the wire and spring radius, E is Young's modulus, ν is Poisson's ratio, and τ is the tangent of helical angle.

The equations of motion, for small deformations of the spring are

$$k_1 U_{xx} + k_2\Theta_{xx} = \gamma U_{tt}, \qquad k_3 U_{xx} + k_4\Theta_{xx} = \mu\Theta_{tt}, \tag{1,2}$$

where $U(x,t)$ and $\Theta(x,t)$ are the axial and rotational displacements at time t with γ and μ the mass and moment of inertia (about the axis) per unit length. These equations are solved for free and forced vibrations of helical springs and rods.

Free vibration of a helical spring. The spring is fixed at one end and free at the other. The initial and boundary conditions are

$$
\begin{array}{llll}
U(x,0) = u_0\frac{x}{h} & \Theta(x,0) = \theta_0\frac{x}{h} & U_t(x,0) = 0 & \Theta_t(x,0) = 0 \\
U(0,t) = 0 & \Theta(0,t) = 0 & U_x(h,t) = 0 & \Theta_x(h,t) = 0
\end{array} \tag{3-6}
$$

where x is axial distance from the fixed end $(x = 0)$, u_0 and θ_0 are the initial axial and rotational displacements at the free end, and h is the spring length.

143

Equations (1) and (2) can be rewritten as

$$A^2 U_{xx} - U_{tt} = D, \qquad B^2 \Theta_{xx} - \Theta_{tt} = D, \qquad (7,8)$$

where $A^2 = (k_1/\gamma) - (k_3/\mu)$ & $B^2 = (k_4/\mu) - (k_2/\gamma)$ and D is an arbitrary constant. We consider solutions of (7,8) of the form

$$U(x,t) = Y(x,t) + \Phi(x) \qquad \text{and} \qquad \Theta(x,t) = Z(x,t) + \Psi(x), \qquad (9,10)$$

where $Y(x,t)$ and $Z(x,t)$ are new independent variables and $\Phi(x)$ and $\Psi(x)$ are unknown functions of x to be determined. Substituting (9,10) into (7,8), and eliminating D gives:

$$\Phi(x) = (Dx^2/2A^2) + a_1 x + a_2 \qquad \text{and} \qquad \Psi(x) = (Dx^2/2B^2) + b_1 x + b_2. \quad (11,12)$$

The arbitrary constants a_j and b_j $(j=1,2)$ are determined through the choices $\Phi(0) = 0$, $\Phi'(h) = 0$, $\Psi(0) = 0$, and $\Psi'(h) = 0$, as $a_1 = -Dh/A^2$, $b_1 = -Dh/B^2$, $a_2 = b_2 = 0$. This reduces to

$$A^2 Y_{xx} = Y_{tt} \qquad \text{and} \qquad B^2 Z_{xx} = Z_{tt}, \qquad (13,14)$$

$$Y(x,0) = u_0 \frac{x}{h} + \frac{D(2hx - x^2)}{2A^2} \qquad Z(x,0) = \theta_0 \frac{x}{h} + \frac{D(2hx - x^2)}{2B^2} \qquad (15,16)$$

$$Y_t(x,0) = Z_t(x,0) = Y(0,t) = Z(0,t) = Y_x(h,t) = Z_x(h,t) = 0 \qquad (17-19)$$

Separation of variables gives the general solution of the BVP (13)–(19)

$$Y = \sum_{n=1}^{\infty} \left[\frac{8(-1)^n u_0}{(2n+1)^2 \pi^2} - \frac{16Dh^2}{(2n+1)^3 A^2 \pi^3} \right] \sin\left(\frac{(2n+1)\pi x}{2h} \right) \cos\left(\frac{A(2n+1)\pi t}{2h} \right), \quad (20)$$

$$Z = \sum_{n=1}^{\infty} \left[\frac{8(-1)^n u_0}{(2n+1)^2 \pi^2} - \frac{16Dh^2}{(2n+1)^3 B^2 \pi^3} \right] \sin\left(\frac{(2n+1)\pi x}{2h} \right) \cos\left(\frac{B(2n+1)\pi t}{2h} \right). $$

$$(21)$$

Free vibrations of a rod. For the free vibration of the rod, we have

$$k_1 = \pi E R^2, \quad k_4 = G I_p, \quad \gamma = \pi R^2 \rho, \quad \mu = G I_p \rho, \quad \text{and} \quad k_2 = k_3 = 0,$$

G is the shear modulus, I_p the polar moment of inertia, and ρ the density of the rod material. The above parameters give $a^2 = E/\rho$ and $B^2 = G/\rho$. Putting $D = 0$ in (7,8) and solving gives the form for free vibrations of the rod

$$U = \sum_{n=1}^{\infty} \frac{(8(-1)^n u_0}{(2n+1)^2 \pi^2} \sin\left(\frac{(2n+1)\pi x}{2h} \right) \cos\left(\frac{(E/\rho)^{1/2}(2n+1)\pi t}{2h} \right) \qquad (22)$$

$$\Theta = \sum_{n=1}^{\infty} \frac{8(-1)^n \theta_0}{(2n+1)^2 \pi^2} \sin\left(\frac{(2n+1)\pi x}{2h} \right) \cos\left(\frac{(G/\rho)^{1/2}(2n+1)\pi t}{2h} \right) \qquad (23)$$

Forced vibrations of a rod. The forced vibration of the rod means a moment M is applied accompanied by an axial force F and the rod released with no initial velocity. The general solution is as in (22,22) except with initial displacement and moment: $u_0 = hF/AE$, $\theta_0 = hM/GI_p$.

144

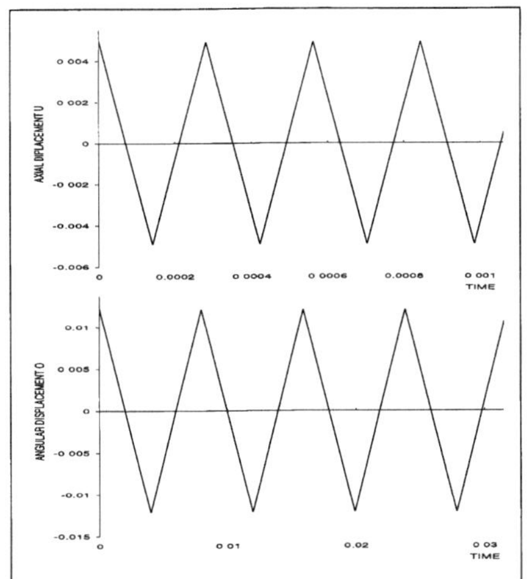

Fig. 1. Free end vibration of the spring.

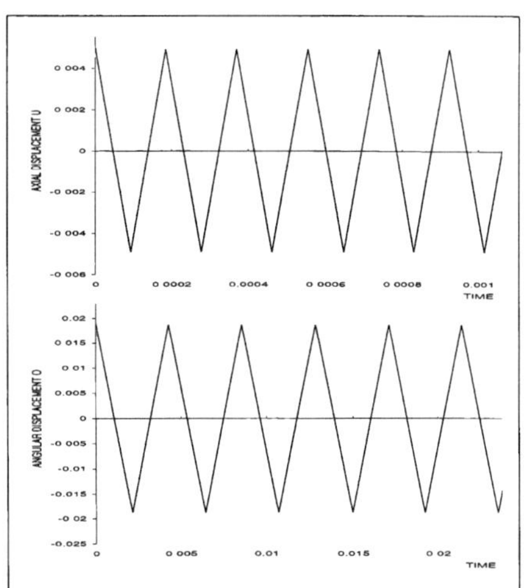

Fig. 3a. Effect of helix angle on spring: $\alpha = 2.5\,\mathrm{deg}$.

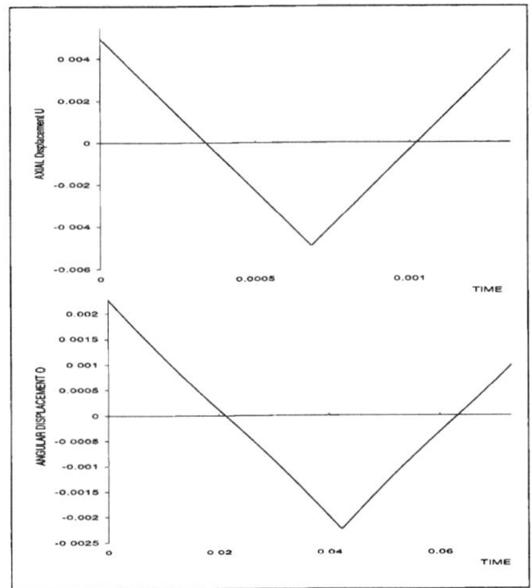

Fig. 2. Free end vibration of the rod.

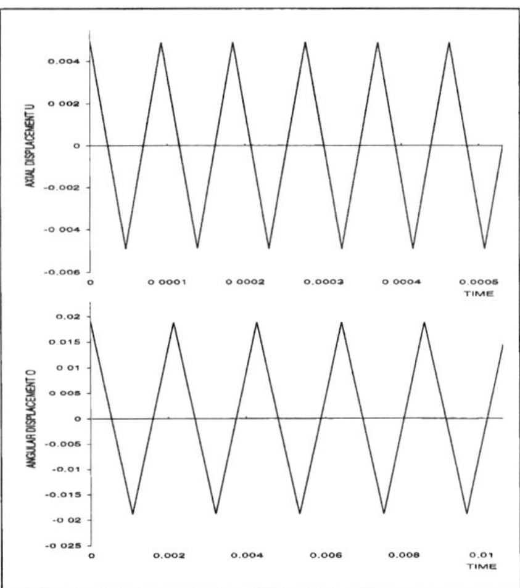

Fig. 3b. Effect of helix angle on spring: $\alpha = 20.0\,\mathrm{deg}$.

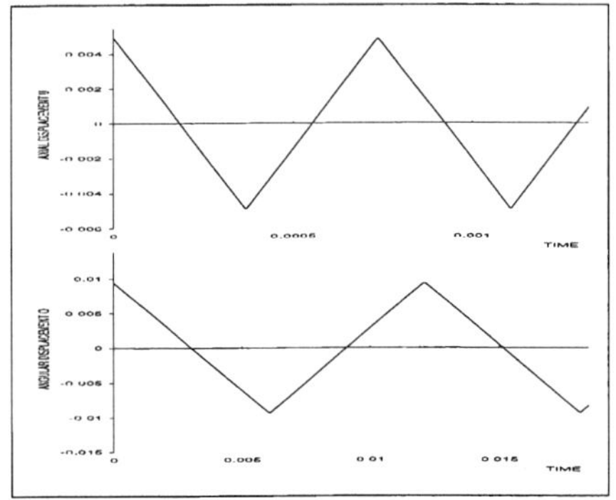

Fig. 4a. Effect of coil diameter:
$2r = 0.4$ cm and $2R = 4.0$ mm.

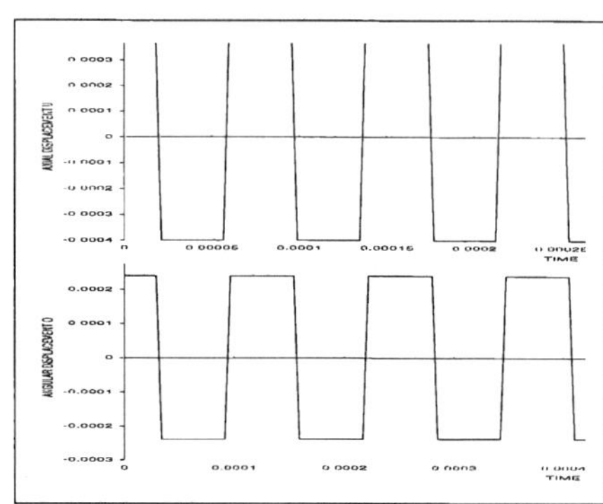

Fig. 6a. Distance ratio $x/h = 0.1$.

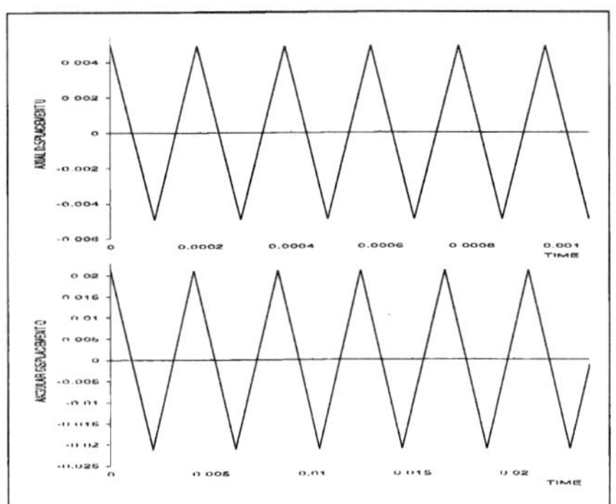

Fig. 4b. Effect of coil diameter:
$2r = 0.8$ cm and $2R = 2.0$ mm.

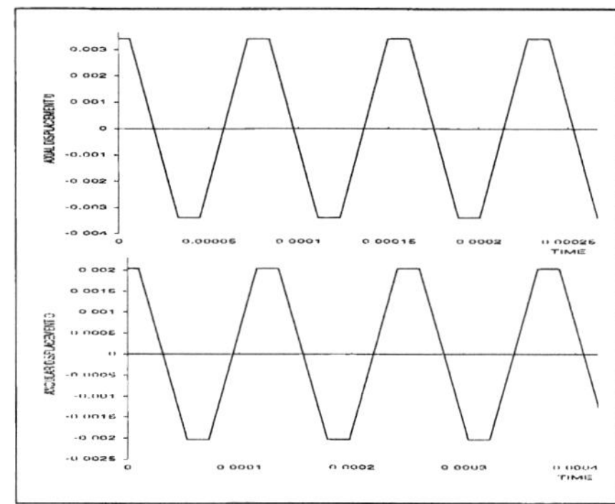

Fig. 6b. Distance ratio $x/h = 0.7$.

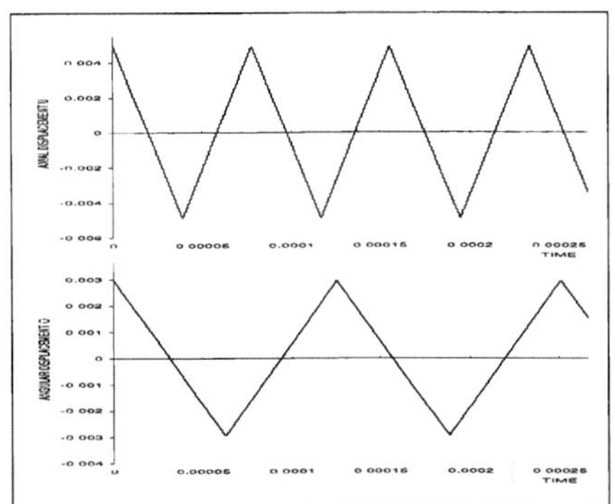

Fig. 5. Effect of spring material:
aluminum spring.

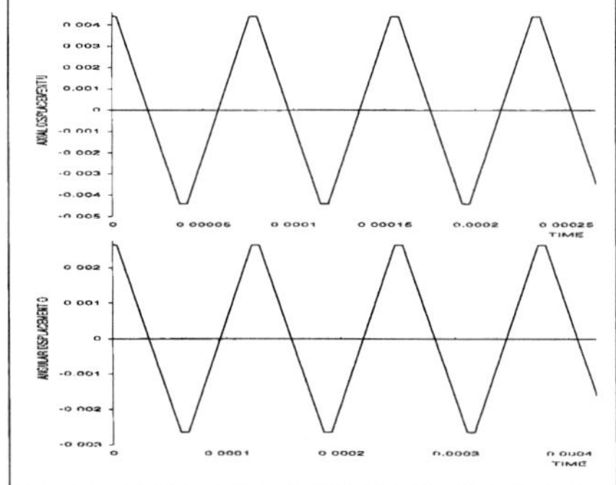

Fig. 6c. Distance ratio $x/h = 0.9$.

146

3. Results and discussion

A finite difference code generated output for the following parameters: $h = 1.0$ cm, $r = 2.0$ cm, $R = 1.0$ mm, $\alpha = 12.5^0$, $U_0 = 0.005$ m, and $\Theta_0 = 0.011$ ($\Theta_0 = 0.003$ for the rod). Figures 1 and 2 show the wave profile for axial and angular modes of vibration of the free end of the spring and rod, respectively. Both modes are triangular with constant amplitude and frequency. For the spring the two vibrational modes are dependent and for the rod independent.

Effects of the helix angle α of the spring. The helix angle α was varied from 2.5 deg to 20 deg with all other parameters constant. Figures 1 and 3 show the amplitude and frequency increase with helix angle due to decreases in both weight and moment of inertia

Effects of coil and spring radii. The values $R = 1$ mm, 2 mm, and 4 mm and $r = 20$ mm, 40 mm, and 80 mm were investigated. Figures 4a and 4b show that the period decreases as the radii are increased.

Effects of the spring material. Increasing the Young's modulus decreases the period. Figures 1 and 5 show steel and aluminum respectively.

Effects of the distance ratio (x/h). x/h was assigned values $0.1, 0.2, \cdots 1.0$ as x/h approaches the fixed end, the mode approaches a rectangular profile, as x/h approaches the free end, the mode becomes more trapezoidal eventually becoming triangular.

4. Conclusions

The free vibration of helical springs and rods with fixed-free boundary conditions are complex. The main difference between rods and springs is that for springs the two modes of vibration are dependent, but for rods they are independent. Separation of variables is a powerful method of solving boundary value problems for helical springs and rods. The only drawback is that the equations may need to be reformulated and the assumptions required for the forced helical spring.

A numerically investigation of the two vibrating modes for the helical spring and rod show that their mode shapes are trapezoidal. This means that the motion of the particles comprising the spring or rod during each mode is not continuous, but is frozen for a certain duration of time at the maximum amplitude of the oscillations. Also, it was found that the amplitude and time duration of oscillations are functions of the distance ratio x/h of the plane along the spring length or rod.

Where possible the results were compared to those published in [9]. The results agree for the free vibration of the helical spring and rod.

References

1. W. Jiang, W. Jones, K. Wu and T. Wang, Nonlinear and linear static dynamics analysis of helical springs, in *Proc. 30th AIAA/ASME/AHS/ASC structures*, 1989, 386–395.

2. J.W. Philips and G.A. Castello, Large deflection of impacted helical springs, *J. Acoustic Soc. Amer.* **51** (1972), 127–131.

3. J.H. Michell, The deflection of curves and surfaces with applications to the vibrations of a helix and a circular ring, *Mess. Math.* **19** (1890), 68–82.

4. Y. Yoshimura and Y. Murata, On the elastic waves propagated along coil springs, *Rep. Inst. Sci. and Tech.* **6** (1952), Tokyo University, 27–35.

5. W.H. Wittrick, On elastic wave propagation in helical springs, *Internat. J. Mech. Sci.* **8** (1966), 25–47.

6. W.H. Wittrick and D. Pearson, An exact solution for the vibration of helical springs using Bernoulli-Euler method, *Internat. J. Mech. Sci.* **28** (1986), 83–96.

7. G.A. Castello, Radial expansion of impacted helical springs, *J. Appl. Mech.* **42** (1975), 789–792.

8. G.A. Castello, Stresses in multilayered cables, *J. Energy Resources Technology* **105** (1983), 337–340.

9. W. Jiang, W. Jones, T. Wang and K. Wu, Free vibration of helical springs, *J. Appl. Mech.* **58** (1991), 222–228.

10. F. Verhulst, *Nonlinear differential equations and dynamical systems*, Springer-Verlag, Berlin-Heidelberg, 1990.

11. R. Nogle and E. Staff, *Fundamentals of differential equations and boundary value problems*, Addison-Wesley, New York, 1993.

12. G.F. Gerald and P.O. Wheatly, *Applied numerical analysis*, Addison-Wesley, New York, 1993.

Department of Engineering Mathematics and Physics, Faculty of Electronic Engineering, Menoufia University, Menouf 32952, Egypt

M.W. FISACKERLY and B.J. McCARTIN

A two-dimensional numerical model of chemotaxis

1. Introduction

Chemotaxis is the process by which cells aggregate under the influence of a secreted chemical substance. This cellular aggregation underlies such physiological processes as cartilage formation and the acceptance or rejection of donated organs. The Oster-Murray model [7] of chemotaxis consists of a nonlinear convection-reaction-diffusion system that cannot be solved analytically. Hence, we appeal to numerical simulations in order to gain insight into this important biological process. Mathematical analysis [3] shows that, in order for aggregation to occur, diffusion must not dominate the process. However, if diffusion is small then the problem becomes singularly perturbed and standard spatial approximations such as central differences are unsuitable [5, 8]. In the following, we will employ an alternative method of spatial approximation, exponential fitting [4], show that it permits larger spatial and temporal increments, and apply it to chemotactic cellular aggregation.

2. Biological model

The model for chemotaxis that is employed herein is due to Oster and Murray [7] and the interested reader is referred there for a full biological justification of its assumptions. The primary variables of this model are cellular concentration, u, and chemoattractant concentration, c. The cells are presumed to be randomly motile and to "convect" in response to the gradient of c. In addition, the cells are presumed to secrete the chemoattractant in a fashion described by Michaelis-Menten saturation. Finally, the chemoattractant is assumed to diffuse and also to decay according to first-order kinetics.

3. Mathematical model

The above biological model, when codified mathematically, yields the following nonlinear parabolic system for the time evolution of the concentrations, u and c [3,7]:

$$\frac{\partial u}{\partial t} = \underbrace{M \Delta u}_{\text{motility}} - \underbrace{a \nabla \cdot (u \nabla c)}_{\text{chemotaxis}}; \qquad \frac{\partial c}{\partial t} = \underbrace{D \Delta c}_{\text{diffusion}} + \underbrace{\frac{bu}{u+h}}_{\text{secretion}} - \underbrace{\mu c}_{\text{degradation}} \qquad (1)$$

Note that the cellular concentration equation is of convection-diffusion type while that for the chemoattractant concentration is of reaction-diffusion type [8].

Before proceeding, we nondimensionalize the system as follows:

$$\begin{aligned} u' &= u/h, & c' &= c\mu/b, & x' &= x\mu/\sqrt{ab}, & y' &= y\mu/\sqrt{ab}, \\ t' &= t\mu, & D' &= D\mu/ab, & M' &= M\mu/ab, \end{aligned} \qquad (2)$$

yielding the dimensionless system

$$\frac{\partial u}{\partial t} = M \Delta u - \nabla \cdot (u \nabla c); \qquad \frac{\partial c}{\partial t} = D \Delta c + \frac{u}{u+1} - c, \qquad (3)$$

The authors thank Dr. David Green, Jr. for encouragement and support of this work.

149

where we have dropped the primes for the sake of simplicity. Thus, we have reduced the number of parameters in the problem from six to two.

The key mathematical question for the above system of equations is, can small perturbations to a constant state evolve into patterns of aggregation? That is, under what conditions will constant solutions become unstable? An extension to two dimensions of the linearized stability analysis of [3, 7] produces the aggregation condition

$$M(Dq^2 + 1) \leq \bar{u}/(\bar{u} + 1)^2 \leq 1/4, \tag{4}$$

where $\bar{u}$ is the constant state and q is essentially the frequency of the perturbation.

Inequality (4) essentially says that M and D must be "small enough" for aggregation to occur. In fact, the smaller they are the larger q may be and, consequently, the more elaborate the resulting aggregation pattern. However, if they become small, then our problem becomes singularly perturbed and special care must be taken in its numerical approximation [4, 5, 8]. The remainder of this paper will be concerned with this issue.

4. Numerical model

We next extend our one-dimensional algorithm [1] to two-dimensions [2]. Consider (3) on the rectangular domain $(0, L_x) \times (0, L_y)$ with initial conditions $u = u_0(x, y)$ and $c = c_0(x, y)$ at $t = 0$ and subject to homogeneous Neumann boundary conditions $u_n = 0 = c_n$, where n is the direction normal to the boundary. Consequently, there is no flux of cells through the boundary, so that we have conservation of cells. The numerical approximations developed below will maintain a discrete version of this conservation law. A constant discretization spacing will be used throughout for each of the variables t, x, and y with Fig. 1 showing the spatial grid topology at spatial location (x_i, y_j) and temporal station t_k. Complete details are available in [2].

Integrating (3) over the two-dimensional control region, R, and applying the divergence theorem reduces the computation to the evaluation of the flux of u and c through $\Gamma := \partial R$. Rather than using central differencing to approximate these fluxes, we instead utilize exponential fitting [4, 5, 8]. In this mode of approximation, we treat the flux through each side of Γ as an ordinary differential equation with constant coefficients that can be solved analytically. This process produces exponential basis functions, hence the name of this procedure.

Finally, approximating the transient terms by forward differences in time, we arrive at the discrete equations for cell density:

$$
\begin{aligned}
u_{i,j}^{k+1} = u_{i,j}^k \\
+ \frac{\Delta t}{\Delta x} M \left\{ B(-p_{i-\frac{1}{2},j}^k)u_{i-1,j}^k + \left[B(p_{i-\frac{1}{2},j}^k) + B(-p_{i+\frac{1}{2},j}^k) \right] u_{i,j}^k + B(p_{i+\frac{1}{2},j}^k)u_{i+1,j}^k \right\} \\
+ \frac{\Delta t}{\Delta y} M \left\{ B(-p_{i,j-\frac{1}{2}}^k)u_{i,j-1}^k + \left[B(p_{i,j-\frac{1}{2}}^k) + B(-p_{i,j+\frac{1}{2}}^k) \right] u_{i,j}^k + B(p_{i,j+\frac{1}{2}}^k)u_{i,j+1}^k \right\};
\end{aligned}
$$

$$p_{i+\frac{1}{2},j}^k := (c_{i+\frac{l+1}{2},j}^k - c_{i+\frac{l-1}{2},j}^k)/M, \qquad p_{i,j+\frac{1}{2}}^k := (c_{i,j+\frac{l+1}{2}}^k - c_{i,j+\frac{l-1}{2}}^k)/M, \tag{5}$$

where $B(z) := z/(e^z - 1)$ is the Bernoulli function, and chemoattractant

$$c_{i,j}^{k+1} = (1 - \Delta t)c_{i,j}^k + \Delta t \left[\frac{u_{i,j}^k}{u_{i,j}^k + 1} \right]$$

$$+\frac{\Delta t}{\Delta x \Delta y}D\cdot\frac{z^2}{2(\cosh z - 1)}\cdot\left[\begin{array}{c}\frac{\Delta y}{\Delta x}(c^k_{i+1,j}-c^k_{i,j})+\frac{\Delta x}{\Delta y}(c^k_{i,j+1}-c^k_{i,j})\\ -\frac{\Delta y}{\Delta x}(c^k_{i,j}-c^k_{i-1,j})-\frac{\Delta x}{\Delta y}(c^k_{i,j}-c^k_{i,j-1})\end{array}\right]$$

$$z := \sqrt{\frac{(\Delta x)^2+(\Delta y)^2}{D}}. \tag{6}$$

The approximations (5) and (6) are explicit in the values for the cellular and chemoattractant concentrations. Hence, starting from the initial values of the cell and chemical concentrations, we may "march in time" to simulate their evolution. The local truncation error of these approximations is $O(\Delta t + (\Delta x)^2 + (\Delta y)^2)$, i.e., they are first-order accurate in time and second-order accurate in space.

Method of Discretization

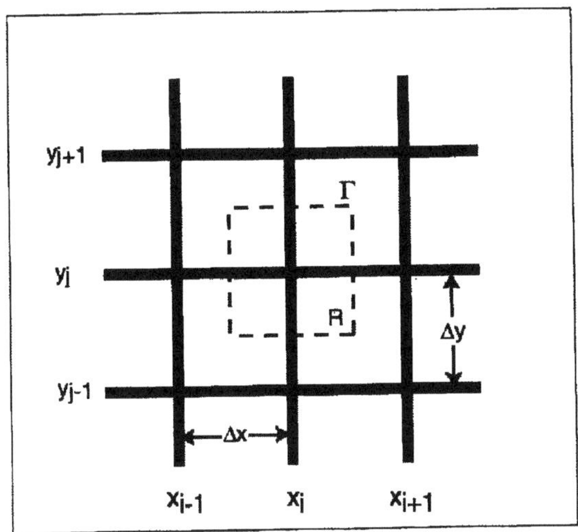

Fig. 1. Two-dimensional spatial discretization.

It must be remembered that we intend to use these approximations for small values of M and/or D. In that case, equations (3) become singularly perturbed and may possess solutions with boundary and interior layers. Since exponential fitting produces the exact solution for such steady-state, one-dimensional, linear problems, it tends to provide superior spatial approximation to that of central differencing even in multi-dimensional, nonlinear problems, as is widely appreciated in the literature [4, 5, 8].

What is not so widely appreciated is that the use of exponential fitting in space also permits the use of larger time increments. Performing a linearized stability analysis [9] on equation (5) yields the time-step restriction

$$\Delta t < \left\{M\left[\frac{B(-p_x)+B(p_x)}{(\Delta x)^2}+\frac{B(-p_y)+B(p_y)}{(\Delta y)^2}\right]\right\}^{-1};$$

$$p_x := \Delta x \cdot c_x/M, \qquad\qquad p_y := \Delta y \cdot c_y/M \tag{7}$$

for the convection-diffusion equation, while a corresponding analysis of (6) yields the time-step restriction

$$\Delta t < \left[\frac{2D}{(\Delta x)^2} \frac{z_x^2}{2(\cosh z_x - 1)} + \frac{2D}{(\Delta y)^2} \frac{z_y^2}{2(\cosh z_y - 1)} \right]^{-1} ;$$
$$z_x := \Delta x / \sqrt{D}, \qquad z_y := \Delta y / \sqrt{D} \tag{8}$$

for the reaction-diffusion equation. As long as these restriction are adhered to, the numerical error should not become unbounded as we step in time.

Inequalities (7) and (8) coincide with the numerical stability criteria for spatial central differences in the limits $M \to \infty$, $D \to \infty$:

$$\Delta t < \left\{ 2M \left[(\Delta x)^{-2} + (\Delta y)^{-2} \right] \right\}^{-1} ; \qquad \Delta t < \left\{ 2D \left[(\Delta x)^{-2} + (\Delta y)^{-2} \right] \right\}^{-1} . \tag{9}$$

However, as $M \to 0$ and $D \to 0$, inequalities (7) and (8) approach

$$\Delta t < \left[\frac{|c_x|}{\Delta x} + \frac{|c_y|}{\Delta y} \right]^{-1} ; \Delta t < \left[\frac{2D}{(\Delta x)^2} \underbrace{\frac{z_x^2}{2(\cosh z_x - 1)}}_{\longrightarrow 0} + \frac{2D}{(\Delta y)^2} \underbrace{\frac{z_y^2}{2(\cosh z_y - 1)}}_{\longrightarrow 0} \right]^{-1} , \tag{10}$$

respectively, while for central differences

$$\Delta t < 2 \left[\underbrace{|p_x|}_{\to \infty} \frac{|c_x|}{\Delta x} + \underbrace{|p_y|}_{\to \infty} \frac{|c_y|}{\Delta y} \right]^{-1} ; \qquad \Delta t < \left[\frac{2D}{(\Delta x)^2} + \frac{2D}{(\Delta y)^2} \right]^{-1} \tag{11}$$

Thus, in the small diffusion limit, exponential fitting permits a larger time-step than does central differencing while, in the large diffusion limit, the time-step restrictions coincide. Yet, as has been previously noted, it is the small diffusion limit that is of most direct interest for cellular aggregation within the Oster-Murray model. Hence, exponential fitting is to be preferred and will be employed in the ensuing examples.

5. Numerical examples

The numerical examples of [1] showed the limited types of aggregation patterns possible in one-dimension. The two-dimensional examples shown in Fig. 2 clearly demonstrates the more intricate and complex aggregation patterns possible in two-dimensions [6]. In all four of these cases, $M = .001$, $D = 1$, $L_x = 1 = L_y$, $\Delta x = .025$, $\Delta y = .025$, and $\Delta t = .000125$. The examples differ only in the initial conditions for the chemical concentration which were derived from [10]. Full numerical details appear in [2]. Despite the rich structure evident in these steady-states, they in no way exhaust the variety of two-dimensional aggregation patterns hidden within the Oster-Murray model.

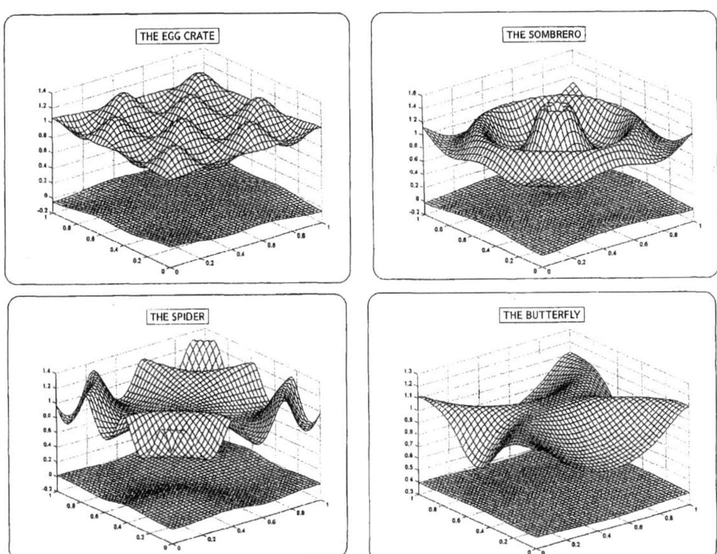

Fig. 2. Steady-state aggregation patterns.

6. Conclusion

In the preceding sections, we have developed a numerical approximation of the Oster-Murray model of chemotaxis in two-dimensions. Our scheme is comprised of exponential fitting in space and forward differencing in time. The resulting explicit algorithm permits larger spatial and temporal increments than would otherwise be afforded by central differencing in space. This increase in efficiency permits us to explore more effectively such exotic aggregation patterns as those presented above. Thus, our employment of exponential fitting may lead to a better understanding of the underlying biological process of chemotaxis.

References

1. M.W. Fisackerly and B.J. McCartin, A one-dimensional numerical model of chemotaxis, in *Proceedings of the Third Biennial Symposium on Mathematical Modeling in the Undergraduate Curriculum*, University of Wisconsin-LaCrosse, 1998.

2. M.W. Fisackerly, Chemotaxis: A Two-Dimensional Numerical Model, B.S. thesis, Kettering University, 1998.

3. P. Grindrod, J.D. Murray and S. Sinha, Steady-state spatial patterns in a cell-chemotaxis model, *IMA J. Maths. Appl. Med. Biol.*, **6** (1989), 69–79.

4. J.J.H. Miller, E. O'Riordan and G.I. Shiskin, *Fitted numerical methods for singular perturbation problems*, World Scientific, 1996.

5. K. Morton, *Numerical solution of convection-diffusion problems*, Chapman & Hall, 1996.

6. V. Nanjundiah, Chemotaxis, Signal Relaying and Aggregation Morphology, *J. Theoret. Biol.* **42** (1973), 63–105.

7. G.F. Oster and J.D. Murray, Pattern formation models and developmental constraints, *J. Experimental Zool.* **251** (1989), 186–202.

8. H. Roos, M. Stynes and L. Tobiska, *Numerical methods for singularly perturbed differential equations: convection-diffusion and flow problems*, Springer, 1996.

9. G. Sewell, *The numerical solution of ordinary and partial differential equations*, Academic Press, 1988.

10. D.H. von Seggern, *Handbook of mathematical curves and surfaces*, CRC Press, 1990.

Department of Applied Mathematics, Kettering University, Flint, Michigan, USA

I.A. GUZ and C. SOUTIS

Asymptotic analysis of fracture theory for layered composites in compression (the plane problem)

1. Introduction

There are two different mechanical approaches to describe the deformation behavior of composites. One of them is based on the piecewise-homogeneous medium model, where the behavior of each material component is described by the equations of three-dimensional solid mechanics, provided certain kinematic boundary conditions are satisfied at the interfaces (exact solution). This enables us to investigate the mechanical response of the material in the most rigorous way at microscopic level. However, due to its complexity, this method is restricted to a very small group of problems. The other approach, or the continuum theory, involves many significant simplifications. With the continuum theory, a composite is modeled as a homogeneous anisotropic material with effective elastic constants, by means of which the physical properties of the original material and the shape and volume fraction of the constituents are taken into account. The continuum theory may be applied when the scale of the investigated phenomenon (for example, the wavelength of the mode of stability loss l) is considerably larger than that of the material structure (say, the fibre diameter or layer thickness h), that is, $l \gg h$. The method based on the model of a piecewise-homogeneous medium is free from such restrictions and is, therefore, an exact one.

The wide usage of the continuum model, due to its simplicity in comparison to the piecewise-homogeneous medium model, raises the question of its accuracy and of its domain of applicability. The answer to this question may be given only by comparing the results delivered by both the continuum theory and the exact approach. The latter imposes no restrictions on the scale of the investigated phenomena and, therefore, has a much larger domain of applicability than the former. The results obtained by means of the continuum theory must follow from those derived using the model of a piecewise-homogeneous medium if the ratio of the scale of the phenomena and the scale of the structure tends to zero, that is,

$$hl^{-1} \to 0. \tag{1}$$

If this is the case, then the continuum theory may be considered as an asymptotically accurate one.

The present investigation is devoted to the substantiation of the continuum theory applied to predict the fracture of a laminated composite material with a periodic structure undergoing large deformations in uniaxial compression. Within the scope of this theory, the moment of stability loss in the structure of the material (internal instability according to Biot [1]) is being treated as the beginning of the fracture process [2]. At present, investigations of the continuum theory accuracy in relation to the model of a piecewise homogeneous medium have been carried out only for the problems of statics and wave propagation (see [3] and [4]). By contrast, problems of stability loss in composite structures undergoing finite (large) deformations have not so far been studied.

155

2. Problem statement for the piecewise-homogeneous medium model

We consider very briefly the statement of the stability problems for layered composites in compression. Suppose that (i) the composite consists of alternating layers with thicknesses $2h_r$ and $2h_m$, which are simulated by incompressible nonlinear elastic isotropic or orthotropic solids with general constitutive equations; (ii) the material is compressed in the plane of the layers by "dead" loads applied at infinity in such a manner that equal deformations along each layer are provided (uniaxial compression, plain strain problem); and (iii) the most accurate approach is used for the investigation of stability—the model of a piecewise-homogeneous medium and the three-dimensional linearized theory of deformable bodies stability (TLTDBS) [5].

Within the exact problem statement on internal instability of laminated composites undergoing large deformations we have to solve the eigenvalue problem described below. In what follows, all values corresponding to the precritical state are marked by the superscript 0 to distinguish them from perturbations of the same values. The indices r and m show that the value is relevant to fiber or matrix, respectively. The elongation factor λ_j in the direction of the OX_j-axis for the given loading is

$$u_i^0 = (\lambda_i - 1)x_i, \quad \lambda_i = \text{const}, \quad \varepsilon_{ij}^0 = (\lambda_i - 1)\delta_{ij}. \tag{2}$$

The stability equations for each of the layers are

$$\frac{\partial}{\partial x_i} t_{ij}^r = 0, \quad \frac{\partial}{\partial x_i} t_{ij}^m = 0, \tag{3}$$

where t_{ij} is the nonsymmetric stress tensor referred to the unit area of the relevant surface elements in the undeformed state (in the reference configuration). More precisely, t_{ij} is the stress component acting in the direction of OX_j on the elementary area with normal OX_i. This is the nonsymmetric Piola-Kirchhoff stress tensor, or nominal stress tensor, according to Hill [6]. Next, we also consider the symmetric stress tensor S_{ij} which reduces to σ_{ij} in the case of small precritical deformations. For incompressible solids, the stresses are connected with the displacements by the formulae (p is hydrostatic pressure)

$$t_{ij} = \kappa_{ij\alpha\beta}\frac{\partial u_\alpha}{\partial x_\beta} + \delta_{ij}q_j p, \quad q_j = \lambda_j^{-1}, \tag{4}$$

with the incompressibility condition $\lambda_1^0\lambda_2^0\lambda_3^0 = 1$. The components of the tensor κ depend on the material properties and loading (that is, on the precritical state). This loading parameter (stress or strain) is the parameter in respect to which the eigenvalue problem is to be solved. In the most general case,

$$\kappa_{ij\alpha\beta} = \lambda_j^0\lambda_\alpha^0[(\delta_{ij}\delta_{\alpha\beta}A_{\beta i} + (1 - \delta_{ij})(\delta_{i\alpha}\delta_{j\beta}\mu_{ij} + \delta_{i\beta}\delta_{j\alpha}\mu_{ji})] + \delta_{i\beta}\delta_{j\alpha}S_{\beta\beta}^0. \tag{5}$$

The particular expressions of $\kappa_{ij\alpha\beta}$ for numerous kinds of constitutive equations were obtained in [2] and [5]. For example, for general elastic solids

$$A_{\beta\alpha} = A_{\beta\alpha}(f,\varepsilon_{nl}^0), \quad \mu_{\beta\alpha} = \mu_{\beta\alpha}(f,\varepsilon_{nl}^0), \quad S_{ij}^0 = f(\varepsilon_{11}^0,\varepsilon_{12}^0,\ldots,\varepsilon_{33}^0). \tag{6}$$

For hyperelastic solids, if Φ is the strain energy density function, then

$$A_{\beta\alpha} = A_{\beta\alpha}(\Phi, \varepsilon_{nl}^0), \quad \mu_{\beta\alpha} = \mu_{\beta\alpha}(\Phi, \varepsilon_{nl}^0). \tag{7}$$

To complete the problem statement, boundary conditions should be written on each matrix-fibre interface:

$$t_{22}^r = t_{22}^m, \quad t_{12}^r = t_{12}^m, \quad u_2^r = u_2^m, \quad u_1^r = u_1^m. \tag{8}$$

It should be emphasised that the approximate approaches do not describe the phenomenon under consideration even on the qualitative level. They turned out (see [2] and [5]) to give a huge discrepancy in comparison to the exact approach and experimental data. For example, even for the simplest case of all linear elastic compressible layers of a composite undergoing small precritical deformations and considered within the framework of the geometrically linear theory, Rosen's formulae [7] for the shear deformation mode $\sigma_{cr} = G_m(1 - V_r)^{-1}$, where $V_r = h_r(h_r + h_m)^{-1}$, $V_m = h_m(h_r + h_m)^{-1}$, and for the extension mode $\sigma_{cr} = 2V_r(E_r E_m V_r V_m^{-1}/3)^{1/2}$, are not worth applying. Indeed, Rosen's model [7] may give not only critical strains that are significantly higher than those obtained by the exact approach, but sometimes, as shown in [2], it may even predict a mode of stability loss different from that obtained within the exact approach. For more complicated models (such as those considered in this investigation), one can expect even bigger discrepancies between the exact and approximate approaches.

The detailed problem statement and solution within the exact approach (as applied to axisymmetric and non-axisymmetric modes, biaxial and uniaxial compression) have been presented in [8] and [9] for the above materials, and in [2], [9]–[11] and other publications for materials with other properties.

3. Asymptotic analysis

To perform an asymptotic analysis, we apply the condition of applicability of the continuum theory (1) to all formulae and calculate the limits analytically; this yields

$$\cosh \alpha_r \eta_j^r \to 1, \quad \cosh \alpha_m \eta_j^m \to 1,$$
$$\sinh \alpha_r \eta_j^r \to \alpha_r \eta_j^r, \quad \sinh \alpha_m \eta_j^m \to \alpha_m \eta_j^m, \tag{9}$$
$$\alpha_r \to 0, \quad \alpha_m \to 0, \quad \text{where} \quad \alpha_r = \pi h_r l^{-1}, \quad \alpha_m = \pi h_m l^{-1}, \quad j = 2, 3.$$

On substitution of (9) into the characteristic determinants derived earlier (see [8] and [9]) for the four modes of stability loss, we get (in the case of the plane problem) the following results:

(i) for the 1st (shear) mode of stability loss,

$$\det \|\beta_{rs}\| \equiv [(\eta_2^r)^2 - (\eta_3^r)^2][(\eta_2^m)^2 - (\eta_3^m)^2]\frac{\pi^2}{l^2}\lambda_1^{-2}\left[\frac{h_m}{h_r}(\kappa_{1212}^m - \kappa_{1212}^r)^2\right.$$
$$\left. - \left(\frac{h_m}{h_r}\kappa_{1221}^m + \kappa_{1221}^r\right)\left(\kappa_{2112}^m + \frac{h_m}{h_r}\kappa_{2112}^r\right)\right] = 0; \tag{10}$$

(ii) for the 2nd (extension) mode of stability loss,

$$\det \|\beta_{rs}\| \equiv -[(\eta_2^r)^2 - (\eta_3^r)^2][(\eta_2^m)^2 - (\eta_3^m)^2]\eta_2^m\eta_3^m\lambda_1^2\kappa_{2112}^r\kappa_{2112}^m = 0; \qquad (11)$$

(iii) for the 3rd mode of stability loss,

$$\det \|\beta_{rs}\| \equiv -[(\eta_2^r)^2 - (\eta_3^r)^2][(\eta_2^m)^2 - (\eta_3^m)^2]$$

$$\times \eta_2^r\eta_3^r\eta_2^m\eta_3^m\frac{\pi^2}{l^2}\lambda_1^2\kappa_{2112}^r\kappa_{2112}^m\left(\frac{h_m}{h_r} + 1\right) = 0; \qquad (12)$$

(iv) for the 4th mode of stability loss,

$$\det \|\beta_{rs}\| \equiv -[(\eta_2^r)^2 - (\eta_3^r)^2][(\eta_2^m)^2 - (\eta_3^m)^2]\eta_2^r\eta_3^r\lambda_1^2\kappa_{2112}^r\kappa_{2112}^m = 0. \qquad (13)$$

It has been shown that for the above models of layers, the roots of the characteristic equations (that is, the parameters $(\eta_j^r)^2$ and $(\eta_j^m)^2$ that depend on the components of the tensor κ and, therefore, on the properties of the layers and loads) are always real and positive. This means that

$$\begin{aligned}
\mathrm{Re}(\eta_2^r)^2 > 0, \quad \mathrm{Im}(\eta_2^r)^2 = 0, \quad \mathrm{Re}(\eta_2^m)^2 > 0, \quad \mathrm{Im}(\eta_2^m)^2 = 0, \\
\mathrm{Re}(\eta_3^r)^2 > 0, \quad \mathrm{Im}(\eta_3^r)^2 = 0, \quad \mathrm{Re}(\eta_3^m)^2 > 0, \quad \mathrm{Im}(\eta_3^m)^2 = 0.
\end{aligned} \qquad (14)$$

Also, the solutions corresponding to a particular phenomenon of internal instability must depend on the properties of both alternating layers, in other words, on the ratio h_r/h_m.

Equations (11), (12), and (13), which correspond to the 2nd, 3rd, and 4th modes of stability loss, respectively, do not have such solutions and, therefore, do not describe the internal instability in the long-wave approximation. This also means that the modes of stability loss, with the exception of the 1st (shear) mode, cannot be described by the continuum theory.

For further analysis, the components of κ can be expressed as [5]

$$\begin{aligned}
\kappa_{2112}^r = \lambda_1^2\mu_{12}^r, \quad \kappa_{1221}^r = \lambda_1^{-2}\mu_{12}^r + (S_{11}^0)^r, \quad \kappa_{1212}^r = \mu_{12}^r, \\
\kappa_{2112}^m = \lambda_1^2\mu_{12}^m, \quad \kappa_{1221}^m = \lambda_1^{-2}\mu_{12}^m + (S_{11}^0)^m, \quad \kappa_{1212}^m = \mu_{12}^m.
\end{aligned} \qquad (15)$$

On substitution of (15) into the characteristic equation (10) for the 1st (shear) mode, we find that

$$\frac{h_m}{h_r}(\mu_{12}^r - \mu_{12}^m)^2$$

$$- \left[\mu_{12}^r + \frac{h_m}{h_r}\mu_{12}^m + \lambda_1^2\left((S_{11}^0)^r + \frac{h_m}{h_r}(S_{11}^0)^m\right)\right]\left(\mu_{12}^m + \frac{h_m}{h_r}\mu_{12}^r\right) = 0. \qquad (16)$$

Introducing the effective values of the stresses $\langle S^0_{11} \rangle$ and parameter $\langle \mu_{12} \rangle$ by

$$\langle S^0_{11} \rangle = (S^0_{11})^r V^*_r + (S^0_{11})^m V^*_m, \quad \langle \mu_{12} \rangle = \mu^r_{12} \mu^m_{12} (\mu^r_{12} V^*_m + \mu^m_{12} V^*_r)^{-1} \tag{17}$$

and taking into account that the volume fractions $(V_r,\ V_m)$ and $(V^*_r,\ V^*_m)$ of the components in the deformed and undeformed states, respectively, are

$$V^*_r = \frac{\lambda^r_2 h_r}{\lambda^r_2 h_r + \lambda^m_2 h_m} = \frac{h_r}{h_r + h_m} = V_r,$$

$$V^*_m = \frac{\lambda^m_2 h_m}{\lambda^r_2 h_r + \lambda^m_2 h_m} = \frac{h_m}{h_r + h_m} = V_m, \tag{18}$$

we now find from (16) that

$$(\Pi^-_1)_T \equiv -\langle S^0_{11} \rangle = \lambda^{-2}_1 \langle \mu_{12} \rangle. \tag{19}$$

This coincides with the results derived by means of the continuum theory of nonlinear laminated composites undergoing large deformations, where $(\Pi^-_1)_T$ denotes the theoretical strength limit.

Thus, it is rigorously proved for laminated nonlinear elastic composites undergoing large deformations in uniaxial compression that the results of the continuum theory follow as a long-wave approximation from those for the first mode of stability loss obtained using the model of a piecewise-homogeneous medium. This establishes the asymptotic accuracy of the continuum theory for such composites.

4. Results for particular models

We can now calculate the accuracy of the continuum theory. Thus, the values of the critical strains (or other critical parameters, for example, critical stresses) calculated by means of the most accurate approach (the piecewise-homogeneous medium model and the equations of TLTDBS) may be found for various types of layered media as described in [2] and [8]–[11]. Using the above formulae, we easily obtain the values of the critical strains for the first mode of stability loss under the condition of applicability of the continuum theory. Comparing these values of the critical parameters, we can estimate the asymptotic accuracy of the continuum theory of fracture for layered media in compression and draw proper conclusions about the viability of using this theory.

It should be noted that for the continuum theory, formula (19) gives the theoretical strength limit for nonlinear elastic incompressible composites as a function of the effective values of the parameter $\langle \mu_{12} \rangle$, which is related to the material/ply properties by (17). This theoretical strength limit is written for the general form of the constitutive equations for the layers. If we need a concrete expression for a particular kind of layer properties, then this can be determined by using the formulae in [5], which give μ^r_{12} and μ^m_{12} for various particular constitutive equations. The monograph [2] also contains expressions for the theoretical strength limits for other layered models. Thus, for example, for the simplest case of all linear elastic compressible layers of a composite undergoing small precritical deformations within the geometrically linear theory, we have $(\Pi^-_1)_T \equiv -\langle \sigma^0_{11} \rangle = \langle G_{12} \rangle$, where $\langle G_{12} \rangle$ is the effective shear modulus of the laminated composite.

References

1. M.A. Biot, *Mechanics of incremental deformations*, Wiley, New York, 1965.

2. A.N. Guz, *Mechanics of fracture of composite materials in compression*, Naukova Dumka, Kiev, 1990. (Russian)

3. L.M. Brekhovskikh, *Waves in laminated media*, Nauka, Moscow, 1979. (Russian)

4. S.M. Rytov, Acoustic properties of small-scale-laminated medium, *Acoustic J.* **2** (1956), 71–83. (Russian)

5. A.N. Guz, *Fundamentals of three-dimensional theory of deformable bodies stability*, Vyshcha Shkola, Kiev, 1986. (Russian)

6. R. Hill, A general theory of uniqueness and stability of elastic-plastic solids, *J. Mech. Phys. Solids* **6** (1958), 236–249.

7. B.W. Rosen, Mechanics of composite strengthening, in *Fiber Composite Materials*, Ch. 3, ASM, Metals Park, 1965.

8. I.A. Guz, Spatial nonaxisymmetric problems of the theory of stability of laminar highly elastic composite materials, *Soviet Appl. Mech.* **25** (1989), 1080–1085.

9. A.N. Guz, Ed., *Mechanics of composite materials and constructive elements*, vol. 1, Naukova Dumka, Kiev, 1982. (Russian)

10. I.A. Guz, Three-dimensional nonaxisymmetric problems of the theory of stability of composite materials with metallic matrix, *Soviet Appl. Mech.* **25** (1989), 1196–1201.

11. I.A. Guz, Internal instability of laminated composites with a metal matrix, *Mech. Composite Materials* **26** (1990), 762–767.

Timoshenko Institute of Mechanics, Nesterov Str. 3, 252057 Kiev, Ukraine

Department of Aeronautics, Imperial College, Prince Consort Road, London, UK

S.M. HAIDAR

Existence and regularity of weak solutions to the displacement boundary value problem of nonlinear elastostatics

1. Introduction

Consider a nonhomogeneous isotropic hyperelastic material body which, in its reference configuration, occupies the open connected bounded subset Ω of R^3. The material particle at $x \in \Omega$ is displaced by the deformation $g : \Omega \to R^3$ to $g(x)$. In the absence of external forces, the total energy associated with a deformation $g(\cdot)$ is given by

$$g \to J(g) := \int_\Omega W(x, \nabla g(x)) \, dx \tag{1}$$

where $W : \Omega \times M_+^{3\times3} \to [0, \infty]$ denotes the corresponding stored energy function and $M_+^{3\times3}$ denotes the space of all 3×3 real matrices with positive determinants. The Euler-Lagrange equations associated with Eq. (1) are given by

$$\mathrm{div}\left[\frac{\partial W}{\partial F}(x, F)\right] = 0, \quad F := \nabla \in g(x). \tag{2}$$

For the case of homogeneous material, Ball and Murat [3] discuss the questions of existence and regularity of solutions to the problem

$$\mathrm{Inf}\{J(g) : g \in C_\lambda\}, \quad \lambda > 0 \qquad \text{where} \tag{3}$$
$$C_\lambda := \{g \in W^{1,1}(\Omega; R^3) : g(x) = \lambda x \text{ on } \partial\Omega, \det \nabla g(\cdot) > 0 \text{ a.e.}, J(g) < \infty\}. \tag{4}$$

Like finite-dimensional minimization problems, the framework for studying the typical problem (3), (4) has two vital components: 1) compactness, and 2) lower semicontinuity (in some topological sense). The nonlinear constraint $\det F > 0$ makes convexity of $W(x, \cdot)$ totally unrealistic in more than one space dimension as, among many other reasons, it contradicts the principle of frame indifference. The absence of convexity makes lower semicontinuity of $J(\cdot)$ a serious problem and obstacle. Since 1950, several attempts to treat this issue have been made, but it is yet to be universally settled (refer to [1], [7], and [12]).

In Section 2 we present an existence result that extends to the nonhomogeneous case that of Ball [2] and, like the results of Ball, is partly motivated by the work of Ogden [8]. In addition, our discussion of the regularity of weak solutions of Eq. (3) represents a novel approach that does not require the consideration of the delicate phase plane analysis. In Section 3 we describe the physical significance of the homogeneity property used for the discussion of regularity in Section 2.

This work was partly supported by a grant from the R & D Office of Grand Valley State University.

2. Existence and regularity of solutions

Let $\Omega = B(0,1) := \{x \in R^3 : |x| < 1\}$. The class of admissible deformations for problem (3), (4) that we consider corresponds to radial displacements of the form: $g(x) = \frac{r(R)}{R}x$, where $R := |x|$.

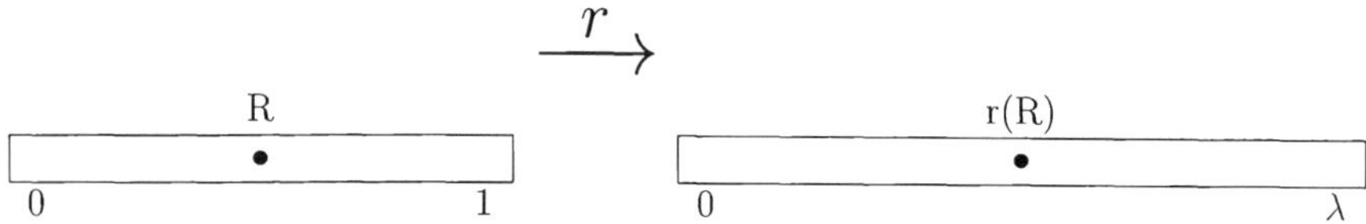

In this case, the stored energy function W can be expressed as a symmetric function $\phi(R,.,.,.)$ of the principal stretches of $\nabla g(x)$. In particular, $W(x, \nabla \in g(x)) = \phi(R, \nu_1, \nu_2, \nu_3)$ where the singular values of $\nabla g(x)$ are $\nu_1 = r'$, $\nu_2 = \nu_3 = r/R$ (refer to [9]). Problem (3), (4) becomes

$$\operatorname{Inf}\{I(r) : r \in \mathcal{U}_{ad}\} \tag{5}$$

$$\mathcal{U}_{ad} := \{r \in W^{1,1}(0,1) : \dot{r}(R) > 0 \text{ a.e.}, \ r(0) = 0, \ r(1) = \lambda, \ I(r) < \infty\} \tag{6}$$

where $J(g) = 4\pi I(r)$ with $I(r) := \int_0^1 R^2 \phi(R; r', r/R, r/R) \, dR$. The Euler-Lagrange equation corresponding to the functional I is

$$\frac{d}{dR} R^2 \phi_{,1} = 2R \phi_{,2}. \tag{7}$$

Following [2, Theorem 4.2], we say that $g \in W^{1,1}(B; R^3)$ is a weak equilibrium solution of (2) if and only if $r'(R) > 0$ a.e. $R \in (0,1)$, $R^2 \phi_{,1}$ and $R^2 \phi_{,2} \in L^1(0,1)$, and $R^2 \phi_{,1} = 2 \int_1^R \rho \phi_{,2}(\rho) \, d\rho + \text{const.}$, where

$$\phi_{,i}(R) := \frac{\partial \phi}{\partial \nu_i}\left(R; r'(R), \frac{r(R)}{R}, \frac{r(R)}{R}\right).$$

To obtain good correlation between theory and experiment, Ogden and J. Ericksen et al. found that expansions of $W(\cdot)$ in symmetric forms such as $\sum \alpha_n \psi(\nu_n)$ do a good job for homogeneous incompressible material. To account for volume change effects one may then consider the form $\sum \alpha_n \psi(\nu_n) + h(\nu_1 \nu_2 \nu_3)$. For the nonhomogeneous case, we present the following result in which $W(x, \cdot)$ admits such expansions.

Theorem 2.1. *Let the functions ϕ, ψ, and h be given such that*

(H0) $\phi(R; \nu_1, \nu_2, \nu_3) \geq \sum_{i=1}^{3} \psi(R, \nu_i) + h(R, \nu_1 \nu_2 \nu_3)$ *for* $R \in (0,1)$, $\quad \nu_i \in (0, +\infty)$;

(H1) $\phi_{,11}(R; \nu_1, \nu_2, \nu_3) > 0$;

(H2) $h : (0,1) \times (0, +\infty) \to R$ *is* C^1 *and strictly convex in its second argument;*

(H3) $h(\cdot, \sigma)$ *is nondecreasing;*

(H4) $\lim_{\sigma \to 0+} h(R, \sigma) = \lim_{\sigma \to \infty} h(R, \sigma)/\sigma^{1+\epsilon} = +\infty$, $\quad \epsilon > 0$;

162

(H5) $\psi : (0,1) \times (0,+\infty) \to [0,+\infty)$ *is C^1 and convex in its second argument with* $\lim_{\nu\to\infty} \psi(R,\nu) = +\infty$.

Then I attains an absolute minimum on $\mathcal{U}_{ad}$.

Proof. Let $\rho = R^3$ and $u(\rho) = r^3(R)$. Then $\dot{u} = du/d\rho = r'(r/R)^2 = \det \nabla g(x)$. In terms of this change of variables, the energy functional takes the form

$$I(u) = \int_0^1 f(\rho, u, \dot{u}) \, d\rho \qquad \text{where}$$

$$f(\rho, u, p) = \phi\left(\rho^{1/3}(rho/u)^{2/3}p, (u/\rho)^{1/3}, (u/\rho)^{1/3}\right) \geq \sum_{i=1}^3 \psi(\rho^{1/3}, \nu_i) + h(\rho^{1/3}, \dot{u}). \tag{8}$$

Put $L := \inf\{I(u) : u \in \mathcal{U}_{ad}\}$ and let $\{u_n\}_{n\in N}$ be a minimizing sequence for I in $\mathcal{U}_{ad}$. For each integer $m > 1$, Eq. (8) and (H3) imply

$$\int_{1/m}^1 h\left((1/m)^{1/3}, \dot{u}_n(\rho)\right) \, d\rho \leq \int_0^1 h(\rho^{1/3}, \dot{u}_n) \, d\rho \leq I(u_n) \leq D,$$

where D is a constant independent of n. Hence, by (H4), the family $\{\dot{u}_n\}_{n\in N}$ is sequentially weakly relatively compact in $L^1(1/m,1)$. Define $\{u_{1,n}\}$ by $u_{1,n} := u_n$ for all n. By De la Valleé Poussin's criterion we can choose the following sequences inductively: $\{u_{m,n}\}_{n=1}^\infty$ to be a subsequence of $\{u_{m-1,n}\}_{n=1}^\infty$ with the property that $\dot{u}_{m,n}$ converging weakly in $L^1(1/m,1)$ to, say, Z_m.

Let Z be the function: $Z : (1/k,1) \to [0,\lambda]$ given by $Z(\rho) := Z_k(\rho)$ where k is chosen so that $\rho \in (1/k,1)$. Following [11], put

$$u(\rho) := \lambda - \int_\rho^1 Z(t) \, dt \text{ and } u_m := u_{m,m} \tag{9}$$

for all $m > 1$. For each $\delta \in (0,1)$, choose m large enough so that $\delta > 1/m$. It is not difficult to show that $u_m \rightharpoonup u$ in $L^1(\delta,1)$ and thereby conclude that

$$u_m \rightharpoonup u \text{ in } W^{1,1}(\delta,1). \tag{10}$$

To establish the (sequentially) weak lower semicontinuity of $I(\cdot)$, we extend the definition of f to $(0,1) \times R \times R$ by setting

$$f(\rho, u, p) = +\infty \text{ if } u \text{ or } p \leq 0. \tag{11}$$

This makes $f(\rho, \cdot, \cdot)$ a continuous function from $R \times R$ to $\bar{R}$ for each $\rho \in (0,1)$. Since $f(\rho, u, \cdot)$ is also convex we may apply a result from [4] to conclude that

$$\int_\delta^1 f(\rho, u, \dot{u}) \, d\rho \leq \liminf_{m\to+\infty} \int_\delta^1 f(\rho, u_m, \dot{u}_m) \, d\rho.$$

From this we see that the nonnegative monotone functions $\delta \mapsto \int_\delta^1 f(\rho, u, \dot{u}) \, d\rho$ have a least upper bound L. The monotone convergence theorem and the definition of L now imply $\int_0^1 f(\rho, u, \dot{u}) \, d\rho = L$.

Hypothesis (H4), (9) and (11) imply that $\dot{u}(\rho) > 0$ for a.e. $\rho \in (0,1)$ and $\dot{u} \in L^1(\delta,1)$ for each $\delta \in (0,1)$. Moreover $\int_\delta^1 \dot{u}(t) \, dt \leq \lambda$. The monotone convergence theorem implies that $u \in W^{1,1}(0,1)$ and Eq. (10) gives $u_m \to u$ uniformly. Thus

$u(0) \geq 0$. By (H4), $u \in W^{1,1+\epsilon}(0,1)$ and the Sobolev Imbedding Theorem yields that $u \in C_b(0,1)$, the space of bounded continuous functions on $(0,1)$ and consequently $u(0) = 0$. It now follows that $u \in \mathcal{U}_{ad}$.

It is natural to ask how regular can this minimizer be? Could it be in a smaller space? Is it Lipschitz or does it satisfy Eq. (7)? Within this context, very little is known about this very complex problem. Our new approach to this problem is based on concepts from the Hamilton-Jacobi theory, field theory of the Calculus of Variations, and the theory of groups of continuous transformations as established in [5].

Artificial example. Consider

$$A(r) := \int_0^1 (r^2 - R)^2 (\dot{r})^6 \, dR$$

with $r(0) = 0$, $r(1) = 1$. Clearly, $r^*(R) := R^{1/2}$ is the absolute minimizer of $A(\cdot)$ in $W^{1,t}(0,1)$ for $t < 2$. If we restrict competing curves to $W^{1,t}(0,1)$ for $t \geq 2$, then r^* is no longer admissible. Let $V(R_0, r_0)$ denote the value of $\mathrm{Inf}\{A(r)\}$ over all $r \in W^{1,1}(0,1)$ satisfying $r(0) = 0$, $r(R_0) = r_0$. The value function $V(\cdot, \cdot)$ provides a lower bound for $A(r)$ at each point (R_0, r_0) and is given by the Hamilton-Jacobi equation

$$V_R(R,r) + f^*(R, r, V_r(R,r)) = 0 \qquad V(0,0) = 0 \tag{12}$$

where $f^*(R,r,z) = \sup_p[pZ - f(R,r,p)]$ with $f(R,r,p) = $ integrand of $A(r)$. Moreover, the solution of the o.d.e. $\dot{r} = \hat{p}(R,r)$ is optimal, where $\hat{p}(R,r)$ is the value of the function $p(\cdot, \cdot)$ at which $f^*(R,r,z)$ is attained. (Refer to [10], p. 18.) For the functional $A(r)$, it is easy to verify that $\hat{p}(R,r) = 1/6^{1/5} \cdot (r^2 - R)^{-2/5} \cdot V_r^{1/5}$ and that Eq. (12) becomes:

$$V_R = \alpha R^{-2/5} \left[\left(\frac{r}{\sqrt{R}} \right)^2 - 1 \right]^{-2/5}, \qquad V_r^{6/5} = 0, \quad V(0,0) = 0 \tag{13}$$

where $\alpha = 5/6^{6/5}$.

The functional $A(r)$ is invariant under the action of the transformations: $\bar{R} = \epsilon R$, $\bar{r} = \epsilon^{1/2} r$, $\epsilon > 0$. Accordingly, we construct (similarity) solutions of Eq. (13) of the form $V(R,r) = h \circ \xi(R,r)$ where $\xi(R,r) = rR^{-1/2}$. Using these solutions, one may easily verify that $\hat{p}(R,r) = 3/5 r R^{-1}$ and that $\hat{r}(R) = R^{3/5}$ is optimal. Moreover, the solution $\hat{r} \in W^{1,t}(0,1)$ for $t \in [2, 5/2)$.

Of course, the integrand of $A(r)$ does not serve as an example for the stored energy function ϕ. However, the work shown above can be generalized (see [5]) to lead in a very natural way to energy functions which are invariant under the action of the group of continuous transformations: $\bar{R} = \epsilon R$, $\bar{r} = \epsilon^\gamma r$ with

$$f(\epsilon R, \epsilon^\gamma r, \epsilon^{\gamma-1} r') = \epsilon^{-1} f(R, r, r'), \epsilon > 0, \gamma \in (0,1). \tag{14}$$

Setting $\epsilon = \frac{1}{R}$ in Eq. (14) yields

$$f(R, r, r') = R^{-1} f(1, r R^{-\gamma}, r' R^{1-\gamma}) := R^{-1} e(P, X) \tag{15}$$

where $X(R,r) = rR^{-\gamma}$ and $P(R,r') = r'R^{1-\gamma}$. We call the attention of the reader to the fact that $e(P,X)$ of Eq. (15) is the restriction to the plane $X_1 = X_2 = X$ of the symmetric quantity $E(P, X_1, X_2)$ associated with $\phi(R; \nu_1, \nu_2, \nu_3)$. For some $\lambda > 0$, observe that an $r(\cdot)$ of the form $r^*(R) = \lambda R^\gamma$ must be an absolute minimizer for $I(r)$ in $W^{1,t}(0, t)$ for $t < (1 - \gamma)^{-1} =: t_0$ because along curves of the form $r^*(R) = \lambda R^\gamma$ and in light of Eq. (15) one has $I(r) = \int_0^1 R^{-1} e(\lambda\gamma, \lambda)\, dR$ which yields the value zero only if there is a zero of $e(\cdot, \cdot)$ of the form $(\lambda\gamma, \lambda)$ in the PX-plane. By restricting the competing curves to $\mathcal{U}_{ad} \cap W^{1,t}(0, 1)$ for $t \geq t_0$, the curve $r^*(\cdot)$ is no longer admissible. In this case, it is possible to produce a more regular solution curve of the form $\hat{r}(R) = \lambda R^\theta$ for $\theta < \lambda$ but we shall not address it here. Singular equilibria of this form corresponding to strongly elliptic isotropic energies are established in [6].

3. Physical interpretation of the homogeneity property

The constitutive property Eq. (14) provides for a plausible model of the onset of fracture in nonhomogeneous elastic materials. Take $e(P, X) = (P - X)^2 (PX)^{-1}$ and assume that the material body $B(0, 1)$ undergoes a rigid deformation. The stored energy function ϕ now behaves like $R^{-2-\gamma}$ and, consequently, $\phi \to \infty$ as $R \to 0^+$. Hence, the material under consideration must be denser (coarser) near the center of the ball than it is elsewhere (e.g., near $R = 1$). In general, when the principal stretches $\nu_1 = \nu_2 = \nu_3 = \nu = $ const, ϕ as a function of R behaves like $R^{-3} q(R)$, where $q(R) := e(\nu R^{1-\gamma}, \nu R^{1-\gamma})$. By the natural growth condition (H4) of Theorem 2.1, $\phi \to +\infty$ as $R \to 0^+$. To account for the possibility of the material to fracture we propose the following argument: Let us expose the unit bar to expansion or compression and think of the intensity of this action as propagating through material layers of different conductivities.

Case 1. Expansion (i.e., $\epsilon > 1$). In this situation, one may arrange the layers as follows:

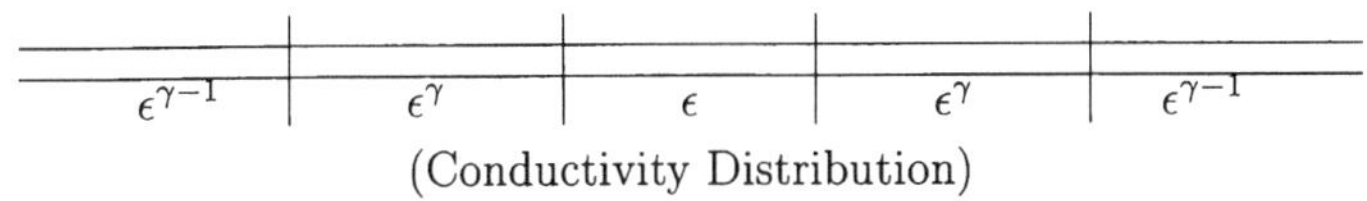

(Conductivity Distribution)

Since $0 < \gamma < 1$ the conductivity near the center of the bar is higher than it is elsewhere. This reflects weaker resistance near the center than elsewhere in the bar.

Case 2. Compression (i.e., $\epsilon < 1$). In this situation, one may arrange the layers as follows:

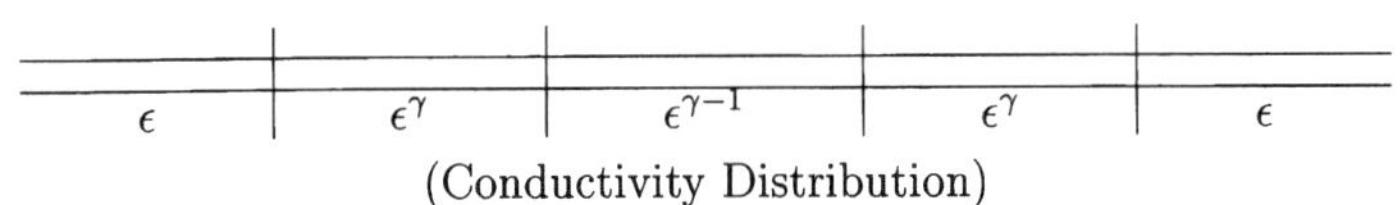

(Conductivity Distribution)

This also reflects weaker resistance near the center of the bar leading to the same outcome as in Case 1.

The above arguments may easily be adapted for the homogeneous case as well as for the incompressible case.

References

1. J. Ball, Constitutive inequalities and existence theorems in nonlinear elasticity, in *Nonlinear analysis and mechanics: Heriot-Watt symposium*, vol. 1, R.J. Knops, Ed., Pitman, London, 1977.

2. J. Ball, Discontinuous equilibrium solutions and cavitation in nonlinear elasticity, *Phil. Trans. Roy. Soc. Lond.* **A306** (1982), 557–611.

3. J. Ball and F. Murat, $W^{1,P}$-quasiconvexity and variational problems for multiple integrals, *J. Functional Anal.* **58** (1984), 225–253.

4. J. Ball, J. Currie and P. Olver, Null Lagrangians, weak continuity, and variational problems of arbitrary order, *J. Functional Anal.* **41** (1981), 135–174.

5. S. Haidar, *The Lavrentiev phenomenon in nonlinear elasticity*, Ph.D. thesis, Carnegie-Mellon University, Pittsburgh, PA, 1989.

6. S. Haidar, An example of J. Ball revisited (to be resubmitted).

7. C. Morrey, Quasiconvexity and the lower semi-continuity of multiple integrals, *Pacific J. Math.* **2** (1952), 25–53.

8. R. Ogden, Large deformation isotropic elasticity: on the correlation of theory and experiment of compressible rubberlike solids, *Proc. Roy. Soc. Lond.* **A326** (1972), 567–583.

9. R. Rivlin and J. Ericksen, Stress-deformation relations for isotropic materials, *J. Rational Mech. Anal.* **4** (1955), 323–425.

10. H. Rund, *The Hamilton-Jacobi theory in the calculus of variations*, Van Nostrand, London, 1966.

11. J. Sivaloganathan, Uniqueness of regular and singular equilibria in radial elasticity, preprint, Heriot-Watt University, 1985.

12. C. Truesdell and W. Noll, The nonlinear field theories of mechanics, in *Handbuch der Physik*, vol. III/3, S. Flügge, ed., Springer, Berlin, 1965.

Departments of Mathematics and Statistics, Grand Valley State University, Allendale, MI 49401, USA

A. HAJI-SHEIKH, F.R. PAYNE and K.J. HAYS-STANG

On convergence and uniqueness of microscale heat transfer equation

1. Introduction

Classical methods of predicting temperature must be modified when attempting to find temperature in thin conductors, nonconductors, and semiconductors subjected to short heat pulses. Known as microscale heat transfer, heat flux and individual body component temperatures can be briefly in local nonequilibrium. For laser heating of metal films, the electrons initially absorb the photon energy, raising the electron temperature, and transfer energy to the colder lattice by electron-phonon collisions [1], reaching temperature equilibrium in picoseconds. Qui and Tien [1,2] studied the heating of the lattice due to electron-phonon collisions. Hays-Stang and Haji-Sheikh [3] showed that the solution for thin metals is directly applicable to nonmetals but with a different set of parameters.

Mathematical formulations of the energy equation in thin films are valid for a broad spectrum of materials from conductors to nonconductors. The solution for different heat conduction models is obtainable from classical solutions of the diffusion equation. The solution exhibits different convergence characteristics depending on the values of the input parameters. The completeness of the solution is discussed. A numerical example is selected to show the solution and to study the convergence of the solution. For metals, the series solution is well behaved and rapid convergence is expected. However, for di-electric materials, a special transformation is needed to achieve the convergence of the series solution.

2. Analysis

The energy equation as derived in [3] is

$$\nabla \cdot [K\nabla T(\mathbf{r}, t)] + \tau_t \frac{\partial \{\nabla \cdot [K\nabla T(\mathbf{r}, t)]\}}{\partial t} + \left[S(\mathbf{r}, t) + \tau_q \frac{\partial S(\mathbf{r}, t)}{\partial t} \right]$$

$$= C\frac{\partial T(\mathbf{r}, t)}{\partial t} + C(\tau_e + \tau_q)\frac{\partial^2 T(\mathbf{r}, t)}{\partial t^2} \tag{1}$$

where T is lattice temperature, K is thermal conductivity, $\mathbf{r}$ is position vector, t is time, S is volumetric heat source, and the parameters τ_q, τ_e, and τ_t can be considered as local disequilibrium times. For metals, τ_q, τ_e, and τ_t can all be nonzero, however τ_q is generally several orders of magnitude less than τ_e and τ_t, and is frequently assumed to be zero. For di-electric materials, $\tau_t = 0$ and $\tau_e = 0$.

Solution. The solution of (1) as given in [3] begins by assuming a series solution for finite bodies of the form

$$T(\mathbf{r}, t) = \sum_{n=1}^{\infty} \psi_n(t) F_n(\mathbf{r}) exp(\gamma_n t) \tag{2}$$

where $F_n(\mathbf{r})$ is the eigenfunction that satisfies equation

$$\nabla \cdot [K\nabla F_n(\mathbf{r})] = -\gamma_n F_n(\mathbf{r}) \tag{3}$$

167

The boundary conditions are linear and homogeneous. Substituting T from (2) in (1), and using the orthogonality condition

$$\int_V F_n(\mathbf{r})F_m(\mathbf{r})dV = 0 \tag{4}$$

leads to the following solution [3]:

$$T(\mathbf{r},t) = T_I(\mathbf{r},t) + T_S(\mathbf{r},t), \tag{5}$$

where the contribution of initial conditions is

$$
\begin{aligned}
T_I(\mathbf{r},t) &= \sum_{n=1}^{\infty} \frac{F_n(\mathbf{r})}{N_n} e^{-(\gamma_n - \beta_n)t} \left\{ \frac{\sinh(\omega_n t)}{\omega_n} \left[(\gamma_n - \beta_n) \int_V F_n(\mathbf{r}')T_i(\mathbf{r}')dV' \right. \right. \\
&\quad \left. \left. + \int_V F_n(\mathbf{r}')T_{ii}(\mathbf{r}')dV' \right] + \cosh(\omega_n t) \int_V F_n(\mathbf{r}')T_i(\mathbf{r}')dV' \right\}
\end{aligned}
\tag{6}
$$

and the contribution of the volumetric heat source is

$$
\begin{aligned}
T_S(\mathbf{r},t) &= \sum_{n=1}^{\infty} \int_{\tau=0}^{t} \int_V \frac{F_n(\mathbf{r})F_n(\mathbf{r}')}{CN_n} e^{-\gamma_n(t-\tau)} \left\{ \frac{e^{\beta_n(t-\tau)} \sinh[\omega_n(t-\tau)]}{\omega_n(\tau_q + \tau_e)} \right\} \\
&\quad \times \left[S(\mathbf{r}',\tau) + \tau_q \frac{\partial S(\mathbf{r}',\tau)}{\partial \tau} \right] dV'd\tau.
\end{aligned}
\tag{7}
$$

The symbols β_n, λ_n, and ω_n stand for $\beta_n = \gamma_n - (1+\gamma_n\tau_t)/2(\tau_q+\tau_e)$, $\lambda_n = \gamma_n[1-\tau_t/(\tau_q+\tau_e)]^{1/2}$, and $\omega_n = (\beta_n^2 - \lambda_n^2)^{1/2}$. Moreover, $T_i(\mathbf{r}) = T(r,0)$ and $T_{ii}(\mathbf{r}) = \partial T(\mathbf{r},t)/\partial t$ as $t \to 0$. The arguments of hyperbolic sine and cosines are real or imaginary but $\psi_n(t)$ is always real. The nonhomogeneous boundary conditions can be accommodated using an appropriate transformation [4,5].

When $t \to 0$, (6) yields the relations

$$T_i(r) = \sum_{n=1}^{\infty} \frac{F_n(\mathbf{r})}{N_n} \int_V F_n(\mathbf{r}')T_i(\mathbf{r}')dV', \tag{8}$$

$$T_{ii}(\mathbf{r}) = \sum_{n=1}^{\infty} \frac{F_n(\mathbf{r})}{N_n} \int_V F_n(\mathbf{r}')T_{ii}(\mathbf{r}')dV'. \tag{9}$$

Both (8) and (9) satisfy the Bessel inequality [6]. Indeed, these relations, in the classical sense, satisfy Parseval's theorem, hence they are closed and complete [6]. They converge if the integrals go to zero as $n \to \infty$. However, for a special case, (7) does not properly converge as will be demonstrated later.

The general (7) exhibits different convergence behavior depending on the values of τ_t, τ_e, and τ_q. One can show that $\beta_n^2 - \lambda_n^2$ is related to $\gamma_n - \beta_n$ by the relation

$$\beta_n^2 - \lambda_n^2 = (\gamma_n - \beta_n)^2 - \gamma_n/(\tau_q + \tau_e) \tag{10}$$

and accordingly, $\gamma_n - \beta_n > (\beta_n^2 - \lambda_n^2)^{1/2}$. Alternatively, one can show that for the arguments of hyperbolic sine and cosine to be real, it is necessary to have $\beta_n^2 - \lambda_n^2 > 0$. This leads to the condition

$$(\tau_t \gamma_n)(\tau_q + \tau_e)/\tau_t < [1 + (\tau_t \gamma_n)]^2/4 \tag{11}$$

When $\beta_n^2 - \lambda_n^2 > 0$, the terms in (6) and (7) decay exponentially, hence the convergence will be achieved. In contrast, when $\beta_n^2 - \lambda_n^2 < 0$, the series may not converge to proper values. The following example discusses the convergence for metals and demonstrates the improper convergence for di-electric materials.

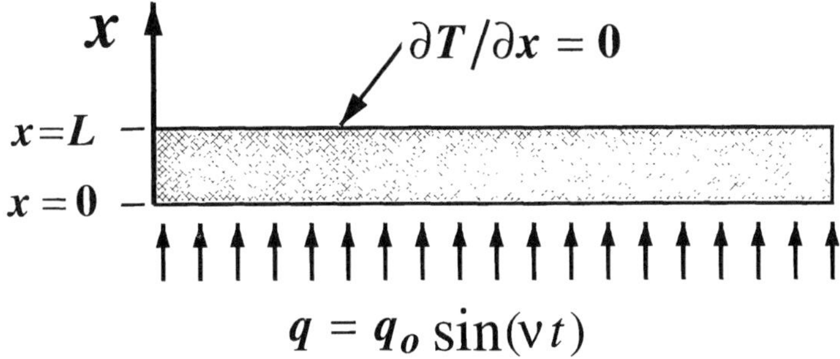

Fig. 1. Schematic of a thin plate and its boundary conditions.

Example. Figure 1 shows a thin plate of thickness L subject to boundary conditions $\partial T/\partial x = 0$ at $x = 0$ and at $x = L$. There is a source of energy at $x = 0$ that is described by the relation $S = q(t)\delta(x - 0)$ where $\delta(x - 0)$ is the Dirac delta function. For a thin metallic plate, $\tau_q << \tau_t$ and $\tau_e << \tau_t$, therefore, the inequality given by (11) is satisfied. The series converge because, at any given time, the coefficients in the Fourier series go to zero as n increases, that is, $\exp[-(\beta_n - \gamma_n)t]\exp[-(\beta_n^2 - \lambda_n^2)^{1/2}t] = \exp[\gamma_n t/(\tau_e + \tau_q)] \to 0$ as $m \to \infty$. In general, when $\beta_n^2 - \lambda_n^2 > 0$, the series solution converges properly.

For di-electric materials such as crystal diamond $\tau_t = \tau_e = 0$. The temperature solution for a di-electric material, when $q = q_o \sin(\nu t)$, after considerable algebra leads to the relation

$$
\begin{aligned}
T(x,t) = {} & \frac{1}{2C\tau_q} \int_{\tau=0}^{t} \sum_{m=1}^{\infty} \frac{2 - \delta_{0m}}{L\omega_m} \cos(m\pi x/L)e^{-(t-\tau)/2\tau_q} \sin[\omega_m(t - \tau)]q(\tau)d\tau \\
& + \frac{1}{C} \int_{\tau=0}^{t} \sum_{m=1}^{\infty} \frac{2 - \delta_{0m}}{L} \cos(m\pi x/L)e^{-(t-\tau)/2\tau_q} \cos[\omega_m(t - \tau)]q(\tau)d\tau
\end{aligned} \tag{12}
$$

The temperature at $x = L$ is computed as a function of time for a frequency of $\nu CL^2/K = 40$ and $K\tau_q/CL^2 = 0.04$ using a nondimensional form of (12). The series solution is evaluated using 1,000, 10,000, 100,000, and 1,000,000 terms. All solutions agree, indicating that convergence may have been achieved. However, a close examination of the data considering the physical phenomenon indicates that the

169

solution does not converge to a proper value. To demonstrate this, the solution of the classical diffusion equation, when $\tau_t = \tau_e = \tau_q = 0$, is also plotted in Fig. 2 using the same boundary conditions and the same heat source. A wave with finite propagation speed, described by (12), should arrive at the $x = L$ surface later than that for Fourier conduction where heat travels at infinite speed; however, this cannot be inferred from the data in Fig. 2.

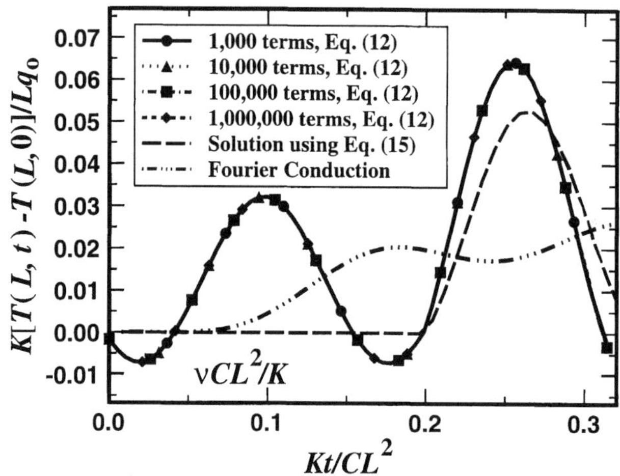

Fig. 2. Solutions using (12) and (15) and comparison with Fourier conduction.

It is possible to improve the convergence of the solution by examining (12) for the specific case of di-electric materials. Because of the parameter $\omega = (m^2\pi^2 K/CL^2\tau_q - 1/4\tau_q^2)^{1/2}$ in the denominator, the first term in (12) properly converges since $\omega_m \to \infty$ as $m \to \infty$. However, the convergence of the second term is not assured because the exponential term is independent of m and $\cos[\omega_m(t - \tau)]$ changes between -1 to 1.

The reason for poor convergence is the appearance of a wave phenomenon in the solution. To account for the wave effect, it is noted that the function $\delta(x - \xi)$ can be expressed in the Fourier series as

$$\delta(x - \xi) = \sum_{m=1}^{\infty} \frac{2 - \delta_{0m}}{L} \cos(m\pi x/L) \cos(m\pi\xi/L) \tag{13}$$

where ξ is a length scale and δ_{0m} is the Kronecker delta; $\delta_{0m} = 1$ when $m = 0$, otherwise $\delta_{0m} = 0$. This relation is subtracted from (18) to obtain a new form of the temperature solution,

$$T(x,t) = \frac{1}{2C\tau_q} \int_{\tau=0}^{t} \sum_{m=1}^{\infty} \frac{2 - \delta_{0m}}{L\omega_m} \cos(m\pi x/L) e^{-(t-\tau)/2\tau_q} \sin[\omega_m(t - \tau)]q(\tau)d\tau$$

170

$$+\frac{1}{C}\int_{\tau=0}^{t}\left\{\delta(x-\xi)-\sum_{m=1}^{\infty}\frac{2-\delta_{0m}}{L}\cos(m\pi x/L)\cos(\pi\xi/L)\right.$$

$$\left.+\sum_{m=1}^{\infty}\frac{2-\delta_{0m}}{L}\cos(m\pi x/L)\cos[\omega_m(t-\tau)]\right\}e^{-(t-\tau)/2\tau_q}q(\tau)d\tau. \qquad (14)$$

The length scale, ξ, is selected arbitrarily. The best choice of ξ is the one that forces $\cos(m\pi\xi/L)$ to approach $\cos[\omega(t-\tau)]$ as $m\to\infty$. Since the parameter ω_n reduces to $m\pi K^{1/2}/L(C\tau_q)^{1/2}$ as $m\to\infty$; this requires ξ to become $\xi=K^{1/2}(t-\tau)/L(C\tau_q)^{1/2}$. This relation implies that ξ is a distance that a wave travels with speed σ if $\sigma=K^{1/2}/(C\tau_q)^{1/2}$; therefore $\tau_q=K/\rho\sigma^2$ and $\xi=\sigma(t-\tau)$. Following integration of the term that contains the Dirac delta, the final form of the solution is

$$\begin{aligned}
T(x,t) &= \frac{1}{C\sigma}e^{-\sigma\xi/2\alpha}+\frac{1}{2C\tau_q}\int_{\tau=0}^{t}\sum_{m=1}^{\infty}\frac{2-\delta_{0m}}{L\omega_m}\cos(m\pi x/L)e^{-(t-\tau)/2\tau_q}\\
&\quad\times\sin[\omega_m(t-\tau)]q(\tau)d\tau+\frac{1}{C}\int_{\tau=0}^{t}\sum_{m=1}^{\infty}\frac{2-\delta_{0m}}{L}\cos(m\pi x/L)\\
&\quad\times\left\{\cos[\omega_m(t-\tau)]-\cos(\pi\xi/L)\right\}e^{-(t-\tau)/2\tau_q}q(\tau)d\tau \qquad (15)
\end{aligned}$$

Note that x in the first term is replaced by ξ to describe multiple reflections of a propagating wave with exponential decay. The energy travels from the surface accepting the heat input to some point in the heat conducting body in a transit time ξ/σ. The distance to that point is ξ. For the first arrival of the wave at the surface opposite the heat input surface, ξ is the coordinate x. If $t>L/\sigma$, the wave is reflected and travels toward the heat input surface. However, the amplitude of the reflected wave is small because of the decaying factor $\exp(-\sigma\xi/2\alpha)$. Equations (12) and (15) are derived using the same partial differential equation and the same boundary conditions. However they yield different temperature values for a given periodic heat flux at $x\to0$; hence, uniqueness is not assured.

In essence, (15) should yield results identical to those of (12). However, the dash line in Fig. 2, using (15), is quite different from the solution using (12). Notice that the dash line indicates a delay of the arrival of energy input from $x=0$ to $x=L$. This is a realistic behavior because the value of $\tau_q=0$, for Fourier conduction, corresponds to $\sigma=\infty$, and therefore the dash line must lag behind the Fourier conduction solution. This wave-like phenomenon in conjunction with the derivation of the Green's function for the thermal wave model is reported in [4].

3. Conclusion

The formulation of a temperature model for microscale systems requires mathematical efforts that can be demanding. The solution for conductors, e.g., gold, silver, copper, and other pure metals, shows satisfactory convergence characteristics. The numerical computations for di-electric materials remain demanding. The appearance of wave-like phenomenon was demonstrated in a simple one-dimensional example; however, (6) and (7) describe generalized solutions for three-dimensional energy transport problems.

There is a paucity of information available on the solution for layered materials. The orthogonality condition that often exists for the Fourier conduction in layered materials cannot be used to obtain solutions for microscale heat conduction problems in layered materials. In the absence of a closed-form solution, one needs to consider a

Volterra-type integral equation to solve for interface conditions of temperature and/or heat flux.

References

1. T.Q. Qui and C.L. Tien, Short-pulse laser heating on metals, *Internat. J. Heat Mass Transfer*, **35** (1992), 719–726.

2. T.Q. Qui and C.L. Tien, Heat transfer mechanisms during short-pulse laser heating on metals, *ASME J. Heat Transfer* **115** (1993), 835–841.

3. K.J. Hays-Stang and A. Haji-Sheikh, A unified solution for heat conduction in thin films, *Internat. J. Heat Mass Transfer* (1999) (to appear).

4. A. Haji-Sheikh and J.V. Beck, Green's function solution for thermal wave equation in finite bodies, *Internat. J. Heat Mass Transfer* **37** (1994), 2615–2626.

5. J.V. Beck, K.D. Cole, A. Haji-Sheikh and B. Litkouhi, *Heat conduction using Green's functions*, Hemisphere, Washington, D.C., 1992.

6. R.V. Churchill, *Fourier series and boundary value problems*, McGraw-Hill, New York, 1941.

Department of Mechanical and Aerospace Engineering, The University of Texas at Arlington, Arlington, Texas 76019-0023, USA

W. HAN and M. SOFONEA

On numerical approximations of a frictionless contact problem for elastic-viscoplastic materials

1. Introduction

In this paper we give numerical analysis of a mathematical model that describes the unilateral quasistatic contact of two elastic-viscoplastic bodies. The contact is without friction and it is modeled by the classical Signorini boundary conditions. Summarizing accounts of contact problems and their numerical approximations can be found in [5] and [7].

We consider elastic-viscoplastic bodies modeled by a constitutive law of the form

$$\dot{\sigma} = \mathcal{E}\dot{\varepsilon} + G(\sigma, \varepsilon) \tag{1}$$

in which σ denotes the stress tensor, ε is the small strain tensor and $\mathcal{E}$, G are given constitutive functions. Here and below the dot above represents the derivative with respect to the time variable.

Rate-type viscoplastic models of the form (1) are used to describe the behavior of real materials like rubbers, metals, pastes, rocks (cf. [2, 6]). A relatively simple one-dimensional example of constitutive law of the form (1) is obtained by taking

$$G(\sigma, \varepsilon) = \begin{cases} -k_1 F_1(\sigma - f(\varepsilon)) & \text{if } \sigma > f(\varepsilon), \\ 0 & \text{if } g(\varepsilon) \leq \sigma \leq f(\varepsilon), \\ k_2 F_2(g(\varepsilon) - \sigma) & \text{if } \sigma < g(\varepsilon), \end{cases}$$

where k_1, $k_2 > 0$ are viscosity constants, f and g are Lipschitz continuous functions with $f(\varepsilon) \geq g(\varepsilon)$, and F_1, $F_2 : \mathbb{R}_+ \to \mathbb{R}$ are increasing functions such that $F_1(0) = F_2(0) = 0$.

The mathematical model describing the quasistatic unilateral contact without friction between two elastic-viscoplastic bodies was analyzed in [9]. This model consists of an evolution equation coupled with a time-dependent variational inequality. The aim of this paper is to present results on numerical approximations of the mechanical problem analyzed in [9]. For detailed proofs of these results, readers are referred to [4].

2. The contact problem

Consider two elastic-viscoplastic bodies whose material particles occupy bounded domains Ω^1 and Ω^2 of $\mathbb{R}^d$ ($d \leq 3$ in applications). In the following, we put a superscript m to indicate that the quantity is related to the domain Ω^m, $m = 1, 2$. For each domain Ω^m, we assume its boundary Γ^m is Lipschitz continuous, and is partitioned into three disjoint measurable parts Γ_1^m, Γ_2^m and Γ_3^m, with $\text{meas}\,(\Gamma_1^m) > 0$. The unit outward normal to Γ^m is denoted by $\nu^m = (\nu_i^m)$. We are interested in the resulting quasistatic process of evolution of the bodies on the time interval $[0, T]$, $T > 0$. Assume that the bodies are clamped on $\Gamma_1^m \times (0, T)$, volume forces of density φ_1^m act on $\Omega^m \times (0, T)$, and surface tractions of density φ_2^m act on $\Gamma_2^m \times (0, T)$. The two bodies are in contact along the common part $\Gamma_3^1 = \Gamma_3^2$, which will be denoted Γ_3 below. The contact is frictionless and it is modeled by the Signorini conditions on $\Gamma_3 \times (0, T)$ with a zero gap function.

The mechanical problem is then the following: For $m = 1, 2$, find the displacement field $\boldsymbol{u}^m = (u_i^m) : \Omega^m \times [0, T] \to \mathbb{R}^d$ and the stress field $\boldsymbol{\sigma}^m = (\sigma_{ij}^m) : \Omega^m \times [0, T] \to S_d$ which satisfy, for $t \in (0, T)$,

$$\dot{\boldsymbol{\sigma}}^m = \mathcal{E}^m \boldsymbol{\varepsilon}(\dot{\boldsymbol{u}}^m) + G^m(\boldsymbol{\sigma}^m, \boldsymbol{\varepsilon}(\boldsymbol{u}^m)) \quad \text{in } \Omega^m, \tag{2}$$

$$\operatorname{Div} \boldsymbol{\sigma}^m + \boldsymbol{\varphi}_1^m = \mathbf{0} \quad \text{in } \Omega^m, \tag{3}$$

$$\boldsymbol{u}^m = \mathbf{0} \quad \text{on } \Gamma_1^m, \tag{4}$$

$$\boldsymbol{\sigma}^m \boldsymbol{\nu}^m = \boldsymbol{\varphi}_2^m \quad \text{on } \Gamma_2^m, \tag{5}$$

$$u_\nu^1 + u_\nu^2 \le 0, \ \sigma_\nu^1 = \sigma_\nu^2 \le 0, \ \sigma_\nu^1(u_\nu^1 + u_\nu^2) = 0, \ \boldsymbol{\sigma}_\tau^m = \mathbf{0} \quad \text{on } \Gamma_3, \tag{6}$$

and the initial value conditions

$$\boldsymbol{u}^m(0) = \boldsymbol{u}_0^m, \ \boldsymbol{\sigma}^m(0) = \boldsymbol{\sigma}_0^m \quad \text{in } \Omega^m. \tag{7}$$

Here and below, $1 \le i, j \le d$, S_d represents the space of second-order symmetric tensors on $\mathbb{R}^d$. The summation convention over a repeated index is adopted, unless stated otherwise. In (2)–(7), $\operatorname{Div} \boldsymbol{\sigma}^m = (\sigma_{ij,j}^m)$ represents the divergence of the tensor-valued function $\boldsymbol{\sigma}^m$, $\boldsymbol{\varepsilon}(\boldsymbol{u}^m) = (\varepsilon_{ij}(\boldsymbol{u}^m))$ denotes the small strain tensor, with $\varepsilon_{ij}(\boldsymbol{u}^m) = \frac{1}{2}(u_{i,j}^m + u_{j,i}^m)$. The index that follows a comma indicates a partial derivative with respect to the corresponding spatial component variable. Moreover,

$$u_\nu^m = u_i^m \nu_i^m, \quad \sigma_\nu^m = \sigma_{ij}^m \nu_i^m \nu_j^m, \quad \boldsymbol{\sigma}_\tau^m = (\sigma_{\tau i}^m) \text{ with } \sigma_{\tau i}^m = \sigma_{ij}^m \nu_j^m - \sigma_\nu^m \nu_i^m$$

denote the normal displacement, normal stress, and tangential stress respectively.

We will use the Hilbert spaces

$$V^m = \{\boldsymbol{v} = (v_i) \mid v_i \in H^1(\Omega^m), \ v_i = 0 \text{ on } \Gamma_1^m, \ 1 \le i \le d\},$$

$$\mathcal{H}^m = \{\boldsymbol{\tau} = (\tau_{ij}) \mid \tau_{ij} \in L^2(\Omega^m), \ 1 \le i, j \le d\},$$

$$\mathcal{H}_1^m = \{\boldsymbol{\tau} \in \mathcal{H}^m \mid \operatorname{Div} \boldsymbol{\tau} \in (L^2(\Omega^m))^d\},$$

and product spaces $V = V^1 \times V^2$, $\mathcal{H} = \mathcal{H}^1 \times \mathcal{H}^2$ and $\mathcal{H}_1 = \mathcal{H}_1^1 \times \mathcal{H}_1^2$. We will use the notation $\boldsymbol{\varepsilon}(\boldsymbol{v}) = (\boldsymbol{\varepsilon}(\boldsymbol{v}^1), \boldsymbol{\varepsilon}(\boldsymbol{v}^2))$ for $\boldsymbol{v} = (\boldsymbol{v}^1, \boldsymbol{v}^2) \in V$ and $\mathcal{E}\boldsymbol{\varepsilon} = (\mathcal{E}^1 \boldsymbol{\varepsilon}^1, \mathcal{E}^2 \boldsymbol{\varepsilon}^2)$, $G(\boldsymbol{\sigma}, \boldsymbol{\varepsilon}) = (G(\boldsymbol{\sigma}^1, \boldsymbol{\varepsilon}^1), G(\boldsymbol{\sigma}^2, \boldsymbol{\varepsilon}^2))$ for $\boldsymbol{\varepsilon} = (\boldsymbol{\varepsilon}^1, \boldsymbol{\varepsilon}^2) \in \mathcal{H}$, $\boldsymbol{\sigma} = (\boldsymbol{\sigma}^1, \boldsymbol{\sigma}^2) \in \mathcal{H}$. The dual of V is denoted V' and $\langle \cdot, \cdot \rangle$ will represent the duality between V' and V. The set of admissible displacement fields is $U = \{\boldsymbol{v} = (\boldsymbol{v}^1, \boldsymbol{v}^2) \in V \mid v_\nu^1 + v_\nu^2 \le 0 \text{ on } \Gamma_3\}$.

Since $\operatorname{meas}(\Gamma_1^m) > 0$, Korn's inequality holds:

$$\|\boldsymbol{\varepsilon}(\boldsymbol{v})\|_{\mathcal{H}^m} \ge c \|\boldsymbol{v}\|_{(H^1(\Omega^m))^d} \quad \forall \boldsymbol{v} \in V^m. \tag{8}$$

Here $c > 0$ is a constant that depends only on Ω^m and Γ_1^m (see, e.g., [8] p. 79).

In the study of the mechanical problem (2)–(7) we make the following assumptions. $\mathcal{E}^m : \Omega^m \times S_d \to S_d$ is a bounded, symmetric, positive definite fourth-order tensor:

$$\begin{cases} \text{(a) } \mathcal{E}_{ijkl}^m \in L^\infty(\Omega^m), \ 1 \le i, j, k, l \le d; \\ \text{(b) } \mathcal{E}^m \boldsymbol{\sigma} \cdot \boldsymbol{\tau} = \boldsymbol{\sigma} \cdot \mathcal{E}^m \boldsymbol{\tau}, \ \forall \boldsymbol{\sigma}, \boldsymbol{\tau} \in S_d, \text{ a.e. in } \Omega^m; \\ \text{(c) There exists an } \alpha^m > 0 \text{ such that} \\ \quad \mathcal{E}^m \boldsymbol{\tau} \cdot \boldsymbol{\tau} \ge \alpha^m |\boldsymbol{\tau}|^2 \ \forall \boldsymbol{\tau} \in S_d, \text{ a.e. in } \Omega^m. \end{cases} \tag{9}$$

$G^m : \Omega^m \times S_d \times S_d \to S_d$ has the properties:

$$\begin{cases} \text{(a) There exists an } L^m > 0 \text{ such that } \forall\, \boldsymbol{\sigma}_1, \boldsymbol{\sigma}_2, \boldsymbol{\varepsilon}_1, \boldsymbol{\varepsilon}_2 \in S_d, \text{ a.e. in } \Omega^m, \\ \quad |G^m(\boldsymbol{x}, \boldsymbol{\sigma}_1, \boldsymbol{\varepsilon}_1) - G^m(\boldsymbol{x}, \boldsymbol{\sigma}_2, \boldsymbol{\varepsilon}_2)| \leq L^m\,(|\boldsymbol{\sigma}_1 - \boldsymbol{\sigma}_2| + |\boldsymbol{\varepsilon}_1 - \boldsymbol{\varepsilon}_2|); \\ \text{(b) For any } \boldsymbol{\sigma}, \boldsymbol{\varepsilon} \in S_d,\ \boldsymbol{x} \mapsto G^m(\boldsymbol{x}, \boldsymbol{\sigma}, \boldsymbol{\varepsilon}) \text{ is measurable;} \\ \text{(c) The mapping } \boldsymbol{x} \mapsto G^m(\boldsymbol{x}, \boldsymbol{0}, \boldsymbol{0}) \in \mathcal{H}^m. \end{cases} \tag{10}$$

$$\boldsymbol{\varphi}_1^m \in W^{1,\infty}(0, T; (L^2(\Omega^m))^d), \quad \boldsymbol{\varphi}_2^m \in W^{1,\infty}(0, T; (L^2(\Gamma_2^m))^d); \tag{11}$$

$$\boldsymbol{u}_0 = (\boldsymbol{u}_0^1, \boldsymbol{u}_0^2) \in U, \ (\boldsymbol{\sigma}_0, \boldsymbol{\varepsilon}(\boldsymbol{u}_0))_{\mathcal{H}} = \langle \boldsymbol{f}(0), \boldsymbol{u}_0 \rangle. \tag{12}$$

For all $t \in [0, T]$ let $\boldsymbol{f}(t)$ denote the element of V' given by

$$\begin{aligned} \langle \boldsymbol{f}(t), \boldsymbol{v} \rangle &= (\boldsymbol{\varphi}_1^1(t), \boldsymbol{v}^1)_{(L^2(\Omega^1))^d} + (\boldsymbol{\varphi}_1^2(t), \boldsymbol{v}^2)_{(L^2(\Omega^2))^d} \\ &\quad + (\boldsymbol{\varphi}_2^1(t), \boldsymbol{v}^1)_{(L^2(\Gamma_2^1))^d} + (\boldsymbol{\varphi}_2^2(t), \boldsymbol{v}^2)_{(L^2(\Gamma_2^2))^d} \qquad \forall\, \boldsymbol{v} = (\boldsymbol{v}^1, \boldsymbol{v}^2) \in V. \end{aligned}$$

The weak formulation of the contact problem is as follows.

Problem P. Find the displacement field $\boldsymbol{u} : [0, T] \to U$ and the stress field $\boldsymbol{\sigma} : [0, T] \to \mathcal{H}_1$ such that

$$\boldsymbol{u}(0) = \boldsymbol{u}_0, \ \boldsymbol{\sigma}(0) = \boldsymbol{\sigma}_0, \tag{13}$$

and for a.a. $t \in (0, T)$,

$$\dot{\boldsymbol{\sigma}}(t) = \mathcal{E}\boldsymbol{\varepsilon}(\dot{\boldsymbol{u}}(t)) + G(\boldsymbol{\sigma}(t), \boldsymbol{\varepsilon}(\boldsymbol{u}(t))), \tag{14}$$

$$(\boldsymbol{\sigma}(t), \boldsymbol{\varepsilon}(\boldsymbol{v} - \boldsymbol{u}(t)))_{\mathcal{H}} \geq \langle \boldsymbol{f}(t), \boldsymbol{v} - \boldsymbol{u}(t) \rangle \quad \forall\, \boldsymbol{v} \in U. \tag{15}$$

The following well-posedness result is proved in [9].

Theorem 2.1. *Under the assumptions* (9), (10), (11) *and* (12), *the problem* P *has a unique solution* $(\boldsymbol{u}, \boldsymbol{\sigma}) \in W^{1,\infty}(0, T; U \times \mathcal{H}_1)$.

3. Spatially semi-discrete approximation

We will discretize the spatial domain by finite elements. Let $\mathcal{T}^h$ be a regular finite element partition of the domain Ω in such a way that if a side of an element lies on the boundary, then the side is entirely on one of the subsets $\overline{\Gamma}_1^m$, $\overline{\Gamma}_2^m$ and $\overline{\Gamma}_3$. We then choose a finite element space $V^h \subset V$ for the approximation of the displacement variable $\boldsymbol{u}$, and another finite element space Q^h for the approximation of the stress variable $\boldsymbol{\sigma}$. The finite element spaces consist of piecewise (images of) polynomials. We will assume that for these finite element spaces, there hold the relations

$$\boldsymbol{\varepsilon}(V^h) \subset Q^h, \quad G(Q^h, Q^h) \subset Q^h. \tag{16}$$

The relation holds if, e.g., we use linear elements for V^h and piecewise constants for Q^h. The discrete admissible set is $U^h = \{\boldsymbol{v}^h = (\boldsymbol{v}^{1,h}, \boldsymbol{v}^{2,h}) \in V^h \mid v_\nu^{1,h} + v_\nu^{2,h} \leq 0 \text{ on } \Gamma_3\}$. Obviously, $U^h \subset U$. Now the spatially semi-discrete scheme is the following.

Problem P^h. Find the displacement field $\boldsymbol{u}^h : [0, T] \to U^h$ and the stress field $\boldsymbol{\sigma}^h : [0, T] \to Q^h$ such that

$$\boldsymbol{u}^h(0) = \boldsymbol{u}_0^h, \ \boldsymbol{\sigma}^h(0) = \boldsymbol{\sigma}_0^h, \tag{17}$$

and for a.a. $t \in (0, T)$,

$$\dot{\boldsymbol{\sigma}}^h(t) = \mathcal{E}\boldsymbol{\varepsilon}(\dot{\boldsymbol{u}}^h(t)) + G(\boldsymbol{\sigma}^h(t), \boldsymbol{\varepsilon}(\boldsymbol{u}^h(t))), \tag{18}$$

$$(\boldsymbol{\sigma}^h(t), \boldsymbol{\varepsilon}(\boldsymbol{v}^h - \dot{\boldsymbol{u}}^h(t)))_{\mathcal{H}} \geq \langle \boldsymbol{f}(t), \boldsymbol{v}^h - \dot{\boldsymbol{u}}^h(t) \rangle \quad \forall \, \boldsymbol{v}^h \in U^h. \tag{19}$$

Here, $\boldsymbol{u}_0^h \in U^h$, $\boldsymbol{\sigma}_0^h \in Q^h$ are appropriate approximations of $\boldsymbol{u}_0$ and $\boldsymbol{\sigma}_0$.

The problem P^h has a unique solution $(\boldsymbol{u}^h, \boldsymbol{\sigma}^h) \in W^{1,\infty}(0, T; U^h \times Q^h)$.

Theorem 3.1. *We have the error estimate*

$$\|\boldsymbol{\sigma} - \boldsymbol{\sigma}^h\|_{L^\infty(0,T;\mathcal{H})} + \|\boldsymbol{u} - \boldsymbol{u}^h\|_{L^\infty(0,T;V)} \tag{20}$$

$$\leq \ c\,(\|\boldsymbol{\sigma}_0 - \boldsymbol{\sigma}_0^h\|_{\mathcal{H}} + \|\boldsymbol{u}_0 - \boldsymbol{u}_0^h\|_V) + c \inf_{\boldsymbol{v}^h \in L^\infty(0,T;U^h)} \Big\{ \|\boldsymbol{u} - \boldsymbol{v}^h\|_{L^\infty(0,T;V)}$$

$$+ (\|\boldsymbol{f}\|_{L^\infty(0,T;V')} + \|\boldsymbol{\sigma}\|_{L^\infty(0,T;\mathcal{H})})^{1/2} \|\boldsymbol{u} - \boldsymbol{v}^h\|_{L^\infty(0,T;V)}^{1/2} \Big\}.$$

The estimate (20) is the basis for convergence analysis, following an argument in [3]. Here we focus on error estimates under suitable regularity assumptions, including

$$\sigma_\nu^1 \in L^\infty(0, T; L^2(\Gamma_3)). \tag{21}$$

Corollary 3.2. *Assume additionally the regularity (21). Then*

$$\|\boldsymbol{\sigma} - \boldsymbol{\sigma}^h\|_{L^\infty(0,T;\mathcal{H})} + \|\boldsymbol{u} - \boldsymbol{u}^h\|_{L^\infty(0,T;V)} \tag{22}$$

$$\leq c\,(\|\boldsymbol{\sigma}_0 - \boldsymbol{\sigma}_0^h\|_{\mathcal{H}} + \|\boldsymbol{u}_0 - \boldsymbol{u}_0^h\|_V)$$

$$+ c \inf_{\boldsymbol{v}^h \in L^\infty(0,T;U^h)} \{ \|\boldsymbol{u} - \boldsymbol{v}^h\|_{L^\infty(0,T;V)} + \|\sigma_\nu^1\|_{L^\infty(0,T;L^2(\Gamma_3))}^{1/2} \|u_\nu - v_\nu^h\|_{L^\infty(0,T;L^2(\Gamma_3))}^{1/2} \}.$$

The inequality (22) is the basis for deriving order error estimates for the semi-discrete solutions. We present one sample result. Assume $\boldsymbol{u} \in L^\infty(0, T; (H^2(\Omega))^d)$ and $u_\nu \in L^\infty(0, T; H^2(\Gamma_3))$. Let us use linear elements for V^h, and piecewise constants for Q^h. We choose the initial values $\boldsymbol{u}_0^h \in U^h$ and $\boldsymbol{\sigma}_0^h \in Q^h$ in such a way that

$$\|\boldsymbol{\sigma}_0 - \boldsymbol{\sigma}_0^h\|_{\mathcal{H}} \leq c\,h, \quad \|\boldsymbol{u}_0 - \boldsymbol{u}_0^h\|_V \leq c\,h.$$

Then using the standard finite element interpolation theory (cf. [1]), we have the optimal order error estimate

$$\|\boldsymbol{\sigma} - \boldsymbol{\sigma}^h\|_{L^\infty(0,T;\mathcal{H})} + \|\boldsymbol{u} - \boldsymbol{u}^h\|_{L^\infty(0,T;V)} \leq c\,h$$

under the regularity assumptions

$$\boldsymbol{u} \in L^\infty(0, T; (H^2(\Omega))^d), \ u_\nu \in L^\infty(0, T; H^2(\Gamma_3)) \text{ and } \sigma_\nu^1 \in L^\infty(0, T; L^2(\Gamma_3)).$$

4. Fully discrete approximation

We use the finite element spaces V^h and Q^h introduced in the last section, with the assumption (16). In addition, we need a partition of the time interval $[0,T] : 0 = t_0 < t_1 < \cdots < t_N = T$. We denote the step-size $k_n = t_n - t_{n-1}$ for $n = 1,\ldots,N$, and let $k = \max_n k_n$ be the maximal step-size. For a sequence $\{w_n\}_{n=0}^N$, we denote $\delta w_n = (w_n - w_{n-1})/k_n$ for the divided difference. No summation is implied over the repeated index n. Then a fully discrete approximation method is the following.

Problem P^{hk}. Find the displacement field $\boldsymbol{u}^{hk} = \{\boldsymbol{u}_n^{hk}\}_{n=0}^N \subset U^h$ and the stress field $\boldsymbol{\sigma}^{hk} = \{\boldsymbol{\sigma}_n^{hk}\}_{n=0}^N \subset Q^h$, such that

$$\boldsymbol{u}_0^{hk} = \boldsymbol{u}_0^h, \ \boldsymbol{\sigma}_0^{hk} = \boldsymbol{\sigma}_0^h, \tag{23}$$

and for $n = 1,\ldots,N$

$$\delta\boldsymbol{\sigma}_n^{hk} = \mathcal{E}\delta\varepsilon(\boldsymbol{u}_n^{hk}) + G(\boldsymbol{\sigma}_{n-1}^{hk}, \varepsilon(\boldsymbol{u}_{n-1}^{hk})), \tag{24}$$

$$(\boldsymbol{\sigma}_n^{hk}, \varepsilon(\boldsymbol{v}^h - \boldsymbol{u}_n^{hk}))_\mathcal{H} \geq \langle \boldsymbol{f}_n, \boldsymbol{v}^h - \boldsymbol{u}_n^{hk} \rangle \quad \forall \boldsymbol{v}^h \in U^h. \tag{25}$$

The fully discrete problem has a unique solution, and the following results hold.

Theorem 4.1. *We have the error estimate*

$$\max_{1 \leq n \leq N} \left(\|\boldsymbol{\sigma}_n - \boldsymbol{\sigma}_n^{hk}\|_\mathcal{H} + \|\boldsymbol{u}_n - \boldsymbol{u}_n^{hk}\|_V \right) \tag{26}$$

$$\leq c \left(\|\boldsymbol{\sigma}_0 - \boldsymbol{\sigma}_0^h\|_\mathcal{H} + \|\boldsymbol{u}_0 - \boldsymbol{u}_0^h\|_V \right) + c k \left(\|\dot{\boldsymbol{\sigma}}\|_{L^\infty(0,T;\mathcal{H})} + \|\dot{\boldsymbol{u}}\|_{L^\infty(0,T;V)} \right)$$

$$+ c \max_{1 \leq n \leq N} \left(\|\boldsymbol{u}_n - \boldsymbol{v}_n^h\|_V + |\langle \boldsymbol{f}_n, \boldsymbol{u}_n - \boldsymbol{v}_n^h \rangle - (\boldsymbol{\sigma}_n, \varepsilon(\boldsymbol{u}_n - \boldsymbol{v}_n^h))|^{1/2} \right).$$

Corollary 4.2. *Assume additionally the regularity (21). Then*

$$\max_{1 \leq n \leq N} \left(\|\boldsymbol{\sigma}_n - \boldsymbol{\sigma}_n^{hk}\|_\mathcal{H} + \|\boldsymbol{u}_n - \boldsymbol{u}_n^{hk}\|_V \right) \tag{27}$$

$$\leq c \left(\|\boldsymbol{\sigma}_0 - \boldsymbol{\sigma}_0^h\|_\mathcal{H} + \|\boldsymbol{u}_0 - \boldsymbol{u}_0^h\|_V \right) + c k \left(\|\dot{\boldsymbol{\sigma}}\|_{L^\infty(0,T;\mathcal{H})} + \|\dot{\boldsymbol{u}}\|_{L^\infty(0,T;V)} \right)$$

$$+ c \max_{1 \leq n \leq N} \inf_{\boldsymbol{v}^h \in L^\infty(0,T;U^h)} \{ \|\boldsymbol{u}(t_n) - \boldsymbol{v}^h\|_V + \|\sigma_\nu^1\|_{L^\infty(0,T;L^2(\Gamma_3))}^{1/2} \|u_\nu(t_n) - v_\nu^h\|_{L^2(\Gamma_3)}^{1/2} \}.$$

The inequalities (26) and (27) are the basis for deriving order error estimates for the fully discrete solution. We present one sample result. Let us use linear elements for the finite element space V^h, and piecewise constants for Q^h. We choose the initial values $\boldsymbol{u}_0^{hk} \in U^h$ and $\boldsymbol{\sigma}_0^{hk} \in Q^h$ in such a way that

$$\|\boldsymbol{\sigma}_0 - \boldsymbol{\sigma}_0^{hk}\|_\mathcal{H} \leq c\,h, \quad \|\boldsymbol{u}_0 - \boldsymbol{u}_0^{hk}\|_V \leq c\,h.$$

Then, under the regularity assumptions

$$\boldsymbol{u} \in L^\infty(0,T;(H^2(\Omega))^d), \ u_\nu \in L^\infty(0,T;H^2(\Gamma_3)) \ \text{and} \ \sigma_\nu^1 \in L^\infty(0,T;L^2(\Gamma_3)),$$

we have the optimal order error estimate

$$\max_{1 \leq n \leq N} \left(\|\boldsymbol{\sigma}_n - \boldsymbol{\sigma}_n^{hk}\|_\mathcal{H} + \|\boldsymbol{u}_n - \boldsymbol{u}_n^{hk}\|_V \right) \leq c\,(k + h).$$

References

1. P.G. Ciarlet, *The finite element method for elliptic problems*, North-Holland, Amsterdam, 1978.

2. N. Cristescu and I. Suliciu, *Viscoplasticity*, Martinus Nijhoff, Editura Tehnica, Bucharest, 1982.

3. W. Han and B.D. Reddy, *Plasticity: mathematical theory and numerical analysis*, Springer-Verlag, New York, 1999.

4. W. Han and M. Sofonea, Numerical analysis of a frictionless contact problem for elastic-viscoplastic materials (submitted for publication).

5. J. Haslinger, I. Hlaváček and J. Nečas, Numerical methods for unilateral problems in solid mechanics, in *Handbook of numerical analysis IV*, P.G. Ciarlet and J.-L. Lions, Eds., Elsevier, Amsterdam, 1996, 313–485.

6. I. Ionescu and M. Sofonea, *Functional and numerical methods in viscoplasticity*, Oxford University Press, Oxford, 1993.

7. N. Kikuchi and J.T. Oden, *Contact problems in elasticity*, SIAM, Philadelphia, 1988.

8. J. Nečas and I. Hlaváček, *Mathematical theory of elastic and elastoplastic bodies: an introduction*, Elsevier, Amsterdam, 1981.

9. M. Rochdi and M. Sofonea, On frictionless contact between two elastic-viscoplastic bodies, *Quart. J. Mech. Appl. Math.* **50** (1997), 481–496.

Department of Mathematics, University of Iowa, Iowa City, IA 52242, USA

Laboratoire de Théorie des Systèmes, University of Perpignan, France

K. HAYAMI and S.A. SAUTER

A panel clustering method for 3-D elastostatics using spherical harmonics

Despite its advantage of boundary-only discretization, the standard boundary element method (BEM) involves huge computational costs for large-scale 3-D problems due to its dense matrix formulation. The situation is even worse for the 3-D elastostatic problem, where the number of unknowns is three times that of the potential problem. In this paper, we will apply the panel clustering method [2] using multipole expansions [1] in order to reduce the computational costs for the 3-D boundary element analysis of elastostatics.

1. The boundary element formulation for 3-D elastostatics

The boundary integral equation for the three-dimensional (linear, isotropic) elastostatic problem is given by

$$c_{lk}(x)u_k(x) + \text{c.p.} \int_\Gamma p_{lk}^*(x,y)u_k(y)\mathrm{d}\Gamma(y) \;=\; \int_\Gamma u_{lk}^*(x,y)p_k(y)\mathrm{d}\Gamma(y)$$
$$(l = 1,2,3), \tag{1}$$

where Γ is the boundary of the domain under consideration, u_k, p_k are the displacement and traction components, respectively, $c_{lk}(x) = \frac{1}{2}\delta_{lk}$ when Γ is smooth at x, and the body force term has been neglected.

$u_{lk}^*(x,y)$ is the fundamental displacement, which is usually given by

$$u_{lk}^* = \frac{1}{16\pi\mu(1-\nu)r'}\{(3-4\nu)\delta_{lk} + r'_{,l}r'_{,k}\} \tag{2}$$

for the three-dimensional case. μ is the shear modulus, ν is the Poisson's ratio, $r' := \|y - x\|$ and $r'_{,k} := \dfrac{\partial r'}{\partial y_k}$, where $x = (x_1, x_2, x_3)^T, y = (y_1, y_2, y_3)^T$. However, it will prove useful to use the alternative expression [3]:

$$u_{lk}^* = \frac{1}{4\pi\mu}\left\{\delta_{lk}\frac{1}{r'} - \frac{1}{4(1-\nu)}r'_{,lk}\right\} = \frac{1}{8\pi\mu}\left\{r'_{,ii}\delta_{lk} - \frac{1}{2(1-\nu)}r'_{,lk}\right\}, \tag{3}$$

where $r'_{,lk} := \dfrac{\partial^2 r'}{\partial y_l \partial y_k}$. The fundamental traction component is given by

$$p_{lk}^*(x,y) = \sigma_{lkj}^* n_j = \lambda u_{li,i}^* n_k + \mu(u_{lk,j}^* + u_{lj,k}^*)n_j, \tag{4}$$

where n_l is the component of the unit outward normal vector at $y \in \Gamma$.

Next, the boundary integral equation (1) is discretized by discretizing the boundary Γ into boundary elements or panels π_α, $(\alpha = 1, ..., n)$. Assuming constant elements, with point x^α representing π_α, we obtain

$$\frac{1}{2}u_l^\alpha + \sum_{\beta=1}^n u_k^\beta \int_{\pi_\beta} p_{lk}^*(x^\alpha, y)\mathrm{d}\Gamma(y) \;=\; \sum_{\beta=1}^n p_k^\beta \int_{\pi_\beta} u_{lk}^*(x^\alpha, y)\mathrm{d}\Gamma(y)$$
$$(\alpha = 1, ..., n; \; l = 1,2,3), \tag{5}$$

where $u_k^\beta = u(x^\beta)$, etc. Given the boundary data, (5) is a system of linear equations for the unknown boundary displacement and traction components u_k^β and p_k^β. Since its matrix is dense and nonsymmetric, it is usually solved using LU-decomposition, which leads to $O(N^3)$ computational work and $O(N^2)$ memory, where $N = 3n$ is the number of unknowns.

Alternatively, one could use iterative solvers for nonsymmetric matrices such as the GMRES. Since the system (5) arising from boundary integral equations such as (1) is usually well conditioned, the method should converge within $M \ll N$ iterations, with the use of a suitable preconditioner when necessary. In this method, the dominant part of the computation is the dense matrix-vector multiplication corresponding to (5) for the unknown boundary data or iteration vector, which costs $O(N^2)$ each. Hence, the amount of computational work is reduced from $O(N^3)$ to $O(MN^2)$, but the memory required is still $O(N^2)$, and it is this memory bottle-neck that hinders the solution of large scale problems using the boundary element method.

2. The panel clustering method

The reason why the matrix in (5) is dense is because the observation point x^α on the element π_α is related to all the elements π_β on the boundary through the kernels $u_{lk}^*(x^\alpha, y)$ and $p_{lk}^*(x^\alpha, y)$.

The panel clustering method [2] and the multipole expansion method [1] overcome this problem by approximating the integral kernels using expansions over clusters of elements which are sufficiently far from x^α, thus reducing the amount of computation and required memory.

In a previous paper [3], we proposed an efficient method for deriving Taylor expansions of the kernels involving u_{lk}^* and p_{lk}^* in 3-D elastostatics. We showed that the work for the approximate matrix-vector multiplication and memory for BEM is $O(N \log^5 N)$ and $O(N \log^4 N)$, respectively, in order that the error due to the approximation by the expansions is consistent with the error coming from the discretization of the boundary integral equation using boundary elements [2].

In this paper, we will consider using multipole expansions of the fundamental displacement u_{lk}^* in order to further reduce the computational costs.

3. Expansion using spherical harmonics

Since u_{lk}^* is a linear combination of $\dfrac{1}{r'}$ and second order spatial derivatives of r' as in (3), the multipole expansion of u_{lk}^* can be derived from those of $\dfrac{1}{r'}$ and r'.

First, we will review the known multipole expansion of $\dfrac{1}{r'}$ given in [1] and then use them to derive the multipole expansions of r', and finally combine them to derive the multipole expansion of the fundamental displacement $u_{lk}^*(x, y)$.

Let x be the observation point, y the field point and y_τ the center of the cluster τ of boundary elements. From now on, $x - y_\tau$ will be expressed in polar coordinates (r, θ, ϕ) centered at y_τ as $x - y_\tau = r(\sin\theta\cos\phi, \sin\theta\sin\phi, \cos\theta)^T$.

Similarly, $y - y_\tau$ will be expressed in polar coordinates (ρ, α, β) centred at y_τ as $y - y_\tau = \rho(\sin\alpha\cos\beta, \sin\alpha\sin\beta, \cos\alpha)^T$. Let γ be the angle between $x - y_\tau$ and $y - y_\tau$.

Then, we have

$$r'^2 = r^2 + \rho^2 - 2r\rho\cos\gamma, \tag{6}$$

where $\cos\gamma = \cos\theta\cos\alpha + \sin\theta\sin\alpha\cos(\phi-\beta)$. Equation (6) gives

$$r' = r(1 - 2u\sigma + \sigma^2)^{\frac{1}{2}},$$

where $\sigma := \dfrac{\rho}{r}$ and $u := \cos\gamma$. Hence, we have

$$\frac{1}{r'} = \frac{1}{r}(1 - 2u\sigma + \sigma^2)^{-\frac{1}{2}}. \tag{7}$$

3.1. Multipole expansion of $1/r'$. For

$$\sigma := \frac{\rho}{r} = \frac{\|y - y_\tau\|}{\|x - y_\tau\|} < 1,$$

we have

$$(1 - 2u\sigma + \sigma^2)^{-\frac{1}{2}} = \sum_{n=0}^{\infty} \sigma^n P_n(u), \tag{8}$$

where $P_n(u)$ is the Legendre polynomial of order n. Then (7) and (8) give the multipole expansion

$$\frac{1}{r'} \approx \left(\frac{1}{r'}\right)_p := \frac{1}{r}\sum_{n=0}^{p}\sigma^n P_n(u), \tag{9}$$

where the truncation error is estimated by

$$\left|\frac{1}{r'} - \left(\frac{1}{r'}\right)_p\right| = \left|\frac{1}{r}\sum_{n=p+1}^{\infty} P_n(u)\sigma^n\right| \leq \frac{1}{r}\sum_{n=p+1}^{\infty}\sigma^n = \frac{1}{r}\frac{\sigma^{p+1}}{1-\sigma},$$

since $|P_n(\cos\gamma)| \leq 1$ for $\gamma \in \mathbf{R}$.

3.2. Addition theorem for Legendre polynomials. Next, making use of the addition theorem for Legendre polynomials

$$P_n(\cos\gamma) = \sum_{m=-n}^{n} Y_n^{-m}(\alpha,\beta)Y_n^m(\theta,\phi), \tag{10}$$

where

$$Y_n^m(\theta,\phi) := \sqrt{\frac{(n-|m|)!}{(n+|m|)!}}P_n^{|m|}(\cos\theta)e^{im\phi}$$

are the spherical harmonics, (9) gives

$$\left(\frac{1}{r'}\right)_p = \frac{1}{r}\sum_{n=0}^{p}\sigma^n\sum_{m=-n}^{n}Y_n^{-m}(\alpha,\beta)Y_n^m(\theta,\phi) = \sum_{n=0}^{p}\sum_{m=-n}^{n}r^{-n-1}Y_n^m(\theta,\phi)\cdot\rho^n Y_n^{-m}(\alpha,\beta)$$

$$\tag{11}$$

so that the contribution from the field point $y - y_\tau : (\rho,\alpha,\beta)$ can be separated from that of the observation point $x - y_\tau : (r,\theta,\phi)$.

Recalling that we want to evaluate $\int_\tau u_{lk}^*(x,y)p_k(y)\mathrm{d}\Gamma_y$ over a cluster of elements $\tau = \cup_{\beta=1}^t \pi_\beta$, where $u_{lk}^*(x,y)$ is given by (3), we obtain

$$\int_\tau \frac{1}{r'}\delta_{lk}p_k(y)\mathrm{d}\Gamma_y = \int_\tau \frac{p_l}{r'}\mathrm{d}\Gamma_y \approx \sum_{n=0}^p \sum_{m=-n}^n r^{-n-1}Y_n^m(\theta,\phi)\sum_{\beta=1}^t p_{l\beta}J_\beta^{n,-m}, \qquad (12)$$

where $J_\beta^{n,-m} := \int_{\pi_\beta} \rho^n Y_n^{-m}(\alpha,\beta)\mathrm{d}\Gamma_y$ are the far-field coefficients, and we have used constant elements.

3.3. Multipole expansion of r'. Next, we derive the multipole expansion for the distance

$$r' := \|y - x\| = r(1 - 2u\sigma + \sigma^2)^{\frac{1}{2}},$$

which is also necessary for the multipole expansion of the fundamental displacement $u_{lk}^*(x,y)$ of 3-D elastostatics given in (3).

In order to make use of (8), we note that

$$\frac{\mathrm{d}}{\mathrm{d}\sigma}\left\{(1 - 2u\sigma + \sigma^2)^{\frac{1}{2}}\right\} = (\sigma - u)(1 - 2u\sigma + \sigma^2)^{-\frac{1}{2}} = (\sigma - u)\sum_{n=0}^\infty \sigma^n P_n(u).$$

Integrating by σ gives

$$\int_0^\sigma \frac{\mathrm{d}}{\mathrm{d}\sigma}\left\{(1 - 2u\sigma + \sigma^2)^{\frac{1}{2}}\right\}\mathrm{d}s = \int_0^\sigma (s - u)\sum_{n=0}^\infty P_n(u)s^n\mathrm{d}s,$$

i.e.,

$$(1 - 2u\sigma + \sigma^2)^{\frac{1}{2}} = 1 - u\sigma + \sum_{n=2}^\infty \frac{\sigma^n}{n}\left\{P_{n-2}(u) - uP_{n-1}(u)\right\}.$$

Further, the recursion

$$(n + 1)P_{n+1}(u) - (2n + 1)uP_n(u) + nP_{n-1}(u) \equiv 0$$

gives

$$(1 - 2u\sigma + \sigma^2)^{\frac{1}{2}} = \sum_{n=0}^\infty \left(\frac{\sigma^2}{2n + 3} - \frac{1}{2n - 1}\right)\sigma^n P_n(u).$$

Hence, we obtain

$$r' = r(1 - 2u\sigma + \sigma^2)^{\frac{1}{2}} \approx r_p' := r\sum_{n=0}^p \left(\frac{\sigma^2}{2n + 3} - \frac{1}{2n - 1}\right)\sigma^n P_n(u),$$

where the truncation error can be estimated by

$$|r' - r_p'| = r\sum_{n=p+1}^\infty \left|\frac{2n + 3 - (2n - 1)\sigma^2}{(2n + 3)(2n - 1)}\sigma^n P_n(u)\right| \leq \frac{r}{2p + 1}\frac{\sigma^{p+1}}{1 - \sigma}.$$

182

Hence, applying the addition theorem of (10), we obtain

$$r_p' = r \sum_{n=0}^{p} \left(\frac{\sigma^2}{2n+3} - \frac{1}{2n-1} \right) \sigma^n \sum_{m=-n}^{n} Y_n^{-m}(\alpha,\beta) Y_n^m(\theta,\phi), \tag{13}$$

so that the contribution from the field point and the observation point can be separated as in (11) for $1/r'$. We note that (11) and (13) have only $\sum_{n=0}^{p}(2n+1) = (p+1)^2$ and $2(p+1)^2$ terms, respectively, whereas the use of three-dimensional Taylor expansions around y_τ as in [3] requires $O(p^3)$ terms in order to obtain p-th order expansions. Thus, the required work and memory for the proposed multipole expansion is expected to be less compared to those of the Taylor expansion.

Then, noting as in [5] that $\dfrac{\partial r'}{\partial y_l} = -\dfrac{\partial r'}{\partial x_l}$, we obtain

$$\int_\tau r'_{,lk} p_k \mathrm{d}\Gamma_y \approx$$

$$-\sum_{n=0}^{p} \frac{1}{2n+3} \sum_{m=-n}^{n} \frac{\partial}{\partial x_l}\{r^{-n-2}Y_n^m(\theta,\phi)\} \sum_{\beta=1}^{t} p_{k\beta} \int_{\pi_\beta} \frac{\partial}{\partial y_k}\{\rho^{n+2}Y_n^{-m}(\alpha,\beta)\}\mathrm{d}\Gamma_y$$

$$+\sum_{n=0}^{p} \frac{1}{2n-1} \sum_{m=-n}^{n} \frac{\partial}{\partial x_l}\{r^{-n}Y_n^m(\theta,\phi)\} \sum_{\beta=1}^{t} p_{k\beta} \int_{\pi_\beta} \frac{\partial}{\partial y_k}\{\rho^{n}Y_n^{-m}(\alpha,\beta)\}\mathrm{d}\Gamma_y. \tag{14}$$

Equations (12) and (14) give

$$\int_\tau u_{lk}^* p_k \mathrm{d}\Gamma \approx \frac{1}{4\pi\mu} \sum_{n=0}^{p} \left[\sum_{m=-n}^{n} r^{-n-1}Y_n^m(\theta,\phi) \sum_{\beta=1}^{t} p_{l\beta} \int_{\pi_\beta} \rho^n Y_n^{-m}(\alpha,\beta)\mathrm{d}\Gamma \right.$$

$$+ \frac{1}{4(1-2n)+3} \sum_{m=-n}^{n} \frac{\partial}{\partial x_l}\{r^{-n-2}Y_n^m(\theta,\phi)\} \sum_{\beta=1}^{t} p_{k\beta} \int_{\pi_\beta} \frac{\partial}{\partial y_k}\{\rho^{n+2}Y_n^{-m}(\alpha,\beta)\}\mathrm{d}\Gamma_y$$

$$\left. - \frac{1}{2n-1} \sum_{m=-n}^{n} \frac{\partial}{\partial x_l}\{r^{-n}Y_n^m(\theta,\phi)\} \sum_{\beta=1}^{t} p_{k\beta} \int_{\pi_\beta} \frac{\partial}{\partial y_k}\{\rho^{n}Y_n^{-m}(\alpha,\beta)\}\mathrm{d}\Gamma_y \right] .$$

The multipole expansions for $\int_\tau p_{lk}^*(x,y) u_k(y)\mathrm{d}\Gamma_y$ can be obtained similarly by using (3) and (4).

By using the proposed multipole expansion, the computational work and memory is expected to be further reduced to $O(N \log^4 N)$ and $O(N \log^3 N)$, from $O(N \log^5 N)$ and $O(N \log^4 N)$ for the Taylor expansion, respectively.

References

1. L. Greengard, *The rapid evaluation of potential fields in particle systems*, MIT Press, Cambridge, 1988.

2. W. Hackbusch and Z.P. Nowak, On the fast matrix multiplication in the boundary element method by panel clustering, *Numer. Math.* **54** (1989), 463–491.

3. K. Hayami and S.A. Sauter, Cost estimation of the panel clustering method applied to 3-D elastostatics, in *Boundary element research in Europe*, Computational Mechanics Publications, Southampton, 1998, 33–42.

4. K. Hayami and S.A. Sauter, Panel clustering for 3-D elastostatics using spherical harmonics, in *Boundary elements XX*, Computational Mechanics Publications, Southampton, 1998, 289–298.

5. K. Yoshida, N. Nishimura and S. Kobayashi, Analysis of three-dimensional elastostatic crack problems with fast multipole boundary integral equation method, *J. Appl. Mech.* **1** (1998) (to appear) (Japanese).

Department of Mathematical Engineering, University of Tokyo, 7-3-1, Hongo, Bunkyo-ku, 113-8656 Tokyo, Japan; email: hayami@simplex.t.u-tokyo.ac.jp

Fakultät für Mathematik und Informatik, Universität Leipzig, Augustusplatz 10/11, Leipzig, Germany; e-mail: sas@mathematik.uni-leipzig.de

184

J.W. HILGERS, B.S. BERTRAM and W.R. REYNOLDS

Extensions of constrained least squares for obtaining regularized solutions to first kind integral equations

1. Introduction

It is well known that many imaging problems require the inversion of a Fredholm integral equation of the first kind,

$$\int K(x - y)f(y)dy = g(x), \tag{1}$$

and that such problems are ill-posed in that the solution, $f(y)$, depends discontinuously on the data $g(x)$. One way to stabilize the problem is to seek a constrained, least squares solution to (1):

$$\text{minimize}_f \ \|Kf - g\| \quad \text{subject to} \quad \|Lf\| \leq r, \tag{2}$$

any solution of which must satisfy

$$(K^*K + \alpha L^*L)f = K^*g. \tag{3}$$

In (2) and (3) L is the regularization operator and α, the regularization parameter, is really a Lagrange multiplier which depends on r in (2). The norm is taken as the usual L_2 norm. Assuming $N(L) \cap N(K) = \{0\}$, (3) can be uniquely solved as

$$f_\alpha = (K^*K + \alpha L^*L)^{-1}K^*g. \tag{4}$$

The actual problem (1) is invariably encountered in noise-contaminated form,

$$Kf = \bar{g} = g + \epsilon$$

where ϵ represents additive error in the measured data as well as round-off error due to finite length computer arithmetic. If the true solution is f_o, satisfying $Kf_o = g$, then a problem some four decades old is how to choose α in (4) so that $\|f_\alpha - f_o\|$ is a minimum, when, of course, f_o is not known.

2. Generalizations and extensions

A number of researchers [1–6] have inquired regarding the optimal choice for L in (4) as well as α. The following theorem proved in [9,10] provides a partial answer to this query.

Consider the discretized version of (1)–(4) where K and L will now denote matrix representations for the kernel, or point spread function, and regularization operator in (1)–(4), respectively. Equation (4) then is

$$f_\alpha = (K^tK + \alpha L^tL)^{-1}K^t\bar{g}. \tag{5}$$

Then the question of how L and α can be chosen for optimal recovery is given, in one sense, by

Theorem 1. *Let the SVD of $L^t L$ be*

$$L^t L = N S^2 N^t$$

where N is orthogonal and S^2 diagonal with positive entries. Then matrices N and S^2 exist such that $f_\alpha|_1 = f_o$ if and only if

$$\langle \epsilon, g \rangle = \epsilon^t g > 0.$$

Note that signal to noise ratio is irrelevant, just the relative orientation of ϵ and g is pertinent. The theorem was proven in [3].

The main importance of the theorem is that a great deal can be gained by varying the singular values of L independently, rather than by the same ratio as when α in (5) is varied.

Exactly, how, or by what strategy, the entries in S^2 should be determined is the obvious problem. Several approaches were examined in [9,10]. These provided some improvement over using (5) directly, but the results were not dramatic, amounting to only a few percent.

What is clear, however, is that each row of L determines the type of regularization applied at the corresponding point in the domain of f_α and the corresponding entry in S^2 dictates the amount of this regularization applied. Depending on the nature of the true solution, f_0, the amount of regularization applied at various places in $[a, b]$ should be quite different. The way to implement this idea is to use (5) to establish an initial construction, then take regions of rapid change, which could indicate an edge, and isolate these points from their complement where f_α is slowly varying or possibly ringing. Thus a two (or more) set partitioning of [a,b] is established, and each component has its own value of α. We define L_i ($i = 1, 2$), to be L on set i and zero elsewhere. We see that (5) now becomes

$$f_{\alpha_1,\alpha_2} = \left[K^t K + \alpha_1 L_1^t L_1 + \alpha_2 L_2^t L_2 \right]^{-1} K^t \bar{g}, \tag{6}$$

and it is clear if $\alpha_1 = \alpha_2$ (6) reverts to (5). But the hope is by choosing the partition wisely, great improvement can be obtained by varying the amount of regularization at different places in [a,b]. To make this point quantitatively, consider

Example 1. We choose the kernel $K(x)$ in (1) to be .02 for $-1/2 < x < 1/2$ and zero otherwise. $f_o(x)$ is a piecewise constant with values one, two, and four (see Fig. 1b). The true image, g, is computed and zero mean Gaussian noise with standard deviation .001 is added. Fig. 1a shows the optimal reconstruction obtained using (5) with a discretized derivative regularizer L of size 101×101. The optimal α is $\alpha_{opt} = 10^{-6}$ and the optimal precision attained is $\|f_{opt} - f_o\| = 1.64$ (the norm being Euclidean). Based on Fig. 1a, the regions around the discontinuities should be partitioned away from the rest of $[-1, 1]$. In particular, $L_1 = L$ except for rows 18–22,38–42,58–62, and 78–82 which are zero. $L_2 = L$ except for the complement of the above rows. Then (6) is used and a minimum sought. Figure 1b shows the solution corresponding to the minimum with $\alpha_1 = 10^{-4}$ and $\alpha_2 = 10^{-16}$ and $\|f_{opt} - f_o\| = .73$, an improvement of over 50%. Note the very small α_2. The regularization has been turned essentially off at the discontinuities and increased two orders of magnitude elsewhere. If α_2 is actually set to zero, the matrix inverted in (6) is singular. Since computations were done with MATLAB, the 10^{-16} is essentially a double precision eps value.

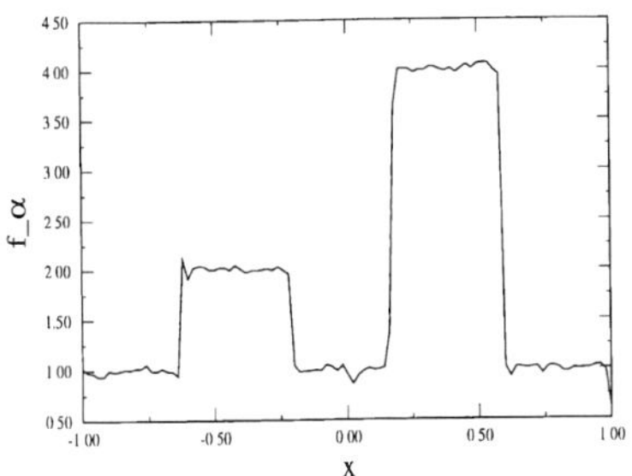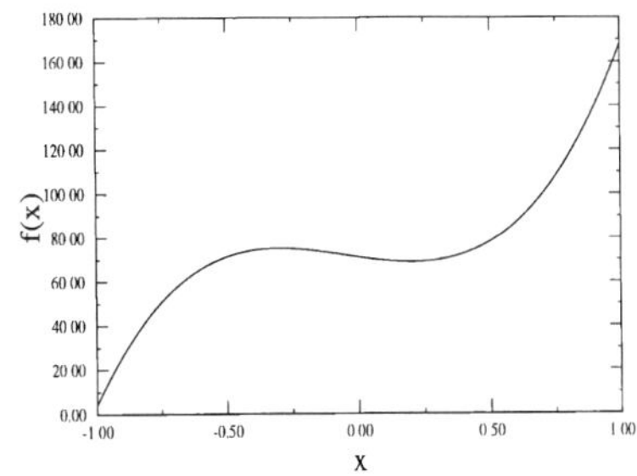

Fig. 1a and 1b.

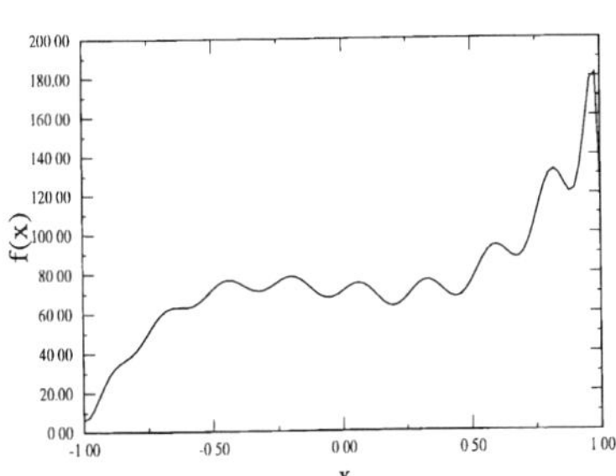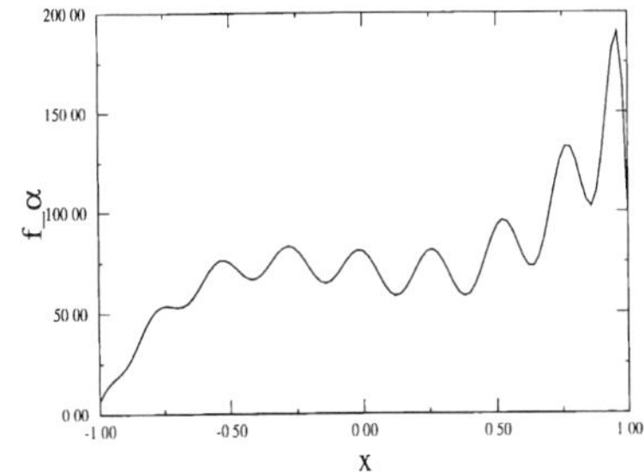

Fig. 2a and 2b.

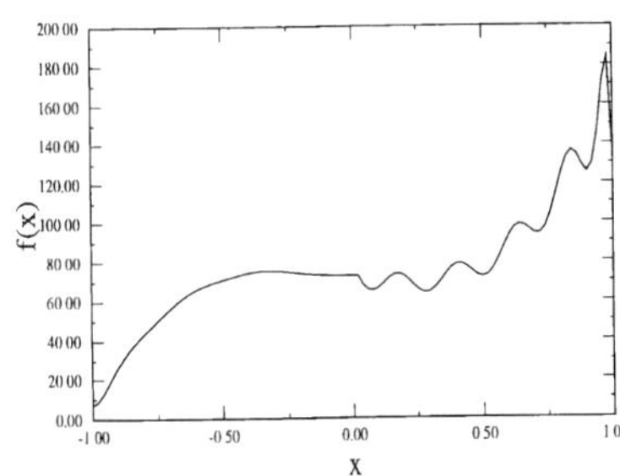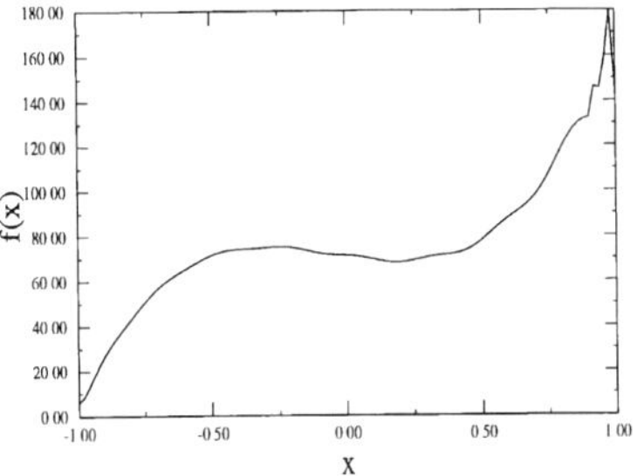

Fig. 3a and 3b.

Example 2. The same setup as above is used, but this time the even more ill-conditioned sinc kernel

$$K(x) = \frac{10\sin(20x)}{\pi x}$$

is employed. The true function is taken as

$$f_o(x) = 100x^3 + 15x^2 - 18x + 71.$$

Fig. 2a and 2b show the optimal reconstructions obtained using the first and second discretized derivatives, respectively. In Fig. 2a $\alpha_{opt} = 10^{-8}$ and $\|f_{opt} - f_o\| = 73$. In Fig. 2b $\alpha_{opt} = 10^{-5}$ and $\|f_{opt} - f_o\| = 124$.

The ringing apparent in Fig. 2a and 2b is caused by the fluctuation at $x = 1$. This, in turn, is due to the way the derivatives are discretized. In the case of the first derivative the last row of L has only a one in the last column, thus introducing $f_\alpha(1)$ into the penalty function rather than a difference.The least square process tries to keep this small, which accounts for the drop at $x = 1$.

To some extent this is endemic to the regularization process. Theoretically, f_α belongs to some Sobolev space which might carry boundary conditions at the end points. If f_o does not satisfy these conditions, ringing can result. These subsidiary conditions can be removed by enlarging the space (see [8]) although this complicates (3) and (4).

Now we have another way to confront this problem. Figure 3a shows the optimal reconstruction using (6) with partitioned rows 1-51 and 52-101. The optimal parameters are $\alpha_1 = 10^{-8}$ and $\alpha_2 = 10^{-5}$ with $\|f_{opt} - f_o\| = 61$. Note the dramatic way the ringing generated at the right boundary is eliminated on $[-1, 0]$. Obviously, it makes sense to move the partition point to the right, isolating the right boundary as much as possible. If the partition is made with rows 1-95 and 96-101, the optimal parameters are $\alpha_1 = 10^{-8}$ and $\alpha_2 = 10^{-6}$ and $\|f_{opt} - f_o\| = 32$. Again, this is over a 50% reduction in the norm of the error. The result is shown in Fig. 3b. If the same partition is used with the second derivative, the optimal parameters are $\alpha_1 = 10^{-5}$ and $\alpha_2 = 1$. The error norm is $\|f_{opt} - f_o\| = 34$, down from 124.

3. Summary and future work

The numerical examples demonstrate that a great deal of improvement in reconstruction quality can be realized by varying the amount of regularization throughout the interval of interest. There are many directions this idea may be taken to generalize (6). For example,

1. The regularization operator, L, rather than just α, could be varied on each component of the partition.

2. Vary the partition itself. In the forgoing the original reconstruction was used to suggest a partition. But it was the only partition tried. Varying the components might lead to significant improvement.

3. Refine the partition. We have used only two components. But more could be tried, although with each new component comes an additional variable.

Finally, in this paper we have used knowledge of f_o to show how much improvement is possible, by varying $\|f_\alpha - f_o\|$. In the practical problem, the parameter choice problem is well known, and many methods exist [11,12]. We have had success with cross-referencing methods [7]. We intend to extend this methodology to the current multivariable case.

References

1. A.K. Katsaggelos, Iterative image restoration, *Optical Engrg.* **28** (1989), 735–748.

2. S.J. Reeves and R.M. Mersereau, Optimal estimation of the regularization parameter and stabilizing functional for regularized image restoration, *Optical Engrg.* **29** (1990), 446–454.

3. J. Hilgers and W. Reynolds, Existence of an optimal regularization operator, SGR TR998, Technical Report, Signature Research, Inc.

4. C.W. Groetsch, *The theory of Tikhonov regularization for Fredholm equations of the first kind*, Pitman, London, 1995.

5. G.H. Golub, M. Heath and G. Wahba, Generalized cross-validation as a method for choosing a good ridge parameter, *Technometrics* **21** (1979), 215–223.

6. J.H. Hilgers and W.R. Reynolds, Instabilities in the optimal regularization parameter relating to image recovery problems, *JOSA-A* **9** (1992), 1273–1279.

7. E.M. Wilcheck, A comparison of cross-referencing and generalized cross validation in the optimal regularization of first kind equations, Masters thesis, Michigan Technological University, 1992.

8. J.W. Hilgers, Non-iterative methods for solving operator equations of the first kind, Ph.D. thesis, University of Wisconsin, 1973 (Math. Research Center TSR # 1413, 1974).

9. J.W. Hilgers, B.S. Bertram, M.M. Alger and W.R. Reynolds, Extensions of the cross referencing method for choosing good regularized solutions to image recovery problems, in *Proc. SPIE*, Barbour et al., Eds., **3171**, 1997, 234–237.

10. M.M. Alger, J.W. Hilgers, B.S. Bertram and W.R. Reynolds, An extension of the Tikhonov regularization based on varying the singular values of the regularization operator, *Proc. SPIE*, Barbour et al., eds., **3171**, 1997, 225–231.

11. C.W. Groetsch, *The theory of Tikhonov regularization for Fredholm equations of the first kind*, Pitman, **105**, London, 1995.

12. M. Hanke, *Conjugate gradient type methods for ill-posed problems*, Pitman Res. Notes Math. Ser. **327**, Longman Scientific and Technical, New York, 1995.

13. A. Neumaier, Solving ill-conditioned and singular linear systems: a tutorial on regularization, *SIAM Rev.* **40** (1998), 636–666.

JWH and WRR: Signature Research Inc., Calumet, Michigan, USA

JWH and BSB: Department of Mathematical Sciences, Michigan Tehcnological University, Houghton, Michigan, USA

R. KANNAN, S. SEIKKALA and M. HIHNALA

Existence and nonexistence results for some boundary value problems at resonance

1. Introduction

We shall consider the differential equation

$$u'' + \pi^2 u + g(x, u') = p(x), \qquad 0 < x < 1, \tag{1}$$

with Dirichlet boundary conditions

$$u(0) = u(1) = 0 \tag{2}$$

and the differential equation

$$u'' + g(x, u') = p(x), \qquad 0 < x < 1, \tag{3}$$

with periodic boundary conditions

$$u(0) = u(1), \quad u'(0) = u'(1) \tag{4}$$

or Neumann boundary conditions

$$u'(0) = u'(1) = 0. \tag{5}$$

In each case we have a boundary value problem at resonance. Indeed, the null space of the linear differential operator $Lu = u'' + \pi^2 u$ with the boundary conditions (2) is spanned by $\phi(x) = \sqrt{2}\sin(\pi x)$ and the null space of $Lu = u''$ with the boundary conditions (4) or (5) by $\phi(x) \equiv 1$. As it is shown in [1], [2], and [3] for the problem (1), (2) the conditions for existence, nonexistence and multiplicity of solutions can be given in terms of the coefficient $\bar{p}$ of the decomposition

$$p(x) = \bar{p}\phi(x) + \tilde{p}(x) \tag{6}$$

where the function $\tilde{p}$ is orthogonal to ϕ. There exists an interval (a, b) such that problem (1), (2) has at least one solution if $\bar{p} \in (a, b)$ and no solution if $\bar{p} \notin [a, b]$, under certain assumptions on g. We shall use a constructive method to find that interval numerically and also derive *a priori* bounds for $|\bar{p}|$ for nonexistence of solutions for the above mentioned boundary value problems.

2. The alternative method

In each of the above problems let $k(x, y)$ be a modified Green's function satisfying, as a function of x, the given boundary conditions and

$$\begin{cases} Lk(x, y) = \delta(x - y) - \phi(x)\phi(y) \\ \int_0^1 k(x, y)\phi(x)dx = 0 \end{cases} \tag{7}$$

where δ is the Dirac's delta function.

Then the problems can be written in the form of a pair of equations,

$$u_\lambda(x) = \lambda\phi(x) + \int_0^1 k(x,y)[\tilde{p}(y) - g(y, u'_\lambda(y))]dy, \qquad (8)$$

$$\bar{\delta}(\lambda) = \bar{p} - \int_0^1 g(x, u'_\lambda(x))dx = 0. \qquad (9)$$

3. The results

Theorem 3.1. [3] *Let $g : [0,1] \times \mathbf{R} \to \mathbf{R}$ be continuous and either be bounded and locally Lipschitz continuous or satisfy the global Lipschitz condition*

$$|g(x,u) - g(x,\bar{u})| \le M|u - \bar{u}|, 0 \le x \le 1, u, \bar{u} \in \mathbf{R}, \qquad (10)$$

where $M \le \frac{3\pi}{2}$. Then there exists an interval $[a,b]$ such that problem (1),(2) has (i) at least one solution if $\bar{p} \in (a,b)$ and (ii) no solution if $\bar{p} \notin [a,b]$.

Theorem 3.2. *Assume that $g : [0,1] \times \mathbf{R} \to \mathbf{R}$ is continuous and satisfies*

$$|g(x,u) - g(x,\bar{u})| \le m_1(x)|u - \bar{u}|, 0 \le x \le 1, u, \bar{u} \in \mathbf{R}, \qquad (11)$$

$$|\tilde{p}(x) - g(x,u)| \le m_0(x), 0 \le x \le 1, u \in \mathbf{R}, \qquad (12)$$

and

$$|\int_0^1 g(x, k\cos(\pi x))sin(\pi x)dx| \le g_0, 0 \le x \le 1, k \in \mathbf{R}, \qquad (13)$$

for some nonnegative functions $m_1, m_0 \in L_1(0,1)$ and for a constant g_0. Then, if

$$|\bar{p}| > \sqrt{2} \int_0^1 \int_0^1 m_0(y)m_1(x)|k_x(x,y)|\sin(\pi x)dx\,dy + \sqrt{2}g_0, \qquad (14)$$

the problem (1),(2) does not have a solution.

Proof. If u_λ is a solution of the integral equation (8),

$$u_\lambda(x) = \lambda\sqrt{2}\sin(\pi x) + \int_0^1 k(x,y)[\tilde{p}(y) - g(y, u_\lambda(y))]dy, \qquad (15)$$

then

$$u'_\lambda(x) = \lambda\sqrt{2}\pi\cos(\pi x) + v(x), \qquad (16)$$

where

$$v(x) = \int_0^1 k_x(x,y)[\tilde{p}(y) - g(y, u_\lambda(y))]dy \qquad (17)$$

satisfies, by (12),

$$|v(x)| \le \int_0^1 |k_x(x,y)|m_0(y)dy. \tag{18}$$

We write

$$g(x, u'_\lambda(x)) = g(x, \lambda\sqrt{2}\pi\cos(\pi x)) + w(x) \tag{19}$$

where

$$w(x) = g(x, \lambda\sqrt{2}\pi\cos(\pi x) + v(x)) - g(x, \lambda\sqrt{2}\pi\cos(\pi x)) \tag{20}$$

satisfies, by the Lipschitz condition (11), the inequality

$$|w(x)| \le m_1(x)|v(x)|. \tag{21}$$

Now, by using (19), inequalities (13), (21), (18) and inequality (14), we obtain

$$\begin{aligned}
|\int_0^1 g(x, u'_\lambda(x))\sin(\pi x)dx| &\le g_0 + \int_0^1 m_1(x)|v(x)|\sin(\pi x)dx \\
&\le g_0 + \int_0^1\int_0^1 m_0(y)m_1(x)|k_x(x,y)|\sin(\pi x)dx\,dy \\
&\le \frac{|\bar p|}{\sqrt{2}}.
\end{aligned}$$

Then

$$\bar\delta(\lambda) = \bar p - \int_0^1 g(x, u_\lambda(x))\sqrt{2}\sin(\pi x)dx \ne 0$$

for all $\lambda \in \mathbf{R}$, which means that problem (8),(9) and hence problem (1), (2) cannot have a solution.

Example 3.1. The boundary value problem

$$\begin{cases} u'' + \pi^2 u + (x - \frac{1}{2})^2\sin u' = \bar p\sqrt{2}\sin(\pi x) + x - \frac{1}{2} \\ u(0) = u(1) = 0 \end{cases} \tag{22}$$

does not have a solution if

$$|\bar p| > \frac{3}{2}\sqrt{2}\int_0^1\int_0^1 (x - \frac{1}{2})^2|y - \frac{1}{2}||k_x(x,y)|\sin(\pi x)dxdy \approx 0.003375.$$

Indeed, we now have $g(x,u) = (x - \frac{1}{2})^2\sin u$, $g_0 = 0$, $m_0(x) = \frac{3}{2}|x - \frac{1}{2}|$ and $m_1(x) = (x-\frac{1}{2})^2$ in Theorem 3.2. We can also apply Theorem 3.1, since g satisfies the Lipschitz condition 3.1 (with $M = \frac{1}{4}$). By solving (22) numerically we have obtained that the interval of existence for $\bar p$ is $(a,b) \approx (-0.0014089, 0.0011733)$.

Next we will consider the differential equation (3),

$$u'' + g(x, u') = p(x), \ 0 < x < 1,$$

with periodic boundary conditions (4) or with Neumann boundary condition (5). In both cases $\phi(x) \equiv 1$ in the decomposition of p.

Theorem 3.3. *Assume that $g : [0,1] \times \mathbf{R} \to \mathbf{R}$ is continuous and satisfies*

$$|g(x,u)| \leq m_1(x)|u|, \ 0 \leq x \leq 1, \ u \in \mathbf{R}, \tag{23}$$

and

$$|\tilde{p}(x) - g(x,u)| \leq m_0(x)|u|, \ 0 \leq x \leq 1, \ u \in \mathbf{R}, \tag{24}$$

for some nonnegative functions $m_1, m_0 \in L_1(0,1)$. If

$$|\bar{p}| > \int_0^1 \int_0^1 m_0(y) m_1(x) |k_x(x,y)| dx dy, \tag{25}$$

then the boundary value problem (3), (4) does not have a solution. The same result holds true for the problem (3), (5).

Proof. In both problems a solution (if it exists) satisfies (8) and (9) and is given by

$$u_\lambda(x) = \lambda + \int_0^1 k(x,y)[\tilde{p}(y) - g(y, u_\lambda'(y))] dy.$$

Then

$$u_\lambda'(x) = \int_0^1 k_x(x,y)[\tilde{p}(y) - g(y, u_\lambda'(y))] dy$$

and, by (24)

$$|u_\lambda'(x)| \leq \int_0^1 |k_x(x,y)| m_0(y) dy.$$

Using (23) and (25) we obtain

$$\begin{aligned}
\left| \int g(x, u_\lambda'(x)) dx \right| &\leq \int_0^1 m_1(x)|u_\lambda'(x)| dx \\
&\leq \int_0^1 m_1(x) \int_0^1 |k_x(x,y)| m_0(y) dy dx \\
&= \int\!\!\!\int_0^1{}_0^1 m_0(y) m_1(x) |k_x(x,y)| dx dy < |\bar{p}|.
\end{aligned}$$

Hence,

$$\bar{\delta}(\lambda) = \bar{p} - \int_0^1 g(x, u_\lambda'(x)) dx \neq 0, \ \lambda \in \mathbf{R},$$

which means that the problems (3), (4) and (3), (5) cannot have a solution.

Example 3.2. For the differential equation

$$u'' + \arctan u' = \bar{p} + x - \frac{1}{2} \tag{26}$$

with Neumann or periodic boundary conditions we have $m_1(x) \equiv 1$ and $m_0(x) \equiv \frac{1+\pi}{2}$. The Landesman-Lazer condition for the differential equations $u'' + \arctan u = \bar{p}$ with Neumann or periodic boundary conditions would allow $|\bar{p}| < \frac{\pi}{2}$ to ensure the existence of a solution. But in the case of Eq. 26, using the estimate (25) we obtain a nonexistence result for the Neumann problem, if $|\bar{p}| > \frac{1+\pi}{6}$ and for the periodic boundary value problem, if $|\bar{p}| > \frac{\pi}{8}$, i.e., both under the Landesman-Lazer condition.

References

1. A. Canada and P. Drabek, On semilinear problems with nonlinearities depending only on derivatives, *SIAM. J. Math. Anal.* **27** (1996), 543–557.

2. P. Habets and L. Sanchez, A two-point problem with nonlinearity depending only on the derivative, *SIAM. J. Math. Anal.* **28** (1997), 1205–1211.

3. R. Kannan and S. Seikkala, Existence of solutions to $u'' + u + g(t, u, u')$ with Dirichlet boundary conditions, preprint.

Department of Mathematics, Faculty of Technology, University of Oulu, Finland

M. KASHTALYAN and C. SOUTIS

Analytic investigation of thick anisotropic plates with undulating surfaces

1. Introduction

Mechanical failure of structures often results from stress concentration at the material surfaces, due to the surface defects inherited from the manufacturing process or acquired from environmental corrosion and impact loadings. The nominally flat surfaces bounding elastically stressed solids appear to be unstable with respect to the formation of surface undulations of wavelengths greater than some critical value [1]. The nature of this instability reflects the nature of surface evolution dictated by the competition between surface and elastic energies.

Although the amplitude of surface undulations is usually small in comparison with their wavelength, they may cause significant stress concentrations near the surface. Stress fields due to surface undulations in isotropic solids (half-planes and half-spaces) under tension were analysed by means of the first-order perturbation approach in [2]. The present paper deals with 3-D elastic stress analysis of anisotropic plates with undulating surfaces under bending loads.

2. Bending of an orthotropic plate (3-D problem)

Suppose an orthotropic plate of finite dimensions $0 \leq x \leq a$, $0 \leq y \leq b$ has undulating bottom and top surfaces S_0, S_1. These surfaces are treated as being slightly perturbed from some reference states in which they are perfectly flat. It is assumed that their geometry can be described by a continuous differentiable (up to any order required) shape function $f(x, y)$ and a small dimensionless parameter ϵ introduced to characterize the magnitude of deviation of an undulating surface S_i from its reference plane $z = h_i$. Shape equations for the perturbed surfaces are given by

$$z = h_i + \epsilon f(x, y), \qquad h_0 = 0, \ h_1 = h. \tag{1}$$

Suppose the plate is loaded by the normal load $Q(x, y)$ at its top surface, while the bottom surface remains free. Boundary conditions at S_1, S_0 are then

$$\sigma_{xt} n_{x,1} + \sigma_{yt} n_{y,1} + \sigma_{zt} n_{z,1} = Q n_{t,1} \tag{2}$$

$$\sigma_{xt} n_{x,0} + \sigma_{yt} n_{y,0} + \sigma_{zt} n_{z,0} = 0 \tag{3}$$

Given the shape equations (1), direction cosines $n_{t,i}$ may be expressed in partial derivatives of the shape function and the small parameter ϵ

$$n_{x,i} = -\frac{\epsilon}{\Delta} \frac{\partial f}{\partial x} \qquad n_{y,i} = -\frac{\epsilon}{\Delta} \frac{\partial f}{\partial y} \qquad n_{z,i} = \frac{1}{\Delta} \qquad i = 0, 1 \tag{4}$$

where

$$\Delta = \pm (1 + \epsilon^2 [(\frac{\partial f}{\partial x})^2 + (\frac{\partial f}{\partial y})^2])^{1/2}. \tag{5}$$

On the edges of the plate ($x = 0, a$ and $y = 0, b$), simple support boundary conditions are assumed to be fulfilled.

3. Perturbation analysis

Due to the complexity of geometrical shape of the surfaces, 3-D boundary-value problem (1)–(5) stated above cannot be solved directly. The boundary shape perturbation method [3] has been applied in order to simplify boundary conditions (2), (3). According to it, stresses and displacements are sought, and load components and direction cosines are represented, in the form of series with respect to the small parameter ϵ,

$$[\sigma_{jt}, u_t, Q, n_{t,i}] = \sum_{p=0}^{\infty} \epsilon^p [\sigma_{jt}^{(p)}, u_t^{(p)}, Q^{(p)}, n_{t,i}^{(p)}]. \tag{6}$$

It is assumed that stresses and displacements in each approximation $\sigma_{jt}^{(p)}$ and $u_t^{(p)}$, if considered at the perturbed surfaces S_0, S_1, can be expanded into Taylor series in the vicinity of the appropriate reference plane $z = h_i$. Then on S_i

$$[\sigma_{jt}, u_t]|_{S_i} = \sum_{p=0}^{\infty} \epsilon^p \sum_{q=0}^{p} \frac{f^q}{q!} \frac{\partial^q}{\partial z^q} [\sigma_{jt}^{(p-q)}, u_t^{(p-q)}]|_{z=h_i}. \tag{7}$$

By means of this approach the initial 3-D boundary-value problem for the plate with perturbed surfaces may be reduced to a sequence of 3-D boundary-value problems for the plate with the perfectly flat faces $z = 0, h$. In the zero-order approximation, i.e., for $p = 0$, boundary conditions at $z = h_i$ take the familiar form

$$\sigma_{zz}^{(0)}|_{z=h} = Q, \quad \sigma_{xz}^{(0)}|_{z=h} = \sigma_{yz}^{(0)}|_{z=h} = 0 \tag{8}$$

$$\sigma_{zz}^{(0)}|_{z=0} = \sigma_{xz}^{(0)}|_{z=0} = \sigma_{yz}^{(0)}|_{z=0} = 0. \tag{9}$$

In the higher order ($p = 1, 2, 3, \ldots$) approximations boundary conditions at the reference planes $z = h_i$ are

$$\sigma_{zt}^{(p)}|_{z=h} = -Q_k^{(p)} + \sum_{q=1}^{p} [N_{x,1}^{(q)} \sigma_{xt}^{(p-q)} + N_{y,1}^{(q)} \sigma_{yt}^{(p-q)} + N_{z,1}^{(q)} \sigma_{zt}^{(p-q)}]_{z=h}, \tag{10}$$

$$\sigma_{zt}^{(p)}|_{z=0} = - \sum_{q=1}^{p} [N_{x,0}^{(q)} \sigma_{xt}^{(p-q)} + N_{y,0}^{(q)} \sigma_{yt}^{(p-q)} + N_{z,0}^{(q)} \sigma_{zt}^{(p-q)}]_{z=0}, \tag{11}$$

where $N_{t,i}^{(q)}$ are differential operators which depend on the shape function f and its derivatives. In the most general form, they are

$$N_{t,i}^{(q)} = - \sum_{s=1}^{q} \frac{f^s}{s!} n_{t,i}^{(q-s)} \frac{\partial^s}{\partial z^s}. \tag{12}$$

In particular, in the first and the second approximations, differential operators $N_{t,i}^{(q)}$ become

$$N_{x,i}^{(1)} = \frac{\partial f}{\partial x} \qquad N_{y,i}^{(1)} = \frac{\partial f}{\partial y} \qquad N_{z,i}^{(1)} = -f \frac{\partial}{\partial z}$$

$$N_{x,i}^{(2)} = f \frac{\partial f}{\partial x} \frac{\partial}{\partial z} \qquad N_{y,i}^{(2)} = f \frac{\partial f}{\partial y} \frac{\partial}{\partial z}$$

$$N_{z,i}^{(2)} = -\frac{1}{2}[f^2 \frac{\partial^2}{\partial z^2} - (\frac{\partial f}{\partial x})^2 - (\frac{\partial f}{\partial y})^2].$$

Since the application of the boundary shape perturbation method transforms only boundary conditions at the perturbed surfaces, equilibrium equations retain their initial form in all approximations, i.e.,

$$C_{11}u_{x,xx}^{(p)} + C_{66}u_{x,zz}^{(p)} + (C_{12} + C_{66})u_{y,xy}^{(p)} + (C_{13} + C_{55})u_{z,xz}^{(p)} = 0$$

$$(C_{12} + C_{66})u_{x,xy}^{(p)} + C_{66}u_{y,xx}^{(p)} + C_{22}u_{y,zz}^{(p)} + (C_{23} + C_{44})u_{z,yz}^{(p)} = 0 \qquad (13)$$

$$(C_{13} + C_{55})u_{x,xz}^{(p)} + (C_{23} + C_{44})u_{y,yz}^{(p)} + C_{55}u_{z,xx}^{(p)} + C_{44}u_{z,yy}^{(p)} + C_{33}u_{z,zz}^{(p)} = 0$$

$$p = 0, 1, 2, \ldots$$

The same is true also for the edge conditions

$$\sigma_{xx}^{(p)}|_{x=0,a} = 0, \quad u_y^{(p)}|_{x=0,a} = u_z^{(p)}|_{x=0,a} = 0$$

$$\sigma_{yy}^{(p)}|_{y=0,b} = 0, \quad u_x^{(p)}|_{y=0,b} = u_z^{(p)}|_{y=0,b} = 0. \qquad (14)$$

4. Solution for wavy surfaces

If the applied load can be expanded in a double trigonometric series

$$Q(x,y) = \sum_{m,n=1}^{\infty} q_{mn} \sin \lambda_m \sin \lambda_n \qquad \lambda_m = \pi m/a, \qquad \lambda_n = \pi n/b \qquad (15)$$

and the undulating surfaces of the plate are described by a shape function

$$f(x,y) = f(y) = h \cos \frac{\omega \pi y}{b} \qquad (16)$$

where ω is a wave generation parameter, then a solution to problem (10)–(14) in any approximation can be found analytically, following [4], in the form

$$u_x^{(p)} = \sum_{s=1}^{p+1} \sum_{m,n=1}^{\infty} \sum_{i=1}^{6} A_{xmni}^{(p,s)} v_{xmni}^{(p,s)}(z/h) \cos \lambda_m x \sin \lambda_n^{(p,s)} y,$$

$$u_y^{(p)} = \sum_{s=1}^{p+1} \sum_{m,n=1}^{\infty} \sum_{i=1}^{6} A_{ymni}^{(p,s)} v_{ymni}^{(p,s)}(z/h) \sin \lambda_m x \cos \lambda_n^{(p,s)} y, \qquad (17)$$

$$u_z^{(p)} = \sum_{s=1}^{p+1} \sum_{m,n=1}^{\infty} \sum_{i=1}^{6} A_{zmni}^{(p,s)} v_{zmni}^{(p,s)}(z/h) \sin \lambda_m x \sin \lambda_n^{(p,s)} y,$$

where (in the first three approximations) $\lambda_n^{(p,s)}$ is

$$\lambda_n^{(0,1)} = \pi n/b \quad \lambda_n^{(1,1)} = (\pi + \omega)n/b \quad \lambda_n^{(1,2)} = (\pi - \omega)n/b \qquad (18)$$

$$\lambda_n^{(2,1)} = \pi n/b \quad \lambda_n^{(2,2)} = (\pi + 2\omega)n/b \quad \lambda_n^{(2,3)} = (\pi - 2\omega)n/b \qquad (19)$$

and $A_{tmni}^{(p,s)}$ are arbitrary constants. The form of functions $v_{tmni}^{(p,s)}(z/h)$ depends on the type of the roots of the characteristic equation associated with the system of simultaneous partial differential equations (13).

ω	$\sigma_{xx}/q\|_{z=h}$ (flat top surface)	$\sigma_{xx}/q\|_{S_1}$ (undulating top surface)	$\delta_{xx}, \%$	$\sigma_{yy}/q\|_{z=h}$ (flat top surface)	$\sigma_{yy}/q\|_{S_1}$ (undulating top surface)	$\delta_{yy}, \%$
2	-16.76	-18.71	11.6	-6.05	-7.58	25.2
6	-16.76	-12.10	-27.8	-6.05	-7.73	27.8
10	-16.76	-11.70	-30.1	-6.05	-7.96	31.5

Table 1. Numerical results.

5. Numerical results

The stress field in the orthotropic plate with the undulating surfaces under bending is illustrated on the following example. The plate dimensions are $h/a = 1/6$, $h/b = 1/12$, and elastic constants of the plate material are taken as

$$c_{22}/c_{11} = 0.543, \quad c_{33}/c_{11} = 0.530,$$

$$c_{12}/c_{11} = 0.233, \quad c_{13}/c_{11} = 0.011,$$

$$c_{23}/c_{11} = 0.098, \quad c_{44}/c_{11} = 0.267,$$

$$c_{55}/c_{11} = 0.160, \quad c_{66}/c_{11} = 0.263.$$

The normal load applied to the plate is

$$Q(x,y) = -q \sin \frac{\pi x}{a} \sin \frac{\pi y}{b}. \tag{20}$$

The plate has wavy top surface S_1 and flat bottom surface S_0. It is assumed that no geometry-induced change in the type of anisotropy has occurred near the undulating surface. The small parameter is $\epsilon = 0.2$, which corresponds to 20% decrease in plate thickness at its center.

Numerical results given in Table 1 show dependence of stress values on the wave generation parameter ω at the undulating top surface in the center of the plate. As a basis of comparison, stress values in the plate of the same dimensions and material, but with the perfectly flat top face $z = h$ are taken. They are provided by the zero-order approximation.

Stress values at the undulating top surface differ considerably from those at the perfectly flat top surface. The difference between stress values at the flat and undulating surfaces can be more than 30%, although surface perturbations are very mild, with the amplitude-to-wavelength ratio less than 0.1.

Two different tendencies for stresses acting in the direction of waving and perpendicularly to it are observed. The former decrease by their absolute value with the increase of the wave generation parameter (or the decrease of the wave-length of perturbations), the latter increase monotonically.

Approximations up to the second order were used in the calculations. In most cases this appeared to suffice to obtain the results with acceptable accuracy. Contribution of the next approximation, estimated using the convergence criterion [3], was not expected to be more than 3.5%.

198

References

1. D.J. Srolovitz, On the stability of surfaces of stresses solids, *Acta Metallurgica* **37** (1989), 621–625.

2. H. Gao, Stress concentration at slightly undulating surfaces, *J. Mech. Phys. Solids* **39** (1991), 443–458.

3. Yu.N. Nemish, *Elements of mechanics of piecewise-homogeneous solids with non-canonical boundary surfaces*, Naukova Dumka, Kiev, 1989 (Russian).

4. N.J. Pagano, Exact solutions for rectangular bidirectional composites and sandwich plates, *J. of Composite Materials* **4** (1970), 20–34.

Timoshenko Institute of Mechanics, Nesterov St. 3, 252057 Kiev, Ukraine

Department of Aeronautics, Imperial College, Prince Consort Road, London, UK

N. KHATIASHVILI

On Stokes's nonlinear integral wave equation

The plane fluid flow problem with a free boundary has been investigated by many authors [1,2,3,4].

In this paper, the planar problem associated with waves in incompressible, heavy fluids is studied, specifically, the model suggested by Lavrent'ev and Shabat [2].

Assume the bottom of the reservoir is planar and the wave moves linearly with constant speed c.

We choose the mobile coordinate system, moving with the wave, with axis oy passing through the maximum point of the wave and the axis ox pointing along the bottom in the direction of movement.

Let us introduce the following notation: $f(z) = \varphi + i\psi$ ($z = x + iy$) is a complex potential, φ is a speed potential, ψ is a stream function, $f'(z)$ a complex speed. In this coordinate system φ does not depend on time.

Problem A. Find the periodic curve $\Gamma : y = y(x)$ such that, if f is a conformal mapping of the area $D = \{0 < y < y(x)\}$ on the strip $\{0 < \psi < q, \ q = \text{const}\}$, $f(\pm\infty) = \pm\infty$, then $\forall z \in \Gamma$ the following condition holds:

$$\frac{1}{2}|f'(z)|^2 + gy = A \quad (A = \text{const}). \tag{1}$$

Problem A has been investigated by numerous authors [2] when $f'(z) \neq 0$.

We consider this problem when $f'(Y) = 0$, where Y is the maximum point of the wave. Such waves are called Stokes's waves: they are peaked at the maximum.

Using a conformal mapping we transform the Problem A to a nonlinear integral equation which we then analyze.

Let x_0 be a period of $f(z)$, and let $f(z)$ be a conformal mapping of one period $OABC$ of the area D on the rectangle $O_1 A_1 B_1 C_1$,

$$
\begin{aligned}
OA &: \quad y = 0, \quad 0 \le x \le x_0; & AB &: \quad x = x_0, \quad 0 \le y \le Y; \\
BC &: \quad y = y(x), \quad 0 \le x \le x_0; & OC &: \quad x = 0, \quad 0 \le y \le Y; \\
O_1 A_1 &: \quad \psi = 0, \quad 0 \le \varphi \le \omega; & A_1 B_1 &: \quad x = \omega, \quad 0 \le \psi \le q; \\
B_1 C_1 &: \quad \psi = q, \quad 0 \le \varphi \le \omega; & O_1 C_1 &: \quad \varphi = 0, \quad 0 \le \psi \le q
\end{aligned}
$$

with ω a constant.

Consider the function $f^{-1}(z_1)$ which is the inverse of $z_1 = f(z)$. From (1) we find that the function $f^{-1}(z_1)$ must satisfy the following conditions:

$$\operatorname{Im} f^{-1}(z_1)|_{\psi=0} = 0, \quad \operatorname{Im} f^{-1}(z_1)|_{\psi=q} = y(x), \tag{2}$$

$$\frac{d}{dz_1} f^{-1}(z_1)\Big|_{\psi=q} = \left(\sqrt{2A - 2g \operatorname{Im} f^{-1}(z_1)}\right)^{-1}. \tag{3}$$

We assume that at the points $z_1 = q + k\omega$, $k = 0, \pm 1, \ldots$, the function $f^{-1}(z_1)$ has an integrable singularity. The function $f^{-1}(z_1)$ has an analytic continuation to the strip $-q < \psi < q$.

Consider the branch of $\ln[f^{-1}(z_1)]'$ for which $\ln 1 = 0$. Equations (2), (3) give the following boundary value problem for $\Phi_1(z_1) = \ln[f^{-1}(z_1)]'$.

Problem B. Find a holomorphic function $\Phi_1(z_1)$ in the strip $-q < y < q$ satisfying the boundary condition

$$\operatorname{Re} \ln[f^{-1}(z_1)]'\big|_{\psi=\pm q} = -\ln\sqrt{2A - 2g\operatorname{Im} f^{-1}(z_1)}; \tag{4}$$

the function $-\ln\sqrt{2A - 2g\operatorname{Im} f^{-1}(z_1)}$ is assumed to belong to the class H_ε^* introduced by N. Muskhelishvili [5].

The function

$$\zeta(z_1) = z_1^{-1}(\zeta), \quad \zeta = \xi + i\eta,$$

$$z_1(\zeta) = C\int_0^\zeta \frac{1}{\sqrt{(a^2 - t^2)(b^2 - t^2)}}\, dt + \frac{\omega}{2} + iq, \qquad C = \text{const}, \tag{5}$$

is a comformal mapping of the rectangle $O_1 A_1 B_1 C_1$ in the upper half-plane, and the rectangle symmetric to $O_1 A_1 B_1 C_1$ by the axix oy on the lower half-plane of the complex variable ζ. $A_1 \leftrightarrow (-b, 0)$, $B_1 \leftrightarrow (-a, 0)$, $C_1 \leftrightarrow (a, 0)$, $O_1 \leftrightarrow (b, 0)$ are the corresponding points, where $\zeta = \xi + i\eta$, a, b, C are the constants.

Let D be a plane cut along the segment $[-a, a]$. By use of (4) and (5) the Problem B transforms to the following problem.

Problem C. Find the piecewise holomorphic function $\Phi(\zeta) = f^{-1}(z_1(\zeta))$, which is bounded at infinity and satisfies the condition

$$\operatorname{Re} \ln \Phi'(\zeta)\big|_{[-a,a]}^{\pm} = \ln|z_1'(\zeta)| \cdot \sqrt{2A - 2g\operatorname{Im}\Phi(\zeta)}^{\,-1}. \tag{6}$$

Assuming that the right-hand side of (6) is known and using the results of the author [5], we can write

$$\ln\Phi'(\zeta) = \frac{1}{\pi i}\int_{-a}^{a} \frac{\sqrt{a^2 - \zeta^2}}{\sqrt{a^2 - t^2}}\left\{\ln|z'(t)|[\sqrt{2A - 2g\operatorname{Im}\Phi(t)}]^{-1}\right\}\frac{1}{t - \zeta}\,dt,$$

$$\int_{-a}^{a} \frac{\ln|z'(t)|[\sqrt{2A - 2g\operatorname{Im}\Phi(t)}]^{-1}}{\sqrt{a^2 - t^2}}\,dt = 0.$$

As $\eta \to 0$, $-a < \xi < a$, we obtain

$$\ln\Phi'(\xi) = \ln|z_1'(\xi)|(\sqrt{2A - 2g\operatorname{Im}\Phi(\xi)})^{-1} +$$
$$+ \frac{1}{\pi i}\int_{-a}^{a} \frac{\sqrt{a^2 - \xi^2}}{\sqrt{a^2 - t^2}}\left\{\ln|z'(t)|\sqrt{2A - 2g\operatorname{Im}\Phi(t)}^{\,-1}\right\}\frac{1}{t - \xi}\,dt. \tag{7}$$

From (7) we get

$$\Phi'(\xi) = |z_1'(\xi)|(\sqrt{2A - 2g\operatorname{Im}\Phi(\xi)})^{-1}$$
$$\times \exp\left\{\frac{1}{\pi i}\int_{-a}^{a} \frac{\sqrt{a^2 - \xi^2}}{\sqrt{a^2 - t^2}}\left\{\ln|z'(t)|\sqrt{2A - 2g\operatorname{Im}\Phi(t)}^{\,-1}\right\}\frac{1}{t - \xi}\,dt\right\}.$$

Then

$$\operatorname{Im} \Phi'(\xi) \;=\; |z_1'(\xi)|(\sqrt{2A - 2g\operatorname{Im}\Phi(\xi)})^{-1}$$
$$\times \sin\left\{\frac{1}{\pi}\int_{-a}^{a}\frac{\sqrt{a^2-\xi^2}}{\sqrt{a^2-t^2}}\left\{\ln|z'(t)|\sqrt{2A-2g\operatorname{Im}\Phi(t)}^{\,-1}\right\}\frac{1}{t-\xi}\,dt\right\}.\tag{8}$$

If we introduce the notation

$$(2A - 2g\operatorname{Im}\Phi(\xi))^{\frac{3}{2}} = u(\xi),$$

from (8) we get the nonlinear integral equation

$$u(\xi) = -3g\int_{a}^{\xi} z_1'(t)\sin A[u(t)]\,dt, \qquad \xi\in[-a,a],\tag{9}$$

where

$$A[u(t)] = \frac{\sqrt{a^2-t^2}}{\pi}\int_{-a}^{a}\frac{\ln|z'(\tau)| - \frac{2}{3}\ln u(\tau)}{(\tau-t)\sqrt{a^2-\tau^2}}\,d\tau.$$

Since three parameters of the conformal mapping can be chosen arbitrarily let a be sufficiently small and we can write

$$\sin A[u(t)] \sim A[u(t)].$$

In the following we will demand that the wave be symmetric, $u(\xi) = u(-\xi)$. Transforming the integral (9) we obtain the integral equation

$$u(\xi) = -\frac{3g}{4\pi}\int_{0}^{a} z'(t)\left[\ln|z'(t)| - \frac{2}{3}\ln u(t)\right]K(t,\xi)\,dt,\tag{10}$$

where

$$K(t,\xi) = 2\ln\frac{\sqrt{b^2-\xi^2}+\sqrt{b^2-t^2}}{\sqrt{b^2-a^2}+\sqrt{b^2-t^2}} + \ln\left|\frac{a^2-t^2}{t^2-\xi^2}\right|, \qquad \xi\in[0,a].$$

Substituting

$$t = a\sin\theta, \qquad \xi = a\sin\theta_0,$$

equation (10) becomes

$$u_0(\theta_0) \;=\; \frac{-3gC}{4\pi}\int_{0}^{\frac{\pi}{2}}\frac{1}{\sqrt{b^2-a^2\sin^2\theta}}$$
$$\left\{\ln\frac{C}{\sqrt{b^2-a^2\sin^2\theta}\,a\cos\theta} - \frac{2}{3}\ln u_0(\theta)\right\}K_0(\theta,\theta_0)\,d\theta\tag{11}$$

$$K_0(\theta,\theta_0) = 2\ln\frac{\sqrt{b^2-a^2\sin^2\theta_0}+\sqrt{b^2-a^2\sin^2\theta}}{\sqrt{b^2-a^2}+\sqrt{b^2-a^2\sin^2\theta}} + \ln\frac{\cos^2\theta}{|\sin(\theta_0-\theta)\sin(\theta_0+\theta)|}$$

where $\theta_0 \in [0,\frac{\pi}{2}]$, $u_0(\theta) = u(a\sin\theta)$, and $b^2 - a^2\sin^2(\theta) > 0$.

202

Using the representations

$$\begin{aligned}
\sin\theta &= \theta \ \prod_{n=1}^{\infty}\left(1 - \frac{\theta^2}{n^2\pi^2}\right), \\
\cos\theta &= \prod_{n=1}^{\infty}\left(1 - \frac{4\theta^2}{(2n-1)^2\pi^2}\right),
\end{aligned}$$

and the behavior of the singular integral near the ends of the line of integration [5,6],

$$\int_0^{\frac{\pi}{2}} \ln a\left(\frac{\pi}{2} - \theta_0\right)\frac{1}{\theta_0 - \theta}\,d\theta = -\frac{1}{2}\ln^2 a\left(\frac{\pi}{2} - \theta_0\right) + C_0 \ln a\left(\frac{\pi}{2} - \theta_0\right) + f^*(\theta_0),$$

where

$$f^*(\theta_0) \in H\left[0, \frac{\pi}{2}\right], \qquad C_0 \text{ is a constant,}$$

we obtain

$$u_0(\theta_0) = -\frac{3gC}{8\pi\sqrt{b^2 - a^2}}\left(\frac{\pi}{2} - \theta\right)\ln^2 a\left(\frac{\pi}{2} - \theta_0\right) + C_0 \ln a\left(\frac{\pi}{2} - \theta_0\right) + f_0^*(\theta_0),$$

where

$$f_0^*(\theta_0) \in H\left[0, \frac{\pi}{2}\right].$$

We want to find $u_0(\theta_0)$ in the following class of functions which we denote by $M^*[0, \frac{\pi}{2}]$:

1. $u_0(\theta_0) \in C[0, \frac{\pi}{2}]$, $u_0(\theta_1) \leq u_0(\theta_2)$, $\theta_1 \geq \theta_2$,

2. $0 < m \leq \dfrac{u_0(\theta_0)}{(\frac{\pi}{2} - \theta_0)\ln^2 a(\frac{\pi}{2} - \theta_0)} \leq M < 1$.

It is clear that $M^*[0, \frac{\pi}{2}] \subset C[0, \frac{\pi}{2}]$. In addition, this set is bounded, closed and convex, and the operator on the right hand side of (11) is completely continuous. Thus applying Schauder's fixed point theorem we get.

Theorem. *There exists a solution of (11) in the class of functions $M^*[0, \frac{\pi}{2}]$.*

If we introduce the notation

$$v(\theta_0) = \frac{u_0(\theta_0)}{(\frac{\pi}{2} - \theta_0)\ln^2 a(\frac{\pi}{2} - \theta_0)}$$

and the approximation $\ln v = \ln v_0 + \frac{v}{v_0} - 1$, where $v_0 = \lim_{\theta \to \frac{\pi}{2}} v$, then we get the following quasi-regular Fredholm approximation of (12)

$$\left(\frac{\pi}{2} - \theta_0\right)\ln^2 a\left(\frac{\pi}{2} - \theta_0\right)v(\theta_0) = -\frac{3gC}{4\pi v_0}\int_0^a v(t)K_0(t, \theta_0)\,dt + f_0(\theta_0),$$

with $f_0(\theta_0)$ a known function.

References

1. U. Heis and S. Tau, *Nonlinear waves*, Cornel University Press, 1974.

2. M.A. Lavrent'ev and B.V. Shabat, *The problems of hydrodynamics and their mathematical models*, Nauka, Moscow, 1977 (Russian).

3. A.E. Sedov, *The plane problems of hydrodynamics and aerodynamics*, Nauka, Moscow, 1980 (Russian).

4. W.F. Ames, *Nonlinear partial differential equations in engineering*, Academic Press, New York, 1965.

5. N.I. Muskhelishvili, *Singular integral equations*, P. Noordhoff, Groningen, 1953.

6. I. Melnik, On Riemann boundary value problem with non-continuous coefficients, *Proc. High Schools (Math.), Kazan*, 1959 (Russian).

I. Vekua Institute of Applied Mathematics, Tbilisi State University, 2 University St., Tbilisi 380043, Georgia

R.W. KOLKKA and D. MALKUS

Dynamics of the spurt phenomenon for single history integral constitutive equations

1. Introduction

The spurt phenomenon for nearly monodisperse, high-molecular-weight, polymer melts and concentrated solutions is of critical importance to the plastics processing industry. These melts and solutions exhibit an abrupt increase in flow rate when the wall shear stress exceeds a critical value in pressure driven shear flow (Fig. 1). Up to the present,

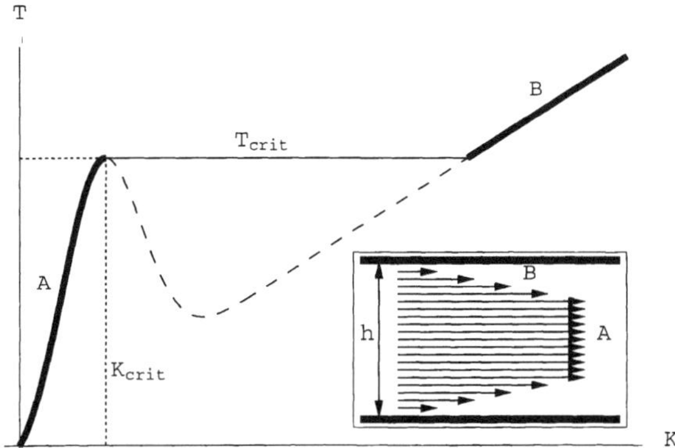

Fig. 1. Steady-state response.

melts and solutions have been modeled by differential constitutive equations for the "extra stress" tensor T, defined by

$$
\begin{aligned}
T &= T_1 + T_2 & D &= \tfrac{1}{2}(\vec{\nabla}\vec{v} + (\vec{\nabla}\vec{v})^T) \\
T_2 &= 2\eta_2 D & T_1\, g(T_1) &+ \lambda\tfrac{\mathcal{D}T_1}{\mathcal{D}t} = 2\eta_1 D.
\end{aligned}
\tag{1}
$$

T_1 is the dominant "polymer" stress contribution and T_2 is a Newtonian viscosity term modeling rapid relaxation effects. The particular constitutive model selects $g(T)$ and the convected derivative $\mathcal{D}/\mathcal{D}t$. The dominant relaxation time λ_1 and the "polymer" viscosity η_1 are experimentally measurable. The Newtonian viscosity η_2 is not. Procedures for determining η_2 are in [1] for Johnson-Segalman models and [2] for differential Larson models.

The key property of the differential models is that their steady shear stress vs. steady shear rate response (Fig. 1) is *non-monotone*. They predict steady flow results [1,2] in very good quantitative agreement with the renowned Vinogradov [3] experimental data for nearly monodispersive polyisoprenes and polybutadienes.

Numerous differential models (Johnson-Segalman, Larson, Giesekus, etc.) match steady-state experimental data with nearly equal precision. The dynamics of the spurt phenomenon must be investigated, to determine which model is superior. This is the focus of [2,4–8] which concluded that all models exhibited similar qualitative dynamics. Supercritical startup of pressure driven shear flow consists of four distinct phases:

1. An initial "Newtonian" phase with viscous stress dominant. A "spike" in the velocity develops on a time-scale $O(\eta_1/\eta_2) \approx 0.01$ with amplitude $\sim \eta_1/\eta_2 \approx 100$.

2. A "latent" phase during which normal stresses build, the polymer stress contributions remains nearly constant, and the velocity profile remains subcritical.

3. A "spurt" phase when the polymer stress contribution drops rapidly to near the steady-state value, the velocity profile develops high shear layers near the wall (with plug-like velocity profile). Throughput increases by orders of magnitude.

4. The final "arrest" phase is a gradual relaxation to the steady state.

We investigate single history integral equation constitutive models as an alternative for modeling spurt. We obtain asymptotics demonstrating that the four dynamic phases of supercritical startup of pressure driven shear flow are ubiquitous.

2. Shear flow dynamics for single history integral constitutive equations

We analyze single integral models of the Rivlin-Sawyers/K-BKZ, which are based on the affine Cauchy and Finger strain measures and are given by

$$T = \int_{-\infty}^{t} m(t-t') \left[\phi_1(I_1, I_2)\gamma_B + \phi_2(I_1, I_2)\gamma_C \right] dt',$$
$$\gamma_B = B - I, \ \gamma_C = I - C, \ B = F\dot{F}^T, \ C = F^T\dot{F}, \ I_1 = tr(B), \ I_2 = tr(C). \tag{2}$$

When $\phi_1 = \partial U/\partial I_1$ and $\phi_2 = \partial U/\partial I_2$ are derived from a viscoelastic potential U, we have the K-BKZ class. We also analyze a generalized integral form Johnson-Segalman model based on a non-affine strain measure γ_{Ba}

$$T = \int_{-\infty}^{t} m(t-t')\gamma_{Ba}dt' \qquad \gamma_{Ba} = E_a \cdot E_a^T - I,$$

where the deformation gradient E_a ($-1 \leq a \leq 1$ is an adjustable parameter) satisfies

$$\frac{E_a}{Dt} - (aD + \Omega)\dot{E}_a = 0 \qquad \text{with} \qquad \Omega = \frac{1}{2}(\vec{\nabla}\vec{v} - (\vec{\nabla}\vec{v})^T).$$

The standard deformation gradient is recovered for $a = 1$. For polymer melts and solutions, the "fading memory" function $m(t-t')$ is of the form

$$m(t-t') = \sum_{i=1}^{n} \frac{G_i}{\lambda_i} e^{\frac{-(t-t')}{\lambda_i}} \quad \text{with relaxation times } \lambda_i \text{ satisfying} \quad \lambda_i > \lambda_{i+1} > 0. \tag{3}$$

The constants G_i are the "shear moduli" of the relaxation modes. For the nearly monodisperse melts we model the first two relaxation modes are well-separated, i.e., $\lambda_1 \gg \lambda_2$ and the other relaxation modes are negligible.

In simple shear flow (Fig. 1), both (2a) and (3a) collapse to

$$T(x,t) = \int_0^\infty m(s)\mathcal{F}(\gamma(t,t-s))ds, \quad \Lambda(x,t) = \int_0^\infty m(s)\mathcal{R}(\gamma(t,t-s))ds,$$
$$\Sigma(x,t) = \int_0^\infty m(s)\mathcal{S}(\gamma(t,t-s))ds, \tag{4}$$

206

where T is the total shear stress, Λ the streamwise normal stress, and Σ the cross-stream normal stress. The relative shear strain is

$$\gamma(t, t-s) = \int_{t-s}^{t} \frac{\partial v}{\partial x}(x, \bar{t})d\bar{t}, \tag{5}$$

where v is the velocity (Fig. 1), and the history variable is $s = t - t'$. The total shear stress and normal stresses can be slightly generalized

$$T(x, t) = \sum_{i=1}^{n} \int_{0}^{\infty} m_i(s)\mathcal{F}_i(\gamma(t, t-s))ds, \quad \text{where} \quad m_i(s) = \frac{G_i}{\lambda_i}e^{\frac{-s}{\lambda_i}}. \tag{6}$$

For the nearly monodispersive melts we model, we simply incorporate the first two modes with $\mathcal{F}_2(\gamma) = \gamma$, and drop the subscript on $\mathcal{F}_1(\gamma) = \mathcal{F}(\gamma)$. We find the model exhibits the spurt phenomenon if $\mathcal{F}(\gamma)$ is non-monotonic. Table 1 gives $\mathcal{F}(\gamma)$ for a

Constitutive Equation	$\mathcal{F} = h(\gamma)\gamma$
J.S.	$\frac{1}{\sqrt{1-a^2}} \sin\left(\sqrt{1 - a^2}\gamma\right)$
Larson, PSM (Papanastasiou-Scriven-Macosko)	$\frac{\gamma}{1+\frac{1}{3}\gamma^2}$
Doi-Edwards (Curtiss-Bird)	$\frac{1}{2\gamma} \int_{0}^{1} \left(1 + \frac{-1 + t^2\gamma}{\sqrt{1 - 2t^2 + t^4\left(\gamma^4 + 4\gamma^2\right)}}\right) dt$
Lodge rubber-like liquid UCM	γ

Table 1. Common constitutive equations.

variety of models. Note, the Johnson-Segalman, Larson integral, and Doi-Edwards (and recent extensions [9]) models all possess this key feature. The Doi-Edwards model is more scientifically appealling than the others because it is based entirely on principles from molecular physics. In fact, MacLeish and Ball [9,10] originally proposed Doi-Edwards as a model for the Vinogradov [3] data. In dimensionless form,

$$T(x, t) = \sigma(x, t) + \sigma_*(x, t) \quad \text{where}$$
$$\sigma(x, t) = \int_{0}^{\infty} e^{-s}\mathcal{F}(\gamma(t, t-s))ds \quad \text{and} \quad \sigma_*(x, t) = \frac{\epsilon}{\nu^2} \int_{0}^{\infty} e^{\frac{-s}{\nu}}(\gamma(t, t-s))ds, \tag{7}$$

with $\epsilon = \eta_2/\eta_1$ and $\nu = \lambda_2/\lambda_1$. The limit $\nu \to 0$, σ_* gives an approximation which we will use in the asymptotic analysis for small ϵ

$$\sigma_*(x, t) \approx \int_{0}^{\infty} se^{\frac{-s}{\nu}}\frac{\partial v}{\partial x}(x, t)ds + \text{h.o.t.} \approx \epsilon\frac{\partial v}{\partial x} + \text{h.o.t.} \tag{8}$$

The dimensionless problem for pressure driven shear flow is described by

$$\frac{1}{\xi}\frac{\partial v}{\partial t} = \frac{\partial T}{\partial x} + f(t), \qquad T(x, t) = \int_{0}^{\infty} e^{-s}\mathcal{F}(\gamma(t, t-s))ds + \epsilon\frac{\partial v}{\partial x}(x, t), \tag{9}$$

with boundary conditions $v(-\frac{1}{2}, t) = \frac{\partial v}{\partial x}(0, t) = T(0, t) = 0$. The prescribed pressure gradient is $f(t)$. For the melts we consider the elasticity number $\xi = \eta_1\lambda_1/\rho h^2 =$

$O(10^{-10})$ is completely negligible. The momentum equation (10) then integrates, giving

$$T(x,t) = -f(t)x, \qquad \frac{\partial v}{\partial x} = \frac{T - \sigma}{\epsilon} \tag{10}$$

In supercritical startup the melt is initially at rest when at $t = 0$ the pressure gradient jumps to a constant value f_a above the model dependant critical value $f_a^{\text{crit}} = 2T_{\text{crit}}$ (Fig. 1). We now solve (10) with $T = T(x) = -f_a x$.

The steady shear stress vs. shear rate response is non-monotonic for ϵ less than a small model dependent critical value, e.g., 0.125 for Johnson-Segalman and 0.0275 for Larson integral. We obtain asymptotic expressions for the stresses and velocity in the small ϵ limit. Motivated by the results from the differential models, we expand in $\tau = t/\epsilon$ near $t = 0$:

$$\sigma = \tilde{\sigma}(x,\tau) \sim \tilde{\sigma}_0(x,\tau) + \epsilon\tilde{\sigma}_1(x,\tau) + \cdots \quad \Lambda = \tilde{\Lambda}(x,\tau) \sim \tilde{\Lambda}_0(x,\tau) + \epsilon\tilde{\Lambda}_1(x,\tau) + \cdots$$
$$\Sigma = \tilde{\Sigma}(x,\tau) \sim \tilde{\Sigma}_0(x,\tau) + \epsilon\tilde{\Sigma}_1(x,\tau) + \cdots$$
$$\tag{11}$$

Substituting (15) into (11) gives the nonlinear Volterra integral equation

$$\tilde{\sigma}_0(x,\tau) = \mathcal{F}(\int_0^\tau (T(x) - \tilde{\sigma}_0(x,\xi))d\xi, \tag{12}$$

Differentiating with respect to τ gives the seperable (and easily solvable for any given model) o.d.e. for the leading order shear stress

$$\frac{d\tilde{\sigma}_0}{d\tau} = \mathcal{F}'(F(\tilde{\sigma}_0))(T - \tilde{\sigma}_0), \tag{13}$$

where $F(\tilde{\sigma}_0) = \mathcal{F}^{-1}(\tilde{\sigma}_0)$. The corresponding normal stress expressions give

$$\tilde{\Lambda}_0(x,\tau) = \mathcal{R}(\int_0^\tau (T - \tilde{\sigma}_0)d\xi), \qquad \tilde{\Sigma}_0 = \mathcal{S}(\int_0^\tau (T - \tilde{\sigma}_0)d\xi). \tag{14}$$

For the Larson integral model we find that

$$F(\tilde{\sigma}_0) = \frac{3}{2l\tilde{\sigma}_0}\left[1 - \sqrt{1 - \frac{4l}{3}\tilde{\sigma}_0^2}\right] \quad \text{and} \quad T_c < T < \sqrt{2}T_c \Longrightarrow \lim_{\tau \to \infty} \tilde{\sigma}_0 = T \tag{15}$$

with a latent phase following. However, if $T > \sqrt{2}T_c$, there is no latent phase with $\tilde{\sigma}_0 = T$ and the solution proceeds directly to spurt. Note $\sqrt{2}T_c$ is the critical stress for instability in step-strain flow for the Larson integral model [11]. Equation (15) shows that this is true for any model, i.e., in general,

$$T_c < T < T_{c,\text{step-strain}} \quad \Longrightarrow \quad \lim_{\tau \to \infty} \tilde{\sigma}_0 = T.$$

When (22) is satisfied the Newtonian phase is followed by a latent phase during which the leading order polymer contribution to the shear stress, σ, nearly equals T and remains almost contant while the normal stresses evolve and continue to increase. For the latent phase, we employ expansions of the form

$$\sigma(x,t) = T(x) + \epsilon\sigma_1(x,t) + \cdots \quad \Lambda(x,t) = \Lambda_0(x,t) + \epsilon\Lambda_1(x,t) + \cdots$$
$$\Sigma(x,t) = \Sigma_0(x,t) + \epsilon\Sigma_1(x,t) + \cdots \quad \sigma(t,t-s) \sim \sigma_0(t,t-s) + \epsilon\sigma_1(t,t-s) + \cdots$$

208

Substituting into (11) gives

$$
\begin{aligned}
T &= \int_0^\infty e^{-s}\mathcal{F}(\gamma_0(t,t-s))ds, & \Lambda_0(x,t) &= \int_0^\infty e^{-s}\mathcal{R}(\gamma_0(t,t-s))ds, \\
\Sigma_0(x,t) &= \int_0^\infty e^{-s}\mathcal{S}(\gamma_0(t,t-s))ds, & \gamma_0(t,t-s) &= -\int_{t-s}^T \sigma_1(x,\xi)d\xi, \\
&\multicolumn{3}{c}{\sigma_1(x,t) = -T/\int_0^\infty e^{-s}\mathcal{F}'(\gamma_0(t,t-s))ds.}
\end{aligned}
$$

Differentiating and integrating by parts generates differential equations governing the evolution of the normal stress Λ_0 and Σ_0. The result for Johnson-Segalman is

$$
\frac{dZ_0}{dt} = \frac{-(1-a^2)T^2}{1+Z_0} - Z_0, \quad \text{where} \quad Z_0 = 1/2(1+a)\Lambda_0 - 1/2(1+a)\Sigma_0 \tag{16}
$$

For the Larson integral model, $\Sigma = 0$, and for Λ_0, we have

$$
\frac{d\Lambda_0}{dt} = \frac{2J(t)T}{1 - \frac{l}{3}\Lambda_0 - \frac{2l}{3}K(t)} - \Lambda_0, \tag{17}
$$

where the integrals $K(t)$ and $J(t)$ are given by,

$$
J(t) = \int_0^\infty \frac{e^{-s}\gamma_0(t,t-s)ds}{(1+\frac{l}{3}\gamma_0^2(t,t-s))^2}, \qquad K(t) = \int_0^\infty \frac{e^{-s}\gamma_0^2(t,t-s)ds}{(1+\frac{l}{3}\gamma_0^2(t,t-s))^2}. \tag{18}
$$

The duration of the latent phase (which is determined from (16–18) [2,4]) depends upon how supercritical the stress T is. The latent phase can often last several relaxation times and would be readily observable in a dynamic experiment.

The latent phase is followed by the spurt phase whose dynamics cannot be readily extracted by ϵ asymptotics. One can, however, readily prove that the time-dependent solution tends to one of the family of spurted steady-state solutions, as was done by Aarts [12]. Aarts also performed numerics for the Larson integral model clearly depicting all four dynamic phases. In addition, "shape memory" feature [4] in unloading of the pressure gradient was demonstrated. Note, equivalent asymptotic results for the Newtonian phase are in [13].

3. Summary and conclusions

In summary, we have shown that the dynamic features for the spurt phenomenon predicted by the single integral constitutive equations are the same as those predicted by differential constitutive equations. This suggests that the differential models do capture the essential underlying molecular physics heretofore solely attributed to the single integral models. Some of these features, in fact, contain universal relationships that hold for all models, and connect the dynamic features with critical stresses in other related shear flows.

References

1. R.W. Kolkka, D.S. Malkus, M.G. Hansen, G.R. Ierley and R.A. Worthing, *J. Non-Newtonian Fluid Mech.* **29** (1988), 303.

2. R.W. Kolkka and H.K. Ganpule, A comparative study of models for the spurt phenomenon, Technical Summary Report, TR-92-43, Fluids Research Oriented Group, Michigan Tech. Univ., 1992.

3. G.V. Vinogradov, A.Y. Malkin, Y.G. Yanovskii, E.K. Borisenkova, B.V. Yarlyko and G.V. Berezhaya, *J. Polymer Sci.* A-2 (1972), 1061.

4. D.S. Malkus, J.A. Nohel and B.J. Plohr, *SIAM J. Appl. Math.* **51** (1991), 899.

5. D.S. Malkus, J.A. Nohel and B.J. Plohr, *J. Comput. Phys.* **87** (1990), 464.

6. D.S. Malkus, J.A. Nohel and B.J. Plohr, Current progress in hyperbolic systems: Riemann problems and computations, in *Contemporary mathematics* **100**, Amer. Math. Sci., 1989, B. Lindquist, Ed., 91–109.

7. J.K. Hunter and M. Slemrod, *Phys. Fluids* **26** (1983), 2345.

8. R.W. Kolkka and G.R. Ierley, *J. Non-Newtonian Fluid Mech.* **33** (1989), 305.

9. T.C.B. McLeish and R.C. Ball, *J. Polymer Sci.* **24** (1986), 1735.

10. T.C.B. McLeish, *J. Polymer Sci.* **25** (1987).

11. R.W. Kolkka, D.S. Malkus and T.R. Rose, *Rheol. Acta* **430** (1991).

12. A.C.T. Aarts, Ph.D. thesis, Eindhoven University of Technology, 1997.

Michigan Technological University, Houghton, Michigan, USA

University of Wisconsin, Madison, Wisconsin, USA

S. LANGDON

A boundary integral equation method for the heat equation

1. Introduction

Consider the positive definite Helmholtz-type problem

$$- \triangle U + \alpha^2 U = 0 \quad \text{in } \Omega, \tag{1}$$

where $\alpha \in R$, subject to either Neumann or Dirichlet boundary conditions, and where Ω is either a bounded or unbounded simply connected two-dimensional domain with smooth boundary Γ. As we shall see in Section 3, some standard boundary integral methods for the heat equation essentially consist of the solution by integral equation methods of triangular elliptic systems, with principle part consisting of the operator in (1), and $\alpha = \mathcal{O}(1/\sqrt{\delta t})$, where δt is the time step. Accurate boundary integral methods for the heat equation essentially require accurate boundary integral methods for (1). This is our goal in this paper.

Using Green's third identity one can reformulate (1) as a first or second kind integral equation, where the kernels of the integrals will typically contain logarithmic singularities. Here we will consider only the second kind integral equations which arise from (1). Assuming that Γ is parameterized by a $\mathcal{C}^\infty$ 2π-periodic bijection $\gamma : [0, 2\pi] \mapsto \Gamma$, with the property $|\gamma'(t)| > 0$ for all $t \in [0, 2\pi]$, these take the form

$$\lambda u(t) + K_\alpha u(t) = f(t), \quad t \in [0, 2\pi]. \tag{2}$$

Here $\lambda \neq 0$ is a constant, f a given function on Γ, and

$$K_\alpha u(t) = \int_0^{2\pi} \left\{ \left. \frac{\partial}{\partial n(y)} \Phi_\alpha(x, y) \right|_{x=\gamma(t), y=\gamma(\tau)} \right\} |\gamma'(\tau)| u(\tau) d\tau, \tag{3}$$

where $\partial/\partial n(y)$ is the derivative with respect to the unit outward normal at $y \in \Gamma$, and $\Phi_\alpha(x, y)$ is the fundamental solution of the Helmholtz equation, which has a logarithmic singularity at $x = y$ and is highly peaked near $x = y$ when α is large. See [1] for more details. It is important to develop numerical methods for (2) that work well when α is large (i.e., $\delta t \to 0$).

When Γ is smooth, standard methods for solving (2) involve splitting the kernel of (3) to remove the logarithmic singularity, and then replacing the operator K_α by the quadrature approximation

$$(K_{\alpha,M} v)(t) = \int_0^{2\pi} \mathcal{P}_M\{k_{1,\alpha}(t, \cdot)v(\cdot)\}(\tau) \log 4 \sin^2 \frac{t - \tau}{2} + \mathcal{P}_M\{k_{2,\alpha}(t, \cdot)v(\cdot)\}(\tau) d\tau, \tag{4}$$

where the operator $\mathcal{P}_M$ is the trigonometric interpolating polynomial of degree M, and $k_{1,\alpha}$, $k_{2,\alpha}$ are smooth 2π-periodic functions that arise when the kernel is split. The integrals in (4) are all known analytically [2].

This work was supported by an EPSRC studentship.

211

The "splitting" approach can be incorporated within a collocation or Nyström method framework. For example the Nyström method of [2] approximates (2) by

$$\lambda u_M(t) + K_{\alpha,M} u_M(t) = f(t), \quad t \in [0, 2\pi]. \tag{5}$$

Using this method we have the error bound

$$\|u - u_M\|_{L_2} \leq C(\alpha) \frac{1}{M^p} \|u\|_{H^p}, \tag{6}$$

for all $p > 1/2$, where $\| \cdot \|_{H^p}$ is the usual Sobolev norm, and $C(\alpha)$ is a constant independent of M, but not of α. So, for fixed α, the convergence is superalgebraic with respect to M. However, the problem is that $k_{1,\alpha}$ and $k_{2,\alpha}$ both blow up exponentially in modulus in α, and so the constant $C(\alpha)$ grows exponentially in α as $\alpha \to \infty$. Thus when α is large, a very large value of M may be needed to achieve an accurate solution. Alternatively, arithmetic overflow may totally destroy the convergence of the method. This is true for all methods that employ the splitting (4). A numerical example demonstrating this can be found in [1].

In this paper we describe in Section 2 a new discrete collocation method for solving (2), which avoids the splitting altogether, and is robust when α is large. We also employ this method to improve a standard boundary integral method for the heat equation (Section 3). Finally, in Section 4 we present some numerical results, demonstrating the properties of the method.

2. A collocation method for (2)

Since Γ is here assumed smooth, it is natural to solve (2) by a global approximation technique. Such methods are able to exploit the smoothness of the solution, yielding superalgebraic or even exponential convergence rates. We use a collocation method in the $2n$-dimensional space

$$T_n := \mathrm{span}\{\phi_k : k = 0, \ldots, 2n - 1\},$$

where

$$\phi_k(t) = \begin{cases} \cos kt, & k = 0, \ldots, n \\ \sin(k - n)t, & k = n + 1, \ldots, 2n - 1. \end{cases}$$

The collocation points will be the equally spaced points $t_j^{(n)} = j\pi/n$, $j = 0, \ldots, 2n - 1$.

The standard collocation method for (2) then seeks a solution

$$u_n = \sum_{k=0}^{2n-1} a_k^{(n)} \phi_k \in T_n,$$

where the coefficients $\{a_k^{(n)}\}_{k=0}^{2n-1}$ are defined by requiring that

$$(\lambda u_n + K_\alpha u_n)(t_j^{(n)}) = f(t_j^{(n)}), \quad j = 0, \ldots, 2n - 1. \tag{7}$$

Equivalently the $a_k^{(n)}$ are found by solving the linear system

$$\sum_{k=0}^{2n-1} \left\{ \lambda \phi_k(t_j^{(n)}) + K_\alpha \phi_k(t_j^{(n)}) \right\} a_k^{(n)} = f(t_j^{(n)}), \quad j = 0, \ldots, 2n - 1.$$

212

This method is only semidiscrete and is not practical until we specify how the integrals $K_\alpha \phi_k(t_j^{(n)})$ should be computed. These are weakly singular integrals, in fact the kernel is of order $|t - \tau|^2 \log|t - \tau|$, as $t \to \tau$ for fixed α. As $\alpha \to \infty$ this singularity occurs in an interval of decreasing size.

To achieve a method which is robust when α is large, we use a rescaling technique to transform the integrals into a more benign form. First note that we can write

$$(K_\alpha v)(t) = \int_0^{2\pi} \theta(\alpha r(t, \tau)) w(t, \tau) v(\tau) d\tau, \tag{8}$$

where $\theta(x)$ has a weak singularity at $x = 0$ and decays exponentially away from $x = 0$, $r(t, \tau) = |\gamma(t) - \gamma(\tau)|$, and $w(t, \tau)$ is a smooth, 2π-periodic function (see [1] for details). Using the periodicity of the integrand we can make the transformation $\tau = t - \tilde{\tau}/\alpha$ in (8) to get

$$(K_\alpha v)(t) = \frac{1}{\alpha} \int_{-\alpha\pi}^{\alpha\pi} \theta\left(\alpha r\left(t, t - \frac{\tilde{\tau}}{\alpha}\right)\right) w\left(t, t - \frac{\tilde{\tau}}{\alpha}\right) v\left(t - \frac{\tilde{\tau}}{\alpha}\right) d\tilde{\tau}. \tag{9}$$

Now as α gets large, the region of integration gets stretched. The integrand has a weak singularity at $\tilde{\tau} = 0$, and decays exponentially away from $\tilde{\tau} = 0$ when α is large, hence we approximate (9) using a graded mesh with more points near $\tilde{\tau} = 0$ (see [1] for details). We then define (dropping the tildes for clarity)

$$(K_{\alpha,m} v)(t) = \frac{1}{\alpha} \int_{-\alpha\pi}^{\alpha\pi} \mathcal{Q}_m \left\{ \theta\left(\alpha r\left(t, t - \frac{(\cdot)}{\alpha}\right)\right) w\left(t, t - \frac{(\cdot)}{\alpha}\right) \right\} (\tau) v\left(t - \frac{\tau}{\alpha}\right) d\tau, \tag{10}$$

where for any $f : (-\infty, \infty) \mapsto R$, $\mathcal{Q}_m f$ is the function which is linear on each mesh interval, and interpolates f at each mesh point. We then have the following theorem, proved in [1].

Theorem 1. *For all $v \in L_2$ there exists a constant C independent of α, m, and v such that*

$$\|K_{\alpha,m} v - K_\alpha v\|_{L_\infty} \leq C \frac{1}{m^2} \|v\|_{L_2}.$$

Crucially, this error estimate is independent of α. Note that in practice v will always be one of the trigonometric basis functions $\phi_k(t)$. Note also that higher order quadrature rules may be applied, with the result that the $1/m^2$ term will be replaced by $1/m^q$ with $q > 2$. Using (10) in (7) we get the fully discrete collocation method

$$(\lambda u_{n,m} + K_{\alpha,m} u_{n,m})(t_j^{(n)}) = f(t_j^{(n)}), \quad j = 0, \ldots, 2n - 1. \tag{11}$$

For further details we refer to [1] and [3] where the proof of the following result will appear.

Theorem 2. *The solution $u_{n,m}$ of the discrete collocation method (11) satisfies*

$$\|u - u_{n,m}\|_{L_2} \leq C \left\{ \frac{1}{n^p} \|u\|_{H^p} + \frac{1}{m^2} \|u\|_{L_2} \right\} \tag{12}$$

for all $p > 1/2$, $\alpha \geq 1$, where C is independent of α, n, m, and u.

In other words, the method is robust to large α and algebraically convergent with respect to m but still superalgebraically convergent with respect to n.

3. Application to the heat equation

The application of boundary integral equation methods to the solution of parabolic PDEs has been widely considered in the literature (see for example [4] and the references therein). One particular method of interest to us is that developed by Chapko and Kress in [5] for solving the heat equation

$$\frac{1}{c}\frac{\partial u}{\partial t} = \Delta u, \quad \text{in } \Omega \times (0,T), \tag{13}$$

for $c, T > 0$, with Dirichlet or Neumann boundary conditions, a given initial condition, and a suitable radiation condition if Ω is unbounded. Here we shall consider the Dirichlet problem, with $u = f$ on Γ. Discretizing (13) in time using the backward Euler method leads to a sequence of inhomogeneous Helmholtz equations of the form

$$-\Delta u_{j+1} + \alpha^2 u_{j+1} = \alpha^2 u_j, \quad \text{in } \Omega, \quad j = 0, \dots, N-1, \text{ where } N\delta t = T, \tag{14}$$

with $u_j = f_j = f(\cdot, j\delta t)$ on Γ, where δt is the time step and $\alpha^2 = 1/c\delta t$. This is the triangular elliptic system mentioned in Section 1. Chapko and Kress construct a fundamental solution $\{\Psi_j(x,y) : j = 0, \dots, N-1\}$ to the sequence (14), and the solution that also satisfies the boundary conditions can then be written as a double layer potential

$$V_j(x) = \frac{1}{\pi} \sum_{k=0}^{j} \int_{\Gamma} q_k(y) \frac{\partial}{\partial \nu(y)} \Psi_{j-k}(x,y) ds(y), \quad x \notin \Gamma, \quad j = 0, \dots, N-1,$$

whose densities $q_j(y)$ can be found by solving the sequence of second kind integral equations

$$q_j(x) + \frac{1}{\pi} \int_{\Gamma} q_j(y) \frac{\partial}{\partial \nu(y)} \Psi_0(x,y) ds(y) =$$

$$f_j(x) - \sum_{k=0}^{j-1} q_k(x) - \frac{1}{\pi} \sum_{k=0}^{j-1} \int_{\Gamma} q_k(y) \frac{\partial}{\partial \nu(y)} \Psi_{j-k}(x,y) ds(y), \quad x \in \Gamma, \tag{15}$$

for $j = 0, \dots, N-1$. Note that on each time level the integral operator remains the same and only the right-hand side changes.

Up to a constant, $\Psi_0 = \Phi_\alpha$, and so for each j, (15) is of the form (2). Each Ψ_j, $j = 0, \dots, N-1$, has a weak singularity at $x = y$, and will be peaked near $x = y$ when α is large. In [5], it is proposed to solve (15) using the Nyström method (5) with a splitting method similar to (4) to deal with the logarithmic singularities. However, this leads to problems as described in §1 when α in (14) is large, i.e., when the time step δt is small. As δt is decreased, one needs to use more and more boundary nodes for the solution of the integral equations, and the method rapidly becomes infeasible. Numerical results demonstrating this are shown in Section 4.

As an alternative, we propose to solve the integral equations (15) using the method described in Section 2, with the obvious extension to the evaluation of the integrals on the right hand side. Numerical results demonstrating the improved convergence rate of this method as δt becomes small are also shown in Section 4.

214

4. Numerical results

We use the same numerical example that appears in [5]. We solve the heat equation (13) with $c = 1$, $T = 1$, on the two-dimensional domain external to the boundary curve

$$\Gamma = \{x(s) = (0.2\cos s, 0.4\sin s - 0.3\sin^2 s)\}, \quad 0 \le s \le 2\pi,$$

with boundary data given by the restriction of the fundamental solution

$$u(x,t) = \frac{1}{4\pi t}\exp\left(-\frac{|x|^2}{4t}\right), \quad |x| > 0, \quad t > 0,$$

to the boundary Γ. We use method (14) with time step $\delta t = T/N$, and thus we expect the solution to converge linearly with respect to δt.

First, we use the Nyström method (5) to solve the integral equations (15), as suggested in [5]. Table 1 shows the error at $T = 1$ between the exact solution and the numerical solution at the two points $x = (0.3, 0)$ and $x = (0.6, 0)$.

	$x = (0.3, 0)$			$x = (0.6, 0)$		
N	M=16	M=32	M=48	M=16	M=32	M=48
10	1.964e-3	1.912e-3	1.902e-3	5.708e-3	5.671e-3	5.663e-3
20	9.284e-4	8.947e-4	8.880e-4	2.575e-3	2.537e-3	2.530e-3
40	4.230e+1	2.961e-4	2.912e-4	1.920e-1	7.181e-4	7.098e-4
80	1.930e+68	1.911e+45	1.102e+31	5.685e+65	1.730e+42	5.566e+27

Table 1. Errors using the "splitting" method to solve (15).

The linear convergence with respect to the time step can initially be seen, but when the time step gets small the constant $C(\alpha)$ in the error bound (6) for the solution of each integral equation gets so large that the error bound is useless, even though the method still seems to be converging superalgebraically with respect to M. Results in [1] demonstrate that all convergence may disappear if the timestep becomes too small.

	$x = (0.3, 0)$			$x = (0.6, 0)$		
N	n=16	n=32	n=48	n=16	n=32	n=48
10	4.629e-4	1.708e-3	1.708e-3	5.450e-3	5.445e-3	5.445e-3
20	6.029e-4	7.834e-4	7.824e-4	2.399e-3	2.400e-3	2.400e-3
40	1.230e-3	2.567e-4	2.558e-4	6.531e-4	6.552e-4	6.552e-4
80	1.471e-3	4.131e-5	4.061e-5	6.248e-5	6.279e-5	6.271e-5

Table 2. Errors using the discrete collocation method of Section 2
to solve (15).

Next, we use the discrete collocation method (11) to solve the integral equations (15). Table 2 again shows the error at $T = 1$ between the exact solution and the numerical solution at the two points $x = (0.3, 0)$ and $x = (0.6, 0)$. Note that here we chose the parameter m in (11) to be sufficiently large that the error would be almost entirely dependent just on N and n. For details of the exact values used see [3].

Provided n is sufficiently large, the linear convergence with respect to the time step can be seen, although the convergence is a little erratic. It is clear that the method continues to work well as the time step gets smaller. Whereas the errors in Table 1 are enormous for $N = 80$ (or equivalently $\alpha \approx 9$ in (14)), the discrete collocation method continues to work well as $\delta t \to 0$.

References

1. S. Langdon and I.G. Graham, Boundary integral methods for singularly perturbed boundary value problems, preprint **98/19** (1998), University of Bath.

2. R. Kress, *Linear integral equation*, Springer-Verlag, Berlin, 1989.

3. S. Langdon, *Domain embedding boundary integral equation methods and parabolic PDEs*, Ph.D. thesis, University of Bath, 1999.

4. C. Lubich and R. Schneider, Time discretization of parabolic boundary integral equations, *Numer. Math.* **63** (1992), 455–481.

5. R. Chapko and R. Kress, Rothe's method for the heat equation and boundary integral equations, *J. Integral Equations Appl.* **9** (1997), 47–69.

Department of Mathematical Sciences, University of Bath, Bath BA2 7AY, UK

Q. LIU and A. NARAIN

Computational simulation and interfacial shear models for downward annular wavy-interface condensing flow in a vertical pipe

1. Introduction

For efficient design and performance of condensers in traditional and modern applications (air-conditioners, looped heat-pipes in space applications, etc.) a need exists—from both fundamental [1] and application [2] perspectives—for tools and techniques that facilitate attainment and accurate predictions of steady annular/stratified internal condensing flows. This paper develops predictive tools for wavy-interface situations involving turbulent vapor/laminar condensate flows in a vertical tube. The flow is quasi-steady—i.e., steady mean flow fields experience wave-induced fluctuations. The flows are sufficiently fast that the entire flow is "parabolic"—i.e., conditions at any point are influenced only by conditions upstream. Situations where the vapor flow is slow and "elliptic" are discussed in the paper [3].

2. 1-D approach: interfacial shear models leading to approximate empirical closure

The steady mean flow under consideration is depicted in Fig. 1. The 1-D integral formulation (see Narain et al. 1997 [4]) leads to the following coupled nonlinear ordinary differential equations for non-dimensional values of average vapor speed (u_{av}), film thickness (δ), average cross-sectional pressure (π), and interfacial shear (f):

$$
\begin{aligned}
\frac{d\mathbf{y}}{dx} &= \mathbf{g}(\mathbf{y}, f), \\
f &= f(x, Re_{in}, (Fr)^{-1}, \frac{Ja}{Pr_1}, \frac{\rho_2}{\rho_1}, \frac{\mu_2}{\mu_1}), \\
\mathbf{y}(0) &= [0, 1, 0]^T,
\end{aligned}
\tag{1}
$$

where $\mathbf{y} \equiv [\pi(x), u_{av}(x), \delta(x)]^T$ and $\mathbf{g} \equiv [g_1, g_2, g_3]^T$ are define in [4]. Resolution of singularity at $x = 0$ requires that interfacial shear f be of the form

$$
\begin{aligned}
f &= f_{\text{model}} \left[\frac{\phi}{\exp(\phi) - 1} \right], \\
\phi &= -\frac{f_{\text{asy}}}{f_{\text{model}}}, \qquad \text{and} \\
f_{\text{asy}} &\equiv f \mid_{x \sim 0} \approx \frac{4}{Re_1 \delta^2(x)} \left(\frac{\rho_1}{\rho_2} \right) \int_0^x \dot{m} \, dx.
\end{aligned}
\tag{2}
$$

The interfacial shear f in (2) has the *required* form [4] at $x \sim 0$ and blends smoothly with f_{model} for large x. The model f_{model} is chosen to give reasonable agreement with experiments [4].

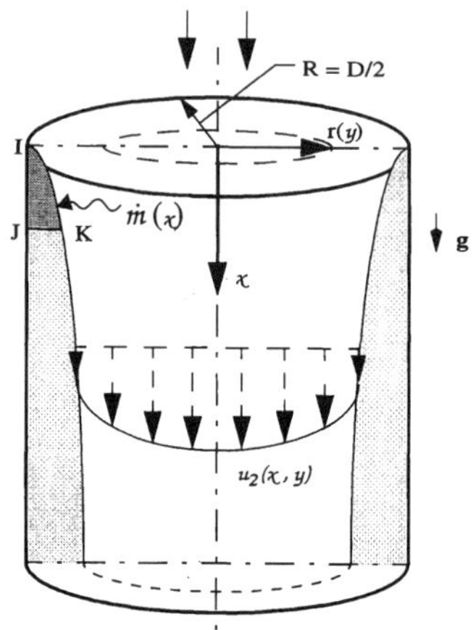

Fig. 1. The physical model for 2-D

The literature contains numerous recommendations for f_{model}: **Models 1 to 6** defined in [4]; **Model 7** Friedel model for air/water two-phase flows [5]; **Model 8** Troniewski and Ulbrich model for two-phase flows [6]; **Model 9** Lockhart, Martinelli and Chisholm model for two-phase flows [7,8].

Figures 2 and 3 (and Table 2 of [4]) show the efficacy of different interfacial shear models and indicate that only the modified Andreussi model $f_{\text{model 6}}$ and $f_{\text{model 5}}$ defined in [4,9] perform well for Experiments I, II & III in [4].

Having selected $f_{\text{model}} = f_{\text{model 5/6}}$, we numerically solve equations (1), (2) for the 1-D flow variables. These interfacial shear models provide the interfacial shear values at the interface for 2-D simulations.

Fig. 2 compares theoretical average heat transfer coefficients (for Models 1–9) with values from Experiment I (see table 2 [4]). The performance of Models 5 and 6 is within experimental uncertainty. Fig. 3 repeats this for Experiment II.

3. 2-D formulation and numerical approach

The mean steady flow problem for Fig. 1 (detailed in [10]) consists of the standard 2-D differential equations (in cylindrical coordinates) of continuity, momentum (in x and y directions), and energy for each phase with interface, inlet, wall, and outlet conditions. The turbulent vapor is described by a $k-\epsilon$ model [11] for the eddy viscosity μ_t ($\mu_t \equiv C_\mu \rho k^2/\epsilon$) in the Boussinesq representation $-\rho \overline{u_i' u_j'} = \mu_t(\partial u_i/\partial x_j + \partial u_j/\partial x_i)$ for the Reynolds stress $-\rho \overline{u_i' u_j'}$. The interior vapor shear stress values ($\approx (\mu + \mu_t)\frac{\partial u_2}{\partial y}$ along planes parallel to the interface) at points P_1, P_2, etc. in Fig. 4 are extrapolated [10] to the near interface "outer-layer" location $P(y = y_p)$. A "good" semi-empirical 1-D interfacial shear model from section 2 provides a value for this extrapolated stress.

A finite volume (SIMPLER [12]) method is used for the computation. The mean film thickness $\Delta(x)$ increases monotonically (the 1-D approach gives an initial estimate) allowing the use of the simple computational domain/grid shown in Fig. 4 at each iteration. See [10] for details.

218

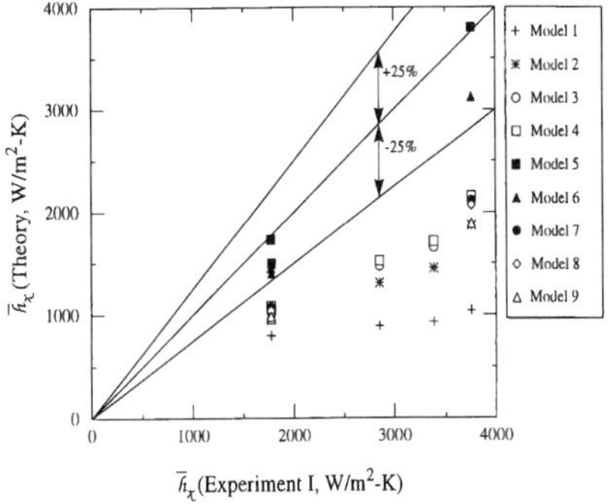

Fig. 2. The figure compares theoretical predictions of average heat transfer coefficient with Experiment I values for the same flow situation. The theoretical predictions employ Models 1 - 9 in (1) - (2). The experimental situations are those listed for Experiment I in Table 2 in [4]. The performance of Model 5 and Model 6 are satisfactory (within experimental uncertainties).

Fig. 3. The figure compares theoretical predictions of local heat transfer co-efficient with Experiment II values for the same flow situation. The theoretical predictions employ Models 1 - 9 in (1) - (2). Perhaps because of the dominance of gravity and the low value of steam to water density ratios ρ_2/ρ_1 involved in these flow situations, the predictions are not sensitive to the choice of interfacial shear models and all the models perform satisfactorily (within experimental uncertainties).

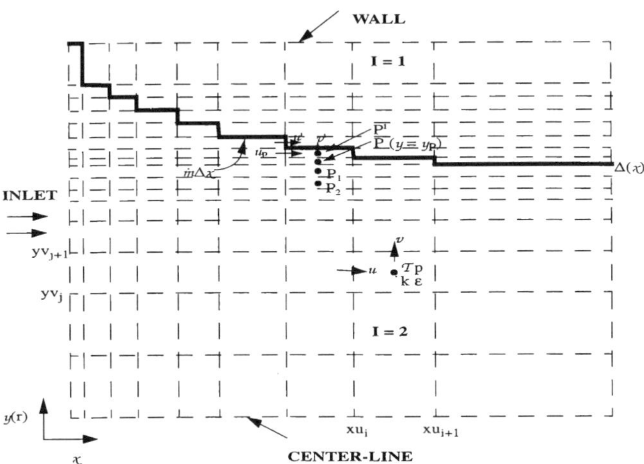

Fig. 4. Sketch depicting the interface and the control volumes

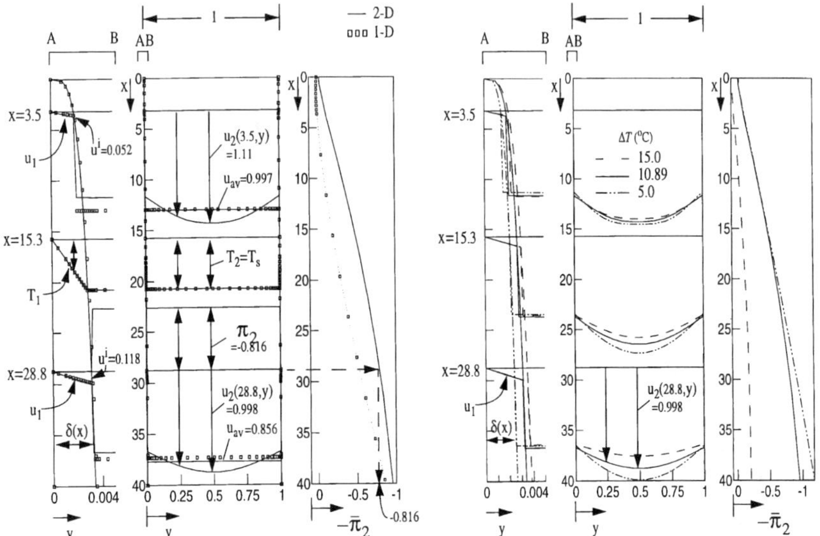

Fig. 5. Compatibility of 2-D solutions with 1-D.   Fig. 6. Effect of changes in ΔT.

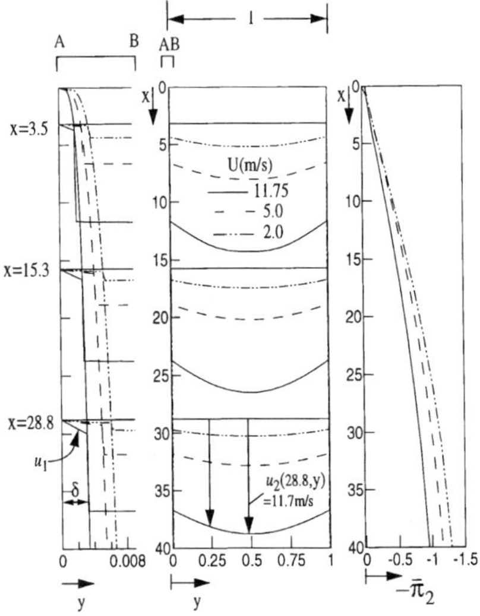

Fig. 7. Effect of changes in inlet speed U

220

4. Simulation results and discussions

Fig. 5 describes downflow of R-11 in a tube. Inlet/outlet vapor-phase Reynolds numbers of, respectively, $Re_{\text{in}} = 1.50 \times 10^5$ and $Re_{\text{out}} = 1.22 \times 10^5$ ensure the entire vapor flow is turbulent since both exceed 2100. The condensate is laminar since the outlet liquid Reynolds number $Re_{\text{liquid}} = 774 < 1000$. The 2-D liquid velocity profile and average vapor speed agree with available experimental heat transfer data. Note that the 2-D average pressure $\overline{\pi_2}$ drops more than the 1-D model because the 2-D model includes vapor friction in the interior and at the interface while the 1-D model neglects vapor friction in the interior.

Fig. 6 shows that film thickness increases with temperature difference ΔT. For the fast turbulent vapor flow situations in Fig. 6, it is apparent that thicker films flow at higher interfacial speeds, higher liquid mass flux, and lower vapor speeds. The lower pressure drop in Fig 6 slows the vapor to the lower speeds associated with higher ΔT.

Fig. 7 shows the effects of changes in inlet speed U. In Fig. 7, increases in inlet speed U with consequent increased interfacial shear cause noticeable thinning of the film. Although the nondimensional pressure drop $\overline{\pi_2}$ decreases as the inlet speed U increases, the physical average pressure drop $\Delta p = \rho_2 U^2 \Delta \overline{\pi}$ increases with U as expected.

5. Conclusions

- A 1-D approach for identifying good interfacial shear models τ^i for wavy-interface annular film condensation in vertical tubes was developed.

- The 1-D approach provides reasonable estimates for the unknown interface location and interfacial shear which enhance the convergence of 2-D models.

- A novel 2-D computational scheme for these condensing flows specifying mean interfacial shear using "good" 1-D models is proposed. The mean interfacial shear specifications renders the model insensitive to the vapor turbulence model.

- The 2-D model predicts pressure variations in the vapor better than the 1-D model.

- The 2-D model can be extended to thick condensate and slow vapor situations described in [3].

References

1. V.P. Carey, *Liquid-vapor phase-change phenomena*, Hemisphere, Washington, DC, 1992.

2. J.W. Palen, G. Breber and J. Taborek, *Heat Transfer Engrg.* **1** (1979), 47–57.

3. G. Yu & A. Narain, Computational simulation and classification of flow domains for laminar/laminar annular/stratified condensing flows, in *Integral methods in science and engineering*, Chapman & Hall/CRC, Boca Raton, FL, 1999.

4. A. Narain, G. Yu and Q. Liu, *Internat. J. Heat Mass Tranfer* **40** (1997), 3559–3575.

5. L. Friedel, New friction pressure drop correlations for upward, horizontal and downward two-phase pipe flow, HTFS symposium, Oxford, (Hoechst AG Ref 372217/24 698), 1978.

6. L. Troniewsky and R. Ulbrich, *Chem. Engrg. Sci.* **39** (1984), 751–765.

7. R.W. Lockhart and R.C. Martinelli, *Chem. Engrg. Prog.* **45** (1949), 39.

8. D. Chisholm, A theoretical basis for the Lockhart-Martinelli correlation for two-phase flow, Report #310 National Engineering Laboratory, East Kilbride, UK. (See also *Internat. J. Heat Mass Transfer* **10** (1967).)

9. A. Narain, *J. Appl. Mech.* **63** (1996), 529.

10. Q. Liu, *Computational simulation and interfacial shear models for wavy annular condensing downward flows in a vertical pipe: turbulent vapor/lamina condensate situation*, Ph.D. thesis, Michigan Technological University, Houghton, Michigan, 1999.

11. D.C. Wilcox, *Turbulence modeling for CFD*, DCW Industries, 1993.

12. S.V. Patankar, *Numerical heat transfer and fluid flow*, Hemisphere, Washington, DC, 1980.

Department of Engineering Mechanics, Michigan Technological University, Houghton, Michigan, USA

B.J. McCARTIN

Compact fourth-order approximation for a nonlinear reaction-diffusion equation arising in population genetics

1. Introduction

Under appropriate assumptions, the transferral of genetic information is governed by a nonlinear reaction-diffusion equation [2, 4, 7]—the generalized Fisher equation [9, 10, 13]. The present paper is concerned with approximating this equation to a high degree of accuracy. The method of approach to the numerical solution of the generalized Fisher equation will be comprised of a hybridization of Mehrstellenverfahren spatial discretization [3] with Runge-Kutta time stepping [14]. In what follows, the relevant biological, mathematical, and numerical models will be outlined. The efficacy of this approach will then be underscored by numerical examples from population genetics.

2. Biological model

We concern ourselves with a single-locus/two-allele genetic trait [10] and seek an equation for the time evolution of the frequency $u(x, t)$ of an advantageous gene (A-allele) within a large, genetically homogeneous population of diploid individuals. We shall assume that the population mates randomly, i.e., there are no genotype-dependent mating preferences, and that there are overlapping generations. The viabilities of the genotypes AA, Aa, aa will be denoted by ρ, σ, τ, respectively.

In [12], it is shown that u satisfies the nonlinear reaction-diffusion equation

$$\frac{\partial u}{\partial t} = \underbrace{D\frac{\partial^2 u}{\partial x^2}}_{\text{diffusion}} + \underbrace{u(1 - u) \cdot [(\rho - 2\sigma + \tau)u + (\sigma - \tau)]}_{\text{reaction}}, \tag{1}$$

where the coefficient, D, is assumed constant. The case of variable D may be accommodated by a change of dependent variable.

3. Mathematical model

Introducing $s := \rho - \tau$; $h := (\sigma - \tau)/(\rho - \tau)$, we arrive at the desired *generalized Fisher equation*

$$\frac{\partial u}{\partial t} = D\frac{\partial^2 u}{\partial x^2} + s(x)u(1 - u)[h(x) + (1 - 2h(x))u], \tag{2}$$

where we have made allowance for spatial variation in the selectivity, $s(x)$, and the heterozygote suitability, $h(x)$. The numerical scheme to be developed below also permits a temporal variation in D, s, and h. However, we shall not have occasion to avail ourselves of this level of generality in the present work.

It is assumed that $\rho > \tau$, so that $s > 0$. Also, we may distinguish the following biologically important cases whose mathematical properties are described in [1, 6, 8]:
- heterozygote-inferiority: $-\infty < h < 0$ $(\sigma < \tau < \rho)$,
- heterozygote-intermediate: $0 \leq h \leq 1$ $(\tau \leq \sigma \leq \rho)$,
- heterozygote-superiority: $1 < h < \infty$ $(\tau < \rho < \sigma)$.

The author thanks Ms. Barbara A. Rowe for assisting in the production of this paper.

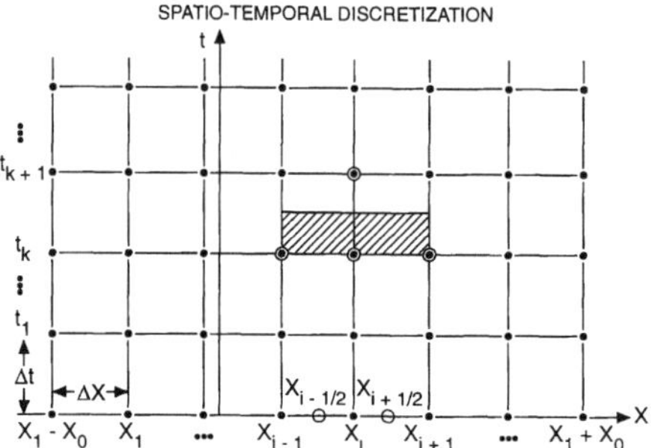

Fig. 1. Spatio-temporal discretization.

4. Numerical model

We next approximate the generalized Fisher equation (2) with homogeneous diffusion, $D = \text{const}$,

$$\underbrace{u_t}_{\text{Runge-Kutta}} = \underbrace{D \cdot u_{xx}}_{\text{Mehrstellenverfahren}} + s(x) \cdot u(1-u)[h(x) + (1 - 2h(x))u], \qquad (3)$$

subject to the initial-boundary conditions

$$u(x,0) = u_I(x); \ u(x_L,t) = u_L(t), \ u(x_R,t) = u_R(t). \qquad (4)$$

We commence by defining $g(x,t) := u_t(x,t)$ and rewriting (3) as

$$D \cdot u_{xx} = g - s \cdot u(1-u)[h + (1-2h)u]. \qquad (5)$$

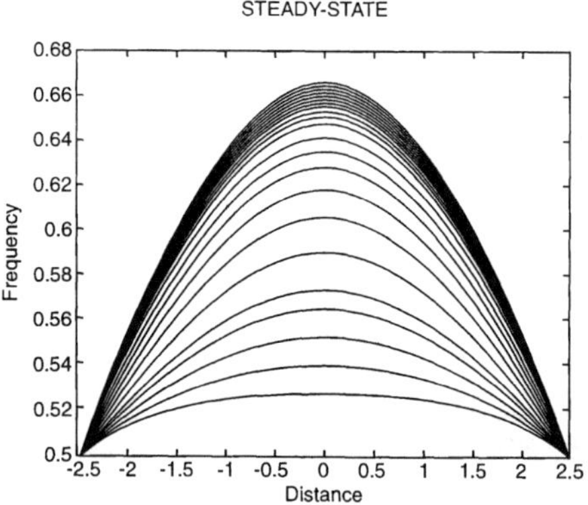

Fig. 2. Steady state.

224

Differentiating (5) twice with respect to x produces

$$D \cdot u_{xxxx} = g_{xx} - 2[s_x u(1-u) + s u_x(1-2u)] \cdot [h_x(1-2u) + (1-2h)u_x]$$
$$- \; [s_{xx}u(1-u) + 2s_x(1-2u)u_x - 2s(u_x)^2 + s(1-2u)u_{xx}] \cdot [h + (1-2h)u]$$
$$- \; [su(1-u)] \cdot [h_{xx}(1-2u) - 4h_x u_x + (1-2h)u_{xx}]. \tag{6}$$

With reference to Fig. 1, consider central difference approximation of the diffusion

$$D \cdot \frac{u_{i+1} - 2u_i + u_{i-1}}{(\Delta x)^2} = (D \cdot u_{xx})_i + \frac{(\Delta x)^2}{12}(D \cdot u_{xxxx})_i + O((\Delta x)^4) \;, \; u_i(t) := u(x_i, t). \tag{7}$$

Applying the Mehrstellenverfahren technique [3] by inserting (6) into (7), rearranging terms, and making additional $O((\Delta x)^4)$ approximations yields

$$(D \cdot u_{xx})_i = - \frac{g_{i+1} - 2g_i + g_{i-1}}{12}$$
$$+ \; \left[D + \frac{(\Delta x)^2}{12} s_i \{(1-2u_i)h_i + (1-2h_i)u_i(2-3u_i)\} \right] \cdot \frac{u_{i+1} - 2u_i + u_{i-1}}{(\Delta x)^2}$$
$$+ \; \frac{(\Delta x)^2}{12} u_i(1-u_i) \left\{ \begin{array}{l} (s_{xx})_i[h_i + (1-2h_i)u_i] \\ +2(s_x)_i(h_x)_i(1-2u_i) + s_i(h_{xx})_i(1-2u_i) \end{array} \right\}$$
$$+ \; \frac{\Delta x}{12}(u_{i+1} - u_{i-1}) \left\{ \begin{array}{l} s_i(h_x)_i(6u_i^2 - 6u_i + 1) \\ (s_x)_i[h_i + 2(1-3h_i)u_i + 3(2h_i - 1)u_i^2] \end{array} \right\}$$
$$+ \; \frac{1}{24}(u_{i+1} - u_{i-1})^2 s_i[(1-3h_i) + 3(2h_i - 1)u_i] + O((\Delta x)^4) \;. \tag{8}$$

Insertion of this last expression into (3) produces (for $i = 1, \ldots, N-1$)

$$\frac{(u_t)_{i+1} + 10(u_t)_i + (u_t)_{i-1}}{12} =$$
$$\left[D + \frac{(\Delta x)^2}{12} s_i \{(1-2u_i)h_i + (1-2h_i)u_i(2-3u_i)\} \right] \frac{u_{i+1} - 2u_i + u_{i-1}}{(\Delta x)^2}$$
$$+ \frac{(\Delta x)^2}{12} u_i(1-u_i) \left\{ \begin{array}{l} (s_{xx})_i[h_i + (1-2h_i)u_i] \\ +2(s_x)_i(h_x)_i(1-2u_i) + s_i(h_{xx})_i(1-2u_i) \end{array} \right\}$$
$$+ \frac{\Delta x}{12}(u_{i+1} - u_{i-1}) \left\{ \begin{array}{l} s_i(h_x)_i(6u_i^2 - 6u_i + 1) \\ +(s_x)_i[h_i + 2(1-3h_i)u_i + 3(2h_i - 1)u_i^2] \end{array} \right\} \tag{9}$$
$$+ \frac{1}{24}(u_{i+1} - u_{i-1})^2 s_i[(1-3h_i) + 3(2h_i - 1)u_i] + s_i u_i(1-u_i)[h_i + (1-2h_i)u_i],$$

where we have neglected $O((\Delta x)^4)$ terms, thus achieving fourth-order spatial accuracy.

Introducing $\vec{u} := [u_1, \ldots, u_{N-1}]^T$, we may write (9) as the matrix ordinary differential equation

$$M \cdot \frac{d\vec{u}}{dt} = F(\vec{u}), \; \vec{u}(0) = \vec{u}_I \;. \tag{10}$$

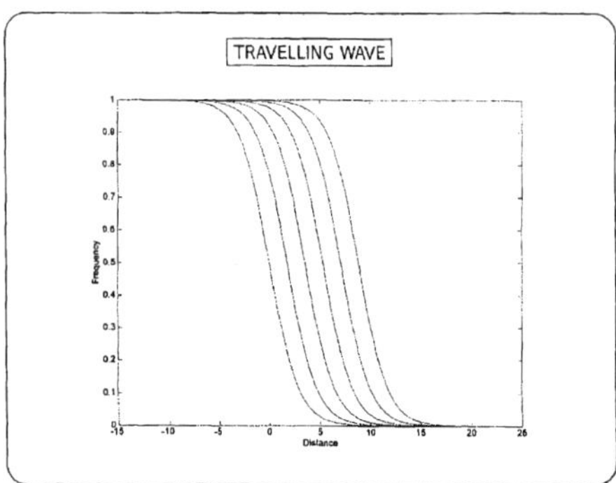

Fig. 3. Travelling wave.

This semidiscretized initial value problem may now be approximated to fourth-order temporal accuracy using the following Runge-Kutta scheme [14]

$$
\begin{aligned}
M \cdot \vec{k}_1 &= F(\vec{u}^k), & M \cdot \vec{k}_2 &= F(\vec{u}^k + \tfrac{\Delta t}{2} \cdot \vec{k}_1), \\
M \cdot \vec{k}_3 &= F(\vec{u}^k + \tfrac{\Delta t}{2} \cdot \vec{k}_2), & M \cdot \vec{k}_4 &= F(\vec{u}^k + \Delta t \cdot \vec{k}_3),
\end{aligned}
\tag{11}
$$

$$
\vec{u}^{k+1} = \vec{u}^k + \frac{\Delta t}{6} \cdot (\vec{k}_1 + 2\vec{k}_2 + 2\vec{k}_3 + \vec{k}_4), \quad \vec{u}^k := \vec{u}(t_k) = \vec{u}(k \cdot \Delta t).
\tag{12}
$$

Note that
$$
\frac{du_0}{dt} = u'_L(t), \ \frac{du_N}{dt} = u'_R(t)
\tag{13}
$$

are specified. Since M is tridiagonal (in fact, symmetric Toeplitz), it can be LU-factored in time $O(N)$ at $t = 0$, while (11) can be solved by forward/backward substitution in time $O(N)$ at each time step. The formal order of accuracy of this new scheme is $O((\Delta x)^4 + (\Delta t)^4)$, i.e., we have achieved fourth-order spatio-temporal accuracy. For Neumann boundary conditions, the local truncation error is of third-order at the boundaries. The scheme is conditionally stable with $\Delta t = O((\Delta x)^2)$.

5. Numerical examples

For the special case $s = .2$, $h = -2.$, $u(-L,t) = .5 = u(L,t)$, it is shown in [11] that there is a unique positive steady-state solution where $L = \sqrt{2.5} \cosh^{-1}(2.5) \approx 2.477327$. This exact solution is given by $u_{exact}(x) = 1.2/[.4 \cosh(\sqrt{.4}x) + 1.4]$, with $u_{max} = u(0) = 2/3$. Figure 2 displays our numerical solution evolving from $u(x, 0) = .5$ to a steady state with $u_{max} = .66651582$. This is accurate to $.02\%$, with $\Delta x = .206444$ ($N = 24$) and $\Delta t = .01$, as compared to $.075\%$ in [11] using twice as many grid points.

For the special case $s = .5$, $h = -.5$, it is shown in [13] that there exists the travelling wave solution $u_{exact}(x,t) = 1/[1 + e^{(x-t/2\sqrt{2})/\sqrt{2}}]$. Figure 3 shows our computed solution for this case propagating to the right with negligible distortion near the predicted speed of $c = 1/2\sqrt{2}$, with $x_L = -15$, $x_R = 25$, $\Delta x = .5$ ($N = 80$), and $\Delta t = .01$.

226

For the case of variable coefficients, we take Fisher's example [5] of a genetic cline with $s(x) = 8x$, $h = .5$ (no dominance, i.e. "additive fitness"). The result is shown in Fig. 4, where we also include the cases $h = -2.$ (heterozygote inferior) and $h = 1.$ (complete dominance). There, we observe the anticipated effect of varying h: by decreasing it, we steepen the cline. Values of $x_L = 0$, $x_R = 6$, $\Delta x = .2$ ($N = 30$), and $\Delta t = .001$ were used.

6. Conclusion

The foregoing sections have successfully extended our fourth-order accurate scheme [12] to the generalized Fisher equation, thus accommodating arbitrary heterozygote fitness. This high order of spatio-temporal accuracy is attained by combining Mehrstellenverfahren spatial discretization [3] with Runge-Kutta time stepping [14]. Numerical computations from population genetics demonstrated the efficacy of this procedure.

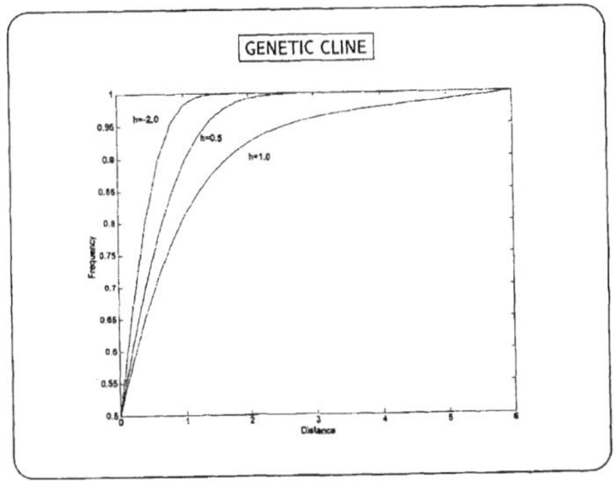

Fig. 4. Genetic clines.

References

1. D. Aronson and H. Weinberger, Nonlinear diffusion in population genetics, in *Partial differential equations and related topics*, J.A. Goldstein, Ed., Springer-Verlag, 1975, 5–49.

2. N.F. Britton, *Reaction-diffusion equations and their applications to biology*, Academic Press, 1986.

3. L. Collatz, Hermitean methods for initial value problems in partial differential equations, in *Topics in numerical analysis*, J.J.H. Miller, Ed., Academic Press, 1972, 41–61.

4. P.C. Fife, *Mathematical aspects of reacting and diffusing systems*, Springer, 1979.

5. R.A. Fisher, Gene frequencies in a cline, *Biometrics* (1950), 353–361.

6. W.H. Fleming, A selection-migration model in population genetics, *J. Math. Biol.* **2** (1975), 219–233.

7. P. Grindrod, *Patterns and waves*, Oxford, 1991.

8. K.P. Hadeler, Nonlinear diffusion equations in biology, in *Ordinary and partial differential equations*, W.N. Everitt and B.D. Sleeman, eds., Springer-Verlag, 1976, 163–206.

9. F.C. Hoppensteadt, *Mathematical methods of population biology*, Cambridge, 1982.

10. F.C. Hoppensteadt and C.S. Peskin, *Mathematics in medicine and the life sciences*, Springer, 1986.

11. V.S. Manoranjan, A.R. Miller, and B.D. Sleeman, Bifurcation studies in reaction-diffusion, *J. Comput. Appl. Maths.* **11** (1984), 27–37.

12. B.J. McCartin, A mathematical model from population genetics: numerical solution of Fisher's equation, in *Proceedings of third biennial symposium on mathematical modeling in the undergraduate curriculum*, University of Wisconsin-La Crosse, 1998.

13. J.D. Murray, *Mathematical biology*, Springer-Verlag, 1989.

14. E.H. Twizell, *Numerical methods, with applications in the biomedical sciences*, Halsted, 1988.

Kettering University, Flint, Michigan, USA

A. MIODUCHOWSKI and P. JANELE

Dynamic deformation of a layered continuum surrounding a cylindrical or spherical cavity

1. Introduction

This paper considers the finite amplitude wave propagation that results when a layered continuum consisting of concentric isotropic compressible hyperelastic layers surrounding a cylindrical, or spherical cavity is subjected to a sudden, spatially uniform application of pressure at the cavity's surface. The layers have different elastic constants, and wave reflection at the cavity's boundary is considered as is wave reflection and transmission at the interface between the layers. Governing equations for the above problems are identical, except for a constant multiplier and are expressed as a system of first order partial differential equations in conservation form. This system of equations is strictly hyperbolic, for the class of strain energy functions used here, with three families of characteristics. The numerical scheme used to obtain the solutions presented here is a finite difference predictor-corrector method which uses the relation along straight characteristics parallel to the t axis in the (R,t) plane. It is a shock finding scheme so that the jump relations are not required for implementation of the scheme. In order to present this method we discuss the finite amplitude cylindrically symmetric wave propagation in a compressible hyperelastic layered solid. Nonlinear elastodynamic solutions for similar homogeneous problems have been addressed in the previous paper [1], in which the hybrid characteristic method—finite difference scheme was introduced. The results shown here are arrived at by the extension of this technique and represent the case of radially varying nonhomogeneity.

2. Governing equations

The cylindrically symmetric, composite hyperelastic medium considered here, is initially in a natural reference configuration. A system of cylindrical coordinates with origin at the cavity center are used to define the time dependent radial deformation

$$r = r(R, t), \tag{1}$$

where r and R are the distances of a particle from the origin in the deformed and reference configurations, respectively. The inner, center, and outer layers occupy the region $R_A \leq R \leq R_B, R_B \leq R \leq R_C$ and $R_C \leq R \leq R_D$, respectively, with $R_D = \infty$ for the unbounded medium. The initial conditions are

$$r(R, 0) = R, \qquad v = \frac{\partial r}{\partial t}(R, 0) = 0 \tag{2}$$

where v is the particle velocity. At time $t = 0^+$, a spatially uniform pressure $p(t) = qH(t)$ is applied at $R = R_A$, where q is a constant and $H(t)$ is the unit step function; consequently $\sigma_r(R_A, t) = -p(t)$ where σ_r is the radial component of Cauchy stress. The principal components of stretch in the tangential, radial, and axial directions are

$$\lambda_\theta = \frac{r}{R}, \qquad \lambda_r = \frac{\partial r}{\partial R}, \qquad \lambda_z = 1, \qquad \lambda_r = \frac{\partial r}{\partial R}, \qquad \lambda_\phi = \lambda_\theta = \frac{r}{R} \tag{3}$$

for spherically symmetric problems. The non-zero components of radial and tangential stress are respectively given by

$$S_r = \frac{\partial W}{\partial \lambda_r}, \quad S_\theta = \frac{1}{n}\frac{\partial W}{\partial \lambda_\theta}, \tag{4}$$

where, as in [1], $W(\lambda_r, \lambda_\theta, \lambda_z)$ is the Gaussian strain energy function per unit volume, with $n = 1$ for cylindrically and $n = 2$ for spherically symmetric problems.

It is convenient to introduce a nondimensionalization scheme that is based on the material properties of the inner layer, identified by the subscript 1,

$$\overline{S}_r = \frac{S_r}{\mu_1}, \quad \overline{S}_\theta = \frac{S_\theta}{\mu_1}, \quad \overline{v} = v\sqrt{\frac{\rho_1}{\mu_1}}, \quad \tau = \frac{t}{R_1}\sqrt{\frac{\mu_1}{\rho_1}}, \quad \overline{R} = \frac{R}{R_1}. \tag{5}$$

The governing equations consist of a nontrivial equation of motion and two compatibility equations. In terms of the reference coordinates (Lagrangian coordinate system), deformation of the inner cylinder is defined by

$$\frac{\partial \overline{v}}{\partial \tau} - \frac{\partial \overline{S}_r}{\partial \overline{R}} - n\frac{\left(\overline{S}_r - \overline{S}_\theta\right)}{\overline{R}} = 0, \qquad \frac{\partial \lambda_r}{\partial \tau} - \frac{\partial \overline{v}}{\partial \overline{R}} = 0, \qquad \frac{\partial \lambda_\theta}{\partial \tau} - \frac{\overline{v}}{\overline{R}} = 0. \tag{6}$$

where all variables are nondimensional. For the other layers, the two compatibility equations are the same as the inner cylinder but the nontrivial equation of motion is

$$\frac{\partial \overline{v}}{\partial \tau} - n\frac{\phi_{il}}{\beta_{il}}\frac{\partial \overline{S}_r}{\partial \overline{R}} - \frac{\phi_{il}}{\beta_{il}}\frac{\left(\overline{S}_r - \overline{S}_\theta\right)}{\overline{R}} = 0, \quad i = 2, 3 \tag{7}$$

where $\beta_{il} = \rho_i/\rho_1$ is the ratio of material densities relative to the reference configuration and $\phi_{il} = \mu_i/\mu_1$. Both β and ϕ exist in the governing equations due to the nature of the nondimensionalization scheme. Note, that the areas of continuously variable nonhomogeneity are replaced by a series of thin cylinders. The material properties for each thin cylinder are then treated as radially varying "constants" with the physical properties averaged for each thin cylinder.

If the concentric cylinders are to remain in contact during deformation, four interface boundary conditions must be satisfied,

$$\begin{aligned}
\lambda_\theta\left(R_B^-, \tau\right) &= \lambda_\theta\left(R_B^+, \tau\right), \\
v\left(R_B^-, \tau\right) &= v\left(R_B^+, \tau\right), \\
\sigma_r\left[\lambda_r\left(R_B^-, \tau\right), \lambda_\theta\left(R_B^-, \tau\right)\right] &= \sigma_r\left[\lambda_r\left(R_B^+, \tau\right), \lambda_\theta\left(R_B^+, \tau\right)\right] \\
\sigma_r \text{ at interface} &< 0, \quad \text{(compression)}.
\end{aligned} \tag{8}$$

where R_B is the radius in the undeformed configuration, at which the inner cylinder and it's adjacent cylinder are in contact. R_B^- and R_B^+ denote the positions at $R = R_B$ just within each of these media respectively. Interface conditions similar to (8) apply at $R = R_C$ where the second cylinder is in contact with its adjacent media.

230

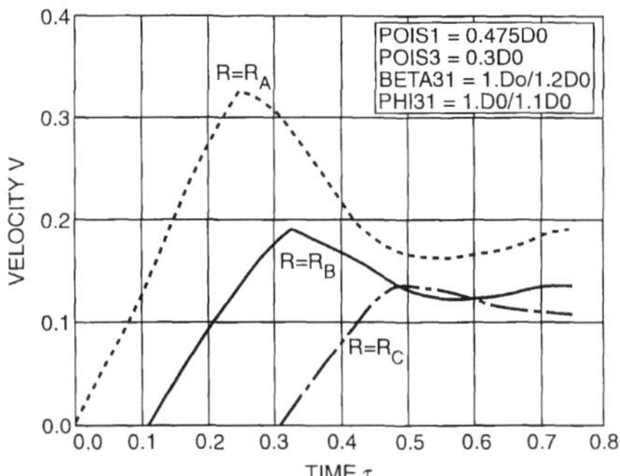

Fig. 1. Velocity vs. time for ramped Heaviside load with
$\nu_1 = 0.475, \nu_3 = 0.3$.

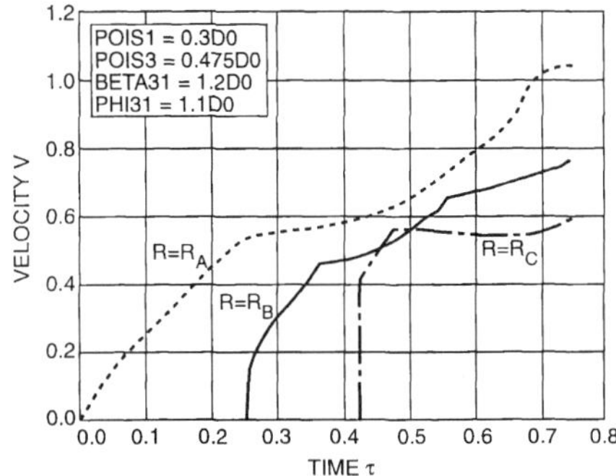

Fig. 2. Velocity vs. time for ramped Heaviside load with
$\nu_1 = 0.475, \nu_3 = 0.3$.

The interface condition (8a) specifies continuity of radial position in the deformed configuration since

$$r_B(\tau) = R_B * \lambda_\theta(R_B, \tau). \tag{9}$$

The velocity interface condition (8b) is based on the constraint that if the cylinders remain in contact during the deformation, the velocity must be continuous across the material boundaries. This is also evident from $v = \partial r / \partial \tau$ since

$$v(R_B, \tau) = \frac{\partial r}{\partial \tau}(R_B, \tau). \tag{10}$$

The stress conditions (8c,d) are based on the physical interpretation that the cylinders remain in contact only if the Cauchy radial stress is continuous (and compressive). These interface conditions are used in the numerical procedure to determine λ_r.

231

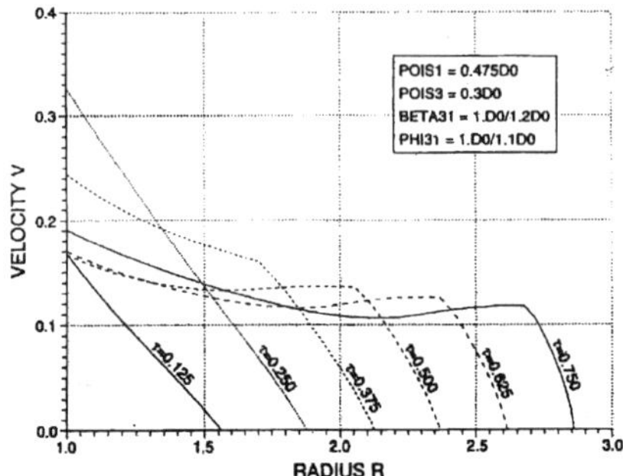

Fig. 3. Velocity vs. radius for ramped Heaviside load with
$\nu_1 = 0.475, \nu_3 = 0.3$.

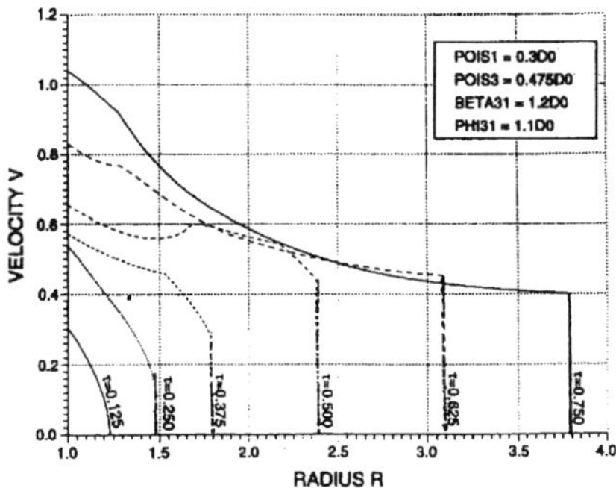

Fig. 4. Velocity vs. radius for ramped Heaviside load with
$\nu_1 = 0.3, \nu_3 = 0.475$.

3. Numerical method and results

The numerical method for which results are presented in this paper is based on a
hybrid finite difference scheme that has been utilized and presented in the previous
paper involving deformation of homogeneous media, [1]. Extension of this method
to handle deformation of multiple concentric cylinders considered here, is relatively
straightforward, with the exception of dealing with the change in material properties
and resulting change in equations of motion at the transition between cylinders. At
these transition radii, the forward/backward finite difference scheme must be modified
using the interface boundary conditions that define the continuity of v, λ_θ, and σ_r. It
should be noted that the application of the interface boundary condition that specifies
continuity of σ_r typically results in a discontinuity in radial stretch λ_r, when applied
to media of differing material properties. Consequently at each time step and at each
of the interface radii, the effective Cauchy radial stress is computed and solved for the

232

discontinuity in λ_r between adjacent cylinders. This is numerically implemented using a Newton-Raphson iterative scheme.

Deformation, and resulting wave propagation is initiated by the uniform application of pressure at the cavity's surface at $\tau = 0$. The case discussed here is when the pressure $p(\tau)$ changes linearly from zero to $1.5\mu_1$ at $\tau = 0.25$, and is maintained thereafter—a ramped Heaviside step function. The outgoing wave passes through the concentric media for which Poisson ratio is constant, but different, in the first and third layer and is changing linearly between these values in the second layer. For numerical calculations it is assumed that $R_A = 1.0, R_B = 1.5$ and $R_C = 2$ with the third layer unbounded, $R_D = \infty$. Figures 1–4 show numerical solutions for two different configurations of material properties: high Poisson ratio, linear transition, low Poisson ratio and vice versa ($v_1 = 0.475, v_3 = 0.3, \beta_{31} = 1/1.2, \phi_{31} = 1/1.1$ or $v_1 = 0.3, v_3 = 0.475, \beta_{31} = 1.2$, and $\phi_{31} = 1.1$).

The results are quantified by two plots: $v(t)$ at $R = R_A, R_B$ and R_C, and $v(R)$ at $t = 0.125, 0.25, \ldots, 0.75$. These velocity profiles clearly show that for the "slow-transient-fast" arrangement of material properties, vs. the "fast-transient-slow," the initial ramp acceleration waves break very quickly and form shock waves. It should be noted that for a spherically symmetric case one is arriving at qualitatively similar results, as can be deduced from (6) and (7).

References

1. P. Janele and A. Mioduchowski, Solution of some non-linear elastodynamic problems, in *Integral methods in science and engineering*, C. Constanda, J. Saranen and S. Seikkala, Eds., Pitman Res. Notes Math. Ser. **375**, Addison Wesley Longman, 1997, 103–108.

Department of Mechanical Engineering, University of Alberta, Edmonton, Alberta, Canada, T6G 2G8

D. MITREA

Boundary value problems for harmonic vector fields on nonsmooth domains

In this paper we undertake a systematic study of natural boundary value problems for the Laplacian acting on vector fields in $\mathbb{R}^3$. Throughout the paper $\Omega \subset \mathbb{R}^3$ denotes a bounded Lipschitz domain, that is, a domain whose boundary is locally described by graphs of Lipschitz functions. Also, $\vec{n}$ stands for its outward unit normal and we set $\Omega_+ = \Omega$, $\Omega_- := \mathbb{R}^3 \setminus \overline{\Omega}$. For a function u defined in $\Omega_\pm$ we let $\mathcal{N}(u)$ be the nontangential maximal function of u (see, for example, [1]). Besides $L^{2,1}(\partial\Omega)$, the Sobolev space of square-integrable functions on $\partial\Omega$ whose tangential gradient is also square-integrable, we also work with $L^2_{\mathrm{tan}}(\partial\Omega)$, the space of tangential vector fields with square-integrable components on $\partial\Omega$. Also of interest are the spaces

$$L^{2,\mathrm{Div}}_{\mathrm{tan}}(\partial\Omega) := \{\vec{f} \in L^2_{\mathrm{tan}}(\partial\Omega);\ \mathrm{Div}\,\vec{f} \in L^2(\partial\Omega)\},$$

where Div is the surface divergence operator, which we equip with the norm

$$\|\vec{f}\|_{L^{2,\mathrm{Div}}_{\mathrm{tan}}(\partial\Omega)} := \|\vec{f}\|_{L^2(\partial\Omega)} + \|\mathrm{Div}\,\vec{f}\|_{L^2(\partial\Omega)},$$

and $L^{2,0}_{\mathrm{tan}}(\partial\Omega) := \{\vec{f} \in L^{2,\mathrm{Div}}_{\mathrm{tan}}(\partial\Omega);\ \mathrm{Div}\,\vec{f} = 0\}$. For a more detailed discussion of these topics we refer the reader to [1]. The reader should be aware that, in what follows, we make no topological assumption on Ω. As a consequence, the following spaces that encode topological information about Ω appear naturally:

$$\mathcal{H}(\Omega) := \{\vec{h} \in L^2(\Omega);\ \mathrm{div}\,\vec{h} = 0,\ \mathrm{curl}\,\vec{h} = 0\ \mathrm{in}\ \Omega\},$$

$$\mathcal{H}_\bullet(\Omega) := \{\vec{h} \in \mathcal{H}(\Omega);\ \mathcal{N}(\vec{h}) \in L^2(\partial\Omega),\ \langle \vec{n}, \vec{h} \rangle = 0\},$$

$$\mathcal{H}_\times(\Omega) := \{\vec{h} \in \mathcal{H}(\Omega);\ \mathcal{N}(\vec{h}) \in L^2(\partial\Omega),\ \vec{n} \times \vec{h} = 0\},$$

$$\mathbb{R}_{\partial\Omega_+} := \mathrm{span}_{\mathbb{R}}\,\{\chi_{\partial\Omega'};\ \Omega'\ \text{bounded connected component of}\ \Omega_+\},$$

$$\mathbb{R}_{\partial\Omega} := \mathrm{span}_{\mathbb{R}}\,\{\chi_\omega;\ \omega\ \text{connected component of}\ \partial\Omega\}.$$

Similar spaces can be defined in Ω_-. Finally, $b_1(\Omega)$ and $b_2(\Omega)$ denote, respectively, the 1st and 2nd Betti numbers of Ω. Also, all constants involved in our estimates will eventually depend only on the constitutive parameters used to describe the Lipschitz character of Ω. We now proceed to describe the main results.

Theorem 1. *Consider the boundary value problem for* $\vec{u} = (u_1, u_2, u_3)$:

$$\Delta\vec{u} = 0 \ \mathrm{in}\ \Omega,$$

$$\mathcal{N}(\vec{u}),\ \mathcal{N}(\mathrm{curl}\,\vec{u}) \in L^2(\partial\Omega), \quad \langle \vec{n}, \vec{u} \rangle = f \in L^2(\partial\Omega), \qquad \text{(BVP1)}$$

$$\vec{n} \times (\mathrm{curl}\,\vec{u})|_{\partial\Omega} = \vec{g} \in L^2_{\mathrm{tan}}(\partial\Omega).$$

234

Then the following assertions hold.

(i) *A solution of* (BVP1) *exists if and only if* $\vec{g}$ *satisfies the compatibility condition* $\vec{g} \in \{\vec{h}|_{\partial\Omega}; \vec{h} \in \mathcal{H}_\bullet(\Omega)\}^\perp$, *where* $\perp$ *denotes the orthogonal complement in* $L^2_{\tan}(\partial\Omega)$.

(ii) *The space of null-solutions is* $\mathcal{H}_\bullet(\Omega)$; *its dimension is* $b_1(\Omega)$. *In particular, the boundary data determine* $\operatorname{curl}\vec{u}$ *and* $\operatorname{div}\vec{u}$ *uniquely and*

$$\|\mathcal{N}(\operatorname{curl}\vec{u})\|_{L^2(\partial\Omega)} \leq \left(\|f\|_{L^2(\partial\Omega)} + \|\vec{g}\|_{L^2(\partial\Omega)}\right).$$

Also, every solution belongs to $H^{1/2,2}(\Omega)$.

Moreover, if the compatibility condition described in (i) *is satisfied, then the following regularity statements are valid.*

(iii) $\mathcal{N}(\operatorname{div}\vec{u}) \in L^2(\partial\Omega)$ *and* $\|\mathcal{N}(\operatorname{div}\vec{u})\|_{L^2(\partial\Omega)} \leq \left(\|f\|_{L^2(\partial\Omega)} + \|\vec{g}\|_{L^2(\partial\Omega)}\right).$

(iv) $\mathcal{N}(\operatorname{curl}\operatorname{curl}\vec{u}) \in L^2(\partial\Omega)$ *if and only if* $\vec{g} \in L^{2,\mathrm{Div}}_{\tan}(\partial\Omega)$; *in this case,*

$$\|\mathcal{N}(\operatorname{curl}\operatorname{curl}\vec{u})\|_{L^2(\partial\Omega)} \leq C\|\vec{g}\|_{L^{2,\mathrm{Div}}_{\tan}(\partial\Omega)}.$$

(v) $\operatorname{curl}\vec{u} = 0$ *in* Ω *if and only if* $\vec{g} = 0$. *Furthermore, when* $\vec{g} = 0$ *we can prescribe periods and have genuine uniqueness. More precisely, for any* $\beta_j \in \mathbb{R}$, $j = 1,\ldots,b_1(\Omega)$, *there exists a unique solution of the problem*

$$\Delta\vec{u} = 0 \ \text{in}\ \Omega, \quad \operatorname{curl}\vec{u} = 0 \ \text{in}\ \Omega,$$
$$\mathcal{N}(\vec{u}) \in L^2(\partial\Omega), \quad \langle\vec{n},\vec{u}\rangle = f \in L^2(\partial\Omega), \qquad\qquad \text{(BVP2)}$$
$$\int_{\gamma_j} \langle\vec{u},\vec{t}_j\rangle = \beta_j, \quad j = 1,\ldots,b_1(\Omega),$$

where $\{[\gamma_j]\}_{j=1,\ldots,b_1(\Omega)}$ *is a basis for* $H^1_{\mathrm{sing}}(\Omega;\mathbb{R})$, *the first singular homology group of* Ω *over the reals, and* $\vec{t}_j$ *are tangential vector fields to the curves* γ_j, $j = 1,\ldots,b_1(\Omega)$.

(vi) $\operatorname{curl}\operatorname{curl}\vec{u} = 0$ *in* Ω *if and only if* $\vec{g} \in L^{2,0}_{\tan}(\partial\Omega)$.

(vii) $\operatorname{div}\vec{u} = 0$ *if and only if* $f \in \mathbb{R}^\perp_{\partial\Omega_+}$. *In particular,* $\vec{g} = 0$ *forces* $\operatorname{curl}\vec{u} = 0$ *and we can prescribe periods, in which case* (BVP1) *becomes*

$$\operatorname{div}\vec{u} = 0 \ \text{in}\ \Omega, \quad \operatorname{curl}\vec{u} = 0 \ \text{in}\ \Omega,$$
$$\mathcal{N}(\vec{u}) \in L^2(\partial\Omega), \quad \langle\vec{n},\vec{u}\rangle = f \in L^2(\partial\Omega), \qquad\qquad \text{(BVP3)}$$
$$\int_{\gamma_j} \langle\vec{u},\vec{t}_j\rangle = \beta_j, \quad j = 1,\ldots,b_1(\Omega).$$

Formulated in this way, the problem (BVP3) *has a solution if and only if* $f \in \mathbb{R}^\perp_{\partial\Omega_+}$, *and the solution is unique.*

Theorem 2. *Consider the following problem for $\vec{u} = (u_1, u_2, u_3)$:*

$$\Delta \vec{u} = 0 \text{ in } \Omega,$$

$$\mathcal{N}(\vec{u}), \; \mathcal{N}(\operatorname{div} \vec{u}) \in L^2(\partial\Omega), \tag{BVP4}$$

$$(\operatorname{div} \vec{u})|_{\partial\Omega} = f \in L^2(\partial\Omega), \quad \vec{n} \times \vec{u} = \vec{g} \in L^2_{\tan}(\partial\Omega).$$

Then the following assertions hold.

(i) *A solution of* (BVP4) *exists if and only if f satisfies the compatibility condition $f \in \{\langle \vec{n}, \vec{h}|_{\partial\Omega}\rangle; \; \vec{h} \in \mathcal{H}_\times(\Omega)\}^\perp$, where $\perp$ denotes the orthogonal complement in $L^2(\partial\Omega)$.*

(ii) *The space of null-solutions for the homogeneous* (BVP4) *is $\mathcal{H}_\times(\Omega)$, which has dimension $b_2(\Omega)$. In particular, $\operatorname{div} \vec{u}$ and $\operatorname{curl} \vec{u}$ are determined uniquely, and*

$$\|\mathcal{N}(\operatorname{div} \vec{u})\|_{L^2(\partial\Omega)} \leq C \left(\|f\|_{L^2(\partial\Omega)} + \|\vec{g}\|_{L^2(\partial\Omega)} \right).$$

Also, every solution belongs to $H^{1/2,2}(\Omega)$.

(iii) *$\mathcal{N}(\operatorname{curl} \vec{u}) \in L^2(\partial\Omega)$ if and only if $\vec{g} \in L^{2,\mathrm{Div}}_{\tan}(\partial\Omega)$. If $\vec{g} \in L^{2,\mathrm{Div}}_{\tan}(\partial\Omega)$, then $\|\mathcal{N}(\operatorname{curl} \vec{u})\|_{L^2(\partial\Omega)} \leq C(\|f\|_{L^2(\partial\Omega)} + \|\vec{g}\|_{L^{2,\mathrm{Div}}_{\tan}(\partial\Omega)})$.*

(iv) *$\mathcal{N}(\nabla \operatorname{div} \vec{u}) \in L^2(\partial\Omega)$ if and only if $f \in L^{2,1}(\partial\Omega)$.*

(v) *$\operatorname{div} \vec{u} = 0$ if and only if $f = 0$. In this case, we can have uniqueness by prescribing periods. More precisely, there exists a unique solution of the problem*

$$\Delta \vec{u} = 0 \text{ in } \Omega, \quad \operatorname{div} \vec{u} = 0 \text{ in } \Omega,$$

$$\mathcal{N}(\vec{u}) \in L^2(\partial\Omega), \quad \vec{n} \times \vec{u} = \vec{g} \in L^2_{\tan}(\partial\Omega), \tag{BVP5}$$

$$\int_{s_j} \langle \vec{u}, \vec{n}_j \rangle = \beta_j, \quad j = 1, \ldots, b_2(\Omega),$$

where $\{[s_j]\}_{j=1,\ldots,b_2(\Omega)}$ is a basis for $H^2_{\mathrm{sing}}(\Omega; \mathbb{R})$, the second singular homology group of Ω over the reals, and $\vec{n}_j$ are unit normal vectors to the surfaces s_j.

(vi) *$\nabla \operatorname{div} \vec{u} = 0$ if and only if $f \in \mathbb{R}_{\partial\Omega}$.*

(vii) *If $\vec{g} \in L^{2,\mathrm{Div}}_{\tan}(\partial\Omega)$, then we can prescribe $\langle \vec{n}, (\operatorname{curl} \vec{u})|_{\partial\Omega} \rangle = \operatorname{Div} \vec{g}$ arbitrarily in the space $\operatorname{Div}(L^{2,\mathrm{Div}}_{\tan}(\partial\Omega))$ instead of $\vec{n} \times \vec{u}$. This leads to the boundary value problem*

$$\Delta \vec{u} = 0 \text{ in } \Omega,$$

$$\mathcal{N}(\vec{u}), \; \mathcal{N}(\operatorname{curl} \vec{u}), \; \mathcal{N}(\operatorname{div} \vec{u}) \in L^2(\partial\Omega), \tag{BVP6}$$

$$(\operatorname{div} \vec{u})|_{\partial\Omega} = f \in L^2(\partial\Omega), \quad \langle \vec{n}, (\operatorname{curl} \vec{u})|_{\partial\Omega} \rangle = \tilde{f} \in L^2(\partial\Omega).$$

Hence, (BVP6) *is solvable if and only if $\tilde{f} \in \operatorname{Div}(L^{2,\mathrm{Div}}_{\tan}(\partial\Omega))$ and $f \in \{\langle \vec{n}, \vec{h}|_{\partial\Omega}\rangle; \vec{h} \in \mathcal{H}_\times(\Omega)\}^\perp$. Moreover, the space of null-solutions is infinitely dimensional and the equivalences* (iv)–(vi) *above hold for this problem.*

236

(viii) $\operatorname{curl}\vec{u} = 0$ *if and only if*

$$\vec{g} \in L_{\tan}^{2,0}(\partial\Omega) \cap \{\vec{h}|_{\partial\Omega}; \vec{h} \in \mathcal{H}_\times(\Omega)\}^\perp \quad \text{and} \quad f \in \mathbb{R}_{\partial\Omega}.$$

Furthermore, if $\vec{g} = 0$, we obtain uniqueness by prescribing periods, namely,

$$\operatorname{div}\vec{u} = 0 \ \text{in} \ \Omega, \quad \operatorname{curl}\vec{u} = 0 \ \text{in} \ \Omega,$$

$$\mathcal{N}(\vec{u}) \in L^2(\partial\Omega), \quad \vec{n} \times \vec{u} = \vec{g} \in L_{\tan}^{2,0}(\partial\Omega), \qquad \text{(BVP7)}$$

$$\int_{s_j} \langle \vec{u}, \vec{n}_j \rangle = \beta_j, \quad j = 1, ..., b_2(\Omega),$$

has a unique solution.

Theorem 3. *Consider the problem for $\vec{u} = (u_1, u_2, u_3)$:*

$$\Delta\vec{u} = 0 \ \text{in} \ \Omega,$$

$$\mathcal{N}(\vec{u}), \ \mathcal{N}(\operatorname{curl}\vec{u}), \ \mathcal{N}(\operatorname{div}\vec{u}) \in L^2(\partial\Omega), \qquad \text{(BVP8)}$$

$$(\operatorname{div}\vec{u})|_{\partial\Omega} = f \in L^2(\partial\Omega), \quad \vec{n} \times (\operatorname{curl}\vec{u})|_{\partial\Omega} = \vec{g} \in L_{\tan}^2(\partial\Omega).$$

Then the following assertions hold.

(i) *There exists solution of* (BVP8) *if and only if $\vec{g} - f\vec{n} \in \{h|_{\partial\Omega}; h \in \mathcal{H}(\Omega)\}^\perp$, where $\perp$ denotes the orthogonal complement in $L^2(\partial\Omega)$.*

(ii) *The space of null-solutions for* (BVP8) *is precisely $\mathcal{H}(\Omega)$. In particular, $\operatorname{div}\vec{u}$ and $\operatorname{curl}\vec{u}$ are determined uniquely and, moreover,*

$$\|\mathcal{N}(\operatorname{div}\vec{u})\|_{L^2(\partial\Omega)} + \|\mathcal{N}(\operatorname{curl}\vec{u})\|_{L^2(\partial\Omega)} \leq \left(\|f\|_{L^2(\partial\Omega)} + \|\vec{g}\|_{L^2(\partial\Omega)} \right).$$

(iii) $\vec{g} \in L_{\tan}^{2,\mathrm{Div}}(\partial\Omega) \Leftrightarrow \mathcal{N}(\operatorname{curl}\operatorname{curl}\vec{u}) \in L^2(\partial\Omega) \Leftrightarrow \mathcal{N}(\nabla\operatorname{div}\vec{u}) \in L^2(\partial\Omega) \Leftrightarrow f \in L^{2,1}(\partial\Omega).$

(iv) $\vec{g} \in L_{\tan}^{2,0}(\partial\Omega) \Leftrightarrow \operatorname{curl}\operatorname{curl}\vec{u} = 0 \Leftrightarrow \nabla\operatorname{div}\vec{u} = 0 \Leftrightarrow f \in \mathbb{R}_{\partial\Omega}.$

(v) $\vec{g} = 0 \Leftrightarrow \operatorname{curl}\vec{u} = 0.$

(vi) $f = 0 \Leftrightarrow \operatorname{div}\vec{u} = 0.$

Theorem 4. *The boundary value problem for $\vec{u} = (u_1, u_2, u_3)$*

$$\Delta\vec{u} = 0 \ \text{in} \ \Omega, \quad \operatorname{curl}\operatorname{curl}\vec{u} = 0 \ \text{in} \ \Omega,$$

$$\mathcal{N}(\vec{u}), \ \mathcal{N}(\operatorname{curl}\vec{u}) \in L^2(\partial\Omega), \qquad \text{(BVP9)}$$

$$\langle \vec{n}, \vec{u} \rangle = f \in L^2(\partial\Omega), \quad \langle \vec{n}, (\operatorname{curl}\vec{u})|_{\partial\Omega} \rangle = g \in L^2(\partial\Omega),$$

is solvable if and only if $g \in \operatorname{Div}\left(L_{\tan}^{2,\mathrm{Div}}(\partial\Omega)\right)$. Moreover, the space of null-solutions is finite dimensional.

Remark. There are appropriate versions of Theorems 1–4 in $\Omega_- = \mathbb{R}^3 \setminus \bar{\Omega}$. Also, similar results can be obtained for the Helmholtz operator $\Delta + k^2$. We postpone the discussion of the latter topic for a separate occasion.

Using suitable layer potential representations, we can reduce (BVP1)–(BVP9) to boundary integral equations involving either scalar potentials, which have been treated in [2], or vector potentials. The latter, together with their relevant properties, are introduced and studied below.

For $\vec{f} = (f_1, f_2, f_3) \in L^2(\partial\Omega)$, the single layer potential operator $\mathcal{S}$ is defined by

$$\mathcal{S}\vec{f}(x) := -\frac{1}{4\pi} \int_{\partial\Omega} \frac{\vec{f}(y)}{|x - y|} \, d\sigma(y), \quad x \in \mathbb{R}^3.$$

Also, the (magnetic dipole) operator M is defined in the principal value sense by

$$M\vec{f} = \vec{n} \times (\text{p.v. curl}\,\mathcal{S}\vec{f}) \quad \text{a.e. on } \partial\Omega.$$

In sharp contrast to the smooth case (cf. [3]), on Lipschitz boundaries M is *no longer a compact operator* on the spaces of interest. Its properties are listed below.

Theorem 5. *If Ω is a bounded Lipschitz domain in $\mathbb{R}^3$, then the following assertions hold.*

(i) The operators $\pm\frac{1}{2}I + M$ are Fredholm with index zero on each of the spaces $L^2_{\tan}(\partial\Omega)$, $L^{2,\mathrm{Div}}_{\tan}(\partial\Omega)$, *and* $L^{2,0}_{\tan}(\partial\Omega)$. *Their kernels on these spaces coincide, and*

$$\mathrm{Ker}\big(\pm\tfrac{1}{2}I + M; L^2_{\tan}(\partial\Omega)\big) = \big\{\vec{n} \times \vec{h}|_{\partial\Omega_\pm}; \vec{h} \in \mathcal{H}_\bullet(\Omega_\pm)\big\}.$$

In particular, $\dim\mathrm{Ker}\big(\pm\tfrac{1}{2}I + M; L^2_{\tan}(\partial\Omega)\big) = b_1(\Omega)$.

(ii) Image$\big(\pm\tfrac{1}{2}I + M; L^2_{\tan}(\partial\Omega)\big) = \big\{\vec{h}|_{\partial\Omega_\mp}; \vec{h} \in \mathcal{H}_\bullet(\Omega_\mp)\big\}^\perp$, where $\perp$ is taken in $L^2_{\tan}(\partial\Omega)$. Similar descriptions are valid for the images of $\pm\tfrac{1}{2}I + M$ when acting on $L^{2,\mathrm{Div}}_{\tan}(\partial\Omega)$ and $L^{2,0}_{\tan}(\partial\Omega)$.

(iii) The operators $\pm\frac{1}{2}I + M$ are isomorphisms on the spaces

$$\{\vec{n} \times \nabla_{\tan} f; \, f \in L^{2,1}(\partial\Omega)\}, \quad L^{2,\mathrm{Div}}_{\tan}(\partial\Omega)/L^{2,0}_{\tan}(\partial\Omega), \quad L^2_{\tan}(\partial\Omega)/L^{2,0}_{\tan}(\partial\Omega),$$

$$L^2_{\tan}(\partial\Omega)/\mathrm{Ker}\big(\pm\tfrac{1}{2}I+M\big), \quad L^{2,\mathrm{Div}}_{\tan}(\partial\Omega)/\mathrm{Ker}\big(\pm\tfrac{1}{2}I+M\big), \quad L^{2,0}_{\tan}(\partial\Omega)/\mathrm{Ker}\big(\pm\tfrac{1}{2}I+M\big),$$

as well as on $(\vec{n} \times L^{2,0}_{\tan}(\partial\Omega))^\perp$, where $\perp$ is taken in $L^2_{\tan}(\partial\Omega)$;

(iv) $\vec{n} \times \mathcal{S} - \big(\pm\tfrac{1}{2}I + M\big)^{-1}\big(\vec{n} \times \nabla_{\tan}\mathcal{S}(\langle\vec{n}, \mathcal{S}(\cdot)\rangle)\big)$ is an isomorphism from the space $\{\vec{n} \times \vec{h}|_{\partial\Omega_\mp}; \vec{h} \in \mathcal{H}_\bullet(\Omega_\mp)\}$ onto $\{\vec{n} \times \vec{h}|_{\partial\Omega_\pm}; \vec{h} \in \mathcal{H}_\bullet(\Omega_\pm)\}$, where the inverse operator $\big(\pm\tfrac{1}{2}I + M\big)^{-1}$ is taken on $\vec{n} \times \nabla_{\tan} L^{2,1}(\partial\Omega)$.

The proof of Theorem 5 follows by combining the results in [4], [5], and [1] together with the techniques in [6].

238

In conclusion, we would like to point out that all our results are valid in the L^p context for arbitrary $1 < p < \infty$ provided that $\partial\Omega \in C^1$ (see [7] for the scalar case).

References

1. D. Mitrea, M. Mitrea and J. Pipher, Vector potential theory on non-smooth domains in $\mathbb{R}^3$ and applications to electromagnetic scattering, *J. Fourier Anal. Appl.* **3** (1997), 131–192.

2. D. Mitrea, The method of layer potentials for non-smooth domains with arbitrary topology, *J. Integral Equations Operator Theory* **29** (1997), 320–338.

3. D. Colton and R. Kress, *Integral equation methods in scattering theory*, Wiley, New York, 1983.

4. D. Mitrea, *Layer potential operators and boundary value problems for differential forms on Lipschitz domains*, Ph.D. thesis, University of Minnesota, Minneapolis, 1996.

5. D. Mitrea and M. Mitrea, Boundary integral methods for harmonic differential forms in Lipschitz domains, *Electronic Research Announcements of A.M.S.* **2** (1996), 92–97.

6. D. Mitrea, M. Mitrea and M. Taylor, *Layer potentials, the Hodge Laplacian, and global boundary problems in nonsmooth Riemannian manifolds*, Mem. Amer. Math. Soc. (to appear).

7. E. Fabes, M. Jodeit and N. Rivière, Potential techniques for boundary value problems on C^1 domains, *Acta Math.* **141** (1978), 165–186.

Department of Mathematics, University of Missouri-Columbia, 202 Math. Sciences Bldg., Columbia, MO 65211, USA; e-mail: dorina@math.missouri.edu

M. MITREA

The oblique derivative problem
for general elliptic systems
in Lipschitz domains

Let M be a smooth, oriented, connected, compact, boundaryless manifold of real dimension m, and let TM and T^*M be its tangent and cotangent bundles, respectively. We assume that M is equipped with a Lipschitzian Riemannian metric tensor $g = g_{jk}dx_j \otimes dx_k$, $g_{jk} \in \mathrm{Lip}\,(M)$. We also denote $\det{(g_{jk})}$ by g, so that if $d\mathrm{Vol}$ stands for the corresponding volume form, then $d\mathrm{Vol} = \sqrt{g(x)}\,dx$ locally.

Consider a smooth vector bundle $\mathcal{E} \to M$ endowed with Lipschitzian Hermitian structures, and let $L(x, D)$ be a second-order differential operator acting on sections of $\mathcal{E}$. Suppose that, about each point, M has a smooth coordinate system and $\mathcal{E}$ has a local trivialization with respect to which L takes the form

$$Lu = \sum_{j,k} \partial_j A^{jk}(x)\partial_k u + \sum_{j} B^j(x)\partial_j u - V(x)u, \tag{1}$$

where A^{jk}, B^j, and V are matrix-valued functions such that

$$A^{jk} \in C^{1+\gamma}, \quad B^j \in H^{1,r}, \quad V \in L^r \tag{2}$$

for some $\gamma > 0$ and $r > m = \dim M$. Here and elsewhere, $H^{s,p}$ denotes the usual scale of Sobolev spaces.

Following [1], for a Lipschitz domain Ω in M (possibly the whole of M) we say that L satisfies the *nonsingularity hypothesis* relative to Ω if

$$u \in H^{1,2}_0(\Omega, \mathcal{E}), \; Lu = 0 \text{ in } \Omega \quad \Rightarrow \quad u = 0 \text{ in } \Omega. \tag{3}$$

Parenthetically, we observe that if L is strongly elliptic, then $L - \lambda$, $\lambda \in \mathbb{R}$, satisfies the nonsingularity hypothesis (3) relative to any subdomain $\Omega \subseteq M$ provided λ is sufficiently large. This is a consequence of Gårding's inequality, which is valid in our setting (even though V may be unbounded). Also, clearly, if L is strongly elliptic and negative semidefinite, then $L - \lambda$ satisfies (3) for any $\lambda > 0$. A concrete example of an elliptic, formally self-adjoint operator satisfying (1)–(3) is the Hodge-Laplacian corresponding to a Riemannian metric with coefficients in $H^{2,r}$, $r > m$.

Let Ω be a Lipschitz subdomain of M, and let $\nu \in T^*M$ be the unit outward conormal to $\partial\Omega$. In order to formalize the partial derivative operator $u \mapsto \nabla_w u + Au$, where $A \in L^\infty(M, \mathrm{Hom}\,(\mathcal{E}, \mathcal{E}))$ and w is a vector field on M transversal to $\partial\Omega$ (that is, $\mathrm{essinf}\,\langle \nu, w \rangle > 0$ on $\partial\Omega$), we work with a first-order differential operator $P = P(x, D)$ on $\mathcal{E}$ such that

$$\sigma(P; \nu), \text{ the principal symbol of } P \text{ (at } \nu\text{), is scalar and } > 0 \text{ on } \partial\Omega. \tag{4}$$

The author was supported in part by NSF Grant DMS-9870018.

240

Before we state the main result, we introduce some notation. With Ω as before, let $\cdot|_{\partial\Omega}$ be the nontangential boundary trace operator. Specifically, for a section $u : \Omega \to \mathcal{E}$ we set

$$u|_{\partial\Omega}(x) := \lim_{\substack{y\in\gamma(x)}} u(y), \quad x \in \partial\Omega, \tag{5}$$

where $\gamma(x) \subseteq \Omega$ is an appropriate nontangential approach region. Finally, $\mathcal{N}$ stands for the nontangential maximal operator defined by

$$\mathcal{N}u(x) := \sup\{|u(y)|;\ y \in \gamma(x)\}, \quad x \in \partial\Omega. \tag{6}$$

The main result of this paper is the theorem below, which addresses the solvability of the oblique derivative problem for general second-order strongly elliptic PDEs in Lipschitz domains. (For earlier results in the flat Euclidean setting and for constant coefficient PDEs, see [2] and [3].)

Our arguments, building on an earlier idea of Calderón [4], make essential use of the results devised in [5] and [1]. This approach is constructive, in the sense that it relies on integral equation methods; indeed, we prove global representation formulas for the solutions in terms of boundary layer potentials. The trend of using such layer potentials "for general elliptic systems" in the nonsmooth context was suggested by A. P. Calderón in [6]. For related developments, the interested reader may consult [5], [1] and the references therein.

In order to state our main result, we recall that $\mathrm{VMO}(\partial\Omega)$, the space of functions of *vanishing mean oscillation*, is the closure of $\mathrm{Lip}\,(\partial\Omega)$ in the BMO "norm"; see [7] for a discussion in the context of spaces of homogeneous type.

Theorem. *Let $\mathcal{E} \to M$ be as above, and let L be a strongly elliptic, (formally) self-adjoint, second-order differential operator acting on sections of $\mathcal{E}$, with coefficients satisfying (2) locally. Suppose that L satisfies the nonsingularity hypothesis (3) relative to any Lipschitz subdomain of M. Also, let P be a first-order differential operator such that (4) is satisfied, whose coefficients belong to $L^\infty(\partial\Omega)$ and whose top coefficients also belong to $\mathrm{VMO}(\partial\Omega)$.*

Then for any fixed Lipschitz domain $\Omega \subset M$ there exists $\varepsilon = \varepsilon(\Omega, L) > 0$ such that for each $p \in (2 - \varepsilon, 2 + \varepsilon)$ the following holds: for any $f \in L^p(\partial\Omega, \mathcal{E})$ satisfying finitely many (necessary) linear conditions, the oblique derivative problem

$$(\mathrm{ODP}) \quad \begin{cases} u \in C^0_{\mathrm{loc}}(\Omega, \mathcal{E}), \\[4pt] Lu = 0 \ \text{in}\ \Omega, \\[4pt] \mathcal{N}(\nabla u) \in L^p(\partial\Omega), \\[4pt] Pu|_{\partial\Omega} = f \ \text{on}\ \partial\Omega, \end{cases} \tag{7}$$

is solvable and its solution is unique modulo a finite-dimensional linear space.

Finally, the index of this problem is zero, that is, the dimension of the space of null-solutions coincides with the number of linearly independent constraints required for the boundary data.

Proof. The construction of a suitable parametrix implies that $L : H^{1,2}(M, \mathcal{E}) \to H^{-1,2}(M, \mathcal{E})$ is Fredholm, while the hypotheses of self-adjointness and strong ellipticity allow us to use a deformation argument to imply that L has index zero. Hence, $L : H^{1,2}(M, \mathcal{E}) \to H^{-1,2}(M, \mathcal{E})$ is invertible.

Let $\Gamma(x,y) \in \mathcal{D}'(M \times M, \mathcal{E} \otimes \mathcal{E})$ be the Schwartz kernel of $L^{-1} : H^{\mu-1,p}(M,\mathcal{E}) \to H^{\mu+1,p}(M,\mathcal{E})$, $0 < \mu < 1$, $1 < p < r$. It follows that $\Gamma \in C^{1+\gamma}(M \times M \setminus \mathrm{diag}\mathcal{E} \otimes \mathcal{E})$ for some $\gamma > 0$. Next, we introduce the *single layer potential* acting on sections $f : \partial\Omega \to \mathcal{E}$ by means of the formula

$$\mathcal{S}f(x) := \int_{\partial\Omega} \langle \Gamma(x,y), f(y) \rangle \, dS(y), \quad x \in M \setminus \partial\Omega, \tag{8}$$

where dS is the surface measure on $\partial\Omega$. Also, we set $Sf := \mathcal{S}f|_{\partial\Omega}$. The idea is to look for a solution to the boundary problem (7) expressed in the form

$$u := \mathcal{S}g \quad \text{in } \Omega, \tag{9}$$

for some $g \in L^p(\partial\Omega, \mathcal{E})$ to be chosen later. Clearly, $Lu = 0$ in Ω and, from [1], we know that $\|\mathcal{N}(\nabla u)\|_{L^p(\partial\Omega)} \leq C\|g\|_{L^p(\partial\Omega,\mathcal{E})}$. Going further, it has been shown in [1] that at almost every $x \in \partial\Omega$ we have

$$Pu(x) = -\tfrac{1}{2}i\sigma(P;\nu(x))\sigma(L;\nu(x))^{-1}g(x) + Tg(x), \tag{10}$$

where T is the principal-value singular integral operator on $\partial\Omega$ (in the sense of removing geodesic balls) given by

$$Tg(x) = \text{p.v.} \int_{\partial\Omega} \langle (P_x \otimes \mathrm{Id}_y)\Gamma(x,y), g(y) \rangle \, dS(y), \quad x \in \partial\Omega. \tag{11}$$

Thus, in order to prove existence for (7), modulo a space of finite dimension, it suffices to show that

$$-\tfrac{1}{2}i\sigma(P;\nu)\sigma(L;\nu)^{-1} + T : L^p(\partial\Omega, \mathcal{E}) \to L^p(\partial\Omega, \mathcal{E}) \tag{12}$$

has a finite codimensional range for p close to 2. What we shall prove is that this operator is Fredholm with index zero.

To begin with, it has been shown in [1] that the operator (11) is bounded for any $1 < p < \infty$. Hence, since the quality of being Fredholm and the index are stable on complex interpolation scales (cf. the discussion in [8]), it is enough to prove this only for $p = 2$. Rewriting the operator (12) in the form

$$\left[-\tfrac{1}{2}\sigma(P;\nu)i\sigma(L;\nu)^{-1} + \tfrac{1}{2}(T - T^*) \right] + \tfrac{1}{2}(T + T^*) =: T_1 + T_2, \tag{13}$$

we see that, by (4) and the strong ellipticity assumption on L, T_1 is a strictly accretive operator on $L^2(\partial\Omega, \mathcal{E})$ and, hence, invertible. Moreover, we claim that

$$T_2 = \tfrac{1}{2}(T + T^*) \text{ is a compact operator on } L^p(\partial\Omega, \mathcal{E}), \quad 1 < p < \infty. \tag{14}$$

We prove this claim first under the stronger assumption that the top coefficients of P are *Lipschitz* continuous on $\partial\Omega$. In this case, the desired conclusion is going to

242

be a consequence of the *weak singularity* of the kernel of T_2 along the diagonal in $\partial\Omega \times \partial\Omega$.

To see this, we work in local coordinates in a small, open neighborhood U of a boundary point $x_0 \in \partial\Omega$. Let $\{\theta_\alpha\}_\alpha$ be a local orthonormal frame of $\mathcal{E}$ over U. In local coordinates, the fundamental solution Γ has the asymptotic expansion

$$\Gamma(x,y) = \frac{1}{\sqrt{g(y)}}\,\Gamma_0(y, x-y) + R(x,y), \tag{15}$$

where, for each $\varepsilon > 0$, the matrix-valued residual term R satisfies

$$|R(x,y)| + |x-y| \cdot |\nabla_x R(x,y)| + |x-y| \cdot |\nabla_y R(x,y)| = \mathcal{O}(|x-y|^{-m+3-\varepsilon}). \tag{16}$$

Also, the symmetric matrix-valued function $\Gamma_0(y,z)$ is even in z and satisfies

$$\Gamma_0(\rho z, y) = |\rho|^{-(m-2)}\Gamma_0(z, y) \ \text{ for any } \rho \in \mathbb{R} \text{ and } z \in \mathbb{R}^m, \tag{17}$$

as well as the estimates

$$(\nabla_y \Gamma_0)(z,y) = \mathcal{O}(|z|^{-(m-2)}) \ \text{ as } z \to 0, \text{ uniformly in } y, \tag{18}$$

and

$$(\nabla_z \Gamma_0)(z,y),\ (\nabla_z \nabla_y \Gamma_0)(z,y) = \mathcal{O}(|z|^{-(m-1)}) \ \text{ as } z \to 0, \text{ uniformly in } y. \tag{19}$$

Identifying U with an open subset of $\mathbb{R}^m$ and P with a matrix of first-order differential operators $P \in \mathrm{Diff}_1(U \times \mathbb{R}^n, U \times \mathbb{R}^n)$ (where $n := \mathrm{rank}\,\mathcal{E}$), we deduce that the main singularity in the kernel of $T + T^*$ is contained in

$$P(x, D_x)g(y)^{-1/2}\Gamma_0(y, x-y) + (P(y, D_y)g(x)^{-1/2}\Gamma_0(x, y-x))^*, \tag{20}$$

where the asterisk denotes matrix transposition. Now, since the principal part of P is scalar and Γ_0 is symmetric, the expression

$$(P(y, D_y)g(x)^{-1/2}\Gamma_0(x, y-x))^* - P(y, D_y)g(x)^{-1/2}\Gamma_0(x, y-x) \tag{21}$$

is actually $\mathcal{O}(|x-y|^{-m+2})$; therefore, it suffices to examine

$$\begin{aligned}
P(x, D_x)&g(y)^{-1/2}\Gamma_0(y, x-y) + P(y, D_y)g(x)^{-1/2}\Gamma_0(x, y-x) \\
&= P(x, D_x)g(y)^{-1/2}\Gamma_0(y, x-y) + P(y, D_y)g(x)^{-1/2}\Gamma_0(x, x-y).
\end{aligned} \tag{22}$$

To this end, since $g \in C^1$ and the commutator $[P, g^{-1/2}]$ is bounded, the above expression can be rewritten as $g^{-1/2}(I + II + III) + \{\text{less singular terms}\}$, where

$$\begin{aligned}
I &:= P(x, D_x)\Gamma_0(y, x-y) + P(x, D_y)\Gamma_0(y, x-y), \\
II &:= P(y, D_y)\Gamma_0(y, x-y) - P(x, D_y)\Gamma_0(y, x-y), \\
III &:= P(y, D_y)\Gamma_0(x, x-y) - P(y, D_y)\Gamma_0(y, x-y).
\end{aligned} \tag{23}$$

On account of the smoothness of the coefficients of P and the estimates (18) and (19), we find that each of the quantities above and, hence, the entire expression (21), is $\mathcal{O}(|x-y|^{-m+2-\varepsilon})$ for any $\varepsilon > 0$. Consequently, the proof of the fact that $T + T^*$ is weakly singular is finished by invoking (16). This concludes the proof of (14) when the (top) coefficients of P are Lipschitz on $\partial\Omega$.

Returning to the general case when the top coefficients of P are just VMO functions, we select a sequence of operators $\{P_j\}_j$ whose top coefficients are Lipschitzian and approximate the top coefficients of P in the BMO-sense arbitrarily well. The idea is that, while the term II in (23) is no longer $\mathcal{O}(|x-y|^{-m+2-\varepsilon})$, this kernel still gives rise to an operator K_{II} that is compact on $L^p(\partial\Omega, \mathcal{E})$, $1 < p < \infty$. This is because K_{II} can be expressed as a limit of commutators between some Calderón-Zygmund type operator K_j (whose kernels are of the form $\partial_{y_k}\Gamma_0(y, x - y)$) and the operators M_{b_j} of multiplication by Lipschitz functions b_j (arising as coefficients of P_j). By construction, the sequence $\{b_j\}_j$ converge to some b in the BMO-sense. In this case,

$$\|K_{II} - [K_j, M_{b_j}]\|_{\mathcal{L}(L^p)} \leq C\|b_j - b\|_{\mathrm{BMO}} \to 0, \tag{24}$$

where $\mathcal{L}(L^p)$ is the space of all bounded operators on L^p; see [5] for a discussion. Since, by the previous piece of reasoning, each $[K_j, M_{b_j}]$ is compact on L^p, it follows that K_{II} is also compact on L^p. Thus, (14) holds. This completes the proof of (14) for operators P with L^∞ coefficients and whose top coefficients are also in VMO.

So far we have proved that, under the assumptions in the theorem, there exists $\varepsilon > 0$ so that the operator (12) is Fredholm with index zero for each $p \in (2-\varepsilon, 2+\varepsilon)$. It follows that the problem (7) is solvable whenever the boundary datum f satisfies

$$f \in \mathrm{Im}\left(-\tfrac{1}{2}i\sigma(P;\nu)\sigma(L;\nu)^{-1} + T : L^p(\partial\Omega, \mathcal{E})\right) \tag{25}$$

and the right side (that is, the range of the operator in (12)) is a closed subspace of finite codimension in $L^p(\partial\Omega, \mathcal{E})$ for each $|2 - p| < \varepsilon$.

To see that (25) is also a necessary condition for the solvability of (7), let us observe that, by the results in [1], any section u satisfying $Lu = 0$ in Ω and $\mathcal{N}(\nabla u) \in L^p(\partial\Omega)$, $|2 - p| < \varepsilon$, is representable in the form $u = \mathcal{S}g$ in Ω for some $g \in L^p(\partial\Omega, \mathcal{E})$. Consequently,

$$\begin{aligned}
Pu|_{\partial\Omega} &= (-\tfrac{1}{2}i\sigma(P;\nu)\sigma(L;\nu)^{-1} + T)g \\
&\in \mathrm{Im}\left(-\tfrac{1}{2}i\sigma(P;\nu)\sigma(L;\nu)^{-1} + T : L^p(\partial\Omega, \mathcal{E})\right).
\end{aligned} \tag{26}$$

Finally, by essentially the same token, any solution u of the homogeneous version of (7) has the form $u = \mathcal{S}h$ for some section h belonging to $\mathrm{Ker}(-\tfrac{1}{2}i\sigma(P;\nu)\sigma(L;\nu)^{-1} + T : L^p(\partial\Omega, \mathcal{E}))$. Consequently, the dimension of the null-space of the oblique derivative problem (7) is

$$\begin{aligned}
\dim \mathrm{Ker}&\left(-\tfrac{1}{2}i\sigma(P;\nu)\sigma(L;\nu)^{-1} + T : L^p(\partial\Omega, \mathcal{E})\right) \\
&= \dim \mathrm{Im}\left(-\tfrac{1}{2}i\sigma(P;\nu)\sigma(L;\nu)^{-1} + T : L^p(\partial\Omega, \mathcal{E})\right).
\end{aligned}$$

This completes the proof of the theorem.

References

1. D. Mitrea, M. Mitrea and M. Taylor, *Layer potentials, the Hodge Laplacian, and global boundary problems in nonsmooth Riemannian manifolds*, Mem. Amer. Math. Soc. (to appear).

2. C. Kenig and J. Pipher, The oblique derivative problem on Lipschitz domains with L^p data, *Amer. J. Math.* **110**, (1988), 715–737.

3. J. Pipher, Oblique derivative problems for the Laplacian in Lipschitz domains, *Revista Mat. Ibero.* **3**, (1987), 455–472.

4. A.P. Calderón, Boundary value problems for the Laplace equation in Lipschitzian domains, "Recent Progress in Fourier Analysis," I. Peral and J. Rubio de Francia, eds., Elsevier/North-Holland, Amsterdam, 1985.

5. M. Mitrea and M. Taylor, Boundary layer methods for Lipschitz domains in Riemannian manifolds, *J. Functional Anal.* **163** (1999), 181–251.

6. A. Calderón, Commutators, singular integrals on Lipschitz curves and applications, Potential techniques for boundary value problems on C^1 domains, *Proc. Int. Congr. Math., Helsinki,* **1**, 1980, 85–96.

7. R. Coifman and G. Weiss, Extensions of Hardy spaces and their use in analysis, *Bull. Amer. Math. Soc.* **83**, (1977), 569–645.

8. N. Kalton and M. Mitrea, Stability results on interpolation scales of quasi-Banach spaces and applications, *Trans. Amer. Math. Soc.* **350** (1998), 3903–3922.

Department of Mathematics, University of Missouri-Columbia, 202 Math. Sciences Bldg., Columbia, MO 65211, USA; e-mail: marius@math.missouri.edu

A. NASTASE

Viscous aerodynamic optimal design of flying configurations via an enlarged variational method

1. Introduction

The evolution of the shape optimal design (OD) is related to the improvement of its two stages, that is, of the *start solutions*—from the reinforced, hyperbolical, potential solutions, to zonal, spectral solutions of the boundary layer (BL) partial differential equations (PDE)—and of the *optimization strategy* itself—from the inviscid aerodynamic OD of the shape of the flying configuration (FC) with given planform, to the global optimization (including also the optimization of the planform similarity parameters) and from inviscid to viscous aerodynamic OD. The author's rapid OD algorithms, used for the inviscid and viscous shape design of a fully optimized Adela wing model, are now enlarged and used for the shape design of a fully optimized, fully integrated wing-fuselage Fadet model. Both models present high values of lift-to-drag ratio.

2. Inviscid outer flow at the BL edge

The flattened, integrated wing-fuselage (IWF) (that is, with a continuous skeleton and with the same tangent planes along the wing/fuselage junction-lines), is considered here as a wing alone, whose surface is discontinuous (in its higher derivatives) along its junction lines. For flattened thick lifting IWF, at moderate angles of attack α, the effect of lift can be separated from the effect of thickness. The thin and the thick-symmetrical IWF components (that is, the skeleton and the thickness distribution of the thick lifting IWF) occur. The skeleton surface $Z(x_1, x_2)$ of the IWF is continuous, but the thickness distributions $Z^*(x_1, x_2)$ of the wing and $Z'^*(x_1, x_2)$ of the central fuselage of the IWF are different. A dimensionless system of coordinates $O\tilde{x}_1\tilde{x}_2\tilde{x}_3$ is used here,

$$\tilde{x}_1 = \frac{x_1}{h_1}, \quad \tilde{x}_2 = \frac{x_2}{\ell_1}, \quad \tilde{x}_3 = \frac{x_3}{h_1}, \tag{1}$$

$$\left(y = \frac{x_2}{x_1}, \quad \tilde{y} = \frac{y}{\ell}, \quad \ell = \frac{\ell_1}{h_1}, \quad B = \sqrt{M_\infty^2 - 1} \right),$$

where ℓ_1, h_1 are the maximal half-span and depth, $\nu = B\ell$, $\bar{\nu} = Bc'$ are the similarity parameters of the planforms of the gross wing and fuselage, and M_∞ is the cruising Mach number. The downwashes $w = \tilde{w}$ on the thin IWF, $w'^* = \bar{w}^*$ on the fuselage $(|\tilde{y}| < \bar{k})$ and $w^* = \tilde{w}^*$ on the wing $(\bar{k} < |\tilde{y}| < 1)$ of the thick-symmetrical IWF, that is,

$$\tilde{w} = \sum_{m=1}^{N} \tilde{x}_1^{m-1} \sum_{k=0}^{m-1} \tilde{w}_{m-k-1,k} \mid \tilde{y} \mid^k, \tag{2}$$

$$\tilde{w}^* = \sum_{m=1}^{N} \tilde{x}_1^{m-1} \sum_{k=0}^{m-1} \tilde{w}^*_{m-k-1,k} \mid \tilde{y} \mid^k, \quad \bar{w}^* = \sum_{m=1}^{N} \tilde{x}_1^{m-1} \sum_{k=0}^{m-1} \bar{w}^*_{m-k-1,k} \mid \tilde{y} \mid^k, \tag{3}$$

246

are superpositions of homogeneous polynomials. The quotient $\bar{k} = \bar{\nu}/\nu$ is taken constant and the axial velocities are coupled: $u = \ell\tilde{u}$, $u^* = \ell\tilde{u}^*$. The coefficients $\tilde{w}_{ij}$, $\tilde{w}_{ij}^*$, $\bar{w}_{ij}^*$ and the similarity parameter ν are unknown and are obtained through the optimization process. The axial disturbance velocities u, u^* on the thin and thick-symmetrical IWF with subsonic leading edges and a central ridge are (see [1]–[3])

$$u = \ell\sum_{n=1}^{N}\tilde{x}_1^{n-1}\left\{\sum_{q=0}^{E(n/2)}\frac{\tilde{A}_{n,2q}\tilde{y}^{2q}}{\sqrt{1-\tilde{y}^2}} + \sum_{q=1}^{E((n-2)/2)}\tilde{C}_{n,2q}\tilde{y}^{2q}\cosh^{-1}\sqrt{\frac{1}{\tilde{y}^2}}\right\}, \qquad (4)$$

$$u^* = \ell\sum_{n=1}^{N}\tilde{x}_1^{n-1}\left\{\sum_{q=0}^{n-1}\tilde{H}_{nq}^*\tilde{y}^q[\cosh^{-1}M_1+(-1)^q\cosh^{-1}M_2] + \sum_{q=0}^{E((n-1)/2)}\tilde{D}_{n,2q}^*\tilde{y}^{2q}\sqrt{1-\nu^2\tilde{y}^2}\right.$$

$$\left. + \sum_{q=0}^{n-1}\tilde{G}_{nq}^*\tilde{y}^q[\cosh^{-1}R_1+(-1)^q\cosh^{-1}R_2] + \sum_{q=1}^{E((n-1)/2)}\tilde{C}_{n,2q}^*\tilde{y}^{2q}\cosh^{-1}\sqrt{\frac{1}{\nu^2\tilde{y}^2}}\right\}, \qquad (5)$$

$$M_{1,2} = \sqrt{\frac{(1+\nu)(1\mp\nu\tilde{y})}{2\nu(1\mp\tilde{y})}}, \quad R_{1,2} = \sqrt{\frac{(1+\bar{\nu})(1\mp\nu\tilde{y})}{2(\bar{\nu}\mp\nu\tilde{y})}}.$$

In (5) the jump of the downwash along the junction lines between the wing and the fuselage was simulated with two artificial ridges located along these lines. As in [1] and [2], the coefficients of u, $\tilde{w}$ and the coefficients of u^* and $\bar{w}^*$ and $\tilde{w}^*$ are connected by homogeneous relations that are linear in the coefficients of the downwashes and strongly nonlinear in ν and $\bar{\nu}$. The lift, pitching moment and inviscid drag coefficients C_ℓ, C_m and C_d of the thin IWF are

$$C_\ell = 8\ell\int_{\bar{O}\bar{A}_1\bar{C}}\tilde{u}\tilde{x}_1 d\tilde{x}_1 d\tilde{y}, \;\; C_m = 8\ell\int_{\bar{O}\bar{A}_1\bar{C}}\tilde{u}\tilde{x}_1^2 d\tilde{x}_1 d\tilde{y}, \;\; C_d = 8\ell\int_{\bar{O}\bar{A}_1\bar{C}}\tilde{u}\tilde{w}\tilde{x}_1 d\tilde{x}_1 d\tilde{y}. \quad (6)$$

C_ℓ and C_m are linear forms and C_d is a quadratic form in the coefficients $\tilde{w}_{ij}$ of the downwash $\tilde{w}$. C_d^* of the thick-symmetric IWF is a quadratic form in the coefficients $\tilde{w}_{ij}^*$ and $\bar{w}_{ij}^*$ of the downwashes $\tilde{w}^*$ and $\bar{w}^*$; specifically,

$$C_d^* = 8\ell\left(\int_{\bar{O}\bar{C}\bar{C}_1}\tilde{u}^*\bar{w}^*\tilde{x}_1 d\tilde{x}_1 d\tilde{y} + \int_{\bar{O}\bar{A}_1\bar{C}_1}\tilde{u}^*\tilde{w}^*\tilde{x}_1 d\tilde{x}_1 d\tilde{y}\right). \qquad (7)$$

3. Spectral solutions for the boundary layer's PDE

Let $\tilde{\delta}^+$, $\bar{\delta}^+$ and $\tilde{\delta}^-$, $\bar{\delta}^-$ be the dimensionless thicknesses of the upper and lower BL on the IWF. The slopes of the BL, taken as superpositions of homogeneous polynomials, and the modified downwashes $\tilde{w}_1$ and $\tilde{w}_1^*$, $\bar{w}_1^*$, due to the solidification of the boundary layer, and their coefficients are of the form

$$\frac{\partial\tilde{\delta}^+}{\partial\tilde{x}_1} = \sum_{m=1}^{N}\tilde{x}_1^{m-1}\sum_{k=0}^{m-1}\tilde{\delta}_{m-k-1,k}^+|\tilde{y}|^k, \quad \frac{\partial\tilde{\delta}^-}{\partial\tilde{x}_1} = \sum_{m=1}^{N}\tilde{x}_1^{m-1}\sum_{k=0}^{m-1}\tilde{\delta}_{m-k-1,k}^-|\tilde{y}|^k,$$

$$\frac{\partial\bar{\delta}^+}{\partial\tilde{x}_1} = \sum_{m=1}^{N}\tilde{x}_1^{m-1}\sum_{k=0}^{m-1}\bar{\delta}_{m-k-1,k}^+|\tilde{y}|^k, \quad \frac{\partial\bar{\delta}^-}{\partial\tilde{x}_1} = \sum_{m=1}^{N}\tilde{x}_1^{m-1}\sum_{k=0}^{m-1}\bar{\delta}_{m-k-1,k}^-|\tilde{y}|^k, \qquad (8)$$

$$\tilde{w}_1 = \sum_{m=1}^{N} \tilde{x}_1^{m-1} \sum_{k=0}^{m-1} \tilde{w}^{(1)}_{m-k-1,k}\,|\tilde{y}|^k, \quad \tilde{w}_1^* = \sum_{m=1}^{N} \tilde{x}_1^{m-1} \sum_{k=0}^{m-1} \tilde{w}^{*\,(1)}_{m-k-1,k}\,|\tilde{y}|^k,$$

$$\bar{w}_1^* = \sum_{m=1}^{N} \tilde{x}_1^{m-1} \sum_{k=0}^{m-1} \bar{w}^{*\,(1)}_{m-k-1,k}\,|\tilde{y}|^k, \quad \tilde{w}^{(1)}_{ij} = \tilde{w}_{ij} + \tfrac{1}{2}(\tilde{\delta}^+_{ij} - \tilde{\delta}^-_{ij}), \tag{9}$$

$$\tilde{w}^{*\,(1)}_{ij} = \tilde{w}^*_{ij} + \tfrac{1}{2}(\tilde{\delta}^+_{ij} + \tilde{\delta}^-_{ij}), \quad \bar{w}^{*\,(1)}_{ij} = \bar{w}^*_{ij} + \tfrac{1}{2}(\bar{\delta}^+_{ij} + \bar{\delta}^-_{ij}).$$

The modified axial disturbance velocities u_1 and u_1^* at the edge of BL are obtained by replacing in (2)–(7) the inviscid coefficients $\tilde{w}_{ij}$ and $\tilde{w}^*_{ij}$, $\bar{w}^*_{ij}$ by the modified ones $\tilde{w}^{(1)}_{ij}$ and $\tilde{w}^{*\,(1)}_{ij}$, $\bar{w}^{*\,(1)}_{ij}$. If u_e, v_e, w_e are the inviscid velocities at the edge of the BL on IWF (which here replace the parallel undisturbed flow of Prandtl) and $\eta = [x_3 - Z(\tilde{x}_1, \tilde{x}_2)]/\delta(\tilde{x}_1, \tilde{x}_2)$ is a new vertical coordinate, then the author's spectral forms of the axial, lateral, and vertical velocities u_δ, v_δ, w_δ in the BL are (see [2], [3])

$$u_\delta = u_e \sum_{i=1}^{N} u_i \eta^i, \quad v_\delta = v_e \sum_{i=1}^{N} v_i \eta^i, \quad w_\delta = w_e \sum_{i=1}^{N} w_i \eta^i. \tag{10}$$

The non-slip conditions on the wing surface ($\eta = 0$) are automatically satisfied. The matching conditions BL/potential flow at the edge ($\eta = 1$) and the continuity equation written at P points are linear relations in u_i, v_i, w_i :

$$\sum_{i=1}^{N} u_i = 1, \quad \sum_{i=1}^{N} i u_i = 0, \quad \sum_{i=1}^{N} i(i-1)u_i = 0, \quad \sum_{i=1}^{N} w_i = 1, \quad \sum_{i=1}^{N} v_i = 1,$$

$$\sum_{i=1}^{N} i v_i = 0, \quad \sum_{i=1}^{N} i(i-1)v_i = 0, \quad \sum_{i=1}^{N} (E_{ip} u_i + F_{ip} v_i + G_{ip} w_i) = 0. \tag{11}$$

In the BL, $p = p_e(\tilde{x}_1, \tilde{x}_2)$. The physical gas equation and an exponential dependence of the viscosity μ on the temperature T are used to eliminate T from the impulse equations. These equations, written at K points of the BL, and the equations (11) form an algebraic system for the coefficients u_i, v_i and w_i; that is,

$$\sum_{i=1}^{N}\sum_{j=1}^{N} u_i\big(A^{(1)}_{ijk} u_j + B^{(1)}_{ijk} v_j + C^{(1)}_{ijk} w_j\big) = D^{(1)}_k + \sum_{i=1}^{N}\big(A^{(1)}_{ik} u_i + B^{(1)}_{ik} v_i + C^{(1)}_{ik} w_i\big),$$

$$\sum_{i=1}^{N}\sum_{j=1}^{N} v_i\big(A^{(2)}_{ijk} u_j + B^{(2)}_{ijk} v_j + C^{(2)}_{ijk} w_j\big) = D^{(2)}_k + \sum_{i=1}^{N}\big(A^{(2)}_{ik} u_i + B^{(2)}_{ik} v_i + C^{(2)}_{ik} w_i\big), \tag{12}$$

$$\sum_{i=1}^{N}\sum_{j=1}^{N} w_i\big(A^{(3)}_{ijk} u_j + B^{(3)}_{ijk} v_j + C^{(3)}_{ijk} w_j\big) = \sum_{i=1}^{N} C^{(3)}_{ik} w_i.$$

The wall shear stress τ_w and the friction drag coefficient $C_d^{(f)}$ of the IWF are

$$\tau_w = \mu \left. \frac{\partial u_\delta}{\partial \eta} \right|_{\eta=0} = \mu u_1 u_e, \quad C_d^{(f)} = 8\nu_f u_1 \int\limits_{\tilde{O}\tilde{A}_1\tilde{C}} u_e \tilde{x}_1 \, d\tilde{x}_1 \, d\tilde{y}. \tag{13}$$

4. Inviscid and viscous OD of IWF Fadet shape

The OD-shape of the Adela model (for $M_\infty = 2$) has a high value L/D, but is too thin for a central fuselage integration. We now optimize the shape of the thin IWF for a given ν, that is, $\tilde{C}_d = \min$, the lift and pitching moment coefficients C_ℓ, C_m are given, and the Kutta condition $u_{\tilde{y} \to 1} = 0$ on subsonic leading edges (LE) (no induced drag due to conturnement of LE at cruise) is fulfilled. The optimal values of $\tilde{w}_{ij}$ and the Lagrange multipliers are, as in [1] and [2], solutions of the linear algebraic system (LAS) (15), (16):

$$C_d \equiv \ell \sum_{n=1}^{N} \sum_{m=1}^{N} \sum_{k=0}^{m-1} \sum_{j=0}^{n-1} \tilde{\Omega}_{nmkj} \tilde{w}_{m-k-1,k} \tilde{w}_{n-j-1,j} = \min, \tag{14}$$

$$\tilde{C}_\ell \equiv \sum_{n=1}^{N} \sum_{j=0}^{n-1} \tilde{\Lambda}_{nj} \tilde{w}_{n-j-1,j} = C_{\ell_0}/\ell, \quad \tilde{C}_m \equiv \sum_{n=1}^{N} \sum_{j=0}^{n-1} \tilde{\Gamma}_{nj} \tilde{w}_{n-j-1,j} = C_{m_0}/\ell,$$

$$\tilde{F}_n \equiv \sum_{j=0}^{n-1} \tilde{\Psi}_{nj} \tilde{w}_{n-j-1,j} = 0, \quad (n = 1, \ldots, N, \ 0 \le j \le (N-1)), \tag{15}$$

$$\sum_{n=1}^{N} \sum_{j=0}^{n-1} \left[\tilde{\Omega}_{n,\theta+\sigma+1,\sigma,j} + \tilde{\Omega}_{\theta+\sigma+1,n,j,\sigma} \right] \tilde{w}_{n-j-1,j} + \lambda_{\theta+\sigma+1} \tilde{\Psi}_{\theta+\sigma+1,\sigma} + \lambda^{(1)} \tilde{\Lambda}_{\theta+\sigma+1,\sigma}$$

$$+ \lambda^{(2)} \tilde{\Gamma}_{\theta+\sigma+1,\sigma} = 0, \quad (1 \le \theta + \sigma + 1 \le N, \ \theta = 0, 1, \ldots, (N-1)). \tag{16}$$

Now the OD of the shape of the thick-symmetrical IWF for a given ν is treated; that is, $C_d^* = \min$ and the relative volumes $\tau_0^* = V_0/S_0^{3/2}$ and $\bar{\tau}_0^* = \bar{V}_0/S_0^{3/2}$ of the wing and fuselage are given, the thickness cancels along the leading edges and the integration conditions along the wing/fuselage junction lines are satisfied as in [1] and [2]. The optimal values of $\tilde{w}_{ij}^*$, $\bar{w}_{ij}^*$ and the Lagrange multipliers are the solutions of the LAS (18)–(21):

$$C_d^* \equiv \ell \sum_{n=1}^{N} \sum_{m=1}^{N} \sum_{k=0}^{m-1} \sum_{j=0}^{n-1} \left[(\tilde{\Omega}_{nmkj}^* \tilde{w}_{n-j-1,j}^* + \tilde{\Omega}_{nmkj}'^* \tilde{w}_{n-j-1,j}^*) \tilde{w}_{m-k-1,k}^* \right.$$

$$\left. + (\bar{\Omega}_{nmkj}^* \tilde{w}_{n-j-1,j}^* + \bar{\Omega}_{nmkj}'^* \bar{w}_{n-j-1,j}^*) \bar{w}_{m-k-1,k}^* \right] = \min, \tag{17}$$

$$\tilde{\tau}^* \equiv \sum_{m=1}^{N} \sum_{k=0}^{m-1} \tilde{\tau}_{mk}^* \tilde{w}_{m-k-1,k}^* = \tau_0^* \sqrt{\ell}, \quad \bar{\tau}^* \equiv \sum_{m=1}^{N} \sum_{k=0}^{m-1} \bar{\tau}_{mk}^* \bar{w}_{m-k-1,k}^* = \bar{\tau}_0^* \sqrt{\ell}, \tag{18}$$

$$\tilde{E}_t^* \equiv \sum_{m=t+1}^{N} \sum_{k=0}^{m-1} \tilde{d}_{mk}^{*(t)} \tilde{w}_{m-k-1,k}^* = 0, \quad \tilde{F}_t^* \equiv \sum_{m=t+1}^{N} \sum_{k=0}^{m-1} \tilde{c}_{mk}^{*(t)} (\tilde{w}_{m-k-1,k}^* - \bar{w}_{m-k-1,k}^*) = 0,$$

$$\tilde{G}_t^* \equiv \sum_{m=t+1}^{N} \sum_{k=0}^{m-1} \tilde{g}_{mk}^{*(t)} \left(\tilde{w}_{m-k-1,k}^* - \bar{w}_{m-k-1,k}^* \right) = 0, \quad t = 0, 1, \ldots, (N-1),$$

$$\tilde{L}_t^* \equiv \sum_{m=t+1}^{N} \sum_{k=0}^{m-1} \tilde{\ell}_{mk}^{*(t)} \left(\tilde{w}_{m-k-1,k}^* - \bar{w}_{m-k-1,k}^* \right) = 0,$$

$$\tag{19}$$

$$\sum_{n=1}^{N} \sum_{j=0}^{n-1} \Big[(\tilde{\Omega}_{n,\theta+\sigma+1,\sigma,j}^* + \tilde{\Omega}_{\theta+\sigma+1,n,j,\sigma}^*) \tilde{w}_{n-j-1,j}^*$$

$$+ (\tilde{\Omega}_{n,\theta+\sigma+1,\sigma,j}^{\prime*} + \bar{\Omega}_{\theta+\sigma+1,n,j,\sigma}^*) \bar{w}_{n-j-1,j}^* \Big] + \mu^{(1)} \tilde{\tau}_{\theta+\sigma+1,\sigma}^*$$

$$+ \sum_{t=0}^{N-1} \left(\mu_t \tilde{d}_{\theta+\sigma+1,\sigma}^{*(t)} + \bar{\mu}_t \tilde{c}_{\theta+\sigma+1,\sigma}^{*(t)} + \eta_t \tilde{g}_{\theta+\sigma+1,\sigma}^{*(t)} + \bar{\eta}_t \tilde{\ell}_{\theta+\sigma+1,\sigma}^{*(t)} \right) = 0, \tag{20}$$

$$\sum_{n=1}^{N} \sum_{j=0}^{n-1} \Big[(\bar{\Omega}_{n,\theta+\sigma+1,\sigma,j}^* + \tilde{\Omega}_{\theta+\sigma+1,n,j,\sigma}^{\prime*}) \tilde{w}_{n-j-1,j}^*$$

$$+ (\bar{\Omega}_{n,\theta+\sigma+1,\sigma,j}^{\prime*} + \bar{\Omega}_{\theta+\sigma+1,n,j,\sigma}^{\prime*}) \bar{w}_{n-j-1,j}^* \Big] + \mu^{(2)} \tilde{\tau}_{\theta+\sigma+1,\sigma}^*$$

$$- \sum_{t=0}^{N-1} \left(\bar{\mu}_t \tilde{c}_{\theta+\sigma+1,\sigma}^{*(t)} + \eta_t \tilde{g}_{\theta+\sigma+1,\sigma}^{*(t)} + \bar{\eta}_t \tilde{\ell}_{\theta+\sigma+1,\sigma}^{*(t)} \right) = 0. \tag{21}$$

Here $1 \leq \theta + \sigma + 1 \leq N$, $\theta = 0, 1, \ldots, N-1$. If now the global optimization of the thick lifting IWF is treated, the optimal values of the free parameters $\tilde{w}_{ij}$, $\tilde{w}_{ij}^*$, $\bar{w}_{ij}^*$ and ν are obtained if the inviscid drag $C_d^{(i)} = C_d + C_d^* = \min$. The above two systems are nonlinear in ν and coupled. If the author's hybrid numerical-analytic method is used, then the two coupled nonlinear systems are replaced by a cascade of LAS obtained by giving ν ($0 < \nu < 1$) several discrete values. Each point of the lower limit curve of the optimal inviscid drag functional $(C_d^{(i)})_{\text{opt}} = f(\nu)$ is obtained analytically. The location of the minimum of this curve ($\nu = \nu_{\text{opt}}$) is determined numerically. If ν_{opt} is introduced in the two systems above, then the best values of $\tilde{w}_{ij}$ and $\tilde{w}_{ij}^*$, $\bar{w}_{ij}^*$ are obtained. The fully optimized, fully integrated shape of the Fadet model (Fig. 2) at $M_\infty = 2.2$ is thicker in its central part than the fully optimized shape of the Adela model (Fig. 1) at $M_\infty = 2$, as required. An iterative OD is proposed in Fig. 3. The inviscid OD-shape of the IWF is now the first step in the iterative viscous shape optimization process. An intermediate computational checking of the inviscid OD-shape is made with the author's spectral, zonal potential/BL viscous solver. The friction drag coefficient $C_d^{(f)}$ of the IWF is 30% of the total drag (at $M_\infty = 2.2$ and $\alpha = 0$). The inviscid OD-shape, checked also for the thermal and structural point of view, can lead to additional auxiliary conditions. In the second step of optimization, the predicted inviscid optimized shape of the IWF is corrected by also including these additional auxiliary conditions in the variational problem and the friction drag coefficient in the drag functional. The corrections of the inviscid optimized shape of the IWF Fadet after the second iteration of the aerodynamic viscous OD are *not significant*, except on the center of the rear part, which is slightly more cambered than the inviscid one. The proposed enlarged, variational method is robust, flexible, evolutive, multidisciplinary, includes friction, and is adapted for the modern subsystems OD strategy with weak interactions, via auxiliary conditions.

250

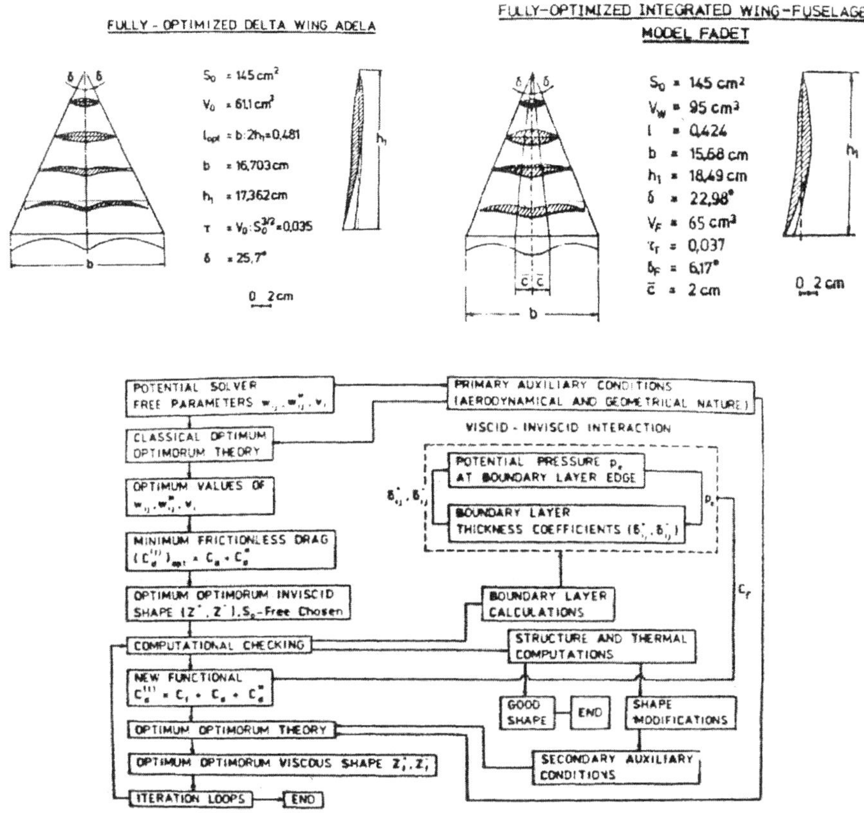

Figs. 1, 2, 3. The fully optimized ADELA and FADET models and the proposed enlarged variational method.

References

1. A. Nastase, Optimum-optimorum integrated wing-fuselage configuration for supersonic transport aircraft of second generation, in *Proc. 15th ICAS Congress*, London, 1986, 324–334.

2. A. Nastase, A new spectral method and its aerodynamic applications, in *Proc. 7th Internat. Symp. on CFD*, Beijing, 1997, 134–139.

3. A. Nastase, The aerodynamic design via iterative optimum-optimorum theory, in *Proc. 4th Internat. Symp. on CFD*, vol. 3, Davis, 1991, 61–66.

Aerodynamik des Fluges, Rhein.-Westf. Technische Hochschule, Aachen, Germany

D. NATROSHVILI and Sh. ZAZASHVILI

Interface crack problem for a piecewise homogeneous anisotropic plane

1. Introduction

We consider the two-dimensional interface crack problem for anisotropic bodies in the Comninou formulation. It is established that, as in the isotropic case, properly incorporating contact zones at the crack tips avoid contradictions connected with the oscillating asymptotic behavior of physical and mechanical characteristics leading to the overlapping of materials. Applying the special integral representation formula for the displacement field the problem in question is reduced to the scalar singular integral equation. The comparison with the results of previous authors shows that the integral equations corresponding to the interface crack problems in the anisotropic and isotropic cases are actually the same from the point of view of the theoretical and numerical analysis. However, there are described some new effects concerning the stress intensity coefficients in the anisotropic case.

2. Formulation of the problem

Denote by $S^{(0)}$ the upper half-plane $x_2 > 0$ and by $S^{(1)}$ the lower half-plane $x_2 < 0$. We assume that the domains $S^{(j)}$ are filled up by anisotropic materials with elastic constants $A_{11}^{(j)}$, $A_{12}^{(j)}$, $A_{13}^{(j)}$, $A_{22}^{(j)}$, $A_{23}^{(j)}$, and $A_{33}^{(j)}$, $j = 0, 1$. The common boundary (x_1-axis) of the above two half-planes will be referred to as *the contact line l*. In what follows we will use the superscript (j) with the physical characteristics corresponding to the domain $S^{(j)}$. We omit the superscript (j) when this causes no confusion.

Consider an interface crack (cut) along the segment $[-L, L]$ on the contact line. Under the action of uniform tension T applied at infinity, in a direction normal to the interface, the crack opens in the interval $(-a, a)$, where a $(0 < a < L)$ is an unknown to be determined in the course of solution. According to the Comninou approach (see [1]) outside of the cut $|x_1| > L$ we have the perfect bond (the rigid contact conditions) between the two materials. We assume that the crack is closed and its two sides are in frictionless contact in the intervals $(-L, -a)$ and (a, L). The boundary conditions on the tractions are that the shear traction must vanish in $(-L, L)$ and that the normal traction must vanish in $(-a, a)$.

The system of equations of elastostatics in the anisotropic case, sans body forces, reads as [2]

$$A_{11}\frac{\partial^2 u_1}{\partial x_1^2} + 2A_{13}\frac{\partial^2 u_1}{\partial x_1 \partial x_2} + A_{33}\frac{\partial^2 u_1}{\partial x_2^2} + A_{13}\frac{\partial^2 u_2}{\partial x_1^2} + A_{12}^*\frac{\partial^2 u_2}{\partial x_1 \partial x_2} + A_{23}\frac{\partial^2 u_2}{\partial x_2^2} = 0,$$

$$A_{13}\frac{\partial^2 u_1}{\partial x_1^2} + A_{12}^*\frac{\partial^2 u_1}{\partial x_1 \partial x_2} + A_{23}\frac{\partial^2 u_1}{\partial x_2^2} + A_{33}\frac{\partial^2 u_2}{\partial x_1^2} + 2A_{23}\frac{\partial^2 u_2}{\partial x_1 \partial x_2} + A_{22}\frac{\partial^2 u_2}{\partial x_2^2} = 0, \qquad (1)$$

where $u = (u_1, u_2)^\top$ is the displacement vector, $A_{12}^* = A_{12} + A_{33}$, and $x = (x_1, x_2) \in S^{(j)} \subset R^2$. Here and in what follows the superscript $\top$ denotes transposition.

The stress components σ_{x_1}, σ_{x_2}, $\tau_{x_1 x_2}$, and the strain components ε_{x_1}, ε_{x_2}, $\varepsilon_{x_1 x_2}$ are related by Hook's law

$$\sigma_{x_1} = A_{11}\varepsilon_{x_1} + A_{12}\varepsilon_{x_2} + A_{13}\varepsilon_{x_1 x_2}, \quad \sigma_{x_2} = A_{12}\varepsilon_{x_1} + A_{22}\varepsilon_{x_2} + A_{23}\varepsilon_{x_1 x_2},$$

$$\tau_{x_1 x_2} = A_{13}\varepsilon_{x_1} + A_{23}\varepsilon_{x_2} + A_{33}\varepsilon_{x_1 x_2}, \qquad (2)$$

252

where

$$\varepsilon_{x_1} = \frac{\partial u_1}{\partial x_1}, \quad \varepsilon_{x_2} = \frac{\partial u_2}{\partial x_2}, \quad \varepsilon_{x_1 x_2} = \frac{\partial u_1}{\partial x_2} + \frac{\partial u_2}{\partial x_1}.$$

The positive definiteness of the potential energy implies that the symmetric matrix $[A_{pq}]_{3\times 3}$ is positive definite [2]. Let $\Delta = \det [A_{pq}]$.

Problem (C). Find regular solutions $u^{(j)}$ to the system (1) in $S^{(j)}$ $(j = 0, 1)$ satisfying the following interface and boundary conditions on the contact line l $(x_2 = 0)$:

$$\text{i)} \quad [u^{(0)}]^+ = [u^{(1)}]^-, \quad [\sigma_{x_2}^{(0)}]^+ = [\sigma_{x_2}^{(1)}]^-, \quad [\tau_{x_1 x_2}^{(0)}]^+ = [\tau_{x_1 x_2}^{(1)}]^-, \quad |x_1| > L, \qquad (3)$$

$$\text{ii)} \quad [u_2^{(0)}]^+ = [u_2^{(1)}]^-, \quad [\sigma_{x_2}^{(0)}]^+ = [\sigma_{x_2}^{(1)}]^-, \quad [\tau_{x_1 x_2}^{(0)}]^+ = [\tau_{x_1 x_2}^{(1)}]^- = 0, \quad a < |x_1| < L, \quad (4)$$

$$\text{iii)} \quad [\sigma_{x_2}^{(0)}]^+ = [\sigma_{x_2}^{(1)}]^- = 0, \quad [\tau_{x_1 x_2}^{(0)}]^+ = [\tau_{x_1 x_2}^{(1)}]^- = 0, \quad |x_1| < a, \qquad (5)$$

where the symbols $[\cdot]^+$ and $[\cdot]^-$ denote limits on l from $S^{(0)}$ and $S^{(1)}$, respectively.

By a regular solution to the system (1) is understood a two-dimensional vector $u^{(j)} = (u_1^{(j)}, u_2^{(j)})^\top$ such that

a) $u^{(j)} \in \mathrm{C}(\overline{S^{(j)}}) \cap \mathrm{C}^2(S^{(j)})$,

b) the corresponding stress components $\sigma_{x_1}^{(j)}$, $\sigma_{x_2}^{(j)}$, and $\tau_{x_1 x_2}^{(j)}$ are continuously extendable on the whole x_1 axis except the points $\{-L; -a; a; L\}$ in the vicinity of which they have integrable singularities,

c) for sufficiently large $|x| = (x_1^2 + x_2^2)^{1/2}$ the following conditions hold:

$$u^{(j)}(x) - u_\infty^{(j)}(x) = O(1), \quad \frac{\partial}{\partial x_k}\left[u^{(j)}(x) - u_\infty^{(j)}(x)\right] = O(|x|^{-2}),$$

$$u_\infty^{(j)}(x) = \frac{T}{A_{22}^{(j)} A_{33}^{(j)} - (A_{23}^{(j)})^2} \begin{bmatrix} -A_{23}^{(j)} \\ A_{33}^{(j)} \end{bmatrix} x_2, \quad k = 1, 2, \quad j = 0, 1. \qquad (6)$$

Theorem. *Let a pair $(u^{(0)}, u^{(1)})$ of vector functions $u^{(0)} = (u_1^{(0)}, u_2^{(0)})^\top$ and $u^{(1)} = (u_1^{(1)}, u_2^{(1)})^\top$ be a regular solution to the homogeneous Problem (C) (i.e., $T = 0$).*

Then $u^{(0)} = (c_1, c_2)^\top$ in $S^{(0)}$ and $u^{(1)} = (c_1, c_2)^\top$ in $S^{(1)}$ where c_1 and c_2 are arbitrary real constants. Moreover, if, in addition, the displacement fields in $S^{(0)}$ and $S^{(1)}$ vanish at infinity, then the homogeneous Problem (C) possesses only the trivial solution.

3. Auxilary problem

In this section by means of potential type integrals we construct a solution to the basic interface problem for the piecewise homogeneous anisotropic elastic plane $\overline{S^{(0)}} \cup \overline{S^{(1)}}$. This problem can be formulated as follows [3]: Find regular solutions $u^{(j)}$ to the system (1) in the domains $S^{(j)}$ $(j = 0, 1)$ satisfying on the interface line l the transmission conditions

$$[u^{(0)}]^+ - [u^{(1)}]^- = f(x_1), \quad [\tau_{x_1 x_2}^{(0)}]^+ = [\tau_{x_1 x_2}^{(1)}]^-, \quad [\sigma_{x_2}^{(0)}]^+ = [\sigma_{x_2}^{(1)}]^-, \quad -\infty < x_1 < +\infty,$$

where $f = (f_1, f_2)^\top \in \mathrm{C}^{1+\alpha}(l)$ is a given vector function with the following asymptotics at infinity $\quad f_k(x_1) = \tilde{c}_k + O\left(\frac{1}{|x_1|^\varepsilon}\right), \quad f_k'(x_1) = O\left(\frac{1}{|x_1|^{1+\varepsilon}}\right), \quad k = 1, 2$; here $\tilde{c}_k$ and $\varepsilon > 0$ are some real constants.

The solution of the above basic interface problem is then representable in the form (for details see [3], [4])

$$u^{(j)}(x) = K^* + \frac{1}{\pi}\mathrm{Im}\sum_{k=1}^{2}\mathcal{E}^{(j)}_{(k)}(X^{(j)} + \mathrm{i}Y^{(j)})\int_{-\infty}^{+\infty}\frac{f(t)dt}{t - z_{kj}},\qquad(7)$$

where K^* is an arbitrary real constant vector, $z_{kj} = x_1 + \alpha_k^{(j)}x_2$, $\alpha_k^{(j)}$ ($k = 1, 2$, $j = 0, 1$) are roots of the so-called characteristic equation of the system (1) (see [2])

$$a_{11}^{(j)}\alpha^4 - 2a_{13}^{(j)}\alpha^3 + (2a_{12}^{(j)} + a_{33}^{(j)})\alpha^2 - 2a_{23}^{(j)}\alpha + a_{22}^{(j)} = 0.$$

Note that this equation possesses only the complex roots. Here $[a_{pq}^{(j)}]_{3\times3}$ denotes the matrix inverse to $[A_{pq}^{(j)}]_{3\times3}$.

Further,

$$\mathcal{E}^{(j)}_{(k)} := -\frac{\mathrm{i}}{m^{(j)}}\begin{bmatrix} A_k^{(j)} & B_k^{(j)} \\ B_k^{(j)} & C_k^{(j)} \end{bmatrix}\begin{bmatrix} B^{(j)} & -A^{(j)} \\ -A^{(j)} & C^{(j)} \end{bmatrix},$$

where

$$A_k = -\frac{2}{\Delta a_{11}}(A_{22}\alpha_k^2 + 2A_{23}\alpha_k + A_{33})d_k, \quad a_{11} = (A_{22}A_{33} - A_{23}^2)\Delta^{-1},$$

$$B_k = \frac{2}{\Delta a_{11}}(A_{23}\alpha_k^2 + (A_{12} + A_{33})\alpha_k + A_{13})d_k,$$

$$C_k = -\frac{2}{\Delta a_{11}}(A_{33}\alpha_k^2 + 2A_{13}\alpha_k + A_{11})d_k, \quad \omega = \frac{a_{12}}{a_{11}} - \mathrm{Re}(\alpha_1\alpha_2),$$

$$d_1^{-1} = (\alpha_1 - \bar\alpha_1)(\alpha_1 - \alpha_2)(\alpha_1 - \bar\alpha_2), \quad d_2^{-1} = (\alpha_2 - \alpha_1)(\alpha_2 - \bar\alpha_1)(\alpha_2 - \bar\alpha_2),$$

$$A = 2\mathrm{i}\sum_{k=1}^{2}\alpha_k d_k, \quad B = 2\mathrm{i}\sum_{k=1}^{2}\alpha_k^2 d_k, \quad C = 2\mathrm{i}\sum_{k=1}^{2}d_k, \quad \kappa_N = \omega(BC - A^2)m^{-1},$$

$$\sum_{k=1}^{2}\begin{bmatrix} A_k & B_k \\ B_k & C_k \end{bmatrix} = \frac{\mathrm{i}m}{BC - A^2}\begin{bmatrix} C & A \\ A & B \end{bmatrix}, \quad m = a_{11}(1 - \omega^2(BC - A^2)).$$

Moreover,

$$X^{(0)} + \mathrm{i}Y^{(0)} = \frac{1}{\Delta_0}\left\{\left(\frac{B^{(1)}C^{(1)} - (A^{(1)})^2}{m^{(1)}a_{11}^{(1)}} + \kappa_N^{(0)}\kappa_N^{(1)}\right)E + \right.$$

$$+\frac{1}{m^{(0)}m^{(1)}}\begin{bmatrix} B^{(1)}C^{(0)} - A^{(0)}A^{(1)} & A^{(0)}C^{(1)} - A^{(1)}C^{(0)} \\ A^{(0)}B^{(1)} - A^{(1)}B^{(0)} & B^{(0)}C^{(1)} - A^{(0)}A^{(1)} \end{bmatrix} -$$

$$\left.-\mathrm{i}\left(\frac{\kappa_N^{(1)}}{m^{(0)}}\begin{bmatrix} A^{(0)} & -C^{(0)} \\ B^{(0)} & -A^{(0)} \end{bmatrix} + \frac{\kappa_N^{(0)}}{m^{(1)}}\begin{bmatrix} A^{(1)} & -C^{(1)} \\ B^{(1)} & -A^{(1)} \end{bmatrix}\right)\right\},$$

$$\Delta_0 = 2\kappa_N^{(0)}\kappa_N^{(1)} + \frac{B^{(0)}C^{(0)} - (A^{(0)})^2}{m^{(0)}a_{11}^{(0)}} + \frac{B^{(1)}C^{(1)} - (A^{(1)})^2}{m^{(1)}a_{11}^{(1)}} +$$

$$+\frac{B^{(1)}C^{(0)} + B^{(0)}C^{(1)} - 2A^{(0)}A^{(1)}}{m^{(0)}m^{(1)}} > 0.$$

The constant matrices $X^{(1)} + \mathrm{i}Y^{(1)}$ is obtained from the above formulas for $X^{(0)} + \mathrm{i}Y^{(0)}$ by the interchange of the superscripts (0) and (1). Note that $X^{(0)} + X^{(1)} = E$,

$Y^{(0)} = Y^{(1)}$, where E stands for the unit matrix. It can be shown that $B > 0$, $C > 0$, $BC - A^2 > 0$, $\omega > 0$, $m > 0$, $\kappa_N > 0$.

From (7) we have the following representation formula for the stress components

$$\begin{bmatrix} \tau^{(j)}_{x_1 x_2} \\ \sigma^{(j)}_{x_2} \end{bmatrix} = -\frac{1}{\pi} \mathrm{Im} \sum_{k=1}^{2} \mathcal{E}'^{(j)}_{(k)} \left(X^{(j)} + \mathrm{i} Y^{(j)} \right) \int\limits_{-\infty}^{+\infty} \frac{f'(t)dt}{t - z_{k_j}}, \tag{8}$$

where $\qquad \mathcal{E}'^{(j)}_{(k)} = \left\{ \kappa_N^{(j)} \begin{bmatrix} 0 & -1 \\ 1 & 0 \end{bmatrix} - \frac{\mathrm{i}}{m^{(j)}} \begin{bmatrix} B^{(j)} & -A^{(j)} \\ -A^{(j)} & C^{(j)} \end{bmatrix} \right\} \mathcal{E}^{(j)}_{(k)}.$

4. Reduction to an integral equation

In this section we will essentially use the representation formulas (7) and (8) to investigate the interface Problem (C).

We look for the solution of Problem (C) in the form

$$u^{(j)}(x) = K^* + u^{(j)}_\infty(x) + \frac{1}{\pi} \mathrm{Im} \sum_{k=1}^{2} \mathcal{E}^{(j)}_{(k)} (X^{(j)} + \mathrm{i} Y^{(j)}) \begin{bmatrix} \int\limits_{-L}^{L} \frac{u_1^{(0)}(t) - u_1^{(1)}(t)}{t - z_{k_j}} dt \\ \int\limits_{-a}^{a} \frac{u_2^{(0)}(t) - u_2^{(1)}(t)}{t - z_{k_j}} dt \end{bmatrix}, \tag{9}$$

where $x \in S^{(j)}$, $j = 0, 1$. Here K^* is again an arbitrary constant vector. Note that in (9) the second summand $u^{(j)}_\infty(x)$, given by (6), represents the solution of (1). Here and throughout this section we use the notations $u^{(0)}(t) := [u^{(0)}(t, 0)]^+$ and $u^{(1)}(t) := [u^{(1)}(t, 0)]^-$ for $-\infty < t < +\infty$.

Clearly, the difference $u_1^{(0)}(t) - u_1^{(1)}(t)$ is unknown in the interval $(-L, L)$, while the difference $u_2^{(0)}(t) - u_2^{(1)}(t)$ is unknown in the interval $(-a, a)$ (see (3)–(4)).

It is evident that the conditions

$$u_1^{(0)}(\pm L) - u_1^{(1)}(\pm L) = 0, \qquad u_2^{(0)}(\pm a) - u_2^{(1)}(\pm a) = 0 \tag{10}$$

are sufficient for the above displacement vectors $u^{(j)}$ to be continuously extendable on the whole contact line l (see [5]). Let

$$u_1^{(0)}(x_1) - u_1^{(1)}(x_1) = -\int\limits_{-L}^{x_1} B_1(t)dt, \qquad u_2^{(0)}(x_1) - u_2^{(1)}(x_1) = -\int\limits_{-a}^{x_1} B_2(t)dt,$$

where $B(x_1) = (B_1(x_1), B_2(x_1))^\top$ is the so-called dislocation vector. We require that

$$\int\limits_{-L}^{L} B_1(t)dt = 0, \qquad \int\limits_{-a}^{a} B_2(t)dt = 0, \tag{11}$$

which guarantee the conditions (10).

From (9) it follows that the stress components can be represented in terms of the dislocation vector

$$\begin{bmatrix} \tau^{(j)}_{x_1 x_2} \\ \sigma^{(j)}_{x_2} \end{bmatrix} = \begin{bmatrix} 0 \\ T \end{bmatrix} + \frac{1}{\pi} \mathrm{Im} \sum_{k=1}^{2} \mathcal{E}'^{(j)}_{(k)} (X^{(j)} + \mathrm{i} Y^{(j)}) \begin{bmatrix} \int\limits_{-L}^{L} \frac{B_1(t)dt}{t - z_{k_j}} \\ \int\limits_{-a}^{a} \frac{B_2(t)dt}{t - z_{k_j}} \end{bmatrix}. \tag{12}$$

The boundary conditions (3)–(5) together with the formulas (9) and (12) lead to the system of integral equations

$$QB_1(x_1) + \frac{R_{21}}{\pi} \int\limits_{-L}^{L} \frac{B_1(t)dt}{t-x_1} + \frac{R_{22}}{\pi} \int\limits_{-a}^{a} \frac{B_2(t)dt}{t-x_1} = T, \quad |x_1| < a,$$

$$Q(H(x_1 - a) - H(x_1 + a))B_2(x_1) + \frac{R_{11}}{\pi} \int\limits_{-L}^{L} \frac{B_1(t)dt}{t-x_1} + \frac{R_{12}}{\pi} \int\limits_{-a}^{a} \frac{B_2(t)dt}{t-x_1} = 0, \qquad (13)$$

$$|x_1| < L,$$

where H stands for the Heviside step function, and

$$Q = \frac{1}{\Delta_0} \left(\kappa_N^{(1)} \frac{B^{(0)}C^{(0)} - (A^{(0)})^2}{m^{(0)}a_{11}^{(0)}} - \kappa_N^{(0)} \frac{B^{(1)}C^{(1)} - (A^{(1)})^2}{m^{(1)}a_{11}^{(1)}} \right),$$

$$\begin{bmatrix} R_{11} & R_{12} \\ R_{21} & R_{22} \end{bmatrix} = \frac{1}{\Delta_0} \left(\frac{B^{(1)}C^{(1)} - (A^{(1)})^2}{m^{(0)}m^{(1)}a_{11}^{(1)}} \begin{bmatrix} B^{(0)} & -A^{(0)} \\ -A^{(0)} & C^{(0)} \end{bmatrix} + \right.$$

$$\left. + \frac{B^{(0)}C^{(0)} - (A^{(0)})^2}{m^{(0)}m^{(1)}a_{11}^{(0)}} \begin{bmatrix} B^{(1)} & -A^{(1)} \\ -A^{(1)} & C^{(1)} \end{bmatrix} \right).$$

Note that $R_{11} > 0$, $R_{22} > 0$, $R_{12} = R_{21}$. It can be easily shown that in the isotropic case the system (13) coinsides with the system of equations which has been obtained and investigated in [6].

Thus, for the unknown functions $B_1(x_1)$ and $B_2(x_1)$ we have the system (13) of integral equations with additional conditions (11).

According to the general theory of singular integral equations (see [5]) we can invert the Cauchy type integral by density $B_1(t)$ from the second equation of the system (13) in the class of functions which are unbounded at the both ends ± 1.

By the change of variables $x = Ls$, $t = Lr$, and retaining the same symbols for the unknown functions in the new variables, we obtain (where $X(s) = (1 - s^2)^{-1/2}$)

$$B_1(s) = -\frac{R_{12}}{R_{11}} H(\gamma^2 - s^2) B_2(s) - \frac{QX(s)}{\pi R_{11}} \int\limits_{-\gamma}^{\gamma} \frac{B_2(r)dr}{X(r)(r-s)}, \quad \gamma = aL^{-1}, \quad |s| < 1. \qquad (14)$$

This equation implies that, if $B_2(t)$ satisfies the second condition of (11), then $B_1(t)$ determined by formula (14) satisfies the first condition of (11) automatically.

Substituting $B_1(s)$ from (14) into the first equation of (13) we obtain

$$\frac{1}{\pi} \int\limits_{-\gamma}^{\gamma} \frac{B_2(r)}{r-s} (\sqrt{1 - s^2} - \nu^2 \sqrt{1 - r^2}) dr = T_* \sqrt{1 - s^2}, \quad |s| < \gamma,$$

$$\nu^2 = Q^2 (R_{11}R_{22} - R_{12}^2)^{-1} > 0, \quad T_* = R_{11}T(R_{11}R_{22} - R_{12}^2)^{-1}. \qquad (15)$$

Thus, we have obtained the scalar integral equation (15) with the second condition in (11), to find the unknown B_2 in the class of functions bounded at the both ends $\pm\gamma$. It is evident that (see [5]), if (15) possesses such a solution, then this solution automatically satisfies the condition $B_2(\pm\gamma) = 0$.

A direct comparison shows that (15) is quite similar to the equation obtained by M.Comninou [1] for isotropic case and investigated by Gautesen and Dundurs in [6]. Therefore, the analysis given in [6] can be applied to our case to construct the explicit solution of (15) and to determine the unknown parameter a.

256

However, a deeper analysis of the stress intensity coefficients yields that we have nonoscillatory singularities at the tip points $\pm L$. Moreover, $\tau_{x_1 x_2}^{(j)}(x_1) = O(|L^2 - x_1^2|^{-1/2})$ as $x_1 \to -L^-$ or as $x_1 \to L^+$, and $\sigma_{x_2}^{(j)}(x_1) = O(|L^2 - x_1^2|^{-1/2})$ as $x_1 \to \pm L$. This latter result is a new effect connected with the anisotropy property of the elastic materials in question. In the isotropic case the normal stresses *are bounded* when a point approaches the end points $\pm L$ along the cut line from outside (cf. [1], [7]).

References

1. M. Comninou, The interface crack, *ASME J. Appl. Mech.* **44** (1977), 631–636.

2. S.G. Lekhnitskii, *Anisotropic plates*, Fizmatgiz, Moscow-Leningrad, 1947.

3. M.O. Basheleishvili, Solution of basic interface problems for piecewise homogeneous anisotropic plane, in *Proceedings of I.Vekua Institute of Applied Mathematics of Tbilisi State University*, **39**, 1990, 13–26.

4. M.O. Basheleishvili, Investigation of plane boundary value problems of statics of anisotropic elastic bodies, in *Proceedings of Computational Centre of Academy of Sciences of Georgian SSR*, **3**, 1963, 93–139.

5. N.I. Muskhelishvili, *Singular integral equations*, Nauka, Moscow, 1968.

6. A.K. Gautesen and J. Dundurs, The interface crack in a tension field, *ASME J. Appl. Mech.* **54** (1987), 93–98.

7. D. Natroshvili and Sh. Zazashvili, The interface crack problem for anisotropic bodies, preprint, Technische Universität Chemnitz, **98-4**, 1998, 1–21.

Department of Mathematics, Georgian Technical University, Tbilisi, Georgia

S. NOMURA and H. EDMISTON

Micromechanics of heterogeneous materials

1. Introduction

The determination of failure conditions for fiber-reinforced composite materials is one of the most critical subjects in structures research, and many empirical as well as semianalytical criteria have been proposed (see [1] for example). However, models to predict composite failure based on micromechanical analysis are few and far between ([2] for example) because of its complicated nature, which calls for more extensive study on failure mechanism of fiber composites.

In this paper, failure envelopes of unidirectionally reinforced composite materials are derived based on micromechanics. Each fiber is modeled as an isotropic cylinder embedded in an isotropic medium that renders the whole composite transversely isotropic. It is assumed that failure of the composite occurs at the interface between the fiber and the matrix due to the yielding at the matrix phase. The stress field at the matrix-fiber interface is estimated using the self-consistent approximation [3,4]. The yield condition in the matrix phase expressed by the local field stress is converted into the yield criterion expressed by the far field stress.

2. Effective medium

The elastic equilibrium equation for heterogeneous materials without body force is expressed as

$$\nabla \cdot (C(r)\nabla u(r)) = 0, \tag{1}$$

where $C(r)$ is the fourth-rank elastic modulus tensor, $u(r)$ is the displacement vector, $\nabla\cdot$ is partial derivative and tensor algebra is assumed throughout. For example, AB represents $A_{ijkl}B_{klmn}$ if both A and B are fourth rank-tensors and Ax represents $A_{ijkl}x_{kl}$ if A is a fourth-rank tensor and x is a second-rank tensor.

The elastic modulus, C, can be decomposed into a reference part and a fluctuating part as

$$C(r) = C^* + \delta C(r), \tag{2}$$

where C^* is a constant elastic modulus of the reference medium yet to be determined and $\delta C(r)$ is a deviation from that of the reference medium. By substituting (2) into (1), one obtains

$$C^*\nabla\nabla u(r) + \nabla \cdot (\delta C(r)\nabla u(r)) = 0. \tag{3}$$

Equation (3) can be solved formally by regarding the second term in the left-hand side of (3) as an imaginary body force as

$$u(r) = u^A(r) + \int \nabla \cdot (\delta C(r')\nabla u(r'))\, g(r - r')dr', \tag{4}$$

where $u^A(r)$ is the displacement field in the homogeneous medium that has the stiffness of C^*. The integral range in (4) is over the whole material points. In (4), $g(r - r')$ is the static Green's function for elasticity defined as

$$C^*\nabla\nabla g(r - r') + \delta\delta(r - r') = 0, \tag{5}$$

This paper is based in part on work supported by the Texas Advanced Technology Program under Grant No. 0036556-044.

258

where δ denotes the Kronecker delta and $\delta(r-r')$ is the Dirac delta function. Equation (4) can be rearranged using integration by parts to

$$u(r) = u^A(r) + \int \delta C(r') \nabla u(r') \nabla g(r - r') dr'. \tag{6}$$

When the composite contains fibers embedded in a homogeneous matrix phase, the quantity, $\delta C(r)$, can be expressed as

$$\delta C(r) = (C^m - C^*) + (C^f - C^m)\psi_f(r), \tag{7}$$

where C^m is the elastic modulus for the the matrix phase, C^f is the elastic modulus for the fiber phase and $\psi_f(r)$ is the characteristic function of the fiber phase defined as

$$\psi_f(r) = \begin{cases} 0 & \text{if } r \in \text{the fiber phase} \\ 1 & \text{if } r \notin \text{the fiber phase.} \end{cases} \tag{8}$$

By taking the ensemble average of (6) after substituting (7), one obtains

$$\begin{aligned} <u(r)> \ = \ & u^A(r) \\ + \ & (C^m - C^*) \int <\nabla u(r')> \nabla g(r - r')dr' \\ + \ & (C^f - C^m) \int <\psi_f(r')\nabla u(r')> \nabla g(r - r')dr'. \end{aligned} \tag{9}$$

The self-consistent approximation is adopted in which C^* is chosen such that the ensemble average of $u(r)$ is equal to $u^A(r)$ for the medium that has the effective elastic modulus of C^* that yields the same ensemble averaged displacement as the composite as

$$<u(r)> = u^A(r). \tag{10}$$

The following replacement of each quantity in (9) is adopted as

$$<\nabla u(r)> \ = \ \nabla u^A(r) \tag{11}$$
$$<\psi_f(r)\nabla u(r)> \ = \ v_f A(C^f, C^*)\nabla u^A(r) \tag{12}$$
$$<\psi_f(r)u(r)> \ = \ v_f u^A(r), \tag{13}$$

where v_f is the volume fraction of the fiber phase, $A(C^f, C^*)$ is the proportionality factor of the displacement gradient inside the fiber when a single fiber is placed in a homogeneous medium of C^* with the applied strain of ∇u^A at remote distance (see [5] for explicit forms).

In (9)–(13), the "ergodic assumption" was adopted that the ensemble average can be replaced by the spatial average. Equations (10)–(13) imply that the displacement gradient field inside each fiber is approximated by the boundary displacement gradient, $u^A(r)$, while the displacement gradient field inside each fiber is approximated by the displacement gradient field when a single fiber is placed in a homogeneous medium.

By combining (9)–(13), the following equation for C^* can be obtained:

$$C^* = C^m + v_f(C^n - C^m)A(C^f, C^*). \tag{14}$$

Equation (14) is a set of nonlinear algebraic equations, since A is a function of C^f and C^*.

3. Composite failure

The stress field, σ^f, inside a single ellipsoidal inclusion in an infinite matrix subject to a far field strain, $\bar{\epsilon}$, is uniform and can be expressed by the Eshelby's method [6] symbolically as

$$\sigma^f = C^f A(C^f, C^m)\bar{\epsilon}, \tag{15}$$

where $A(C^f, C^m)$ is the same A in (12) with C^* replaced by C^m. When there are multiple fibers distributed in an infinite matrix in a statistically homogeneous way, the effective elastic modulus, C^*, can be used in place of C^m in (15) as

$$\sigma^f = C^f A(C^f, C^*)\bar{\epsilon}. \tag{16}$$

At the boundary between the matrix and the fiber, the displacements and tractions must be continuous. However, the strains across the boundary need not be continuous, but may be described in terms of a "jump" parameter λ_i as [7]

$$\epsilon_{ij}^{out} - \epsilon_{ij}^{in} = \lambda_i n_j, \tag{17}$$

where ϵ^{out} is the strain outside the fiber, ϵ^{in} is the strain inside the fiber and n_i is the outward unit normal vector to the surface of the fiber.

Equating the tractions at the boundary results in

$$C_{ijkl}^m \epsilon_{kl}^{out} n_j = C_{ijkl}^f \epsilon_{kl}^{in} n_j. \tag{18}$$

Substituting (17) into (18) results in a set of simultaneous equations for the parameters λ_i:

$$C_{ijkl}^m \lambda_k n_l n_j = (C_{ijkl}^f - C_{ijkl}^m)\epsilon_{kl}^{in} n_j. \tag{19}$$

Solving these equations for λ_i allows the calculation of ϵ_{ij}^{out} in terms of ϵ_{ij}^{in}.

The stresses outside the fiber may then be calculated from the applied far field stress, $\bar{\sigma}$, the elastic moduli of the fiber and the matrix, C^f and C^m, respectively, and the fiber volume fractions, v_f, as

$$\sigma^{out} = C^m \epsilon^{out} = G(C^f, C^m, v_f)\bar{\sigma}, \tag{20}$$

where $G(C^f, C^m, v_f)$ is the proportionality factor to the far stress field. Equation (20) shows that σ^{out} can be expressed as a function of the applied far field stresses.

The interface stresses, σ^{out}, are now substituted into an appropriate failure criterion for the matrix phase as

$$F(\sigma^{out}) = 0. \tag{21}$$

This is translated into a failure criterion in terms of $\bar{\sigma}$ using (20) as

$$F(\bar{\sigma}) = 0. \tag{22}$$

Thus, a failure criterion for the entire composite is written in terms of the applied far field stress.

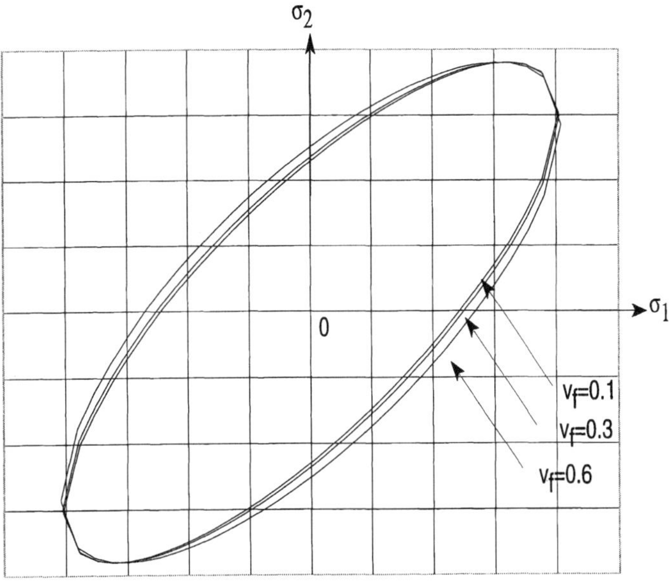

Figure 1:Typical failure envelope

For example, the von Mises failure criterion is expressed using the interface stresses as

$$\sigma'_{ij}\sigma'_{ij} = s^2, \tag{23}$$

where σ'_{ij} is the deviatoric part of the stress and s is a constant.

Substituting (20) into (23) yields the failure criterion in terms of the externally applied stress as

$$g_{ijkl}\bar{\sigma}_{kl}\bar{\sigma}_{ij} = s^2, \tag{24}$$

where g_{ijkl} is a fourth-rank tensor and is a function of the composite constituents, the fiber volume fraction, and the fiber shape (aspect ratio).

4. Results and discussion

Figure 1 represents the failure envelopes computed for a composite reinforced with cylindrical fibers aligned unidirectionally. The material is transversely isotropic and the envelopes represent the allowable states of axial stress that may act in the plane of isotropy. The envelopes presented represent composites with fiber volume fractions of 0.1, 0.3 and 0.6. In the transverse plane the reinforcement is not continuous and the strength of the material is controlled by the strength of the matrix. The failure envelopes represent this characteristic as well as demonstrate an increase in strength with increasing fiber volume fraction.

The assumption of initial failure at the phase boundary implies that this modeling technique is most applicable to composites subject to matrix dominated failure. This technique, however, is not restricted to the von Mises criterion. Any suitable failure theory can be incorporated into the algorithm to generate the failure envelopes.

References

1. S.W. Tsai and E.M. Wu, A general theory of strength for anisotropic materials, *J. Composite Materials* **5** (1971), 58–80.

2. G.J. Dvorak and Y.A. Bahai-El-Din, Elastic-plastic behavior of fibrous composites, *J. Mech. Phys. Solids* **27** (1979), 51.

3. S. Nomura and T.-W. Chou, Bounds for elastic moduli of multiphase short-fiber composites, *J. Appl. Mech.* **51** (1984), 540–545.

4. R. Hill, A Self-consistent mechanics of composite materials, *J. Mech. Phys. Solids* **15** (1965), 79–95.

5. S. Nomura and D. Ball, Micromechanical formulation of effective thermal expansion coefficients of unidirectional fiber composites, *Advanced Composites* **3** (1993), 143–152.

6. J.D. Eshelby, The determination of the elastic field of an ellipsoidal inclusion and related problems, *Proc. Roy. Soc. London* **A241** (1957), 376–396.

7. T. Mura, *Micromechanics of defects in solids*, 2nd ed., Martinus Nijhoff, Dordrecht, 1987.

Department of Mechanical and Aerospace Engineering, The University of Texas at Arlington, Arlington, Texas 76019-0023, USA

Z.C. OKONKWO

Admissibility and optimal control for stochastic difference equations

1. Introduction and statement of the problem

Motivated by results in [1], in this paper we study the problem associated with the stochastic difference equation

$$x(k+1,\omega) = A(k)x(k,\omega) + B(k)u(k,\omega) + f(k,\omega)\eta(k,\omega) \tag{1}$$

with the random initial condition

$$P_0 x(0,\omega) = \zeta(\omega) \in X_0 \subset R^m. \tag{2}$$

The first part of the paper is concerned with the boundedness of the solution process of the stochastic difference equation (1), (2); in the second part, we prove the existence and uniqueness of the solution of the optimal control problem described by (1), (2) and the quadratic cost functional

$$C(x,u) = <x,x>_g + <Qu,u>. \tag{3}$$

All vectors $x = \{x(k,\omega)\}_{k\geq 0}$ take values in R^m; all vectors $u = \{u(k,\omega)\}_{k\geq 0}$ take values in R^q, $q \leq m$; $\{A(k)\}$ is a sequence of invertible $m \times m$ matrices with real entries; all $\{B(k)\}$ are $m \times q$ matrices with bounded entries; $\{f(k,\omega)\}$ are $m \times l$ random matrix-valued functions; and $\{\eta(k,\omega)\}$ is a sequence of R^l normalized discrete processes, independent of the random initial value $\zeta(\omega)$ and with a positive definite covariance matrix $\mathcal{Q}$. The projection P_0 is defined below. $(\Omega, \mathcal{J}, \mathcal{P})$ is the underlying complete probability space, $\mathcal{J}$ is the σ-algebra of the subsets of Ω, $\mathcal{J}_k$ is the natural filtration with $\mathcal{J}_k \subset \mathcal{J}_{k+1}$, and $\mathcal{P}$ is the probability measure. Throughout this paper we assume that $\omega \in \Omega$, where Ω is the sample space. Also, $g = \{g_k\}_{k\geq 0}$, is a weight such that $g_k > 0$ for all k and $\sum_{k=0}^{\infty} g_k < \infty$, and Q is a self-adjoint positive definite linear operator on the sequence space $l^2(Z_+ \times \Omega, R^q)$. If for every $k \geq 0$ the spectral radius of $A(k)$ satisfies $\rho(A(k)) \leq \delta < 1$, then the asymptotic stability of the homogeneous equation is assured. When A is a constant $m \times m$ matrix, it is possible to have d eigenvalues inside the unit circle and $m - d$ eigenvalues outside it (see [2]). In this case, one can expect only partial stability.

Notation. In what follows, $(l^\infty)^m = l^\infty(Z_+ \times \Omega, R^m)$ is the Banach space of m-dimensional stochastic vector sequences $x = \{x(k,\omega)\}_{k\geq 0}$, equipped with the norm

$$\|x\|_{l^\infty} = \sup_{k\geq 0} = E|x(k,\omega)| < \infty. \tag{4}$$

In (4) and in the sequel, $E|\mu|$ denotes the expected value of μ.

The square of the norm on the Hilbert space $(l_g^2)^m$ is

$$\|x\|_g^2 = <x,x>_g = \sum_{k=0}^{\infty} \operatorname{tr} E[x(k,\omega)\cdot(x(k,\omega))^T]g_k < \infty, \tag{5}$$

$Z_+ = \{0,1,2,\ldots\}$, and $M_2(R_+ \times \Omega, R^{m\times l})$ is the Hilbert space of matrix-valued random sequences equipped with the norm

$$\|f\|_{M_2} = \sum_{k=0}^{\infty} \operatorname{tr} E[f(k,\omega)\cdot(f(k,\omega))^T] < \infty. \tag{6}$$

Throughout the paper we assume that (1) has an ordinary dichotomy. This implies that there is a projection P_0 on R^m, a positive constant α, and a positive integer n such that

$$|\Psi(n+1)P_0\Psi^{-1}(j+1)| \le \alpha, \quad n \ge j \ge 0, \tag{7}$$

$$|\Psi(n+1)P_1\Psi^{-1}(j+1)| \le \alpha, \quad j \ge n \ge 0, \tag{8}$$

where $P_1 = I_{m\times m} - P_0$ and

$$\Psi(r+1) = A(r)\cdot A(r-1)\cdots A(1)\cdot A(0), \tag{9}$$

with $\Psi(0) = I_{m\times m}$ (the identity $m \times m$ matrix).

We note that the homogeneous system

$$x(k+1,\omega) = A(k)x(k,\omega), \quad k \ge 0, \tag{10}$$

associated with (1) has the solution

$$x(r+1,\omega) = \Psi(r+1)x(0,\omega). \tag{11}$$

Consider the system

$$x(j+1,\omega) = A(j)x(j,\omega) + g(j,\omega), \quad j \ge 0, \tag{12}$$

and suppose that $g(j,\omega) \in l^1(Z_+ \times \Omega, R^m)$. Let P_0 be the projection of R^m onto its subspace X_0 consisting of the values for $k = 0$ of those vectors that are almost surely bounded solutions of (10) on Z_+. The solution of (12) with the random partial initial condition

$$P_0 x(0,\omega) = \zeta(\omega) \in X_0 \subset R^m \tag{13}$$

is [1]

$$x(n+1,\omega) = \Psi(n+1)\zeta(\omega) + \sum_{j=0}^{n} \Psi(n+1)P_0\Psi^{-1}(j+1)g(j,\omega)$$

$$- \sum_{j=n+1}^{\infty} \Psi(n+1)P_1\Psi^{-1}(j+1)g(j,\omega). \tag{14}$$

Equality (14) is obtained under the assumption of admissibility of the sequence spaces $\{(l^1)^m, (l^\infty)^m\}$, and remains valid if we assume the admissibility of the pair $\{(l^2)^m, (l^\infty)^m\}$. Since $l^1 \subset l^2$, (7) and (8) do not guarantee the convergence of the series in (14). It is essential, therefore, that we strengthen conditions (7) and (8).

2. Admissibility of the pair of sequence spaces $\{(l^2)^m, (l^\infty)^m\}$

The following assertion [1] gives necessary and sufficient conditions for the admissibility with respect to (12) of the pair $\{(l^2)^m, (l^\infty)^m\}$.

Theorem 2.1. *Equation* (12) *has a unique bounded solution almost surely on* Z_+ *(for each* $\omega \in \Omega$*) for every* $g \in (l^2)^m$ *if and only if there is a constant* $\gamma > 0$ *such that for every* $n \geq 0$

$$\sum_{j=0}^{n} |\Psi(n+1)P_0\Psi^{-1}(j+1)|^2 + \sum_{j=n+1}^{n+1} |\Psi(n+1)P_1\Psi^{-1}(j+1)|^2 \leq \gamma^2. \tag{15}$$

Theorem 2.2. *Suppose that the assumptions in Theorem* 2.1, *including inequality* (15), *remain valid, and that*

$$\sum_{k=0}^{\infty} E\{f(j,\omega)\eta(j,\omega)\} = 0, \tag{16}$$

$$\sum_{k=0}^{\infty} E|f((j,\omega)\eta(j,\omega))|^2 = \sum_{j=0}^{\infty} \operatorname{tr} E[f(j,\omega) \cdot (f(j,\omega))^T] < \infty. \tag{17}$$

Then problem (1), (2) *has a unique bounded solution on* Z_+ *almost surely.*

3. The optimal control problem

The optimal control problem is to determine a control $\bar{u}$ satisfying

$$\bar{u} \in \mathcal{U} \subset l^2(Z_+ \times \Omega, R^q), \tag{18}$$

where $\mathcal{U}$ is closed convex set in the control space $l^2(Z_+ \times \Omega, R^q)$ such that

$$\mathcal{C}(\bar{x}, \bar{u}) = \min_{u \in \mathcal{U}} \mathcal{C}(x, u), \tag{19}$$

with

$$\mathcal{C}(x, u) = \langle x, x \rangle_g + \langle Qu, u \rangle. \tag{20}$$

Let us characterize the closed convex set $\mathcal{U}$. The controls in $\mathcal{U}$ are such that

$$E[u(j,\omega)(u(j,\omega))^T] < \infty. \tag{21}$$

Furthermore,

$$\sum_{j=0}^{\infty} E|u(j,\omega)|^2 = \sum_{j=0}^{\infty} \operatorname{tr} E[u(j,\omega) \cdot (u(j,\omega))^T] < \infty. \tag{22}$$

We also assume that $\theta \notin \mathcal{U}$. The continuous version of assumptions (21) and (22) can be found in [3]. We note that choosing the initial condition

$$P_0 x(0,\omega) = \theta \in X_0 \subset R^m \tag{23}$$

does not diminish the generality of the problem (see, for example, [4]). From (14) we deduce that, if $g(j,\omega) = h(j,\omega) + f(j,\omega)\eta(j,\omega)$ and $h(j,\omega) = B(j)u(j,\omega)$, then (12) and (23) become

$$x(n+1,\omega) = \sum_{j=0}^{n} \Psi(n+1)P_0\Psi^{-1}(j+1)\{B(j)u(j,\omega) + f(j,\omega)\eta(j,\omega)\}$$

$$- \sum_{j=n+1}^{\infty} \Psi(n+1)P_1\Psi^{-1}(j+1)\{B(j)u(j,\omega) + f(j,\omega)\eta(j,\omega)\}. \tag{24}$$

One can therefore deal with the problem of minimizing $\mathcal{C}(x,u)$ subject to (1) and the initial condition (23). Let

$$Tu = \sum_{k=0}^{n} \Psi(n+1)P_0\Psi^{-1}(j+1)B(j)u(j,\omega)$$

$$- \sum_{k=n+1}^{\infty} \Psi(n+1)P_1\Psi^{-1}(j+1)B(j)u(j,\omega),$$

$$F = \sum_{k=0}^{n} \Psi(n+1)P_0\Psi^{-1}(j+1)f(j,\omega)\eta(k,\omega)$$

$$- \sum_{k=n+1}^{\infty} \Psi(n+1)P_1\Psi^{-1}(j+1)f(j,\omega)\eta(j,\omega).$$

Theorem 3.1. *Consider the optimal control problem described by (1), (23) and (3) in which $\min \mathcal{C}(x,u)$ is to be found, and suppose that*

(i) *$\{A(k)\}$ is a sequence of $m \times m$ invertible matrices; if we write $\Psi(r+1) = A(r) \cdot A(r-1) \cdots A(1) \cdot A(0)$, $r \geq 1$, and $(\mathcal{W})(r) = \Psi(r+1)\Psi^{-1}(r+1)$, then $\sup\{\|\mathcal{W}(r)\|, \|\mathcal{W}^{-1}(r)\| < \infty; \{B(k)\}$ is a sequence of $m \times q$ matrices whose entries are uniformly bounded on Z_+;*

(ii) *the assumptions in Theorem 2.1, including inequality (15), hold;*

(iii) *$\{f(k,\omega)\}$ is a sequence of $m \times l$ random matrix-valued functions satisfying the conditions in Theorem 2.2, $\{f(k,\omega)\} \in \mathcal{M}_2(R_+ \times \Omega, R^{m \times l})$, and $\{\eta(k,\omega)\}$ is a sequence of R^l-valued normalized random processes;*

(iv) $g \in l^1$ is a weight, while Q is a positive definite self-adjoint operator such that $Q \in L(l^2, l^2)$;

(v) the control is restricted to a closed convex set $\mathcal{U}$ in $(l^2)^q$.

Then there exists a unique control $\bar{u} = \{\bar{u}(k, \omega)\} \in \mathcal{U} \subset (l^2)^q$ such that

$$\mathcal{C}(\bar{x}, \bar{u}) = \min_{u \in \mathcal{U}} \mathcal{C}(x, u).$$

Proof. We express the cost functional in the form

$$\mathcal{C}(x, u) = [Tu + F, Tu + F]_g + [Qu, u], \tag{25}$$

where $[Tu + F, Tu + F]_g = \langle x, x \rangle_g$ and $[Qu, u] = \langle Qu, u \rangle$. Since $\mathcal{C}(x, u)$ is continuous with respect to u in $\mathcal{U}$ and $\mathcal{U}$ is closed and convex, $\mathcal{C}(x, u)$ attains its maximum and minimum values on $\mathcal{U}$. Also, since $[Qu, u] > 0$ and $[Tu + F, Tu + F]_g > 0$, we conclude that $\mathcal{C}(x, u)$ is strictly positive. $\mathcal{C}(x, u)$ is strictly convex, so its maximum and minimum values on $\mathcal{U}$ are unique. Hence, the infimum of $\mathcal{C}(x, u)$ on $\mathcal{U}$ is attained and is unique, and we conclude that there exists a unique control $\bar{u} \in \mathcal{U} \subset (l^2)^q$ such that $\mathcal{C}(\bar{x}, \bar{u}) = \min_{u \in \mathcal{U}} \mathcal{C}(x, u)$, where $\bar{x} = T\bar{u} + \bar{F}$.

References

[1] Z.C. Okonkwo, Admissibility and optimal control for difference equations, *Dynam. Systems Appl.* **5** (1996), 627–634.

[2] W.A. Coppel, Dichotomies in stability theory, Lecture Notes in Math. **62**, Springer, Berlin-Heidelberg-New York, 1978.

[3] Z.C. Okonkwo and G.S. Ladde, Itô-type stochastic differential systems with abstract Volterra operators and their control, *Dynam. Systems Appl.* **6** (1997), 461–468.

[4] Z.C. Okonkwo, Control problems in the class of linear-quadratic systems, Ph.D. thesis, University of Texas at Arlington, 1994.

Department of Mathematics and Computer Science, Albany State University, Albany, GA 31705, USA

F.R. PAYNE and K.R. PAYNE

Linear and sublinear Tricomi via DFI

1. Introduction

An optimum computer methodology [6] for nonlinear DE systems, **direct, formal integration (DFI)**, requires only one numeric approximation (a quadrature) and has had wide success in nonlinear physics. A partial list of applications include: supersonics, nonlinear heat conduction, Laplace, 3-D Navier-Stokes, and 4-D turbulent channel flows. Major DFI modes are "**SIMPLEX**", one integration trajectory, and "**NAD**", a distinct trajectory for each variable. Even second-year students easily calculate NL ODEs by DFI [7]. Prior and current works on Tricomi's "mixed-type" PDE [9] are described.

2. DFI methodology

Motivation arose from the fact that computers solve IEs better than DEs for the following reasons:

1. "Bad" finite arithmetic operations of division and subtraction are avoided.

2. Quadrature accuracy improves to any level, limited only by machine.

Furthermore, one looses no applicability and gains nice features, such as:

1. Any DE converts to a Volterra IE/IDE of the second kind.

2. Uniqueness is guaranteed for technology (Tricomi [12]).

3. DFI needs only second-year calculus [7].

DFI methodology is a three stage process [6, 8–10]:

I. **Formally integrate any DE system** over any path from an initial point to an upper limit; **Volterra IEs** of the second kind result.

II. **Study all IE/IDEs** for new insights; some arise due to the new forms.

III. **IE solution** (manual iterations assist understanding and programming).

Three advantages accruing to DFI usage are as follows:

1. Any DE system yields an IE system optimally compatible with digital machinery as demonstrated by solution of 300 P/ODEs, 1981–98.

2. Under DFI first-order DEs require iteration, whereas high-order DEs often do not. Computers can solve complex DEs faster than simpler ones.

3. DFI solves IVP via a sequence of "micro" IVPs, BVPs by a double IVP sequence and ABVPs by a triple IVP sequence.

3. Two sophomore-graduate class examples

To display DFI and machine compatibility, two NL DEs are considered: 1) a first-order IVP that requires iteration, 2) a second-order IVP that does not require iteration.

1. **Riccati equation (IVP).** One form of this oldest of NLDEs is $u' = 1 - u^2$ with IC as $u(0) = 0$. A single y-SIMPLEX formal integration yields:

$$u(t) = t - \int_0^t u^2(s)\, ds. \tag{1}$$

Trapezoid rules, with linear predictors, reproduce the exact solution, $\tanh(t)$.

2. **Boundary-layer flow ODE model (IVP).** A second example is a flow model with viscosity, inertia and pressure gradient:

$$u''(y) = u^2 - 1 \tag{2}$$

and $u(0) = 0$ ["no slip"]; $u(y \to \infty) \to 1$ [asymptotics]. Dual y-SIMPLEX yields:

$$u(y) = yu'(0) - y^2/2 + \int_0^y (y - s)u^2(s)\, ds. \tag{3}$$

DFI accurately reproduces the solution, $3\tanh^2 \left[y/\sqrt{2} + \tanh^{-1} \left(\sqrt{2/3} \right) \right] - 2$. An advantage is the lag factor $(\eta - s)^k/k!$ [1] which reduces multiple to single integrals. Choose as starter the low order part or all terms outside the integral. "Micro"-Picard iterations [6] converge ~ 10 times faster than standard Picard.

4. On-going DFI studies: Euler flow and Tricomi-type equations

1. **Euler flow on the unit square (incompressible)** [9]

$$\text{Continuity}: \ \nabla \cdot \boldsymbol{u} = 0. \ \ \text{Momentum}: \ \boldsymbol{u} \cdot \nabla \boldsymbol{u} = -\nabla \mathrm{P}/\rho. \tag{4}$$

Natural anti-derivative (DFI/NAD) integrates and eliminates **all** derivatives; continuity and y-momentum become (after 3 integrations each in x and y):

$$\int_0^x [v(t,y) - v(t,0)]dt = \int_0^y [u(x,s) - u(0,s)]ds, \tag{5}$$

$$\int_0^x [v^2(t,y) - v^2(t,0)]dt = \int_0^y [uv(x,s) - uv(x,0)]ds - \int_0^x [P(t,y) - P(t,0)]dt/\rho, \tag{6}$$

and a similar equation for uv. Eliminating all derivatives implies that no finite differences are needed as in SIMPLEX; accuracy and speed are enhanced [9,10].

2. **Tricomi's mixed-type PDE**

Early work [9] used a unit circle D offset in the elliptic region, a smooth non-characteristic boundary. Here we consider, purely for programming expediency, the domain $D = (-1, 1) \times (-1, 1)$ with the same BC, and proceed to add a sublinear nonlinearity to the problem. This second regime has it origins in recent analytical work (Lupo and Payne [2]). More precisely, consider

$$T(u) \equiv yu_{xx} + u_{yy} = 0 \ \text{in} \ D, \ \ u = g \ \text{on} \ \partial D. \tag{7}$$

Mixed type (elliptic-hyperbolic) PDEs are traditionally associated with transonic flows. Dirichlet BC on D have been discussed by Morawetz [5] and Payne [11]; existence of solutions with isolated singularities at parabolic boundary points remains open. As a model problem, data g are taken linear with unit jump at the sonic line endpoint, $y = 0$. Place $f(x) = u(x,0)$ on the sonic line and solve an elliptic BVP, via shooting, in the sector $y > 0$. Applying DFI twice gives

$$u(x,y) = f(x) + y\partial_y u(x,0) - \partial_x^2 \int_0^y s(y - s)u(x,s)\, ds. \tag{8}$$

	σ	Iterations	u–RMS	x–RMS	Δ from the initial field
Elliptic	8^{-9}	8–12	10^{-6}	10^{-6}	small to x = 0.92
Hyperbolic	10^{-13}	0	0	10^{-6}	similar

Table 1. Linear results.

The integrand vanishes at $s = 0$ and $s = y$, making numerics docile. The y-slope of u at the sonic line is "shot". The hyperbolic region BVP converts to an IVP [Volterra IDE (8)]. Solutions were inserted into the PDE, differenced, and squares summed and averaged for RMS deviations. Global residuals were $\sim 10^{-10}$ (elliptic) and $\sim 10^{-13}$ (hyperbolic) (below quadrature error) validating solutions.

In the sublinear case, where we consider

$$T(u) = u|u|^{1-p} \quad \text{where } 0 < p < 1, \tag{9}$$

the coding changes are minor. Dual y-SIMPLEX gives

$$\begin{aligned} u(x,y) &= f(x) + y\partial_y u(x,0) - \partial_x^2 \int_0^y s(y-s)u(x,s)\,ds \\ &\quad - \operatorname{sgn}(u)\int_0^y (y-s)|u|^p\,ds. \end{aligned} \tag{10}$$

The only change from the homogeneous case (8) is the second integral in (10) requiring two new lines of code, initializing and updating an accumulator. The linear case [9,10] was rerun on the square as validation. Sublinear p-values in [0, 0.99] were used. Runs on an Intel PRO200, 200 MHz, took 10 to 60 seconds.

Solution checks were: 1) insert into the PDE, difference, square, and average for RMS; 2) iterate RMS deviations as convergence tests; and 3) compute the final solution's maximal value near the singularity and RMS deviation from the initial (linear) field. Growth of RMS deviation from the inertial field as p increases are shown. All cases are $\sim$ linear far from the singularity. The feedback influence increases with p over a growing area near the "jump". Results are summarized below. DFI allows latitude in field traversing. The algorithm: 1) fills the elliptic field with BC interpolations; 2) shoots for the unknown $\partial u(x,0)/\partial y$; 3) increments in x, repeats until the sonic line end; 4) sweeps until errors become acceptable; 5) inputs elliptic solutions to the hyperbolic solver; 6) solves a series of IVPs from $y = 0$ to Δy; 7) increments in y and solves a new IVP; 8) terminates as y reaches the outer boundary and the solution is complete. The elliptic BVP requires sweeping the entire domain whereas the hyperbolic IVP requires no sweeping at all.

The linear case, $T(u) = 0$, was rerun on a much faster computer than in [9] on a grid of 512 y- and 128 x-points, adequate for this docile case, and yields AR = 8 ("aspect ratio" $\equiv$ ratio of step size, $\Delta x/\Delta y$) as a good numeric compromise. The results are contained in Table 1, (where $\sigma \equiv$ global RMS deviation, and x–RMS = transverse values).

Similar grids (512 × 128) were used for most sublinear calculations; these results are shown in Table 2. Note the poor σ-accuracy vs. Table 1.

p	σ	Iterations	Shoots	u–RMS	x–RMS
0.01	.0219/.044	6/0	4/0	$10^{-4}/0$	$10^{-6}/0$
0.10	.0195/.040	6/0	4/0	$10^{-4}/0$	$10^{-6}/0$
0.25	.018/.036	6/0	4/0	$10^{-4}/0$	$10^{-6}/0$
0.50	.016/.031	6/0	4/0	$10^{-4}/0$	$10^{-6}/0$
0.75	.014/.028	6/0	4/0	$10^{-4}/0$	$10^{-5}/0$
0.90	.013/.026	6/0	4/0	$10^{-4}/0$	$10^{-5}/0$
0.99	.0128/.026	6/0	3/0	$10^{-4}/0$	$10^{-5}/0$
1.00	.0127/.0254	6/0	3/0	$10^{-4}/0$	$10^{-4}/0$

Table 2. Sublinear results; elliptic/hyperbolic ratios.

The authors are not happy with these numerical results in the sublinear case. Questions as to whether solutions may not exist to the sublinear problem (related to the overderminded nature of the linear problem [5,11]), or the presence of multiple solutions (as occurs with a symmetrized verson of the problem with Tricomi boundary conditions [3]) remain. Digital machinery cannot answer such questions.

DFI provides great grid latitudes. The primary pattern began at the sonic line end and was y-swept; then x was incremented and y-sweeps were repeated to cover the field. One can invert any pattern, yielding four choices, by changing only two code lines.

NAD was first implemented a decade ago for Euler. NAD was planned here but the RAM was inadequate; 192 MB were required. The Tricomi/NAD equations are easily obtained; simply integrate (10) twice again but over x:

$$\int_{-1}^{x} (x - z)u(z,y)\, dz = \int_{-1}^{x} (x - z)[f(z) + y\partial_y u(z,0)]\, dz$$
$$+ \int_{0}^{y} s(y - s)[u(-1,s) - u(x,s)]\, ds - \mathrm{sgn}(u) \int_{-1}^{x} \int_{0}^{y} (x - z)(y - s)|u|^p\, ds dz. \quad (11)$$

NAD does require a third quadrature to unravel the implicit u-solution. The only prior cited NAD applications were to Euler (1988) [9] and inviscid Burgers [10].

Planned additions and modifications to the Tricomi solver include: 1) replace SIMPLEX by NAD; 2) utilize a finer grid; 3) develop a Mach line tracer; 4) consider shock wave modeling; 5) return to original geometry [9]; 6) consider new sonic line data; and 7) Utilize romberg quadrature for arbitrary accuracy. In summary, **DFI** is a robust, accurate solver of homogeneous Tricomi $T(u) = 0$, but accuracy problems remain for the sublinear case. Possible causes are: 1) central difference error $[O(10^{-4})]$ is 64 times the quadrature error due to the required aspect ratio > 4; 2) u-jump on the sonic line may be incompatible with Tricomi. Some mystery remains.

4.1. "Micro"-Picard iteration [6], DFI's "life blood"

This Picard modification, similar to Caratheodory iteration cited in [4], makes DFI competitive. One iterates until converged on an interval and creates a new IVP on the

next interval. "Micro-" Picard generates a simple solution hierarchy for DEs; coding
is simple; add a FORTRAN "DO-loop" for each level:

1. IVP = the limit of a sequence of "mini" IVPs
2. BVP = the limit of a sequence of IVPs
3. ABVP = the asymptotic limit of a sequence of BVPs.

4.2. Some new analytical questions concerning Tricomi

1. Why is sublinear accuracy much worse than in the homogeneous case?
2. Is the square geometry incompatible with the Tricomi sublinear case?
3. Why does Tricomi exhibit great sensitivity to the "aspect ratio"?

5. Closure

These examples show DFIs breadth. DE systems yield readily to DFI; integral methods
are more compatible to digital machines. DFI has many major appeals (see [8]): 1)
NAD eliminates all derivatives; 2) optimum machine compatibility; 3) iterations are
often unneeded; 4) uniqueness assured for technology [12]; 5) multiple algorithms; 6)
analytics near IP provide new insights; 7) "micro"-Picard saves CPU time; 8) easy
global and local checks; 9) tiny FORTRAN codes: the elliptic solver has 16 lines and
hyperbolic solver has 13 lines. Validation codes are 300 and 200 lines. Total codes are
345 lines (elliptic) and 247 lines (hyperbolic) in length.

Readers are encouraged to add **DFI Stages I** and **II** to their tool kits. Study of
the new forms [Stage II] provides insights **not** obtainable any other way. These and
recent experiences with Euler, Lorenz chaos, and Burgers sharpen the dichotomy of
discrete vs. continua. Recall Kipling's (paraphrased) words:

"Discrete is discrete and continua are continua,
and never the twain shall meet" (to arbitrary accuracy).

References

1. V.L. Lovitt, Linear integral equations, McGraw-Hill, 1924; Dover, 1950.

2. D. Lupo and K.R. Payne, A dual variational approach to a class of nonlocal
   semilinear Tricomi problems, *Nonlinear Diff. Equations Appl.* (to appear).

3. D. Lupo, A.M. Micheletti and K.R. Payne, Existence of eigenvalues for reflected
   Tricomi operators and applications to multiplicity of solutions for sublinear and
   asymptotically linear Tricomi problems, *Adv. Diff. Equations* (to appear).

4. R.K. Miller, *Non-linear Volterra equations*, W.A. Benjamin, 1971.

5. C.S. Morawetz, The Dirichlet problem for the Tricomi equation, *Comm. Pure
   Appl. Math.* **23** (1970), 587–601.

6. F.R. Payne, Lecture notes, 1980; oral paper, 1981; AIAA Symposium, UTA.

7. F.R. Payne, Seven sophomore FORTRAN classes (181 students), 1996–98, UTA.

8. F.R. Payne, A non-linear PDE system solver optimal for computer, in *Integral
   methods in science and engineering*, C. Constanda, Ed., Pitman Res. Notes Math.
   Ser. **317**, Longman, Harlow, 1994, 61–71.

9. F.R. Payne and K.R. Payne, New facets of DFI, a DE solver for all seasons, in *Integral methods in science and engineering*, C. Constanda, J. Saranen, and S. Seikkala, Eds., Pitman Res. Notes Math. Ser. **375**, Addison Wesley Longman, Harlow-New York, 1997, 176–180.

10. F.R. Payne, Euler and inviscid Burger high-accuracy solutions, in *Noninear problems in aerospace and aviation*, S. Sivasundaram, Ed., Gordon & Breach (in press).

11. K.R. Payne, Interior regularity of the Dirichlet problem for the Tricomi equation, *J. Math. Anal. Appl.* **199** (1996), 271–292.

12. E.G. Tricomi, *Integral equations*, Interscience, Dover, 1985.

Department of Mechanical and Aerospace Engineering, University of Texas at Arlington, Arlington, Texas, USA; e-mail: frpdfi@airmail.net

Dipartimento di Matematica, Politecnico di Milano, Milano, Italy; e-mail: paynek@mate.polimi.it

R.M. PEAT and A.C. McBRIDE

Multidimensional fractional integrals on spaces of smooth functions

1. Introduction

In [1], Rubin discussed the multidimensional fractional integrals $B^\alpha_\pm$ and I^α which are related to the Riesz potential. In particular he showed that by making use of expansions in terms of spherical harmonics, we can relate these operators to the 1-dimensional Erdélyi-Kober operators, which were studied by McBride in [2]. Mapping properties of the n-dimensional operators are then inherited from the corresponding properties of the 1-dimensional operators, which are well known.

We shall study our operators on spaces that consist of smooth functions and are n-dimensional analogues of the test function spaces $F_{p,\mu,\sigma}(\mathbb{R}_+)$ introduced by McBride and Spratt in [2] and [3]. These spaces provide an appropriate setting for the study of such operators.

Notation. $\mathbb{N}$ is the set of all nonnegative integers. $\mathbb{R}$ (respectively $\mathbb{R}_+$) is the set of all real numbers (respectively all positive real numbers). $\mathbb{R}^n = \{x = (x_1,\ldots,x_n) : x_j \in \mathbb{R} \text{ for } j = 1,\ldots,n\}$. $\dot{\mathbb{R}}^n = \mathbb{R}^n \setminus \{0\}$. We denote by Σ the unit sphere given by $\Sigma = \{x \in \mathbb{R}^n : |x| = 1\}$, and $|\Sigma| = 2\pi^{n/2}/\Gamma(n/2)$ is the surface area of Σ.

For $1 \le p \le \infty$ and $\mu \in \mathbb{C}$ we define the weighted spaces $L_{p,\mu}(\mathbb{R}_+)$ and $L_{p,\mu}(\dot{\mathbb{R}}^n)$ by $L_{p,\mu}(\mathbb{R}_+) = \{f : \|f\|_{p,\mu} < \infty\}$ where for $1 \le p < \infty$,

$$\|f\|_{p,\mu} \equiv \left[\int_0^\infty |r^\mu f(r)|^p \frac{dr}{r}\right]^{1/p} \quad \text{and} \quad \|f\|_{\infty,\mu} \equiv \operatorname*{ess\,sup}_{r \in \mathbb{R}_+} |r^\mu f(r)|, \tag{1}$$

and $L_{p,\mu}(\dot{\mathbb{R}}^n) = \{\phi : \|\phi\|_{p,\mu} < \infty\}$, where for $1 \le p < \infty$,

$$\|\phi\|_{p,\mu} \equiv \left[\int_{\dot{\mathbb{R}}^n} ||x|^\mu \phi(x)|^p \frac{dx}{|x|^n}\right]^{1/p} \quad \text{and} \quad \|\phi\|_{\infty,\mu} \equiv \operatorname*{ess\,sup}_{x \in \dot{\mathbb{R}}^n} ||x|^\mu \phi(x)|. \tag{2}$$

The reader should note that we have used the same notation for the norms on the spaces defined on $\mathbb{R}_+$ and $\dot{\mathbb{R}}^n$.

2. Preliminaries

The operators that we shall study are n-dimensional analogues of 1-dimensional operators associated with the Mellin transform. We call an operator a Mellin multiplier transform if it satisfies the relation

$$[\mathcal{M}(Tf)](s) = m(s)(\mathcal{M}f)(s) \tag{3}$$

274

where f is a suitable function defined on the positive half-line, s is a suitably restricted complex number, and $\mathcal{M}$ is the Mellin transform given by

$$(\mathcal{M}f)(s) = \int_0^\infty r^s f(r) \frac{dr}{r}. \tag{4}$$

We call the function $m(s)$ the multiplier of the operator T, and many properties of T can be obtained by studying the multiplier m.

In order to extend this to operators defined on $\dot{\mathbb{R}}^n$ we expand functions defined on $\dot{\mathbb{R}}^n$ in the following way.

We can write any suitable function ϕ defined on $\dot{\mathbb{R}}^n$ in the form

$$\phi(x) = \sum_{j,\nu} \phi_{j,\nu}(r) Y_{j,\nu}(x') \tag{5}$$

where the numbers $\phi_{j,\nu}$ defined by

$$\phi_{j,\nu}(r) = \int_\Sigma \phi(rx') Y_{j,\nu}(x') dx' \tag{6}$$

are called the Fourier-Laplace coefficients of ϕ and are defined on $\mathbb{R}_+$.

$\{Y_{j,\nu}(x')\}$ is an orthonormal basis of spherical harmonics on Σ. Here $j \in \mathbb{N}$ and $\nu = 1, \ldots, d_n(j)$, where $d_n(j)$ is the dimension of the subspace of the spherical harmonics of order j. Also, $x' \equiv x/|x|$ is a point of the unit sphere and $|x| \equiv r$.

Finally, we introduce the spaces upon which we shall study our operators. These spaces are subspaces of the weighted spaces $L_{p,\mu}(\mathbb{R}_+)$ and $L_{p,\mu}(\dot{\mathbb{R}}^n)$ and for $1 \le p \le \infty$ and $\mu \in \mathbb{C}$ are given by by

$$F_{p,\mu,\sigma}(\mathbb{R}_+) = \{f : \gamma_k^{p,\mu,\sigma}(f) < \infty\}$$

where for $\sigma = 0$

$$\gamma_k^{p,\mu,0}(f) = \|\delta_r^k f\|_{p,\mu} = \|r^\mu \delta_r^k f\|_p, \tag{7}$$

and

$$F_{p,\mu,\sigma}(\dot{\mathbb{R}}^n) = \{\phi : \lambda_k^{p,\mu,\sigma}(\phi) < \infty\}$$

where for $\sigma = 0$

$$\lambda_l^{p,\mu,0}(\phi) = \sup_{|\gamma| \le l} \|\mathcal{D}_\gamma \phi\|_{p,\mu}, \quad l \in \mathbb{N}. \tag{8}$$

Here $\delta_r = r(d/dr)$ and, for any multi-index $\gamma = (\gamma_1, \ldots, \gamma_n)$, the operator $\mathcal{D}_\gamma$ is given by $\mathcal{D}_\gamma = r^{|\gamma|}(\partial^{|\gamma|}/\partial x_1^{\gamma_1} \ldots \partial x_n^{\gamma_n})$ where $|\gamma| = \gamma_1 + \cdots + \gamma_n$. The seminorms and norms for $\sigma > 0$ and $\sigma < 0$ can be found in [4] and [5]. We note that the spaces $F_{p,\mu,\sigma}(\mathbb{R}_+)$ and $F_{p,\mu,\sigma}(\dot{\mathbb{R}}^n)$ are Fréchet spaces under the topology generated by the relevant seminorms. These spaces provide an appropriate setting for the study of many operators that occur in applied mathematics.

3. Erdélyi-Kober operators

The 1-dimensional operators that we shall study are homogeneous modifications of
the transforms I_M^α and K_M^α defined for $\operatorname{Re}\alpha > 0$, $M > 0$ and suitable functions f by

$$(I_M^\alpha f)(r) = \frac{M}{\Gamma(\alpha)} \int_0^r (r^M - t^M)^{\alpha-1} t^{M-1} f(t)dt \tag{9}$$

$$(K_M^\alpha f)(r) = \frac{M}{\Gamma(\alpha)} \int_r^\infty (t^M - r^M)^{\alpha-1} t^{M-1} f(t)dt. \tag{10}$$

These operators map $L_{p,\mu}(\mathbb{R}_+)$ into $L_{p,\mu-M\alpha}(\mathbb{R}_+)$ under appropriate conditions.

Definition 1. Let $\operatorname{Re}\alpha > 0$ and $\eta \in \mathbb{C}$. Then we define

$$(I_M^{\eta,\alpha} f)(r) = r^{-M\eta-M\alpha} I_M^\alpha r^{M\eta} f(r) \tag{11}$$

$$(K_M^{\eta,\alpha} f)(r) = r^{M\eta} K_M^\alpha r^{-M\eta-M\alpha} f(r). \tag{12}$$

The operators $I_M^{\eta,\alpha}$ and $K_M^{\eta,\alpha}$ are the so-called Erdélyi-Kober operators and were
studied by McBride in [2]. We shall find it convenient to study these operators when
$M = 2$. The reason behind this will become clear in the next section.

We shall first give results for these operators on the spaces $L_{p,\mu}(\mathbb{R}_+)$.

Theorem 2. *Let* $1 \le p \le \infty$, $\mu \in \mathbb{C}$ *and* $\operatorname{Re}\alpha > 0$. *Then*
  (i) $I_2^{\eta,\alpha}$ *is a continuous linear mapping of* $L_{p,\mu}(\mathbb{R}_+)$ *into itself provided that*
$\operatorname{Re}(\eta + 1 - \mu/2) > 0$.
  (ii) $K_2^{\eta,\alpha}$ *is a continuous linear mapping of* $L_{p,\mu}(\mathbb{R}_+)$ *into itself provided that*
$\operatorname{Re}(\eta + \mu/2) > 0$.

We now exhibit the Mellin multipliers of these operators.

Theorem 3. *Let* $f \in L_{2,\mu}(\mathbb{R}_+)$, $\operatorname{Re}\alpha > 0$, μ, $\eta \in \mathbb{C}$ *and let* s *be a suitably restricted
complex number. Then*

$$[\mathcal{M}(I_2^{\eta,\alpha} f)](s) = \frac{\Gamma(\eta + 1 - s/2)}{\Gamma(\eta + \alpha + 1 - s/2)}(\mathcal{M}f)(s) \tag{13}$$

with the multiplier existing if $\operatorname{Re}(\eta + 1 - \mu/2) > 0$, *and*

$$[\mathcal{M}(K_2^{\eta,\alpha} f)](s) = \frac{\Gamma(\eta + s/2)}{\Gamma(\eta + \alpha + s/2)}(\mathcal{M}f)(s) \tag{14}$$

with the multiplier existing if $\operatorname{Re}(\eta + \mu/2) > 0$.

We aim to obtain the mapping properties of these operators relative to the spaces
$F_{p,\mu,\sigma}(\mathbb{R}_+)$. Since we are working in spaces of smooth functions we can extend the
range of admissible parameters. We do this by using analytic continuation of the
Gamma functions in the multipliers of the operators $I_2^{\eta,\alpha}$ and $K_2^{\eta,\alpha}$ and we obtain
the following.

276

Theorem 4. *Let $1 \leq p \leq \infty$, $\mu \in \mathbb{C}$, $\alpha \in \mathbb{C}$ and $\sigma \in \mathbb{R}$.*

(i) For $-\operatorname{Re}(\eta+1-\mu/2) \notin \mathbb{N}$, $I_2^{\eta,\alpha}$ is a continuous linear mapping of $F_{p,\mu,\sigma}(\mathbb{R}_+)$ into itself. If, in addition, $-\operatorname{Re}(\eta+\alpha+1-\mu/2) \notin \mathbb{N}$, then $I_2^{\eta,\alpha}$ is a homeomorphism of $F_{p,\mu,\sigma}(\mathbb{R}_+)$ onto itself and $(I_2^{\eta,\alpha})^{-1} = I_2^{\eta+\alpha,-\alpha}$.

(ii) For $-\operatorname{Re}(\eta+\mu/2) \notin \mathbb{N}$, $K_2^{\eta,\alpha}$ is a continuous linear mapping of $F_{p,\mu,\sigma}(\mathbb{R}_+)$ into itself. If, in addition, $-\operatorname{Re}(\eta+\alpha+\mu/2) \notin \mathbb{N}$, then $K_2^{\eta,\alpha}$ is a homeomorphism of $F_{p,\mu,\sigma}(\mathbb{R}_+)$ onto itself and $(K_2^{\eta,\alpha})^{-1} = K_2^{\eta+\alpha,-\alpha}$.

4. Multidimensional fractional integrals

The n-dimensional operators that we shall study were discussed by Rubin in [1] and are multidimensional modifications of the usual Riemann-Liouville fractional integrals.

Definition 5. Let $\operatorname{Re}\alpha > 0$ and $\gamma_{n,\alpha} = 2/(\Gamma(\alpha)|\Sigma|)$. Then for suitable functions ϕ defined on $\dot{\mathbb{R}}^n$ we define the operators B_+^α and B_-^α by

$$(B_+^\alpha \phi)(x) = \gamma_{n,\alpha} \int_{|y|<|x|} \frac{(|x|^2 - |y|^2)^\alpha}{|x-y|^n} \phi(y)dy, \tag{15}$$

$$(B_-^\alpha \phi)(x) = \gamma_{n,\alpha} \int_{|y|>|x|} \frac{(|y|^2 - |x|^2)^\alpha}{|x-y|^n} \phi(y)dy \tag{16}$$

and the operator I^α by

$$(I^\alpha \phi)(x) = c_{n,\alpha} \int_{\dot{\mathbb{R}}^n} \frac{\phi(y)}{|x-y|^{n-\alpha}} dy \tag{17}$$

where $c_{n,\alpha} = 2^{-\alpha}\pi^{-n/2}\Gamma[(n-\alpha)/2]/\Gamma(\alpha/2)$, $0 < \operatorname{Re}\alpha < n$.

We note that I^α is the usual Riesz potential (see [6]) and that $B_\pm^\alpha$ were introduced in [1] in connection with the inversion problem for Riesz potentials in a ball.

In the first instance we shall consider the modified operators $\mathcal{B}_\pm^\alpha$ and $\mathcal{I}^\alpha$ where

$$\mathcal{B}_\pm^\alpha \phi = r^{-2\alpha} B_\pm^\alpha \phi \quad \text{and} \quad \mathcal{I}^\alpha \phi = r^{-\alpha} I^\alpha \phi. \tag{18}$$

These modified operators commute with rotations and dilations and map the space $L_{p,\mu}(\dot{\mathbb{R}}^n)$ into itself under appropriate conditions.

Theorem 6. *Let $1 \leq p \leq \infty$, $\operatorname{Re}\alpha > 0$ and $\mu \in \mathbb{C}$. Then*

(i) $\mathcal{B}_+^\alpha$ is bounded on $L_{p,\mu}(\dot{\mathbb{R}}^n)$ provided that $\operatorname{Re}\mu < n$.

(ii) $\mathcal{B}_-^\alpha$ is bounded on $L_{p,\mu}(\dot{\mathbb{R}}^n)$ provided that $\operatorname{Re}(\mu - 2\alpha) > 0$.

(iii) If $\operatorname{Re}\alpha < \operatorname{Re}\mu < n$ then $\mathcal{I}^\alpha$ is bounded on $L_{p,\mu}(\dot{\mathbb{R}}^n)$ and the following factorisation holds for $\phi \in L_{p,\mu}(\dot{\mathbb{R}}^n)$:

$$\mathcal{I}^\alpha \phi = 2^{-\alpha} \mathcal{B}_+^{\alpha/2} \mathcal{B}_-^{\alpha/2} \phi = 2^{-\alpha} \mathcal{B}_-^{\alpha/2} \mathcal{B}_+^{\alpha/2} \phi. \tag{19}$$

To obtain the mapping properties of these operators relative to the spaces $F_{p,\mu,\sigma}(\dot{\mathbb{R}}^n)$ we need to find an explicit form for the Fourier-Laplace coefficients of $\mathcal{B}_\pm$. In [1], Rubin proved that $(\mathcal{B}_+^\alpha \phi)_{j,\nu}$ and $(\mathcal{B}_-^\alpha \phi)_{j,\nu}$ have the form

$$(\mathcal{B}_+^\alpha \phi)_{j,\nu} = I_2^{(n+j)/2-1,\alpha} \phi_{j,\nu} \tag{20}$$

$$(\mathcal{B}_-^\alpha \phi)_{j,\nu} = K_2^{j/2-\alpha,\alpha} \phi_{j,\nu}. \tag{21}$$

We now see how our multidimensional fractional integrals are related to the Erdélyi-Kober operators presented in the previous section. In particular we can use Theorem 4 and (19) to obtain the following results.

Theorem 7. *Let $1 \leq p \leq \infty$, $\alpha \in \mathbb{C}$, $\sigma \in \mathbb{R}$ and $\mu \in \mathbb{C}$. Then*

(i) $\mathcal{B}_+^\alpha$ is a continuous linear mapping of $F_{p,\mu,\sigma}(\dot{\mathbb{R}}^n)$ into itself if $\operatorname{Re}\mu - n \notin \mathbb{N}$. If also $\operatorname{Re}(\mu - 2\alpha) - n \notin \mathbb{N}$, then $\mathcal{B}_+^\alpha$ is a homeomorphism of $F_{p,\mu,\sigma}(\dot{\mathbb{R}}^n)$ onto itself.

(ii) $\mathcal{B}_-^\alpha$ is a continuous linear mapping of $F_{p,\mu,\sigma}(\dot{\mathbb{R}}^n)$ into itself if $\operatorname{Re}(2\alpha-\mu) \notin \mathbb{N}$. If, in addition, $-\operatorname{Re}\mu \notin \mathbb{N}$, then $\mathcal{B}_-^\alpha$ is a homeomorphism of $F_{p,\mu,\sigma}(\dot{\mathbb{R}}^n)$ onto itself.

(iii) $\mathcal{I}^\alpha$ is a continuous linear mapping of $F_{p,\mu,\sigma}(\dot{\mathbb{R}}^n)$ into itself if $\operatorname{Re}(\alpha-\mu) \notin \mathbb{N}$ and $\operatorname{Re}(\mu-n) \notin \mathbb{N}$. If, in addition, $\operatorname{Re}(\mu-\alpha-n) \notin \mathbb{N}$ and $-\operatorname{Re}\mu \notin \mathbb{N}$, then $\mathcal{I}^\alpha$ is a homeomorphism of $F_{p,\mu,\sigma}(\dot{\mathbb{R}}^n)$ onto itself.

We are now in a position to study the original operators $B_\pm^\alpha$ and I^α.

Definition 8. *For $1 \leq p \leq \infty$, $\alpha \in \mathbb{C}$ and $\mu \in \mathbb{C}$ such that $\operatorname{Re}\mu - n \notin \mathbb{N}$ and $\operatorname{Re}(\alpha - \mu) \notin \mathbb{N}$, define $B_\pm^\alpha$ and I^α on $F_{p,\mu,\sigma}(\dot{\mathbb{R}}^n)$ by*

$$B_\pm^\alpha = r^{2\alpha} \mathcal{B}_\pm^\alpha \quad \text{and} \quad I^\alpha = r^\alpha \mathcal{I}^\alpha. \tag{23}$$

The above definition and the fact that r^λ maps $F_{p,\mu,\sigma}(\dot{\mathbb{R}}^n)$ onto $F_{p,\mu-\lambda,\sigma}(\dot{\mathbb{R}}^n)$ leads us to the following result.

Theorem 9. *Let $1 \leq p \leq \infty$, $\alpha \in \mathbb{C}$ and $\mu \in \mathbb{C}$. Then*

(i) B_+^α is a continuous linear mapping of $F_{p,\mu,\sigma}(\dot{\mathbb{R}}^n)$ into $F_{p,\mu-2\alpha,\sigma}(\dot{\mathbb{R}}^n)$ if $\operatorname{Re}\mu - n \notin \mathbb{N}$. If also $\operatorname{Re}(\mu - 2\alpha) - n \notin \mathbb{N}$ then B_+^α is a homeomorphism of $F_{p,\mu,\sigma}(\dot{\mathbb{R}}^n)$ onto $F_{p,\mu-2\alpha,\sigma}(\dot{\mathbb{R}}^n)$ with inverse $B_+^{-\alpha}$.

(ii) B_-^α is a continuous linear mapping of $F_{p,\mu,\sigma}(\dot{\mathbb{R}}^n)$ into $F_{p,\mu-2\alpha,\sigma}(\dot{\mathbb{R}}^n)$ if $\operatorname{Re}(2\alpha - \mu) \notin \mathbb{N}$. If also $-\operatorname{Re}\mu \notin \mathbb{N}$ then B_-^α is a homeomorphism of $F_{p,\mu,\sigma}(\dot{\mathbb{R}}^n)$ onto $F_{p,\mu-2\alpha,\sigma}(\dot{\mathbb{R}}^n)$ with inverse $B_-^{-\alpha}$.

(iii) I^α is a continuous linear mapping of $F_{p,\mu,\sigma}(\dot{\mathbb{R}}^n)$ into $F_{p,\mu-\alpha,\sigma}(\dot{\mathbb{R}}^n)$ if $\operatorname{Re}(\mu - n) \notin \mathbb{N}$ and $\operatorname{Re}(\alpha - \mu) \notin \mathbb{N}$. If also $\operatorname{Re}(\mu - \alpha - n) \notin \mathbb{N}$ and $-\operatorname{Re}\mu \notin \mathbb{N}$ then I^α is a homeomorphism of $F_{p,\mu,\sigma}(\dot{\mathbb{R}}^n)$ onto $F_{p,\mu-\alpha,\sigma}(\dot{\mathbb{R}}^n)$ with inverse $I^{-\alpha}$.

The corresponding theory for the adjoints of the operators $B_{\pm}^{\alpha}$ and I^{α} relative to the dual spaces $F'_{p,\mu,\sigma}(\dot{\mathbb{R}}^n)$ can be obtained and used to find distributional solutions of equations in potential theory. This work will be presented in a future paper.

References

1. B.S. Rubin, Fractional integrals and weakly singular integral equations of the first kind in the n-dimensional ball, *J. Anal. Math.* **63** (1994), 19–102.

2. A.C. McBride, *Fractional calculus and integral transforms of generalised functions*, Pitman Res. Notes Math. Ser. **31**, Pitman Press, 1979.

3. W.J. Spratt, *A classical and distributional theory of Mellin multiplier transforms*, Ph.D. thesis, University of Strathclyde, 1985.

4. R.M. Peat and A.C. McBride, Spaces of smooth functions for n-dimensional fractional integrals. I, *Strathclyde Math. Research Report* **40**, 1998.

5. R.M. Peat and A.C. McBride, Spaces of smooth functions for n-dimensional fractional integrals. II, *Strathclyde Math. Research Report* **41**, 1998.

6. E.M. Stein, *Singular integrals and differentiability properties of functions*, Princeton University Press, Princeton, NJ, 1970.

Department of Mathematics, University of Strathclyde, Glasgow, U.K.

A. PEDAS and G. VAINIKKO

Piecewise polynomial projection methods for nonlinear multidimensional weakly singular integral equations

1. Introduction

Second kind Fredholm integral equations with weakly singular kernels typically have solutions that are nonsmooth near the boundary of integration. In this paper, on the basis of certain regularity properties of the exact solution, the piecewise polynomial collocation and Galerkin methods on graded grids are discussed to solve nonlinear multidimensional weakly singular integral equations. The superconvergence effect at collocation points for the collocation method is studied (Section 3) and global convergence estimates for Galerkin approximations are derived (Section 4).

2. Integral equation

We consider the nonlinear integral equation

$$u(x) = \int_G K(x, y, u(y))dy + f(x), \qquad x \in G, \tag{1}$$

where $G \equiv \{(x_1, \dots, x_n) : 0 < x_k < b_k, \ k = 1, \dots, n\} \subset \mathbb{R}^n$ is an n-dimensional parallelepiped. We shall make the following assumptions (A1)–(A3).

(A1) The kernel $K(x, y, u)$ is m times $(m \geq 1)$ continuously differentiable with respect to x, y and u for $x, y \in G$, $x \neq y$, $u \in \mathbb{R}$, whereby there exists a real number $\nu \in (-\infty, n)$ such that, for any nonnegative integer $l \in \mathbb{Z}_+$ and multi-indices $\alpha = (\alpha_1, \dots, \alpha_n) \in \mathbb{Z}_+^n$ and $\beta = (\beta_1, \dots, \beta_n) \in \mathbb{Z}_+^n$ with $l + |\alpha| + |\beta| \leq m$, the following estimates hold:

$$\left| D_x^\alpha D_{x+y}^\beta \left(\partial/\partial u\right)^l K(x, y, u) \right| \leq \psi_1(|u|)\tau_{\nu+|\alpha|}(|x - y|), \tag{2}$$

$$\begin{aligned}
\left| D_x^\alpha D_{x+y}^\beta \left(\partial/\partial u\right)^l K(x, y, u_1) - D_x^\alpha D_{x+y}^\beta \left(\partial/\partial u\right)^l K(x, y, u_2) \right| \\
\leq \psi_2(\max\{|u_1|, |u_2|\})|u_1 - u_2|\tau_{\nu+|\alpha|}(|x - y|).
\end{aligned} \tag{3}$$

Here $|\alpha| = \alpha_1 + \dots + \alpha_n$ for $\alpha \in \mathbb{Z}_+^n$, $|x| = (x_1^2 + \dots + x_n^2)^{\frac{1}{2}}$ for $x \in \mathbb{R}^n$,

$$D_x^\alpha = (\partial/\partial x_1)^{\alpha_1} \cdots (\partial/\partial x_n)^{\alpha_n}, \quad D_{x+y}^\beta = (\partial/\partial x_1 + \partial/\partial y_1)^{\beta_1} \cdots (\partial/\partial x_n + \partial/\partial y_n)^{\beta_n},$$

$$\tau_s(t) = \begin{cases} 1 & \text{for } s < 0 \\ 1 + |\log t| & \text{for } s = 0 \\ t^{-s} & \text{for } s > 0 \end{cases} \qquad (s, t \in \mathbb{R}, \ t > 0), \tag{4}$$

and the functions $\psi_1 \colon [0, \infty) \to [0, \infty)$ and $\psi_2 \colon [0, \infty) \to [0, \infty)$ are assumed to be monotonically increasing.

This investigation was partly supported by the Estonian Science Foundation (Grant No. 2999).

280

(A2) $f \in C^{m,\nu}(G)$. The space $C^{m,\nu}(G)$ is defined as the collection of all m times continuously differentiable functions $u \colon G \to \mathbb{R}$ such that the estimates

$$|D_x^\alpha u(x)| \le c\tau_{|\alpha|-n+\nu}(\rho(x)), \qquad \left|(\partial/\partial x_k)^l u(x)\right| \le c'\tau_{l-n+\nu}(\rho_k(x)),$$

hold for $x \in G$, $|\alpha| \le m$, $l = 1, \ldots, m$; $k = 1, \ldots, n$, where c and c' are positive constants, $\rho_k(x) = \min\{x_k, b_k - x_k\}$, $k = 1, \ldots, n$, and $\rho(x) = \min_{1 \le k \le n} \rho_k(x)$ is the distance from x to ∂G, the boundary of G. Note that $C^m(\overline{G}) \subset C^{m,\nu}(G)$.

(A3) Equation (1) has a solution $u_0 \in L^\infty(G)$ and the linearized integral equation

$$v(x) = \int_G K_0(x,y)v(y)dy, \qquad K_0(x,y) = [\partial K(x,y,u)/\partial u]_{u=u_0(y)},$$

has only the trivial solution $v = 0$ in $L^\infty(G)$.

Remark 1. The assumption (A1) holds, for example, for the kernels $K(x,y,u) = K_1(x,y,u)|x-y|^{-\nu}$ $(0 < \nu < n)$ and $K(x,y,u) = K_1(x,y,u)\log|x-y|$ $(\nu = 0)$ where $K_1(x,y,u)$ is a $m+1$ times continuously differentiable function with respect to x, y, u for $x, y \in \overline{G}$, $u \in \mathbb{R}$.

Remark 2. From (A1)–(A3) it follows that, for the solution $u_0 \in L^\infty(G)$ of (1), we actually have $u_0 \in C^{m,\nu}(G)$ [4].

3. Collocation method

To define the partition of $\overline{G} = \{(x_1, \ldots, x_n) : 0 \le x_k \le b_k,\ k = 1, \ldots, n\}$ into cells we choose a vector $N = (N_1, \ldots, N_n)$ of natural numbers $N_1, \ldots, N_n$ and introduce in each of intervals $[0, b_k]$, $k = 1, \ldots, n$, the following $2N_k+1$ grid points $x_{k,N}^i$, $i = 0, 1, \ldots, 2N_k$:

$$x_{k,N}^i = (b_k/2)\,(i/N_k)^r, \quad i = 0, 1, \ldots, N_k; \qquad x_{k,N}^{N_k+i} = b_k - x_{k,N}^{N_k-i}, \ i = 1, \ldots, N_k. \quad (5)$$

Here $r \in \mathbb{R}$, $r \ge 1$, characterizes the nonuniformity of the grid. If $r = 1$ then the grid points (5) are uniformly located. Using (5) we introduce the partition of $\overline{G}$ into closed cells $G_N^{j_1,\ldots,j_n} \equiv \{(x_1, \ldots, x_n) : x_{k,N}^{j_k-1} \le x_k \le x_{k,N}^{j_k},\ k = 1, \ldots, n\} \subset \overline{G}$, $j_k = 1, \ldots, 2N_k$, $k = 1, \ldots, n$.

We determine the collocation points in following way. Choose m points $\eta_1, \ldots, \eta_m$ in the interval $[-1,1]$: $-1 \le \eta_1 < \ldots < \eta_m \le 1$. By affine transformations we transfer them into the interval $\left[x_{k,N}^{j_k-1}, x_{k,N}^{j_k}\right]$ $(j_k = 1, \ldots, 2N_k;\ k = 1, \ldots, n)$: $\xi_{k,N}^{j_k,q_k} = x_{k,N}^{j_k-1} + (\eta_{q_k} + 1)(x_{k,N}^{j_k} - x_{k,N}^{j_k-1})/2$, $q_k = 1, \ldots, m$. We assign the collocation points

$$\left(\xi_{1,N}^{j_1,q_1}, \ldots, \xi_{n,N}^{j_n,q_n}\right), \quad j_k = 1, \ldots, 2N_k; \quad q_k = 1, \ldots, m; \quad k = 1, \ldots, n, \qquad (6)$$

to the cells $G_N^{j_1,\ldots,j_n}$. To a continuous function $u : \overline{G} \to \mathbb{R}$ we introduce a piecewise polynomial interpolation function $P_N u : \overline{G} \to \mathbb{R}$ as follows: 1) on every cell $G_N^{j_1,\ldots,j_n}$ $(j_k = 1, \ldots, 2N_k;\ k = 1, \ldots, n)$ $(P_N u)(x_1, \ldots, x_n)$ is a polynomial of degree not exceeding $m - 1$ with respect to any of arguments $x_1, \ldots, x_n$; 2) $P_N u$ interpolates u at points (6). Thus, the interpolation function $P_N u$ is uniquely defined in every cell separately and may have jumps on the hyperplains $x_k = x_{k,N}^i$, $i = 1, \ldots, 2N_k - 1$;

$k = 1, \ldots, n$. We may treat $P_N u$ as a multivalued function on these hyperplanes. In the case $\eta_1 = -1$, $\eta_m = 1$, $P_N u$ is a continuous function on $\overline{G}$.

Let us denote by E_N the range of the operator $\mathcal{P}_N$. This is a finite dimensional space of all functions $u_N \in L^\infty(G)$ that are polynomials of degree$\leq m - 1$ with respect to any of arguments $x_1, \ldots, x_n$ on each of cells $G_N^{j_1, \ldots, j_n} \subset \overline{G}$ ($j_k = 1, \ldots, 2N_k$; $k = 1, \ldots, n$). If $\eta_1 > -1$ or $\eta_m < 1$ then the dimension of E_N is $(2m)^n N_1 \cdots N_n$. If $\eta_1 = -1$, $\eta_m = 1$, then $\dim E_N = (2N_1(m - 1) + 1) \cdots (2N_n(m - 1) + 1)$.

We look for an approximate solution $u_N \in E_N$ to integral equation (1) determing it from the following conditions:

$$\left[u_N(x) - \int_G K(x, y, u_N(y))dy - f(x) \right]_{x = (\xi_{1,N}^{i_1, p_1}, \ldots, \xi_{n,N}^{i_n, p_n})} = 0,$$

$$i_k = 1, \ldots, 2N_k; \quad p_k = 1, \ldots, m; \quad k = 1, \ldots, n. \tag{7}$$

We obtain a system of algebraic equations from (7) choosing a basis for E_N. For instance, we can present $u_N \in E_N$ in the form

$$u_N(x_1, \ldots, x_n) = \sum_{j_1=1}^{N_1} \cdots \sum_{j_n=1}^{N_n} \sum_{q_1=1}^{m} \cdots \sum_{q_n=1}^{m} c_{j_1, \ldots, j_n}^{q_1, \ldots, q_n} \varphi_{1,N}^{j_1, q_1}(x_1) \ldots \varphi_{n,N}^{j_n, q_n}(x_n) \tag{8}$$

where

$$\varphi_{k,N}^{j_k, q_k}(x_k) = \begin{cases} \displaystyle\prod_{\substack{i=1 \\ i \neq q_k}}^{m} \frac{\left(x_k - \xi_{k,N}^{j_k, i}\right)}{\left(\xi_{k,N}^{j_k, q_k} - \xi_{k,N}^{j_k, i}\right)}, & x_{k,N}^{j_k - 1} \leq x_k \leq x_{k,N}^{j_k} \\ 0, & \text{otherwise} \end{cases} \tag{9}$$

for $q_k = 1, \ldots, m$; $j_k = 1, \ldots, 2N_k$; $k = 1, \ldots, n$. Now the collocation conditions (7) will take a form of a (nonlinear) system that determines the coefficients $\{c_{j_1, \ldots, j_n}^{q_1, \ldots, q_n}\}$.

It is shown in [4] how to choose the scaling parameter r in (5) so that the method (7) might have the best convergence rate in supremum-norm on conditions (A1)–(A3):

$$\sup_{x \in \overline{G}} |u_N(x) - u_0(x)| \leq \text{const } h_N^m \qquad \text{for } \begin{cases} r > \dfrac{m}{n - \nu} & \text{if } n - \nu \leq m \\ r \geq 1 & \text{if } n - \nu > m \end{cases}, \tag{10}$$

where

$$h_N = \max\{b_1/N_1, \ldots, b_n/N_n\}. \tag{11}$$

The following theorem shows how to choose r and collocation points so that for this method the superconvergence phenomenon at collocation points would take place.

Theorem 1. *Let assumption (A3) and the following conditions be fulfilled.*

(A1') The kernel $K(x, y, u)$ and $\tilde{K}(x, y, u) \equiv \partial K(x, y, u)/\partial u$ are $m + \mu + 1$ times ($0 \leq \mu \leq m - 1$) continuously differentiable with respect to x, y and u for $x, y \in G$, $x \neq y$, $u \in \mathbb{R}$, whereby there exists a real number $\nu \in (-\infty, n)$ such that $K(x, y, u)$ and $\tilde{K}(x, y, u)$ satisfy (2) and (3) for $|\alpha| + |\beta| + l \leq m + \mu + 1$.

(A2') $f \in C^{m+\mu+1, \nu}(G)$.

(A4) The scaling parameter $r = r(m, n, \nu, \mu)$ in (5) satisfies the following conditions:

$$r > m/(n - \nu), \quad r \geq (m + n - \nu)/(n - \nu + 1) \quad if \quad n - \nu < \mu + 1;$$
$$r > m/(n - \nu), \quad r > (m + \mu + 1)/(n - \nu + 1) \quad if \quad \mu + 1 \leq n - \nu \leq m;$$
$$r \geq 1 \qquad\qquad r > (m + \mu + 1)/(n - \nu + 1) \quad if \quad m < n - \nu.$$

(A5) The collocation points (6) are generated by the nodes $\eta_1, \ldots, \eta_m$ of a quadrature formula

$$\int\limits_{-1}^{1} g(\xi)d\xi \approx \sum_{q=1}^{m} w_q g(\eta_q), \qquad -1 \leq \eta_1 < \ldots < \eta_m \leq 1,$$

which is exact for all polynomials of degree $m + \mu$, $0 \leq \mu \leq m - 1$.

Then there exist N_k^0 $(k = 1, \ldots, n)$ and $\delta_0 > 0$ such that, for $N_k \geq N_k^0$ $(k = 1, \ldots, n)$, the collocation method (7) determines a unique approximation $u_N \in E_N$ to u_0 satisfying $\|u_N - u_0\|_{L^\infty(G)} \leq \delta_0$. The following estimate holds:

$$\varepsilon_N \leq \text{const } h_N^m[h_N^{\mu+1}\tau_{\mu+1+\nu-n}(h_N) + h_N^n\tau_\nu(h_N)], \tag{12}$$

where

$$\varepsilon_N = \max_{j_k=1,\ldots,n;\ q_k=1,\ldots,m;\ k=1,\ldots,n} |u_N(x) - u_0(x)|_{x=(\xi_{1,N}^{j_1,q_1},\ldots,\xi_{n,N}^{j_n,q_n})}$$

and h_N and $\tau_s(t)$ are defined in (11) and (4), respectively.

Proof. We outline the basic idea on which the proof will be based (cf. [1,3]). Let $u_0 \in L^\infty(G)$ be a solution to (1). The hypotheses (A1') and (A2') for K and f imply that $u_0 \in C^{m+\mu+1,\nu}(G)$ (see Remark 2). Further, we have (see [4, p. 144]),

$$\varepsilon_N \leq c_1 \sup_{x \in G}\left|\int\limits_G \frac{\partial K(x, y, u_0(y))}{\partial u}[(P_N u_0)(y) - u_0(y)]dy\right| + c_2\|u_0 - P_N u_0\|_{L^2(G)}^2, \tag{13}$$

where c_1 and c_2 are positive constants not depending on N (on h_N). The second term on the right in the last inequality can be estimated on the basis of Lemma 7.2 in [4, p. 116]: due to $u_0 \in C^{m+\mu+1,\nu}(G)$ and (A4), $\|u_0 - P_N u_0\|_{L^\infty(G)}^2 \leq c_3 h_N^{2m}$, $c_3 = \text{const} > 0$. To establish the estimate (12) we have to prove that the first term on the right in the inequality (13) can be estimated by $c_4 h_N^m[h_N^{\mu+1}\tau_{\mu+1+\nu-n}(h_N) + h_N^n\tau_\nu(h_N)]$, $c_4 = \text{const} > 0$. This can be done on the basis of properties of $u_0 \in C^{m+\mu+1,\nu}(G)$ and assumptions (A1') and (A5). We refer to [3] for details. $\square$

Remark 3. If $\mu + 1 \leq n$ or if $\nu > 0$ then under conditions of Theorem 1,

$$\varepsilon_N \leq \text{const } h_N^{m+\mu+1}\tau_{\mu+1+\nu-n}(h_N). \tag{14}$$

Theorem 2. [3] *Let the conditions of Theorem 1 be fulfilled. We assume additionally that $\nu \leq 0$, $\mu + 1 > n$ and for $|\alpha| \leq \min\{\mu+1-n, -\nu\}$, $0 \leq k \leq \min\{\mu+1-n, -\nu\}$, the derivatives $D_y^\alpha \partial^{k+1} K(x, y, u)/\partial u^{k+1}$ are bounded and continuous on $G \times G \times (-\rho, \rho)$ with any $\rho > 0$, including the diagonal $x = y$. Then the error estimate (14) holds.*

Corollary. *Under the conditions of Theorem 1, the iterated approximation $\tilde{u}_N(x) = \int_G K(x, y, u_N(y))dy + f(x)$, $x \in G$, satisfies*

$$\sup_{x \in \overline{G}} |\tilde{u}_N(x) - u_0(x)| \leq \text{const } h_N^m [h_N^{\mu+1} \tau_{\mu+1+\nu-n}(h_N) + h_N^n \tau_\nu(h_N)].$$

Under the conditions of Theorem 2,

$$\sup_{x \in \overline{G}} |\tilde{u}_N(x) - u_0(x)| \leq \text{const } h_N^{m+\mu+1} \tau_{\mu+1+\nu-n}(h_N).$$

4. Galerkin method

Let $(\cdot, \cdot)$ denote the inner product for $L^2(G)$ and let T be the integral operator in equation (1): $(Tu)(x) = \int_G K(x, y, u(y))dy$. We look for an approximate solution $u_N \in E_N$ (see Section 2) to (1), determining it now from the following conditions:

$$(u_N - Tu_N - f, v_N) = 0 \qquad \forall v_N \in E_N. \tag{15}$$

Choosing a basis for E_N, we obtain from (15) a system of algebraic equations. In particular, we may use the Lagrange functions (9).

Theorem 3. *Assume (A1)–(A3) and let the grid points (5) be used.*

Then there exist N_k^0 ($k = 1, \ldots, n$) and δ_0 such that, for $N_k \geq N_k^0$ ($k = 1, \ldots, n$), the conditions (15) determine a unique approximation $u_N \in E_N$ to the solution u_0 of (1) satisfying $\|u_N - u_0\|_{L^\infty(G)} \leq \delta_0$. The following error estimates hold:

$$\text{if } m < n - \nu \text{ then } \sup_{x \in \overline{G}} |u_N(x) - u_0(x)| \leq c\, h_N^m \text{ for any } r \geq 1; \tag{16}$$

$$\text{if } m = n - \nu \text{ then } \sup_{x \in \overline{G}} |u_N(x) - u_0(x)| \leq c \begin{cases} h_N^m(1 + |\log h_N|), & r = 1, \\ h_N^m, & r > 1; \end{cases} \tag{17}$$

$$\text{if } m > n - \nu \text{ then } \sup_{x \in \overline{G}} |u_N(x) - u_0(x)| \leq c \begin{cases} h_N^{r(n-\nu)}, & 1 \leq r \leq m/(n - \nu), \\ h_N^m, & r \geq m/(n - \nu). \end{cases} \tag{18}$$

Here r is the parameter characterizing the nonuniformity of the grid (5), h_N is defined in (11) and the constant c is independent of N (of h_N).

Proof. We consider (1) as the equation $u = Tu + f$ in the space $L^\infty(G)$. Let $u_0 \in L^\infty(G)$ be a solution to this equation. Due to Remark 2, $u_0 \in C^{m,\nu}(\overline{G})$. Let $Q_N : L^2(G) \to E_N$ be the orthogonal projection of $L^2(G)$ onto E_N, which we will also regard as a projection operator on $L^\infty(\overline{G})$ to E_N. With Q_N, we can rewrite (15) as $u_N = Q_N Tu_N + Q_N f$. Further, we can check that the operators $T : L^\infty(G) \to L^\infty(G)$ and $Q_n T : L^\infty(G) \to L^\infty(G)$ satisfy the conditions (i)–(v) of a general convergence theorem in [4, p. 58], where $E = E_h = L^\infty(G)$, $h = h_N$, $p_h = Q_N$, $T_h = Q_N T$, and $f_h = Q_N f$. Therefore (see [4, p. 58]), there exist $N_k^0 > 0$ ($k = 1, \ldots, n$) and $\delta_0 > 0$ such that, for $N_k \geq N_k^0$ ($k = 1, \ldots, n$), equation $u_N = Q_N Tu_N + Q_N f$ has a unique solution $u_N \in E_N$ satisfying $\|u_N - u_0\|_{L^\infty(G)} \leq \delta_0$ and $\|u_N - u_0\|_{L^\infty(G)} \leq c\, \|u_0 - Q_N u_0\|_{L^\infty(G)}$, where the constant c is independent of N. Let P_N denote the interpolation operator, introduced in Section 2. Then $u_0 - Q_N u_0 = u_0 - P_N u_0 + Q_N(P_N u_0 - u_0)$. Now, using the corresponding estimates for $\|P_N u_0 - u_0\|_{L^\infty(G)}$ from [4, p. 115], and the uniformly boundedness of Q_N, $\|Q_N\| \leq$ const, where $\|Q_N\|$ denotes the norm of Q_N when it is regarded as an operator on $L^\infty(G)$ to $L^\infty(G)$, we obtain the estimates (16)–(18). $\square$

Theorem 4. *Let the conditions of Theorem 3 be fulfilled with $\psi_i(t) = c_1 + c_2 t^{p-1}$ $(i = 1, 2; t \in [0, \infty); c - 1, c_2 = const$ where $1 < p < n/\nu$ in the case $0 < \nu < n$ and $1 < p < \infty$ in the case $\nu \leq 0$. Assume that the scaling parameter $r = r(m, n, \nu, p)$ is chosen as follows: $r \geq 1$ if $m < n - \nu + (1/p)$ and $r > m/(n - \nu + (1/p))$ if $m \geq n - \nu + (1/p)$. Then for the Galerkin approximation u_N we have*

$$\|u_N - u_0\|_{L^p(G)} \leq c h_N^m. \tag{19}$$

Proof. In analogy to the proof of Theorem 3, the proof of (19) is based on the corresponding estimates for $\|u_0 - P_N u_0\|_{L^p(G)}$ from [4, p. 115] $(1 < p < \infty)$. $\square$

References

1. A. Pedas, Superconvergence of the spline collocation method for nonlinear two dimensional weakly singular integral equations, *Differentsialnye Uravnen.* **33** (1997), no. 9, 1–8 (Russian).

2. A. Pedas and G. Vainikko, Superconvergence of piecewise polynomial collocations for nonlinear weakly singular integral equations, *J. Integral Equations Appl.* **9** (1997), 379–406.

3. A. Pedas and G. Vainikko, Spline collocation method for weakly singular integral equations, *Proc. Estonian Acad. Sci. Phys. Math.* **48** (1999), no. 2, 69–78.

4. G. Vainikko, *Multidimensional weakly singular integral equations*, Lect. Notes Math. **1549**, Springer, Berlin-Heidelberg, 1993.

Institute of Applied Mathematics, University of Tartu, J.Liivi 2-206, 50409 Tartu, Estonia; e-mail: arvet.pedas@ut.ee

Institute of Mathematics, Helsinki University of Technology, Otakaari 1, 02150 Espoo, Finland; e-mail: gennadi.vainikko@hut.fi

O.G. RUEHR and B.S. BERTRAM

Asymptotics and inequalities for a class of infinite sums

1. Introduction

We begin by defining, for $2\beta - 1 > \alpha \geq 0$ and $x \geq 0$,

$$S(\alpha, \beta) = \sum_{n=1}^{\infty} \frac{n^\alpha}{(n^2 + x)^\beta}. \tag{1}$$

This sum is a generalization of one encountered in an extension of Mathieu's inequality [1], where, in particular, the inequality $S^2(1, 2) \geq S(1, 3)$ was stated as an unproved lemma. Later, two proofs of the lemma were given in [2]. Our goal is to establish the asymptotic behavior for large, positive x of $S(\alpha, \beta)$. As a byproduct we will discover a generalization of the Alzer-Brenner inequality, namely, $(\beta - 1)S^2(1, \beta) \geq S(1, 2\beta - 1)$. Finally, we present a general criterion for inequalities of this type.

2. Asymptotics for S for large x

We use a simple Laplace transform to obtain the representation

$$S(\alpha, \beta) = \int_0^\infty \frac{t^{\beta-1} R_\alpha(t) e^{-xt}}{\Gamma(\beta)} dt, \tag{2}$$

where

$$R_\alpha(t) = \sum_{n=1}^{\infty} n^\alpha e^{-n^2 t} \tag{3}$$

We will give two derivations for the algebraic terms in the asymptotic behavior of R_α as $t \to 0^+$,

$$R_\alpha(t) \approx \frac{\Gamma(\frac{\alpha+1}{2})}{2t^{\frac{\alpha+1}{2}}} - 2\sin(\frac{\pi\alpha}{2}) \sum_{j \geq 0} \frac{t^j \Gamma(2j + \alpha + 1)\zeta(2j + \alpha + 1)}{j!(2\pi)^{2j+\alpha+1}}. \tag{4}$$

Term by term integration then yields, for S:

Theorem 1. *We have*

$$S_{\alpha,\beta}(x) \approx \frac{\Gamma(\frac{\alpha+1}{2})\Gamma(\beta - \frac{\alpha+1}{2})}{2\Gamma(\beta)x^{\beta - \frac{\alpha+1}{2}}} - 2\sin(\frac{\pi\alpha}{2}) \sum_{j \geq 0} \frac{\Gamma(2j + \alpha + 1)\Gamma(\beta + j)\zeta(2j + \alpha + 1)}{j!(2\pi)^{2j+\alpha+1}\Gamma(\beta)x^{\beta+j}}. \tag{5}$$

Here also we are ignoring terms which are exponentially small as $x \to \infty$.

286

3. Transform method for R_α asymptotics

Calculate the Mellin transform of R_α term by term as follows:

$$\bar{R}_\alpha(s) = \int_0^\infty t^{s-1} R_\alpha(t)dt = \sum_{n\geq 1} \frac{n^\alpha \Gamma(s)}{n^{2s}} = \Gamma(s)\zeta(2s - \alpha). \tag{6}$$

We invert the transform using the standard contour

$$R_\alpha(t) = \frac{1}{2\pi i}\int t^{-s}\Gamma(s)\zeta(2s - \alpha)ds. \tag{7}$$

Taking account of the poles of the gamma function and that of the zeta function, we obtain formally

$$R_\alpha(t) \approx \frac{\Gamma(\frac{\alpha+1}{2})}{2t^{\frac{\alpha+1}{2}}} + \sum_{j\geq 0} \frac{(-1)^j t^j \zeta(-\alpha - 2j)}{j!}. \tag{8}$$

It is important to note that by ignoring the integral on part of the contour we are not getting the exponentially small terms in the asymptotic expansion. Finally, with the aid of the reflection formula for the zeta function, [4, p. 19]

$$\zeta(1 - z) = 2(2\pi)^{-z}\Gamma(z)\zeta(z)cos(\frac{\pi z}{2}), \tag{9}$$

we have

$$R_\alpha(t) \approx \frac{\Gamma(\frac{\alpha+1}{2})}{2t^{\frac{\alpha+1}{2}}} - 2\sin(\frac{\pi\alpha}{2})\sum_{j\geq 0} \frac{t^j \Gamma(2j + \alpha + 1)\zeta(2j + \alpha + 1)}{j!(2\pi)^{2j+\alpha+1}}. \tag{10}$$

4. R_α asymptotics using an identity

We begin by quoting an identity of Euler

$$\sum_{n=-\infty}^{\infty} e^{-n^2 t} = \sqrt{\frac{\pi}{t}}\sum_{n=-\infty}^{\infty} e^{-n^2\pi^2/t}. \tag{11}$$

and its generalization, which we rediscovered (see [2],[3]),

$$\sum_{n=-\infty}^{\infty} |n|^{2\beta}e^{-n^2 t} = \frac{\Gamma(\beta + 1/2)}{t^{\beta+1/2}}\sum_{n=-\infty}^{\infty} {}_1F_1[\beta + 1/2; 1/2; \frac{-n^2\pi^2}{t}]. \tag{12}$$

We outline our proof of (12) as follows. Define the Weyl fractional integral operator [5, p. 201] by

$$L_\alpha\{B(t)\} = \int_t^\infty \frac{B(\tau)d\tau}{\Gamma(\alpha)(\tau - t)^{1-\alpha}}.$$

Then we have, with appropriate restrictions, the following results:

$$L_\alpha\{e^{-xt}\} = x^{-\alpha}e^{-xt} \qquad [5,\ p.\ 202\ (11)] \tag{13}$$

$$L_\alpha\{t^{-p}e^{-\frac{x}{t}}\} = \frac{\Gamma(p-\alpha)t^{\alpha-p}}{\Gamma(p)} \, {}_1F_1[p-\alpha;p;-\frac{x}{t}] \qquad [5, \text{ p. } 203 \ (15)] \qquad (14)$$

The generalization can be established with the aid of these identities and an appeal to analytic continuation. Use (13) on the left side of (11) ($x = n^2$) and use (14) on the right side of (11) ($x = n^2\pi^2, p = 1/2$). Then simplify and take $\alpha = -\beta$ (by analytic continuation) to get (12).

Rearranging (12) yields

$$R_\alpha(t) = \frac{\Gamma(\frac{\alpha+1}{2})}{t^{\frac{\alpha+1}{2}}} \left\{ \frac{1}{2} + \sum_{n=1}^{\infty} \, {}_1F_1[\frac{\alpha+1}{2};1/2;\frac{-n^2\pi^2}{t}] \right\}. \qquad (15)$$

Now insert the well-known algebraic asymptotic behavior of the confluent hypergeometric functions, [4, p. 289],

$${}_1F_1(a;c;-z) \approx \frac{\Gamma(c)}{\Gamma(c-a)z^a} \, {}_2F_0(a,a-c+1;1/z). \qquad (16)$$

Interchanging the order of summation, identifying the Riemann zeta function, and simplifying the various factorial functions, we obtain again the result (10) of the previous paragraph.

5. A one-parameter generalization of the Alzer-Brenner inequality

Taking $\alpha = 1$ and looking at the first term of the asymptotic expansion we attempt to extend the Alzer-Brenner inequality, $S^2(1,2) \geq S(1,3)$. Clearly we must compare $S^2(1,\beta)$ with $S(1,2\beta - 1)$. In order that the dominant terms on both sides coincide, as $x \to \infty$, we need the factor $\beta - 1$. This yields the conjecture, $(\beta - 1)S^2(1,\beta) \geq S(1,2\beta - 1)$, i.e.,

Theorem 2. *We have*

$$(\beta - 1)\left[\sum_{n\geq 1} \frac{n}{(n^2+x)^\beta}\right]^2 \geq \sum_{n\geq 1} \frac{n}{(n^2+x)^{2\beta-1}}.$$

Proof. We will outline a proof subject to the validity of a simple lemma that we have not yet established. From the integral representation (2), we can write the inequality as $\Delta \geq 0$ where

$$\Delta = \Gamma(\beta)\Gamma(\beta - 1)[(\beta - 1)S^2(1,\beta) - S(1,2\beta - 1)] \qquad (17)$$

$$= \int_0^\infty t^{\beta-1}R_1(t)dt \int_0^\infty s^{\beta-1}R_1(s)e^{-x(s+t)}ds - Q\int_0^\infty w^{2\beta-2}R_1(w)e^{-xw}dw$$

where

$$Q = \frac{\Gamma(\beta)\Gamma(\beta - 1)}{\Gamma(2\beta - 1)} = \int_0^1 t^{\beta-1}(1-t)^{\beta-2}dt.$$

Letting $s+t = w$ in the first pair of integrals and interchanging the order of integration we get, with a change of scale $t \to wt$, the following representation

$$\Delta = \int_0^\infty e^{-xw}w^{2\beta-2}\int_0^1 [t(1-t)]^{\beta-2}\{wt(1-t)R_1(wt)R_1(w(1-t)) - tR_1(w)\}dtdw. \qquad (18)$$

288

We pause to note that it would be sufficient that the bracketed quantity be non-negative. However, since $R_1(w) \approx 1/2w$ (as $w \to 0^+$), we get, for the bracket, the approximation $1/4w - t/2w$, which fails. We remedy this situation as follows. Employing the symmetry in the inner integral, we replace t by $1 - t$ and add this result to the original integral and multiply by two to get

$$4\Delta = \int_0^\infty e^{-xw} w^{2\beta-3} \int_0^1 [t(1-t)]^{\beta-2} \{4w^2 t(1-t) R_1(wt) R_1(w(1-t)) - 2w R_1(w)\} dt\, dw.$$

(19)

Defining $F(x) = 2x R_1(x)$, we see that it is sufficient for our proof to establish the following:

Lemma 1. *For $0 \le t \le 1$ and for $x \ge 0$,*

$$F(xt) F[x(1-t)] \ge F(x). \tag{20}$$

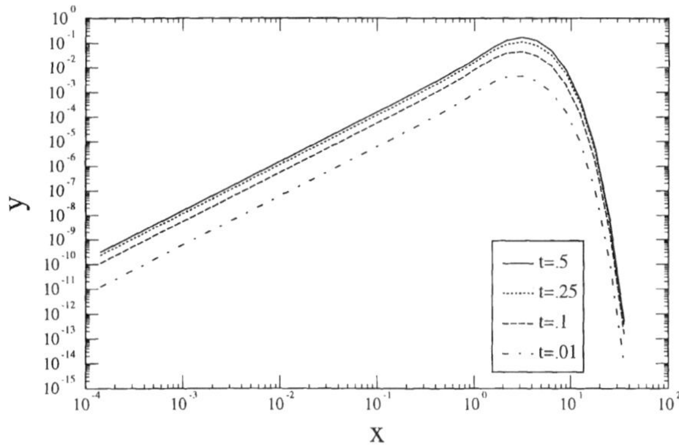

Fig. 1. Evidence for Lemma 1.

From considerable numerical evidence, as indicated in Fig. 1 (where for fixed t, $y = F(xt)F(x(1-t) - F(x))$, we have no doubt that the lemma is true. It is useful to recall that, from previous sections, we have the (divergent) asymptotic expansion for $F(x)$ as $x \to 0^+$.

$$F(x) \approx 1 - \sum_{n \ge 1} |B_{2n}| \frac{x^n}{n!}. \tag{21}$$

(Here we are using the familiar formula for $\zeta(2n)$, [4, p. 19] together with the expansion obtained previously (4).)

6. General integral representation for Alzer-Brenner type inequalities

The inequalities we seek have the pattern $S(\alpha, \beta)^2 \geq QS(a, b)$. In order that Q be constant and that the inequality should hold as $x \to \infty$, we require that

$$b = 2\beta - \alpha + \frac{a}{2} - \frac{1}{2} \tag{22}$$

and

$$Q = \frac{\Gamma^2(\frac{\alpha+1}{2})\Gamma^2(\beta - \frac{\alpha+1}{2})\Gamma(b)}{2\Gamma^2(\beta)\Gamma(2\beta - \alpha - 1)\Gamma(\frac{a+1}{2})}. \tag{23}$$

Following the procedure used in the previous section we can write

$$[\Gamma(\beta)S(\alpha, \beta)]^2 = \int_0^\infty w^{2\beta-1}e^{-xw}\int_0^1 [t(1-t)]^{\beta-1}R_\alpha(wt)R_\alpha(w(1-t))dtdw. \tag{24}$$

To simplify the notation we introduce the definitions

$$F_\alpha(x) = \frac{2x^{\frac{\alpha+1}{2}}R_\alpha(x)}{\Gamma(\frac{\alpha+1}{2})} \tag{25}$$

$$G(\alpha, a, x, t) = F_\alpha(xt)F_\alpha(x(1-t)) - F_a(x). \tag{26}$$

Omitting details, we obtain, finally, the representation

$$S^2(\alpha, \beta) - QS(a, b) = \left[\frac{\Gamma(\frac{\alpha+1}{2})}{2\Gamma(\beta)}\right]^2 \int_0^\infty w^{2\beta-\alpha-2}e^{-xw}\int_0^1 (t(1-t))^{\beta-\frac{\alpha}{2}-\frac{3}{2}}G(\alpha, a, w, t)dtdw. \tag{27}$$

For those values of α and a (and, of course, $w \geq 0$ and $0 \leq t \leq 1$) for which G is non-negative, we obtain the inequality $S^2(\alpha, \beta) \geq QS(a, b)$, b and Q as defined above. To summarize, we state the following:

Lemma 2. *For $0 \leq t \leq 1$ and $x \geq 0$*

$$F_\alpha(xt)F_\alpha(x(1-t)) - F_a(x) \geq 0, \tag{28}$$

where

$$F_\alpha(x) = \frac{2x^{\frac{\alpha+1}{2}}}{\Gamma(\frac{\alpha+1}{2})}\sum_{n=1}^\infty n^\alpha e^{-n^2 x}. \tag{29}$$

Preliminary computations indicate that the lemma is true for some pairs (α, a) and *false* for others. In this connection, it is appropriate to examine prospective inequalities near $x = 0$ (as opposed to x near infinity). For $x = 0$, $S(\alpha, \beta) = \zeta(2\beta - \alpha)$, so that from Theorem 2 we would deduce $(\beta - 1)\zeta^2(2\beta - 1) \geq \zeta(4\beta - 3)$. Note that this result follows from Lemma 1 and is entirely consistent with the behavior of the zeta function near its pole. More generally, we would have

$$\zeta^2(2\beta - \alpha) \geq Q\zeta(4\beta - 2\alpha - 1). \tag{30}$$

Whenever (30) fails to be true, so must Lemma 2 for the same parameter values. While a study of this circumstance is incomplete at this writing, it appears that (30) is false when a and α are equal and between zero and one-half. In conclusion, we have both Lemma 2 and (30) to aid in the construction of possible inequalities of the Alzer-Brenner type.

References

1. H. Alzer, J.L. Brenner and O.G. Ruehr, On Mathieu's inequality, *J. Math. Anal. Appl.* **218** (1998), 606–610.

2. R.L. Curry, B. Bertram, J.E. Wilkins, Jr., C.C. Rousseau and O.G. Ruehr, Problem 97-1, *SIAM Rev.* **40** (1998), 714–718.

3. V. Kovalenko, N.E. Frankel, M.L. Glasser and T. Taucher, *Generalized Euler-Jacobi inverse formula and asymptotics beyond all orders*, Math. Lecture Notes Ser. **214**, London, 1995, 45.

4. W. Magnus, F. Oberhettinger and R.P. Soni, *Formulas and theorems for the special functions of mathematics and physics*, Springer-Verlag, New York, 1966.

5. A. Erdélyi et al., *Tables of integral transforms*, vol. 2, Bateman Manuscript Project, McGraw-Hill, New York, 1954.

Department of Mathematical Sciences, Michigan Technological University, Houghton, MI 49931, USA

C. SAMOILA, R. TEODORESCU and D. CENTEA

A new theorem concerning the Buckingham Rayleigh methods and general dimensional analysis

1. Fundamental theorem of dimensional analysis (D.A.)

Considering any quantity, say G, in the units length L, mass M, time T, we can write

$$g = f(l, m, t).$$ (1)

Let us now consider two states G_1 and G_2 of the quantity G:

$$g_1 = f(l_1, m_1, t_1) \quad \text{and} \quad g_2 = f(l_2, m_2, t_2).$$ (2)

The exchanges of the measure units with the factor attached: λ, μ, τ, made as two state functions become

$$g_1' = f(\lambda l_1, \mu m_1, \tau t_1) \quad \text{and} \quad g_2' = f(\lambda l_2, \mu m_2, \tau t_2).$$ (3)

Obviously, the ratio of the state functions is constant, because, independently of the unit system adopted, we have

$$\frac{g_1}{g_2} = \frac{g_1'}{g_2'} \quad \frac{f(l_1, m_1, t_1)}{f(l_2, m_2, t_2)} = \frac{f(\lambda l_1, \mu m_1, \tau t_1)}{f(\lambda l_2, \mu m_2, \tau t_2)} = \dots$$ (4)

Therefore, in general,

$$\frac{f(\lambda l, \mu m, \tau t)}{f(l, m, t)} = K(\lambda, \mu, \tau).$$ (5)

It is clear that permuting the symbols l, m, t, and λ, μ, τ, respectively, will in no way affect the first member of (5) and hence

$$f(l, m, t) \cdot K(\lambda, \mu, \tau) \equiv f(\lambda, \mu, \tau) \cdot K(l, m, t),$$ (6)

or

$$\frac{f(l, m, t)}{K(l, m, t)} = \frac{f(\lambda, \mu, \tau)}{K(\lambda, \mu, \tau)} = \text{const} = x.$$ (7)

As a consequence, (5) becomes

$$f(\lambda l, \mu m, \tau t) \equiv f(l, m, t) \cdot \frac{1}{x} \cdot f(\lambda, \mu, \tau).$$ (8)

Langhaar demonstrated that a conventional unity value can be assumed for x, hence

$$f(\lambda l, \mu m, \tau t) \equiv f(l, m, t) \cdot f(\lambda, \mu, \tau).$$ (9)

For simplicity, only the derivation with respect to "l" will be considered:

$$\frac{\partial f}{\partial u}(\lambda l, \mu m, \tau t) \cdot \frac{du}{dl} = \frac{\partial f}{\partial u}(l, m, t) \cdot \frac{du}{dl} \cdot f(\lambda, \mu, \tau),$$ (10)

292

where $\lambda l = u$ for the first member and $l = u$ for the second member. Multiplying relation (10) by "l", in both members and taking account of (9), we have successively

$$f(\lambda, \mu, \tau) \equiv \frac{f(\lambda l, \mu m, \tau t)}{f(l, m, t)},$$

$$\left(u \cdot \frac{\partial f}{\partial u}\right)(\lambda l, \mu m, \tau t) = \left(u \cdot \frac{\partial f}{\partial u}\right)(l, m, t) \cdot f(\lambda, \mu, \tau), \tag{11}$$

$$\left(\frac{u}{f} \cdot \frac{\partial f}{\partial u}\right)(\lambda l, \mu m, \tau t) = \left(\frac{u}{f} \cdot \frac{\partial f}{\partial u}\right)(l, m, t), \tag{12}$$

$$\frac{u}{f} \cdot \frac{\partial f}{\partial u} = \text{const} = \lg. \tag{13}$$

Separating variables and integrating, (C_1 is an integration constant) we obtain

$$\frac{\partial f}{f} = \lg \cdot \frac{\partial u}{u}, \tag{14}$$

$$\ln f = C_1 \lg \cdot \ln u \text{ or } f = C_1 \cdot u^{\lg}. \tag{15}$$

Proceeding similarly for variables m and t, we get the complete shape of the function

$$f = C_1 \cdot l^{\lg} \cdot m^{mg} \cdot t^{tg}. \tag{16}$$

2. Rayleigh's method

Applied in 1899 for studying one problem of heat transfer, Rayleigh's method settled functional dependence between a dependent quantity P and independent variables which influenced them, expressed by

$$P = F(A, B, C, \ldots, M, N, \ldots). \tag{17}$$

Function "F" is expanded into an infinite series with respect to the powers of the independent variables, as follows:

$$P = \sum K_i \cdot A^{ai} \cdot B^{bi} \cdot C^{ci} \cdot \ldots \cdot M^{mi} \cdot N^{ni} \cdot \ldots, \tag{18}$$

from which Rayleigh keeps consideration of only one product of powers, because all terms of the infinite series should have the same dimensionality, namely that of the physical quantity P:

$$P = K \cdot A^a \cdot B^b \cdot C^c \cdot \ldots \cdot M^m \cdot N^n \cdot \ldots, \tag{19}$$

either a product of powers, or under similitude criteria product of powers

$$\Pi = K \cdot \Pi_m^{km} \cdot \Pi_n^{kn}. \tag{20}$$

3. Buckingham's method (π theorem)

Let us consider the same physical phenomena as in Rayleigh's method (17). This relation must be independent of choice of the measurement unit's de. If, for example, A and B are fundamental physical quantities of the unit system adopted, the other quantities in (17) are derived, and can be expressed, in terms of A and B, in accordance with the dimensional matrix, as follows:

<table>
<tr><td></td><td>P</td><td>C</td><td>$D \ldots M$</td><td>$N \ldots$</td></tr>
<tr><td>A</td><td>a_p</td><td>a_c</td><td>$a_d \ldots a_m$</td><td>$a_n \ldots$</td></tr>
<tr><td>B</td><td>b_p</td><td>b_c</td><td>$b_d \ldots b_m$</td><td>$b_n \ldots$</td></tr>
</table>

$$(21)$$

On the base of matrix (21) new nondimensional quantities for all these secondary quantities are defined:

$$\Pi_P = \frac{P}{A^{a_p} \cdot B^{b_p}}; \quad \Pi_C = \frac{C}{A^{a_c} \cdot B^{b_c}}; \quad \Pi_D = \frac{D}{A^{a_d} \cdot B^{b_d}}; \quad \Pi_M = \frac{M}{A^{a_m} \cdot B^{b_m}}. \tag{22}$$

Replacing these secondary quantities in (17) by the new dimensionless ones (22), we have

$$\Pi_P = f\left(A, B, \Pi_c, \Pi_d, \ldots, \Pi_m, \Pi_n, \ldots\right). \tag{23}$$

The change of measurement units A and B, to A$'$ and B$'$ does not modify dimensionless Π, and relation (23) becomes

$$\Pi_P = f\left(A', B', \Pi_c, \Pi_d, \ldots, \Pi_m, \Pi_n, \ldots\right). \tag{24}$$

Therefore we find that quantities A, B can take any value, while $\Pi_c, \Pi_d, \ldots, \Pi_m, \Pi_n, \ldots$ do not change. Hence, (23) is independent of A and B, and can be written as

$$\Pi_P = \varphi\left(\Pi_c, \Pi_d, \ldots, \Pi_m, \Pi_n, \ldots\right). \tag{25}$$

4. General dimensional analysis

C.I. Staicu started from the findings of the Rayleigh and Buckingham methods, which need laborious experiments for identification of the numerical values of the exponents and of the proportionality constant value. Staicu proposed to determine values for all exponents through calculation and to determine the proportionality constant, based on experimental data.

For the same physical phenomenon (17), dimensional analysis imposes

$$f_1\left(M, N, P, \ldots\right) = f_2\left(A, B, C, \ldots\right) \tag{26}$$

and the final relation:

$$P = K \frac{A^{a_1} \cdot B^{b_1} \cdot C^{c_1} \cdot \ldots}{M^{m_1} \cdot N^{n_1} \cdot \ldots}. \tag{27}$$

This paper tries to demonstrate the compatibility of the Rayleigh, Buckingham, and general dimensional (G.D.A.) methods.

5. Background expositions

Consider the pushing force of a ship's propeller, expressed in accordance with (16), as a product of powers:

$$Q = K \cdot \rho^{z_1} \cdot V^{z_2} \cdot d^{z_3} \cdot n^{z_4} \cdot g^{z_5} \cdot \eta^{z_6}, \tag{28}$$

where Q is the thrust of the propeller, [kg·m/s^2]; ρ, the water density [kg/m^3]; V, the ship speed, [m/s]; d, the propeller diameter, [m]; n, the propeller rotation, [1/s]; g, the gravitational acceleration, [m/s^2]; η, the dynamic viscosity, [kg/ms]; and K, a constant of proportionality.

The dimensional equation of (28) will be

$$\left[\frac{Kg \cdot m}{s^2}\right] = K \left[\frac{Kg}{m^3}\right]^{z_1} \cdot \left[\frac{m}{s}\right]^{z_2} \cdot [m]^{z_3} \cdot \left[\frac{1}{s}\right]^{z_4} \cdot \left[\frac{m}{s^2}\right]^{z_5} \cdot \left[\frac{Kg}{m \cdot s}\right]^{z_6} \tag{29}$$

and dimensional homogeneous conditions become

$$\left\{ \begin{array}{l} z_1 + z_6 = 1, \\ -3z_1 + z_2 + z_3 + z_5 - z_6 = 1, \\ -z_2 - z_4 - 2z_5 - z_6 = -2. \end{array} \right\} \tag{30}$$

Having six unknowns and only three equations, system (30) yields arbitrary election for three of the quantities:

$$\left\{ \begin{array}{l} z_1 = 1 - z_6, \\ z_3 = 4 - z_2 - z_5 - 2z_6, \\ z_4 = 2 - z_2 - 2z_5 - z_6. \end{array} \right\} \tag{31}$$

With solutions (31), equation (28) takes the form

$$\frac{Q}{\rho \cdot d^4 \cdot n^2} = K \cdot \left[\frac{V}{d \cdot n}\right]^{z_2} \cdot \left[\frac{g}{d \cdot n^2}\right]^{z_5} \cdot \left[\frac{\eta}{\rho \cdot d^2 \cdot n}\right]^{z_6} \tag{32}$$

Solution (32) shows the existence of four nondimensional complexes, of magnitude exponents: z_1, z_5, z_6, and of a proportionality constant K, all being unknowns.

The same example, using the Π theorem, starts from the implicit relation

$$\varphi\left[Q, \rho, V, d, n, g, \eta\right] = 0. \tag{33}$$

If the number of variables is $x = 7$, and if the S.I. system of measurement units is utilized, fundamental units are M, L, T, ($k = 3$). From the Π theorem, we get $x - k = 4$ nondimensional complexes. Changing the fundamental units, L/L', M/M', T/T', we see that (33) becomes

$$\varphi\left[Q\frac{M'L'}{T'}, \rho\frac{M'}{L'^3}, V\frac{L'}{T'}, d \cdot L', n \cdot \frac{1}{T'}, \eta\frac{M'}{L'T'}, g\frac{L'}{T'^2}\right] = 0. \tag{34}$$

Because function (33) is independent of the given fundamental measurement units, we can try this elimination by expressing others as a function of three selected, namely, ρ, d, n:

$$\rho\frac{M'}{L'^3} = 1; \quad d \cdot L' = 1; \quad n \cdot \frac{1}{T'} = 1.$$

From here, $L' = \frac{1}{d}$, $T' = n$, $M' = \frac{1}{\rho \cdot d^3}$, and

$$\varphi\left[\frac{Q}{\rho \cdot d^4 \cdot n^2}, 1, 1, 1, \frac{V}{n \cdot d}, \frac{\eta}{\rho \cdot d^2 \cdot n}, \frac{g}{d \cdot n^2}\right] \tag{35}$$

From G.D.A. we find that

$$d^{z_1} \cdot n^{z_2} \cdot \rho^{z_3} \cdot Q^{z_4} = K \cdot V^{z_5} \cdot g^{z_6} \cdot \eta^{z_6}. \tag{36}$$

With the same three fundamental measurement units M, N, T, the dimensional matrix of (36) will be

	Q	ρ	V	D	n	g	η
L	1	-3	1	1	0	1	-1
M	1	1	0	0	0	0	1
T	-2	0	-1	0	-1	-2	-1

$$\tag{37}$$

The condition of dimensional homogeneity of (36) requires that

$$\left\{\begin{array}{l} Z_4 - 3Z_3 - Z_5 + Z_1 - Z_6 + Z_7 = 0, \\ Z_4 + Z_3 - Z_7 = 0, \\ -2Z_4 + Z_5 - Z_2 + 2Z_6 + Z_7 = 0. \end{array}\right\} \tag{38}$$

We present eight solutions, in integers, whose exponents have acceptable values:

	D	n	ρ	Q	V	g	η
	Z_1	Z_2	Z_3	Z_4	Z_5	Z_6	Z_7
R_1	3	4	1	1	2	1	2
R_2	1	3	1	1	1	1	2
R_3	1	3	1	2	2	1	3
R_4	4	4	2	1	1	1	3
R_5	2	3	2	2	1	1	4
R_6	3	4	2	2	2	1	4
R_7	1	3	2	3	2	1	5
R_8	4	4	3	2	1	1	5

$$\tag{39}$$

Employing R4 in accordance with G.D.A., we obtain from (36)

$$d^4 \cdot n^4 \cdot \rho^2 \cdot Q = K \cdot V^1 \cdot g^1 \cdot \eta^3, \tag{40}$$

from which the pushing force of a propeller results as

$$Q = K \frac{V \cdot g \cdot \eta^3}{n^4 \cdot d^4 \cdot \rho^2}. \tag{41}$$

Does this difference of expression cause us to suspect (41)? To answer this question, we note that from (32) and (35) we have

$$\Pi_1 = \frac{Q}{\rho \cdot d^4 \cdot n^2}; \quad \Pi_2 = \frac{V}{n \cdot d}; \quad \Pi_3 = \frac{g}{d \cdot n^2}; \quad \Pi_4 = \frac{\eta}{\rho \cdot d^2 \cdot n}. \tag{42}$$

It will be seen that the fact that in these nondimensional complexes, besides the selected fundamental quantities ρ, d, n, the other variables enter only one time, namely Q in Π_1, v in Π_2, g in Π_3, and η in Π_4. These variables will be called "nonfundamental", their roll in obtaining the required answer being illustrated by the following assertion.

Theorem. *Nondimensional complexes reduced through restricted dimensional analysis (Rayleigh's and Buckingham's methods) distributed directly or inversely proportional to the dependent variable (according to the distribution of nonfundamental quantities) and raised to the appropriate powers of the same quantities, yield a solutions matrix in conformity with the G.D.A. method. This leads to functional relations identical to those established through G.D.A.*

Applying this theorem to example (28) we obtain for the nondimensional complex (42) the power-product function

$$\Pi_1^1 = K \cdot \Pi_2^1 \cdot \Pi_3^1 \cdot \Pi_4^3, \tag{43}$$

in which the non-fundamental quantity Q from Π_1 has exponent $Z_4 = 1$, V from Π_2 has exponent $Z_5 = 1$, g from Π_3 has exponent Z_6, and η from Π_4 has exponent $Z_7 = 3$. Therefore we have to take account of the distribution of these nonfundamental quantities in connection with the dependent quantity Q, in accordance with relation (36).

In accordance with the enunciated theorem, relations (41) and (43) must be identical. Replacing nondimensional complexes (42) in (43) and simplifying yields

$$\frac{Q}{\rho \cdot d^4 \cdot n^2} = K \cdot \frac{V}{n \cdot d} \cdot \frac{g}{d \cdot n^2} \cdot \frac{\eta^3}{\rho^3 \cdot d^6 \cdot n^3}, \quad Q = K \frac{V \cdot g \cdot \eta^3}{n^4 \cdot d^4 \cdot \rho^2}, \tag{44}$$

as required.

Faculty of Science of Materials, "Transilvania" University , Brasov, Romania;
e-mail: csam@unitbv.ro

V.Al. SAVA

Energy decay for a weak solution of the non-Newtonian fluid equations with slowly varying external forces

1. Introduction

We are concerned with the asymptotic behavior of weak solutions to the system of equations governing the flow of a non-Newtonian fluid [10], [11]

$$\frac{\partial \mathbf{v}}{\partial t} + \mathbf{v} \cdot \nabla \mathbf{v} = -\nabla p + \nu \Delta \mathbf{v} + k \frac{\partial(\Delta \mathbf{v})}{\partial t} + f,$$

$$\nabla \cdot \mathbf{v} = 0, \quad \text{in } t > 0, \ \mathbf{x} \in I\!\!R^3, \tag{1}$$

$$\mathbf{v}(0, \mathbf{x}) = \mathbf{v}^0(\mathbf{x}), \quad \mathbf{x} \in I\!\!R^3, \tag{2}$$

where

$$\mathbf{v} = \mathbf{v}(\mathbf{x}, t) = (v_1(\mathbf{x}, t), v_2(\mathbf{x}, t), v_3(\mathbf{x}, t)) \quad \text{and} \quad p = p(\mathbf{x}, t)$$

denote unknown velocity vector and pressure at point $(\mathbf{x}, t) \in I\!\!R^3 \times (0, \infty)$, while

$$\mathbf{v}^0 = \mathbf{v}^0(\mathbf{x}) = (v_1^0(\mathbf{x}), v_2^0(\mathbf{x}), v_3^0(\mathbf{x})) \quad \text{and} \quad \mathbf{f} = \mathbf{f}(\mathbf{x}, t) = (f_1(\mathbf{x}, t), f_2(\mathbf{x}, t), f_3(\mathbf{x}, t))$$

are given initial velocity and external force. The material coefficients ν and k are constants and positive.

We shall discuss the energy decay problem of the weak solution to (1) and (2): $\|\mathbf{v}(t)\|_2^2 \to 0$ as $t \to \infty$. Detailed studies on the initial-boundary value problem for (1) have been given in [2] and [3].

In this paper, we first extend these results to more general external forces and establish nonuniform decay. For that purpose, we construct a weak solution that satisfies a generalized energy inequality (see Theorem 3). The Fourier splitting method ([4], [5]) combined with an argument due to Masuda [6] on the generalized energy inequality for the Navier-Stokes equations, yields nonuniform decay.

Throughout this paper, the following notations will be used. Let $\mathcal{L}^2(I\!\!R^3)$ and $\mathcal{H}_0^1(I\!\!R^3)$ denote the completions of $\mathbf{C}_0^\infty$ ($\mathcal{C}_0^\infty$ functions with divergence free) in the $L^2(I\!\!R^3)$-norm $\|\cdot\|_2$ and the Dirichlet (homogeneous H^1) norm $\|\nabla \cdot\|_2$, respectively. We denote

$$L^p(a, b; L^q) = \left\{ \mathbf{f} : (a, b) \times I\!\!R^3 \to I\!\!R^3; \ \|\mathbf{v}\|_{L^p(a,b;L^q)} = \left(\int_a^b \|\mathbf{f}(\tau)\|_p^q d\tau \right)^{1/q} < \infty \right\},$$

and $\mathcal{H}^1 = \mathcal{H}_0^1 \cap \mathcal{L}^2$. The symbol $< \cdot, \cdot >$ denotes the inner product in L^2. Moreover, we consider $\|v(t)\|_k^2 \equiv \|v(t)\|_2^2 + k\|\nabla v(t)\|_2^2$.

2. Results

We first give the definition of the weak solution.

298

Definition 1. For $\mathbf{v}^0 \in \mathcal{H}_0^1$ and $T > 0$, we call $\mathbf{v}$ a weak solution of the initial-value problem (1), (2) if and only if

(i) $\mathbf{v} \in L^\infty(0, T; \mathcal{L}^2) \cap L^2(0, T; \mathcal{H}_0^1)$;

(ii) For any $0 \leq s < t \leq T$, $\mathbf{v}$ satisfies

$$
\begin{aligned}
\int_s^t & \left[-\langle \mathbf{v}(\tau), \boldsymbol{\Phi}_\tau(\tau) \rangle + \nu \langle \nabla \mathbf{v}(\tau), \nabla \boldsymbol{\Phi}(\tau) \rangle \right. \\
& \left. - k \langle \nabla \mathbf{v}(\tau), (\nabla \boldsymbol{\Phi}(\tau))_\tau \rangle + \langle \mathbf{v}(\tau) \cdot \nabla \mathbf{v}(\tau), \boldsymbol{\Phi}(\tau) \rangle \right] d\tau \\
= & -\langle \mathbf{v}(t), \boldsymbol{\Phi}(t) \rangle + \langle \mathbf{v}(s), \boldsymbol{\Phi}(s) \rangle \\
& - k \langle \nabla \mathbf{v}(t), \nabla \boldsymbol{\Phi}(t) \rangle + k \langle \nabla \mathbf{v}(s), \nabla \boldsymbol{\Phi}(s) \rangle + \int_s^t \langle \mathbf{f}(\tau), \boldsymbol{\Phi}(\tau) \rangle d\tau,
\end{aligned} \tag{3}
$$

for all test functions $\boldsymbol{\Phi} \in C^1([0, T]; \mathcal{C}_0^\infty)$.

The external force $\mathbf{f}$ is assumed to satisfy the following condition.

A. For $\mathbf{x}_0 \in I\!\!R^3$, let $\rho = \rho_{x_0}(\mathbf{x}) = \mid \mathbf{x} - \mathbf{x}_0 \mid$. Suppose that for $0 < \gamma \leq 1$, $2 \leq p < 6/(1 + 2\gamma)$ and $\theta = 4p/(2p\gamma + 3p - 6)$ we have $\rho^\gamma \mathbf{f} \in L^{\theta'}(0, \infty; L^{p'})$, where p' and θ' are the conjugates of p and θ.

Remark 1. These conditions ensure that $\mathbf{f}$ is in the dual of the space to which the weak solution belongs.

Now we establish the existence of a weak solution that satisfies a certain energy inequality. To this end, we need some a priori estimates for the solution. First we recall the following lemma, due to T. Ogawa [7].

Lemma 2. *Let* $0 \leq \gamma \leq 1$ *and* $2 \leq p \leq 6/(1 + 2\gamma)$, *and suppose that* $\mathbf{f}$ *satisfies A. Then for* $\mathbf{v} \in L^\infty(0, T; \mathcal{L}^2) \cap L^2(0, T; \mathcal{H}_0^1)$ *and* $0 \leq s < t < \infty$ *we have*

$$
\int_s^t \mid \langle \mathbf{f}, \mathbf{v} \rangle d\tau \leq C \varepsilon^{(1-\lambda)(1-\gamma)}(t) \left(\int_s^t \|\nabla \mathbf{v}(\tau)\|_2^2 d\tau \right)^{1/\theta} \|\rho^\gamma(\mid \mathbf{x} - \mathbf{x}_0 \mid) \mathbf{f}\|_{L^{\theta', p'}}, \tag{4}
$$

where $\mathbf{x}_0 \in I\!\!R^3$, $\varepsilon(t) = \sup_{\tau < t} \|\mathbf{v}(\tau)\|_2^2$ *and* $\lambda = [3/(1 - \gamma)] \cdot (1/2 - 1/p)$.

Theorem 3. *Let* $0 \leq s \leq t < \infty$, *and suppose thet* $\mathbf{f}$ *satisfies assumption A above. Then for all* $\mathbf{v} \in L^\infty(0, T; \mathcal{L}^2) \cap L^2(0, T; \mathcal{H}_0^1)$ *satisfying the strong energy inequality*

$$
\|\mathbf{v}(t)\|_k^2 + 2\nu \int_s^t \|\nabla \mathbf{v}(\tau)\|_2^2 d\tau \leq \|\mathbf{v}(s)\|_k^2 + 2 \int_s^t \langle \mathbf{f}(\tau), \mathbf{v}(\tau) \rangle d\tau, \tag{5}
$$

for almost all $s > 0$ *and all* $t \geq s$, *we have the following a priori estimate on* $\mathbf{v}$:

$$
\sup_{0 \leq \tau \leq t} \|\mathbf{v}(\tau)\|_k < c_1, \tag{6}
$$

$$
\int_s^t \|\nabla \mathbf{v}(\tau)\|_2^2 d\tau \leq c_1, \tag{7}
$$

where c_1 *is a constant that only depends on* $\|\mathbf{v}^0\|_2$, $\|\nabla \mathbf{v}^0\|_2$, $\|\rho^\gamma \mathbf{f}\|_{L^{\theta', p'}}$, ν *and* k.

Proof. Note that $2 \leq \theta$ and $(1 - \lambda)(1 - \gamma) \leq 1$ in Lemma 2. Hence the a priori estimates (6) and (7) can be obtained by applying Lemma 2 to (5).

Remark 2. Interpolating (6) and (7) for $1 \leq q \leq 6$ and $3/q + 2/\sigma = 3/2$, we have $\|\mathbf{v}\|_{L^{\sigma,q}} \leq c_2$.

Theorem 4. *Let $\mathbf{v}^0 \in \mathcal{H}_0^1$, and suppose that $\mathbf{f}$ satisfies assumption A above. Then there exists a weak solution $\mathbf{v}$ to problem (1), (2) in $L^\infty(0, \infty; \mathcal{L}^2) \cap L^2(0, \infty, \mathcal{H}_0^1)$ satisfying the strong energy inequality (5).*

Moreover, for $E \in C^1(R_+, R)$ with $E(t) \geq 0$ and $\psi \in C^1(\mathbb{R}; C^1 \cap L^2)$, the weak solution satisfies

$$E(t)\|\psi(t) * \mathbf{v}(t)\|_k^2 \leq E(s)\|\psi(s) * \mathbf{v}(s)\|_k^2 + \int_s^t E_\tau(\tau)\|\Psi(\tau) * \mathbf{v}(\tau)\|_k^2 d\tau$$

$$+ 2\int_s^t E(\tau) \, | \, \langle \psi_\tau(\tau) * \mathbf{v}(\tau), \psi(\tau) * \mathbf{v}(\tau) \rangle - \nu\|\nabla\psi(\tau) * \mathbf{v}(\tau)\|_2^2 \, | \, d\tau$$

$$+ 2\int_s^t (|\, k\langle \nabla\psi_\tau(\tau) * \mathbf{v}(\tau), \nabla\psi(\tau) * \mathbf{v}(\tau)\rangle| + E(\tau) \,|\, \langle \mathbf{f}(\tau), \psi(\tau) * \psi(\tau) * \mathbf{v}(\tau)\rangle\,|)d\tau$$

$$+ 2\int_s^t E(\tau) \, | \, \langle \mathbf{v}(\tau) \cdot \nabla\mathbf{v}(\tau), \psi(\tau) * \psi(\tau) * \mathbf{v}(\tau) \rangle \, | \, d\tau \tag{8}$$

for $0 \leq s < t < \infty$. (Here $(\mathbf{v} * \varphi)(\mathbf{x}) = \int_{-\infty}^\infty \mathbf{v}(\mathbf{x} - \mathbf{y})\varphi(\mathbf{y})d\mathbf{y}$ is the convolution.)

Theorem 5. *Let $E \in C^1(\mathbb{R}_+, \mathbb{R})$ and $\tilde{\psi} \in C^1(0, \infty; L^\infty)$ such that $(1 - \tilde{\psi}^2) \in L^\infty(0, \infty; L^2)$ and $\nabla\mathcal{F}^{-1}(1 - \tilde{\psi}^2) \in L^\infty(0, \infty; L^2)$. Then the weak solution constructed in Theorem 4 also satisfies*

$$E(t)(\|\tilde{\psi}(t)\hat{\mathbf{v}}(t)\|^2 + k\|\xi\tilde{\psi}(t)\hat{\mathbf{v}}(t)\|_2^2)$$

$$\leq E(s)(\|\tilde{\psi}(s)\hat{\mathbf{v}}\|^2 + k\|\xi\tilde{\psi}(s)\hat{\mathbf{v}}(s)\|_2^2) + \int_s^t E_\tau(\|\tilde{\psi}(\tau)\hat{\mathbf{v}}(\tau)\|^2 + k\|\xi\tilde{\psi}(\tau)\hat{\mathbf{v}}(\tau)\|_2^2)d\tau$$

$$+ 2\int_s^t E(\tau) \, |\langle\tilde{\psi}_\tau(\tau)\hat{\mathbf{v}}(\tau), \tilde{\psi}(\tau)\hat{\mathbf{v}}(\tau)\rangle + k\langle\xi\tilde{\psi}_\tau(\tau)\hat{\mathbf{v}}(\tau), \xi\psi(\tau)\hat{\mathbf{v}}(\tau)\rangle - \nu\|\xi\tilde{\psi}(\tau)\hat{\mathbf{v}}(\tau)\|_2^2 \, | \, d\tau$$

$$+ 2\int_s^t E(\tau)[|\langle\mathbf{v}\cdot\nabla\mathbf{v}(\tau), (1 - \tilde{\psi}^2(\tau))\hat{\mathbf{v}}(\tau)\rangle| + |\langle\hat{\mathbf{f}}(\tau), \tilde{\psi}^2(\tau)\hat{\mathbf{v}}(\tau)\rangle|]d\tau \tag{9}$$

for almost all $s > 0$ and all $t \geq 0$. (A superposed hat denotes the Fourier transform $\hat{f} = \mathcal{F}(f) = \int f(\mathbf{x})e^{-i\langle\xi\cdot\mathbf{x}\rangle}d\mathbf{x}$.) In particular, the weak solution satisfies

$$E(t)\|\mathbf{v}(t)\|_k^2 \leq E(s)\|\mathbf{v}(s)\|_k^2 + \int_s^t E_\tau(\tau)\|\mathbf{v}(\tau)\|_k^2 d\tau$$

$$- 2\int_s^t E(\tau)(\|\nabla\mathbf{v}(\tau)\|_2^2 - |\langle\mathbf{f}(\tau), \mathbf{v}(\tau)\rangle|)d\tau. \tag{10}$$

Corollary 6. *For a weak solution* $\mathbf{v}$ *and* $\varphi \in L^2(\mathbb{R}^3)$ *such that* $\nabla \mathcal{F}^{-1}(\varphi) \in L^2(\mathbb{R}^3)$, *we have*

$$\|\check{\varphi} * \mathbf{v}(t)\|_k^2 \leq \|e^{\frac{\nu\Delta}{1-k\Delta}(t-s)}\check{\varphi} * \mathbf{v}(s)\|_k^2$$
$$+ 2\int_s^t \left[| \langle \mathbf{v} \cdot \nabla\mathbf{v}, e^{\frac{2\nu\Delta}{1-k\Delta}(t-\tau)}\check{\varphi}^2 * \mathbf{v}(\tau)\rangle | + | \langle \mathbf{f}(\tau), e^{\frac{2\nu\Delta}{1-k\Delta}(t-\tau)}\check{\varphi}^2 * \mathbf{v}(\tau)\rangle | \right] d\tau, \quad (11)$$

where $\check{\varphi} = \mathcal{F}^{-1}(\varphi)$ *is the inverse Fourier transform of* φ.

Theorem 7. *Let* $\mathbf{v}^0 \in \mathcal{L}^2$, $\nu > \kappa$ *and* $\mathbf{f}$ *satisfies A. Then the weak solution* $\mathbf{v}$ *of* *(1), (2) constructed in Theorem 4 satisfies the energy decay:*

$$\|v(t)\|_k \to 0 \ \ as \ t \to \infty.$$

Proof. We first show the decay of the low frequency part of the energy. Taking $\varphi(\xi) = e^{-\nu|\xi|^2/(1+\kappa|\xi|^2)}$ we have, by Corollary 6 and Plancherel's identity,

$$\|\varphi^1 v(t)\|_2^2 + k\|\varphi\xi\hat{v}\|_2^2 \leq \|e^{\frac{-\nu|\xi|^2}{1+k|\xi|^2}(t-s)}\varphi(s)\hat{\mathbf{v}}(s)\|_2^2 + k\|\xi e^{\frac{-\nu|\xi|^2}{1+k|\xi|^2}(t-s)}\varphi(s)\hat{\mathbf{v}}(s)\|_2^2$$
$$+ 2\int_s^t | \langle \mathbf{v} \cdot \nabla\mathbf{v}, e^{\frac{2\nu\Delta}{1-k\Delta}(t-\tau)}\check{\varphi}^2(\tau) * \mathbf{v}(\tau)\rangle | \, d\tau$$
$$+ 2\int_s^t | \langle \mathbf{f}(\tau), e^{\frac{2\nu\Delta}{1-k\Delta}(t-\tau)}\check{\varphi}^2(\tau) * \mathbf{v}(\tau)\rangle | \, d\tau. \quad (12)$$

Since $\check{\varphi}$ is a rapidly decreasing function, from the Hausdorff-Young, Hölder and Sobolev inequalities it follows that

$$| \langle \check{\varphi}^2 * \mathbf{v} \cdot \nabla\mathbf{v}, e^{\frac{2\nu\Delta}{1-k\Delta}(t-\tau)}\mathbf{v}\rangle | \leq C(\varphi)\|\mathbf{v}(\tau)\|_2\|\nabla\mathbf{v}(\tau)\|_2. \quad (13)$$

Hence, by (12), (13) and Lemma 2,

$$\|\varphi\hat{\mathbf{v}}(t)\|_k^2 \leq \|e^{\frac{-\nu|\xi|^2}{1+k|\xi|^2}(t-s)}\varphi(s)\hat{\mathbf{v}}(s)\|_2^2 + k\|\varphi e^{\frac{-\nu|\xi|^2}{1+k|\xi|^2}(t-s)}\varphi(s)\hat{\mathbf{v}}(s)\|_2^2$$
$$+ 2\int_s^t C(\varphi)\varepsilon(\tau)\|\nabla\mathbf{v}\|_2^2 d\tau + C\varepsilon^{(1-\lambda)(1-\gamma)}(t)\left(\int_s^t \|\rho^\gamma\mathbf{f}\|_{p'}^{\theta'} d\tau\right)^{1/\theta'}\left(\int_s^t \|\nabla\mathbf{v}\|_2^2 d\tau\right)^{1/\theta} \quad (14)$$

Since

$$\varlimsup_{t\to\infty}\|e^{\frac{-\nu|\xi|^2}{1+k|\xi|^2}(t-s)}\varphi(s)\hat{\mathbf{v}}(s)\|_k^2 = 0, \quad \varlimsup_{t\to\infty}\|\xi e^{\frac{-\nu|\xi|^2}{1+k|\xi|^2}(t-s)}\varphi(s)\hat{\mathbf{v}}(s)\|_2^2 = 0,$$

we have, by taking a limit $t \to \infty$ in (14), that

$$\varlimsup_{t\to\infty}\left(\|\varphi(t)\hat{\mathbf{v}}\|_2^2 + k\|\xi\varphi\hat{v}(t)\|_2^2\right) \leq C \sup_t \mathcal{E}(t)\int_s^\infty \|\nabla\mathbf{v}(\tau)\|_2^2 d\tau$$

$$+ C \sup_{t} \mathcal{E}^{(1-\lambda)(1-\gamma)}(t) \left(\int_s^t \|\rho^\gamma \mathbf{f}\|_{p'}^{\theta'} d\tau \right)^{1/\theta'} \left(\int_s^\infty \|\nabla \mathbf{v}(\tau)\|_2^2 d\tau \right)^{1/\theta}. \quad (15)$$

By Theorem 3, the right-hand side of (15) converges 0 as $s \to \infty$.

Next, we shall estimate the high frequency part of the energy. Choose $\tilde{\psi} = 1 - \varphi$, $\varphi = e^{-\nu|\xi|^2/(1+\kappa|\xi|^2)}$ and let $\chi(t) = \{\xi \in I\!\!R^3; \ |\ \xi \ | \leq G(t)\}$. Then (9) yields

$$E(t)(\|(1-\varphi)\hat{\mathbf{v}}\|_2^2 + k\|\xi(1-\varphi)\hat{\mathbf{v}}\|_2^2) \leq E(s) \left(\|(1-\varphi)\hat{\mathbf{v}}(s)\|_2^2 \right.$$

$$+ k\|\xi(1-\varphi)\hat{\mathbf{v}}(s)\|_2^2) + \int_s^t E'(\tau)(1 + kG^2(\tau)) \int_{\chi(\tau)} |\ (1-\varphi)\hat{\mathbf{v}}(\tau)\ |^2 \ d\xi d\tau$$

$$+ \int_s^t \left[(E'(\tau) - 2\nu E(\tau)G^2(\tau)) \int_{I\!\!R^3 - \chi(\tau)} |\ (1-\varphi)\hat{\mathbf{v}}(\tau)\ |^2 \ d\xi \right] d\tau$$

$$+ 2\int_s^t \left[kE'(\tau) \int_{I\!\!R^3 - \chi(\tau)} |\ (1-\varphi)\hat{\mathbf{v}}(\tau)\ |^2 \ d\xi - \nu E(\tau) \int_{\chi(\tau)} |\ (1-\varphi)\hat{v}(\tau)\ |^2 \ d\xi \right] d\tau$$

$$+ 2\int_s^t E(\tau) \left[|\ \langle \mathbf{v}\cdot\widehat{\nabla}\mathbf{v}, (1-(1-\varphi)^2)\hat{\mathbf{v}}(\tau)\rangle \ | + |\ \langle \hat{\mathbf{f}}, (1-\varphi)^2\hat{\mathbf{v}}(\tau)\rangle \ | \right] d\tau. \quad (16)$$

Since $\mathcal{F}^{-1}[1 - (1-\varphi)^2] \equiv \psi$ is a rapidly decreasing function, we can estimate

$$\int_s^t E(\tau) |\ \langle \mathbf{v}\cdot\nabla\mathbf{v}, \psi * \mathbf{v}(\tau)\rangle \ | \ d\tau \leq C(\varphi)\mathcal{E}(t)\int_s^t E(\tau)\|\nabla\mathbf{v}(\tau)\|_2^2 d\tau. \quad (17)$$

If we choose $E(t) = (1+t)^\alpha$, $G^2(t) = \alpha/2\nu(1+t)$ in (16) and (17), then

$$\|(1-\varphi)\hat{\mathbf{v}}(t)\|_2^2 + k\|\xi(1-\varphi)\hat{\mathbf{v}}(t)\|_2^2$$

$$\leq \left(\frac{1+s}{1+t} \right)^\alpha \left(\|(1-\varphi)\hat{\mathbf{v}}(s)\|_2^2 + k\|\xi(1-\varphi)\hat{\mathbf{v}}(s)\|_2^2 \right)$$

$$+ \frac{\alpha}{2\nu(1+t)^\alpha} \int_s^t (1+\tau)^{\alpha-2}[k\alpha + 2\nu(1+\tau)] \int_{\chi(\tau)} |\ (1-\varphi)\hat{\mathbf{v}}(\tau)\ |^2 \ d\xi d\tau$$

$$+ C(\varphi)\frac{1}{(1+t)^\alpha}\mathcal{E}(t)\int_s^t (1+\tau)^\alpha\|\nabla\mathbf{v}(\tau)\|_2^2 d\tau$$

$$+ C\sup_{t,\xi} |\ (1-\varphi)^2\ | \frac{1}{(1+t)^\alpha}\int_s^t (1+\tau)^\alpha |\ \langle \hat{\mathbf{f}}(\tau), (1-\varphi)^2\hat{\mathbf{v}}(\tau)\rangle \ | \ d\tau. \quad (18)$$

Observing that $|\ 1-\varphi\ | \leq \nu\xi^2$ if $|\ \xi\ | < \sqrt{2/\nu}$, we have

$$\overline{\lim}_{t\to\infty}(\|(1-\varphi)\hat{\mathbf{v}}(t)\|_2^2 + k\|\xi(1-\varphi)\hat{\mathbf{v}}(t)\|_2^2) \leq C\sup_{t}\mathcal{E}(t)\int_s^\infty |\ \nabla\mathbf{v}(\tau)\|_2^2 d\tau$$

$$+ C \sup_t \mathcal{E}^{(1-\lambda)(1-\gamma)}(t) \cdot \|\rho^\gamma \mathbf{f}\|_{L^{\theta',p'}} \left(\int_s^\infty \|\nabla \mathbf{v}(\tau)\|_2 d\tau \right)^{1/\theta}, \qquad (19)$$

where γ, p and θ are defined in assumption A. As $s \to \infty$, the terms on the right-hand side of (19) tend to 0. This proves the theorem.

References

1. A.P. Oskolkov, On some model nonstationary systems in the theory of non-Newtonian fluids, in *Boundary value problems of mathematical physics, IX, Proc. Steklov Inst. Math.* **127**, Providence, RI, Amer. Math. Soc. 1977, 37–66.

2. A.P. Oskolkov, The uniqueness and solvability in the large of boundary value problems for the equations of motion of aqueous solutions of polymers, *Zap. Nauchn. Sem. LOMI* **38**, 98–136.

3. A.P. Oskolkov, A certain nonstationary quasilinear system with a small parameter that regularizes the system of N.S. equations, in *Probl. Mat. Anal.*, no. 4, Izdat. Leningrad. Univ., 1973, 78–87.

4. M.E. Schonbek, Large time behaviour of solutions to the Navier-Stokes equations, *Comm. Partial Diff. Equat.* **11**, 733–763.

5. M.E. Schonbek, L^2-decay for weak solutions of the Navier-Stokes equations, *Arch. Rational Mech. Anal.* **88**, 209–222.

6. K. Masuda, Weak solutions of the Navier-Stokes equations, *Tohoku Math. J.* **36**, 623–646.

7. T. Ogawa, Energy decay for a weak solution of the Navier-Stokes equation with slowly varying external forces, *J. Functional Anal.* **144**, 325–358.

8. J. Leroy, Sur le mouvement d'un liquide visqeux emplissant l'espace, *Acta Math.* **63**, 193–248.

9. L. Caffareli, R. Kohn and L. Nirenberg, Partial regularity of suitable weak solutions of the Navier-Stokes equations, *Comm. Pure Appl. Math.* **35**, 971–831.

10. W.L. Wilkinson, *Non-Newtonian fluids. Fluid mechanics, mixing and heat transfer*, Pergamon Press, New York, 1960.

11. K. Walters, Non-Newtonian effects in some elastico-viscous liquids whose behaviour at small rates of shear is characterized by a general linear equation of state, *Quart. J. Mech. Appl. Math.* **15**.

Department of Mathematics, Technical University "Gh.Asachi", Iasi-6600, Romania

V.Al. SAVA

An integral representation for the solution of 2D nonstationary flow of micropolar fluids

1. Introduction

It is desirable for some aspects of hydrodynamical problems and for the proof of existence of solutions to have integral representations of solutions. With this main objective in view, we proceed to examine the system of coupled differential equations governing the motion of slow, nonstationary flow of micropolar fluids.

The integral representation for the solution of equations that govern steady flow of incompressible micropolar fluids was considered by Ramkissoon and Majumdar [1,2] and Dragos and Homentcovschi [3].

The plan of this paper is straightforward. We state the basic equations governing the flow, and, using the method given in [5], we derive a reciprocal theorem. The fundamental singular solution is then derived in two-dimensional regions and the integral representation of the solution of the motion is given.

2. Basic equations and the reciprocal theorem

For slow flow of incompressible micropolar fluids, we have, respectively, the following equations of motion, linear constitutive law, and continuity equation [4]:

$$t_{ij,i} + \rho f_j = \rho \frac{\partial v_j}{\partial t},$$
$$m_{ij,i} + \varepsilon_{jik} t_{ik} + \rho l_j = \rho a^2 \frac{\partial w_j}{\partial t}; \tag{1}$$

$$t_{ij} = -p\delta_{ij} + (2\mu + \kappa)d_{ij} + \kappa \varepsilon_{ije}(\omega_e - w_e),$$
$$m_{ij} = \alpha w_{e,e}\delta_{ij} + \beta w_{i,j} + \gamma w_{j,i}; \tag{2}$$

$$v_{i,i} = 0, \tag{3}$$

where t_{ij}, m_{ij}, v_i, w_i, f_i, l_i, denote, respectively, the components of the stress tensor, couple stress, velocity vector, microrotation vector, body force, and body couple; p is the pressure, ε_{ijk} is the alternating tensor, δ_{ij} is the Kronecker delta; α, β, γ, μ, κ are constants characteristic of particular fluid under consideration; ρ is the density and a^2 is the microinertia coefficient. Moreover,

$$\omega_e = \tfrac{1}{2}\varepsilon_{ekr}(v_{k,r} - v_{r,k}), \quad d_{ke} = \tfrac{1}{2}(v_{k,e} + v_{e,k}). \tag{4}$$

The Clausius-Duhem inequality implies [3] that κ, μ, γ and $3\alpha + \beta + \gamma$ are each nonnegative.

Equations (1)–(3) reduce to the following system of coupled vector equations in the case of incompressible slow flows:

$$(\nu + k)\Delta \mathbf{v} + k\nabla \times \mathbf{w} - \frac{1}{\rho}\nabla p + \mathbf{f} = \frac{\partial \mathbf{v}}{\partial t},$$

$$(\alpha + \beta)\nabla\nabla\mathbf{w} + \gamma\Delta\mathbf{w} + k\nabla \times \mathbf{v} - 2k\mathbf{w} + \rho\mathbf{l} = \rho a^2 \frac{\partial \mathbf{w}}{\partial t}, \tag{5}$$

$$\nabla \cdot \mathbf{v} = 0.$$

To (5) we append the initial conditions

$$\mathbf{v}(\mathbf{x},0) = \mathbf{v}^0(\mathbf{x}), \quad \mathbf{w}(\mathbf{x},0) = \mathbf{w}^0(\mathbf{x}), \tag{6}$$

where $\mathbf{v}^0$ is solenoidal.

It is easy to obtain the following assertion.

Theorem 1. *The functions* v_i, w_i, t_{ij}, m_{ij}, $(i,j = 1,2,3)$ *satisfy (1) and the initial conditions (6) if and only if*

$$g * t_{ij,i} + F_j = \rho v_j,$$
$$g * (m_{ij,i} - \varepsilon_{jer}t_{er}) + L_j = \rho a^2 w_j \quad \text{on } \Omega \times [0,\infty), \tag{7}$$

where

$$g(t) = 1, \quad F_i = \rho(g * f_i + v_i^0), \quad L_i = \rho(g * l_i + a^2 w_i^0), \quad u * v = \int\limits_0^t u(\mathbf{x}, t - \tau)v(\mathbf{x}, \tau)d\tau,$$

and Ω *is region of space occupied by the fluid.*

We shall now give the following reciprocal theorem.

Theorem 2. *Let* $(\mathbf{v}^{(\alpha)}, \mathbf{w}^{(\alpha)}, p^{(\alpha)}, t_{ij}^{(\alpha)}, m_{ij}^{(\alpha)}, F^{(\alpha)}, L^{(\alpha)})$, $\alpha = 1,2$ *represent any two motions of same fluid which conform to (1)–(3) and (6). Let* Σ *be a closed surface bounding any fluid volume* Ω *and* $\mathbf{v}^{(\alpha)}, \mathbf{w}^{(\alpha)} \in C^1([0,\infty); \Omega \cup \Sigma)$. *Then the following reciprocal relation holds:*

$$\int\limits_\Omega (F_j^{(1)} * v_j^{(2)} + L_j^{(1)} * w_j^2)d\mathbf{x} + \int\limits_\Sigma g * (t_j^{(1)} * v_j^{(2)} + m_j^{(1)} * w_j^{(2)})d\sigma$$

$$= \int\limits_\Omega (F_j^{(2)} * v_j^{(1)} + L_j^{(2)} * w_j^{(1)})d\mathbf{x} + \int\limits_\Sigma g * (t_j^{(2)} * v_j^{(1)} + m_j^{(2)} * w_j^{(1)})d\sigma, \tag{8}$$

where

$$t_j^{(\alpha)} = t_{ij}^{(\alpha)}n_i, \quad m_j^{(\alpha)} = m_{ij}^{(\alpha)}n_i, \quad \alpha = 1,2,$$

are, respectively, the surface tractions and surface moments acting on Σ *and* n_i *the component of exterior unit-normal to* Σ.

Proof. Using the relations (4), (7) and the divergence theorem, we get

$$\int_\Omega g * (t_{ij}^{(\alpha)} * v_{j,\alpha}^{(\beta)} + m_{ij}^{(\alpha)} * w_{j,i}^{(\beta)} - \varepsilon_{jer} t_{er}^{(\alpha)} * w_j^{(\beta)}) d\mathbf{x}$$

$$= \int_\Omega (F_j^{(\alpha)} * v_j^{(\beta)} + L_j^{(\alpha)} * w_j^{(\beta)}) d\mathbf{x} + \int_\Sigma g * (t_j^{(\alpha)} * v_j^{(\beta)} + m_j^{(\alpha)} * w_j^{(\beta)}) d\sigma$$

$$+ \rho \int_\Omega (v_j^{(\alpha)} * v_j^{(\beta)} + a^2 w_j^{(\alpha)} * w_j^{(\beta)}) d\mathbf{x}.$$

But

$$g * (t_{ij}^{(1)} * v_{j,i}^{(2)} + m_{ij}^{(1)} * w_{j,i}^{(2)} - \varepsilon_{jer} t_{er}^{(1)} * w_j^{(2)})$$

$$= g * (t_{ij}^{(2)} * v_{j,i}^{(1)} + m_{ij}^{(2)} * w_{j,i}^{(1)} - \varepsilon_{jer} t_{er}^{(2)} * w_j^{(1)}),$$

and the reciprocal relation (8) follows.

3. Fundamental solutions

Let us consider the 2D slow flows. The equations (5) reduce to

$$\frac{\partial v_\alpha}{\partial t} - (\nu + k) v_{\alpha,\beta\beta} - k \varepsilon_{\alpha\beta} \varphi_{,\beta} + \frac{1}{\rho} p_{,\alpha} = f_\alpha, \quad \alpha = 1, 2,$$

$$a^2 \frac{\partial \varphi}{\partial t} - \gamma \varphi_{,\beta\beta} + k \varepsilon_{\alpha\beta} v_{\alpha,\beta} + 2k\varphi = l, \qquad (9)$$

$$v_{\alpha,\alpha} = 0,$$

where $\varepsilon_{\alpha\beta}$ is the alternating tensor and φ is the microrotation.

To obtain the fundamental solution, we first seek the solution

$$\{\mathbf{v}^j(\mathbf{x}, t), \varphi^j(\mathbf{x}, t)\} = (v_1^j(\mathbf{x}, t), v_2^j(\mathbf{x}, t), \varphi^j(\mathbf{x}, t)), \quad j = 1, 2, 3,$$

of

$$\frac{\partial v_\alpha^j}{\partial t} - (\nu + k) v_{\alpha,\beta\beta}^j - k \varepsilon_{\alpha\beta} \varphi_{,\beta}^j + \frac{1}{\rho} p_{,\alpha}^j = e_\alpha^j \delta(\mathbf{x}, t), \quad \alpha = 1, 2,$$

$$a^2 \frac{\partial \varphi^j}{\partial t} - \gamma \varphi_{,\beta\beta}^j + k \varepsilon_{\alpha\beta} v_{\alpha,\beta}^j - 2k\varphi^j = e_3^j \delta(\mathbf{x}, t), \qquad (10)$$

$$v_{\alpha,\alpha}^j = 0,$$

where $\delta(\mathbf{x}, t)$ stand for the Dirac measure at the origin and $\mathbf{e}^j = (e_1^j, e_2^j, e_3^j)$, $j = 1, 2, 3$, are the unit vectors whose j-th components are equal to 1.

If we apply to (10) the Fourier transform with respect to variables x_α, $\alpha = 1, 2$, and then the Laplace transform with respect to the variable t, that is,

$$(\tilde{v}_\alpha, \tilde{\varphi}, \tilde{p}) = \int\!\!\!\int_{-\infty}^{\infty} (v_\alpha, \varphi, p) e^{iy \cdot x} dx_1 dx_2, \quad (\hat{\tilde{v}}_\alpha, \hat{\tilde{\varphi}}, \hat{\tilde{p}}) = \int_0^{\infty} (\tilde{v}_\alpha, \tilde{\varphi}, \tilde{p}) e^{-ist} dt,$$

we obtain (after some calculations)

$$\hat{\tilde{v}}_\alpha^\beta(\mathbf{y}, s) = \frac{1}{s + (\nu + k) \mid y \mid^2} \left[\delta_{\alpha\beta} + k^2 e_{\alpha\gamma} e_{\beta\theta} \frac{y_\gamma y_\theta}{Q(s, \mathbf{y})} - \frac{y_\alpha y_\beta}{\mid \mathbf{y} \mid^2} \right], \qquad \alpha = 1, 2,$$

$$\hat{\tilde{v}}_\alpha^3(\mathbf{y}, s) = kie_{\beta\alpha} \frac{y_\beta}{Q(s, \mathbf{y})}, \qquad \hat{\tilde{p}}^\alpha(\mathbf{y}, s) = i \frac{y_\alpha}{\mid \mathbf{y} \mid^2}, \qquad \alpha, \beta = 1, 2, \tag{11}$$

$$\hat{\tilde{\varphi}}^\alpha(\mathbf{y}, s) = kie_{\alpha\beta} \frac{y_\beta}{Q(s, \mathbf{y})}, \qquad \hat{\tilde{\varphi}}^3 = \frac{s + (\nu + k) \mid \mathbf{y} \mid^2}{Q(s, \mathbf{y})}, \qquad \alpha = 1, 2, \quad \hat{\tilde{p}}^3(\mathbf{y}, s) = 0,$$

where

$$\begin{aligned}
Q(s, \mathbf{y}) =& a^2 s^2 + s[a^2(\nu + k) \mid \mathbf{y} \mid^2 + \gamma \mid \mathbf{y} \mid^2 + 2k] \\
&+ k(2\nu + k) \mid \mathbf{y} \mid^2 + \gamma(\nu + k) \mid \mathbf{y} \mid^4 .
\end{aligned} \tag{12}$$

To determine the properties of roots of the quadratic expression $Q(s, \mathbf{y})$ in (12), it is necessary to specify a value for a^2, the microinertia coefficient. At present, we have no physical basis on which to arbitrarily assign a value to a^2. However, consider the limiting case in which

$$\varphi = \tfrac{1}{2} \varepsilon_{\alpha\beta} v_{\beta,\alpha}. \tag{13}$$

Equation (13) implies that the spherical particles present in the fluid rotate at an angular velocity equivalent to the regional angular velocity of the flowing fluid. This is, admittedly, a stringent restriction on the microrotation, but it does represent a physical possibility and will enable us to calculate a value for a^2 at least as realistic as one that might arbitrarily be selected.

Substituting the expression for φ in (13) into (9) yields, after simplification,

$$a^2 = \frac{2\gamma}{2\mu + k}. \tag{14}$$

Introducing the value for a^2 given by (14) into the quadratic expression $Q(s, \mathbf{y})$ and taking inverse Fourier and Laplace transforms of (11), we obtain

$$v_\alpha^\beta(\mathbf{x}, t) = \frac{1}{2\pi} \left[H\delta_{\alpha\beta} + e_{\alpha\gamma} e_{\beta\theta} \frac{\partial^2 \Phi}{\partial x_\gamma \partial x_\theta} - \frac{\partial^2 \Psi}{\partial x_\alpha \partial x_\beta} \right], \qquad \alpha, \beta = 1, 2,$$

$$v_\alpha^3(\mathbf{x}, t) = -\varphi^\alpha(\mathbf{x}, t) = \frac{2\nu + k}{2\pi\gamma} e_{\beta\alpha} \frac{\partial M}{\partial x_\beta}, \qquad p^\alpha(\mathbf{x}, t) = -\frac{\rho}{2\pi} \frac{\partial}{\partial x_\alpha} (\ln \frac{1}{r}) \delta(t), \tag{15}$$

$$\varphi^3(\mathbf{x}, t) = \frac{1}{2\pi\gamma} \left(G + \frac{(2\nu + k)}{\gamma} M \right), \qquad p^3(\mathbf{x}, t) = 0,$$

where

$$\Phi(r, t) = H(r, t) + M(r, t) + N(r, t), \qquad \Psi(r, t) = \ln \frac{1}{r} - I + \int_0^\infty \frac{Y_0(ru) e^{-(\nu + k)u^2 t}}{u} du,$$

$$H(r,t) = \frac{r}{2t\sqrt{\nu+k}}e^{-r^2/(4t(\nu+k))}, \quad G(r,t) = \frac{r}{t\sqrt{2(2\nu+k)}}e^{-r^2/(2t(2\nu+k))},$$

$$M(r,t) = \left(e^{-k(2\nu+k)t/\gamma} - 1\right)\int_0^\infty \frac{uY_0(ru)\,e^{-(2\nu+k)u^2t/2}}{u^2 - 2(2\nu+k)/\gamma}du,$$

$$N(r,t) = \int_0^\infty \frac{Y_0(ru)\,e^{-(2\nu+k)u^2t/2}}{u}du,$$

$$-I = PF\int_0^\infty \frac{J_0(u)}{u}du = \int_0^1 \frac{J_0(u)-1}{u}du + \int_1^\infty \frac{J_0(u)}{u}du,$$

$J_0(u)$ and $Y_0(u)$ denote the Bessel functions of zeroth order and first and second kind, respectively, and $r = \sqrt{x_1^2 + x_2^2}$.

The fundamental solution is

$$v_\alpha = v_\alpha^\beta f_\beta + v_\alpha^3 l, \quad \omega = \omega^\alpha f_\alpha + \omega^3 l, \quad \alpha = 1,2, \ \beta = 1,2.$$

4. Integral representation

With the aid of reciprocal relation (8), we shall now derive the integral representation of the basic solutions.

For the 2D flows, the reciprocal relation (8) reduces to

$$\iint_D (F_\alpha^{(1)} * v_\alpha^{(2)} + L^{(1)} * \varphi^2)d\mathbf{x} + \oint_\Gamma g * (t_\alpha^{(1)} * v_\alpha^{(2)} + m^{(1)} * \varphi^2)ds$$

$$= \iint_D (F_\alpha^{(2)} * v_\alpha^{(1)} + L^{(2)} * \varphi^1)d\mathbf{x} + \oint_\Gamma g * (t_\alpha^{(2)} * v_\alpha^{(1)} + m^{(2)} * \varphi^1)ds, \quad (16)$$

where D is a region of (x_1, x_2) plane, Γ its boundary and $L = \rho(g * l + a^2\varphi^0)$.

Assuming in (17) that $F_\alpha^{(1)} = \delta_{\alpha\beta}\delta(\mathbf{x} - \mathbf{y}, t - \tau)$, $L^{(1)} = 0$, we obtain

$$v_\alpha(\mathbf{y},t) = \iint_D [F_\beta(\mathbf{x},t) * v_\alpha^\beta(\mathbf{x} - \mathbf{y},t) + L(\mathbf{x},t) * \varphi^\alpha(\mathbf{x} - \mathbf{y},t)]d\mathbf{x}$$

$$+ \oint_\Gamma g * (t_\beta^\alpha(\mathbf{x},t) * v_\alpha^\beta(\mathbf{x} - \mathbf{y},t) + m(\mathbf{x},t) * \varphi^\alpha(\mathbf{x} - \mathbf{y},t)ds_x$$

$$- \oint_\Gamma g * (t_\beta^\alpha(\mathbf{x} - \mathbf{y},t) * v_\beta(\mathbf{x},t) + m^\alpha(\mathbf{x} - \mathbf{y},t) * \varphi(\mathbf{x},t))ds_x, \quad (17)$$

where

$$t_\beta^\alpha(\mathbf{x} - \mathbf{y}, t) = [-p^\alpha(\mathbf{x} - \mathbf{y}, t)\delta_{\beta\gamma} + (2\mu + \kappa)v_{\beta,\gamma}^\alpha(\mathbf{x} - \mathbf{y}, t)$$
$$+ \mu(v_{\gamma,\beta}^\alpha(\mathbf{x} - \mathbf{y}, t) - v_{\beta,\gamma}(\mathbf{x} - \mathbf{y}, t)) + \kappa\varepsilon_{\beta\gamma}\varphi^\alpha(\mathbf{x} - \mathbf{y}, t)]n_\gamma, \quad (18)$$
$$m^\alpha(\mathbf{x} - \mathbf{y}, t) = \gamma\varphi_{,\beta}^\alpha(\mathbf{x} - \mathbf{y}, t)n_\beta.$$

Letting now $F_\alpha^{(1)} = 0$ and $L^{(1)} = \delta(\mathbf{x} - \mathbf{y}, t - \tau)$, we find that

$$\varphi(\mathbf{y}, t) = \iint_D F_\alpha(\mathbf{x}, t) * v_\alpha^3((\mathbf{x} - \mathbf{y}, t) + L(\mathbf{x}, t) * \varphi^3(\mathbf{x} - \mathbf{y}, t)]dx$$

$$+ \oint_\Gamma g * (t_\alpha(\mathbf{x}, t) * v_\alpha^3(\mathbf{x} - \mathbf{y}, t) + m(x, t) * \varphi^3(\mathbf{x} - \mathbf{y}, t))ds_x$$

$$- \oint_\Gamma g * (t_\alpha^3(\mathbf{x} - \mathbf{y}, t) * v_\alpha(x, t) + m^3(\mathbf{x} - \mathbf{y}, t) * \varphi(x, t))ds_x, \quad (19)$$

where

$$t_\alpha^3(\mathbf{x} - \mathbf{y}, t) = [(2\mu + \kappa)v_{\alpha,\beta}^3(\mathbf{x} - \mathbf{y}, t) - \mu(v_{\beta,\alpha}^3(\mathbf{x} - \mathbf{y}, t) - v_{\alpha,\beta}^3(\mathbf{x} - \mathbf{y}, t))$$
$$+ \kappa\varepsilon_{\alpha\beta}\varphi^3(\mathbf{x} - \mathbf{y}, t)] \cdot n_\beta,$$
$$m^3(\mathbf{x} - \mathbf{y}, t) = \gamma\varphi_{,\alpha}^3(\mathbf{x} - \mathbf{y}, t) \cdot n_\alpha.$$

The formulas (18) and (19) give the integral representation for the solution of the system of equations which govern the nonstationary slow 2D motion of incompressible micropolar fluids for arbitrary forces $\mathbf{f}$, couples l, initial conditions $\mathbf{v}^0$, φ^0 and for bounded domains D in plane.

References

1. M. Ramkissoon and S.R. Majumdar, Representations and fundamental singular solution in micropolar fluid, *Z. Angew. Math. Mech.* **56**, 197–203.

2. M. Ramkissoon and S.R. Majumdar, Potential and Green's functions in micropolar fluid theory, *Z. Angew. Math. Mech.* **60**, 249–255.

3. L. Dragos and D. Homentcovschi, Fundamental matrices in micropolar fluids, *Z. Angew. Math. Mech.* **63**, 389–390.

4. A.C. Eringen, Theory of micropolar fluids, *J. Math. Mech.* **16**, 1–18.

5. D. Iesan, On some reciprocity theorems and variational theorems in linear dynamics, *Mem. Acad. Sci. Torino. Cl. Sci. Fis. Mat. Nat. Ser. 4* **17**, 1–20.

Department of Mathematics, Technical University "Gh.Asachi", Iasi-6600, Romania

Q. SHENG

A monotonically convergent adaptive method for nonlinear combustion problems

1. Introduction

Given that α, $\beta > 0$, $t_0 \geq 0$, we consider the following nonlinear reaction-diffusion problem with a singular source term,

$$u_t = \alpha^2 u_{xx} + p(x)u_x + q(x)/(1-u)^\beta, \quad 0 < x < 1, \ t_0 < t, \tag{1}$$

$$u(x, t_0) = 0, \ 0 < x < 1; \quad u(0, t) = u(1, t) = 0, \ t_0 \leq t, \tag{2}$$

where $0 \leq p(x) \leq b$, $0 < a \leq q(x) \leq b$, $0 \leq x \leq 1$. Problem (1), (2) provides a regularized mathematical model for the combustion of two gases meeting in a gap between porous walls at distance $1/\alpha$ apart [1–3,8].

It is observed [4–6] that, for given p, q, and β, there exists a *critical reference* $\alpha^* > 0$ such that for $\alpha > \alpha^*$, the unique solution of the combustion problem exists globally. However, for any $\alpha \leq \alpha^*$, there is a finite *quenching time* $T_\alpha > t_0$ such that

$$\lim_{t \to T_\alpha} u(1/2, t) = 1, \quad \lim_{t \to T_\alpha} u_t(1/2, t) = +\infty. \tag{3}$$

Most of existing numerical procedures for computing solutions of (1), (2) are indirect, fixed-grid and based on Kawarada's original framework [4]. Needless to say, such computations are often inefficient and unreliable due to the blow-up singularity (3) involved.

This paper is concerned with a modified adaptive method dealing with problem (1), (2). Moving mesh mechanisms are achieved both in space and time. Special monitor functions based on arc-lengths of u_t are derived. We show that, under proper smoothness constraints, the numerical solution generated converges monotonically to the physical solution.

2. The fully adaptive scheme

Let $h = 1/N$ be the mathematical grid size. Based on the physical coordinate $x_k = x_{k-1} + h_k$, $k = 1, \ldots, N$, $x_0 = 0$, where

$$h_k = \int_0^1 \mu(u_t(x,t), x)dx \int_{s_{k-1}}^{s_k} (\mu(u_t(x,t), x))^{-1}dx, \quad \sum_{k=1}^N h_k = 1,$$

$$\mu(u_t(x,t), x) = \sqrt{1 + u_{xt}^2(x,t)}, \quad t_0 \leq t, \ s_k = s_{k-1} + h, \ k = 1, \ldots, N, \ s_0 = 0,$$

we consider the following system of nonlinear differential equations generated via the semi-discretization of (1), (2),

$$\frac{du_k}{dt} = \sigma^2 D_+ D_- u_k + p_k D_0 u_k + g_k(u_k), \quad 1 \leq k \leq N-1, \ t_0 < t, \tag{4}$$

$$u_k(t_0) = 0, \ 1 \leq k \leq N-1, \quad u_0 = u_N = 0, \tag{5}$$

This work was supported in part by the State of Louisiana under Grant LEQSF-(1997-00)-RD-B-15.

where $u_k = u(x_k, t)$, $p_k = p(x_k)$, $g_k = q(x_k)/(1 - u_k)^\beta$ and D_+, D_-, D_0 are forward, backward, and central difference operators, respectively. System (4), (5) may be conveniently written in the matrix form

$$u_t = Au + g(u), \ t_0 < t; \ \ u(t_0) = u^0,$$

for which the formal solution can be expressed as

$$u(t) = E(tA)u^0 + \int_0^t E((t - \tau)A)g(u)d\tau, \ \ t_0 < t,$$

with $E(zA)$ being the analytic semigroup generated.

Let $R_0^{(i)}$, $i = 2, 3$, be L-acceptable rational approximations to E and $u^0 = 0$. Based on the above solution formula, we construct the following fully adaptive scheme,

$$c^m = R_0^{(2)}(\tau_m A)u^m + \tau_m R_1^{(2)}(\tau_m A)g(u^m); \tag{6}$$

$$
\begin{aligned}
u^{m+1} = {} & R_0^{(3)}(\tau_m A)u^m + \tau_m \left\{ \left(\kappa R_1^{(3)}(\tau_m A) + (1 - 2\kappa)R_2^{(3)}(\tau_m A) \right) g(u^m) \right. \\
& \left. + \left((1 - \kappa)R_1^{(3)}(\tau_m A) + (2\kappa - 1)R_2^{(3)}(\tau_m A) \right) g(c^m) \right\}, \ \ m = 0, 1, \dots, \tag{7}
\end{aligned}
$$

where $0 \leq \kappa \leq 1$ and u^m are approximations of $u(t_m)$, $t_{m+1} = t_m + \tau_m$, and

$$\tau_m = \max\left\{ \tau_0, c_0\tau \min_k \left[\tau^2 + (u_t(h_k k, t) - u_t(h_k k, t - \tau))^2 \right]^{-1/2} \right\},$$

$$R_j^{(i)}(\tau_m A) = \left(R_{j-1}^{(i)}(\tau_m A) - I \right)(\tau_m A)^{-1}, \ \ i = 2, 3, \ j = 1, 2, \ m = 0, 1, \dots, \tag{8}$$

for controlling parameters $0 < \tau_0, \ \tau, \ c_0 << 1$.

3. Accuracy and convergence

Denote $I - \frac{\tau_m}{4}A = A_1$, $I - \frac{\tau_m}{3}A = A_2$ and $I + \frac{5\tau_m}{12}A = A_3$. Consider rational approximations $R_0^{(i)}(\tau_m A) = A_1^{-1}A_2^{-1}A_3$, $i = 2, 3$. In this case, recalling (8), algorithm (6), (7) can be reformulated to

$$A_1\delta_1^m = p^m, \ A_2 c^m = \delta_1^m; \ A_1\delta_2^m = q^m, \ A_2 u^{m+1} = \delta_2^m, \ \ m = 0, 1, \dots, \tag{9}$$

where

$$
\begin{aligned}
p^m = {} & A_3 u^m + \frac{\tau_m}{12}(12I - \tau_m A)g(u^m), \ \ q^m = A_3 u^m + \tau_m \gamma^m; \\
\gamma^m = {} & A_1 A_2 \left(\kappa R_1^{(3)}(\tau_m A) + (1 - 2\kappa)R_2^{(3)}(\tau_m A) \right) g(u^m) \\
& + \left((1 - \kappa)R_1^{(3)}(\tau_m A) + (2\kappa - 1)R_2^{(3)}(\tau_m A) \right) g(c^m).
\end{aligned}
$$

Lemma 1. *Let $u(x, t)$ be the solution of (1), (2) and e_k be the local truncation error between (1), (2) and (9) at x_k, $k = 1, \dots, N - 1$. If $|u_{xxx}| \leq M$, then*

$$e_k \leq M(h_k + h_{k+1})\left(\alpha^2 + p_k h_{k+1} \right).$$

311

The conclusion implies that, even if approximations used in (4), (6), (7) and $R_j^{(i)}$, $i = 2, 3$, $j = 1, 2$, can be of second order accuracy, the highest order of accuracy of (11) may not exceed one due to comprehensive mesh-moving mechanisms involved.

Lemma 2. *Given that*

$$h_k, \ h_{k+1} \le 2\alpha^2/p_k \quad whenever \ p_k \ne 0. \tag{10}$$

Then A_1, A_2 are monotone and nonsingular. Their inverse matrices are nonnegative and monotone.

Further, let $\tilde{h}_k = h_k(h_k + h_{k+1})$, $\hat{h}_k = h_{k+1}(h_k + h_{k+1})$, $k = 1, 2, \ldots, N - 1$, and $\omega = (1, 1, \ldots, 1)^T$ be the vector with $N - 1$ unit components.

Lemma 3. *Let*

$$R = (1 - c_1\tau_m)I + \frac{\tau_m^2}{12}A(c_2 I + A),$$

where c_1, c_2 are real constants and $c_1 > 0$ therewith. If

$$h_1 \le \frac{2\alpha^2}{p_1} \quad whenever \ p_1 \ne 0 \ and \ h_1 h_2, \ h_{N-1}h_N \le \frac{2\alpha^2}{c_2} \quad whenever \ c_2 > 0,$$

$$c_1\tau_m + \tau_m^2 \max\left\{ \frac{\alpha^4}{3\tilde{h}_1\tilde{h}_2}, \frac{(2\alpha^2 + h_{N-1}p_{N-2})(2\alpha^2 + h_N p_{N-1})}{12\hat{h}_{N-2}\hat{h}_{N-1}} \right\} < 1,$$

then $R\omega > 0$.

Based on preliminary results given by Lemmas 2 and 3, we can state the following assertion.

Theorem 1. *Let condition (10) hold. If $0 \le u^m < 1$, $Au^m + g(u^m) \ge 0$, then the solution of the fully adaptive scheme (9), $\{u^m\}_{m=0}^{\infty}$,*

(i) *forms a monotonically increasing positive sequence;*

(ii) *increases monotonically until unity is exceeded by an element of the solution vector, or converges to the steady solution of the problem (1), (2).*

Further, for $0 < \xi < 1$, $f(\xi) = 2$, if

$$h_1 \le \frac{2\alpha^2}{p_1} \quad whenever \ p_1 \ne 0, \ h_1 h_2, \ h_{N-1}h_N \le 2\xi\alpha^2;$$

$$\tau_0 \max\{1/\xi, 3/2\} + \tau_0^2\sigma^4 \max\left\{ \frac{1}{3\tilde{h}_1\tilde{h}_2}, \frac{1}{3\hat{h}_{N-2}\hat{h}_{N-1}} \right\} < 1,$$

then $0 \le u^0 < u^1 < 1$.

The proof of the theorem can be divided into three stages. During the first stage, we may show that $c^m \ge u^m \ge 0$ for $m = 0, 1, \ldots$ Further, we derive $u^{m+1} \ge u^m$ through inequalities

$$\tau_m\left(I - \frac{\tau_m}{12}A\right)(Au^m + g(u^m)) \ge 0, \quad m = 0, 1, \ldots$$

Finally, our conclusions can be achieved through a study of properties of functions $u^1 - \omega$ and R that are defined in Lemma 3.

312

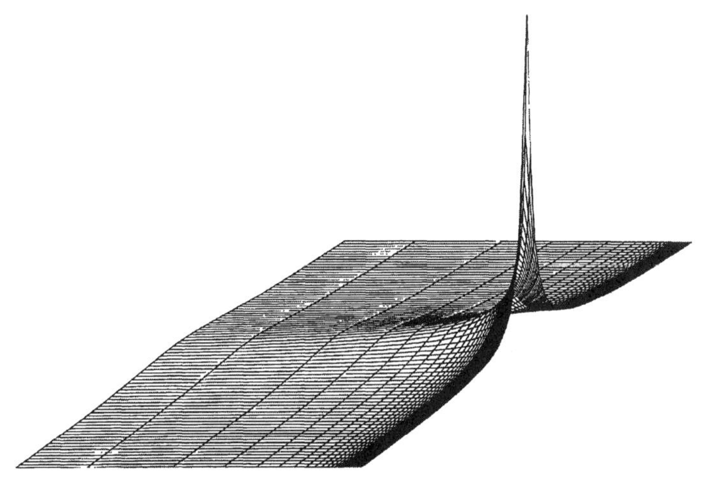

Fig. 1. The explosive rate function u_t with $\alpha = 0.5$, $\beta = 1$.

4. Numerical demonstrations

Let $p \equiv 0$, we consider the following nondegenerate combustion problem:

$$u_t = \alpha^2 u_{xx} + 2/(1-u)^\beta, \quad 0 < x < 1, \ 0 < t < T, \tag{11}$$

$$u(x,0) = 0, \ 0 < x < 1; \quad u(0,t) = u(1,t) = 0, \ 0 < t < T. \tag{12}$$

The above thermal explosion model has been studied by several authors (see [1,3–5] and references therein) and an explosive solution is defined as $\max_{0<x<1}\{u\} \geq 1$, $t < \infty$. Our computations well match and significantly improve the classical results.

Without loss of generality, we set $\kappa = 0$ in the computation. The mathematical grid size in space is chosen to be 0.01, while the initial time step is chosen between 0.001 and 0.01. We observe that, when the numerical solution of (11), (12) is advanced to be near the thermal explosive point, should quenching exist, function u_t becomes very sensitive and increases rapidly with respect to time. The latter leads to fast decays of the grid sizes in the space and time. To see more clearly the phenomenon, we let the spatial adaptiveness be frozen in Fig. 1 which gives a three-dimensional profile of the numerical solution u_t prior to quenching.

In Fig. 2, we further show the computed explosive rate function profile at time levels $t = 0.6508$, 0.7696, 0.7760, 0.7780, 0.7784, 0.7786, 0.7787, 0.7788, 0.77883, and 0.77888, respectively.

We plot the convergence or explosion/quenching time, T_α, corresponding to $\beta = 0.5, 1, 2$ in Fig. 3. It may also be interesting to observe that $T_\alpha \to +\infty$ as $\alpha \to \alpha^*$ and $T_\alpha \to c_\beta$ as $\alpha \to +\infty$, where c_β is a constant depending on β. The later case, together with other important problems such as relations between the quenching phenomena and singular perturbation theories while $\alpha \to 0$, and post-quenching characters, still demand extensive investigations.

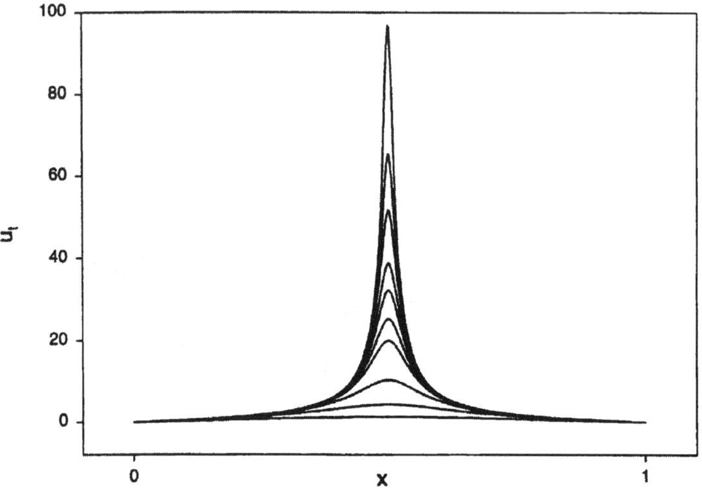

Fig. 2. Profile of the rate function u_t with $\alpha = 0.5$, $\beta = 1$.

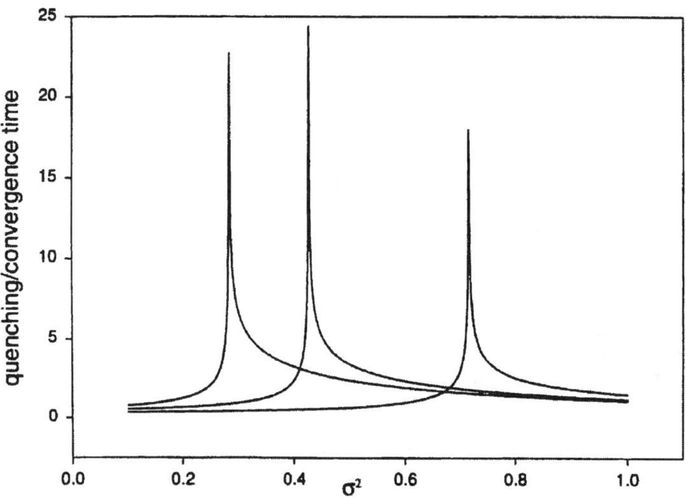

Fig. 3. The profiles of T_α with $\beta = 0.5$, 1, and 2 from left to right, respectively.

314

The author would like to thank IMSE-1998 conference organizers for their enthusiastic discussions and encouragements in the line of the study.

References

1. A. Acker and W. Walter, The quenching problem for nonlinear parabolic differential equations, in *Lecture Notes in Math.* **564**, Springer-Verlag, 1–12.

2. M. Branch, M. Beckstead, T. Litzinger, M. Smooke and V. Yang, Nonsteady combustion mechanisms of advanced solid propellants, *Annual Technical Report*, 1994-05, Center for Combustion and Environmental Research, University of Colorado, Boulder, CO.

3. C. Chen, The method of fundamental solutions for nonlinear thermal explosions, *Comm. Numer. Methods Engrg.* **11**, 675–681.

4. H. Kawarada, On solutions of initial-boundary value problem for $u_t = u_{xx} + 1/(1 - u)$, *Publ. Res. Inst. Math. Sci.* **10**, 729–736.

5. Q. Sheng and A. Khaliq, A compound adaptive approach to degenerate nonlinear quenching problems, *Numer. Methods Partial Diff. Equat.* (to appear).

6. Q. Sheng and A. Khaliq, On modified arc-length adaptive algorithms for degenerate reaction-diffusion equations (submitted for publication).

7. D. Voss and A. Khaliq, Parallel LOD method for second order time dependent PDEs, *Computers Math. Appl.* **30**, 25–35.

8. W. Walter, Parabolic differential equations with a singular nonlinear term, *Funkcial Ekvac.* **19**, 271–277.

Department of Mathematics, University of Southwestern Louisiana, Lafayette, LA 70504-1010, USA

G.R. THOMSON and C. CONSTANDA

Stationary oscillations of elastic plates with Robin boundary conditions

1. Introduction

In [1] and [2], classical boundary integral equation methods were used to solve problems in the theory of time-harmonic plate oscillations with Dirichlet and Neumann boundary data. We now consider applying these techniques to the corresponding Robin boundary value problems. More detailed discussions of the results presented here can be found in [3]–[5].

In what follows, Greek and Latin subscripts and superscripts take the values 1,2 and 1,2,3, respectively, summation over repeated indices is understood, a superscript ^{T} denotes matrix transposition, $x = (x_1, x_2)$ and $y = (y_1, y_2)$.

We consider a homogeneous and isotropic elastic plate of density ρ and Lamé constants λ and μ, which occupies the region $\bar{S} \times [-h_0/2, h_0/2]$ in $\mathbb{R}^3$, where S is a domain in $\mathbb{R}^2$ bounded by a closed C^2-curve ∂S, and h_0 is the (constant) thickness of the plate. The bounded domain enclosed by ∂S is denoted by S^+, and we write $S^- = \mathbb{R}^2 \backslash \bar{S}^+$.

The stationary oscillations of frequency ω of an elastic plate with transverse shear deformation are governed by the system [6]

$$A^\omega(\partial_x)u(x) = H(x), \quad x \in S, \tag{1}$$

where $u = (u_1, u_2, u_3)^{\mathrm{T}}$ characterizes the displacements, $H = (H_1, H_2, H_3)^{\mathrm{T}}$ is related to the forces and moments acting on the plate and its faces, and $A^\omega(\partial_x) = A^\omega(\partial/\partial x_1, \partial/\partial x_2)$ is the matrix operator defined by

$$A^\omega(\xi_\alpha) = \begin{pmatrix} h^2\{\mu(\Delta + k_3^2) + (\lambda + \mu)\xi_1^2\} & h^2(\lambda + \mu)\xi_1\xi_2 & -\mu\xi_1 \\ h^2(\lambda + \mu)\xi_1\xi_2 & h^2\{\mu(\Delta + k_3^2) + (\lambda + \mu)\xi_2^2\} & -\mu\xi_2 \\ \mu\xi_1 & \mu\xi_2 & \mu(\Delta + k^2) \end{pmatrix};$$

here $h^2 = h_0^2/12$, $\Delta = \xi_1^2 + \xi_2^2$, $k^2 = \rho\omega^2/\mu$, and $k_3^2 = k^2 - 1/h^2$.

A particular solution of system (1) is constructed in [7]; therefore, without loss of generality, we may consider the homogeneous system

$$A^\omega(\partial_x)u(x) = 0, \quad x \in S. \tag{2}$$

The boundary stress operator $T(\partial_x) = T(\partial/\partial x_1, \partial/\partial x_2)$ is given by [6]

$$T(\xi_\alpha) = \begin{pmatrix} h^2\{(\lambda + 2\mu)\nu_1\xi_1 + \mu\nu_2\xi_2\} & h^2(\mu\nu_2\xi_1 + \lambda\nu_1\xi_2) & 0 \\ h^2(\lambda\nu_2\xi_1 + \mu\nu_1\xi_2) & h^2\{\mu\nu_1\xi_1 + (\lambda + 2\mu)\nu_2\xi_2\} & 0 \\ \mu\nu_1 & \mu\nu_2 & \mu\nu_\alpha\xi_\alpha \end{pmatrix},$$

where $\nu = (\nu_1, \nu_2)^{\mathrm{T}}$ is the unit outward normal to ∂S.

316

We assume throughout that

$$\lambda + \mu > 0, \quad \mu > 0, \quad \rho \omega^2 h^2 > \mu.$$

The reasons for these restrictions are explained in [8].

We denote by $\mathcal{B}^\omega$ the class of functions defined in S^- which, as $|x| \to \infty$, satisfy the radiation conditions formulated in [9].

Let $\mathcal{F}$ and $\mathcal{G}$ be (3×1)-vector functions defined on ∂S, and let $\sigma \in C^{1,\alpha}(\partial S)$, $\alpha \in (0,1)$, be a symmetric (3×3)-matrix function. The interior and exterior Robin problems are formulated as follows.

$(\mathrm{R}^{\omega+})$ Find $u \in C^2(S^+) \cap C^1(\bar{S}^+)$ that satisfies (2) in S^+ and

$$(Tu + \sigma u)|_{\partial S} = \mathcal{F}. \tag{3}$$

$(\mathrm{R}^{\omega-})$ Find $u \in C^2(S^-) \cap C^1(\bar{S}^-) \cap \mathcal{B}^\omega$ that satisfies (2) in S^- and

$$(Tu + \sigma u)|_{\partial S} = \mathcal{G}. \tag{4}$$

We denote the homogeneous interior Robin problem by $(\mathrm{R}_0^{\omega+})$. A function u is said to be *regular* in S if $u \in C^2(S) \cap C^1(\bar{S})$. If the homogeneous interior Robin problem possesses a nontrivial regular solution u for a particular ω, then that value of ω is called an *eigenfrequency* of $(\mathrm{R}_0^{\omega+})$ with corresponding *eigensolution u*.

2. Uniqueness of solution

The properties of the coupling matrix σ in the boundary conditions (3) and (4) are important when considering the uniqueness of solutions of $(\mathrm{R}^{\omega+})$ and $(\mathrm{R}^{\omega-})$. We consider three separate cases:

(C1) $\mathrm{Re}\,\sigma$ positive definite and $\mathrm{Im}\,\sigma = 0$;
(C2) $\mathrm{Im}\,\sigma$ either positive definite or negative definite;
(C3) $\mathrm{Im}\,\sigma$ positive semidefinite.

The following assertions were proved in [3].

Theorem 1. (i) *If* (C1) *holds, then there exists an infinite set of real eigenfrequencies of* $(\mathrm{R}_0^{\omega+})$.

(ii) *If* (C2) *holds, then* $(\mathrm{R}^{\omega+})$ *has at most one regular solution.*

(iii) *If* (C3) *holds, then* $(\mathrm{R}^{\omega-})$ *has at most one regular solution.*

The proof of Theorem 1(iii) relies on various far-field estimates of solutions of (2) in S^- which were derived in [9]. The result follows from an application of Rellich's Lemma.

We remark that in [10] it is shown that the homogeneous interior Dirichlet problem $(\mathrm{D}_0^{\omega+})$ has an infinite set of eigenfrequencies.

3. Single layer and double layer potentials

Let $D^\omega(x,y)$ be the matrix of fundamental solutions for the matrix operator $A^\omega(\partial_x)$ constructed in [8]. We introduce the single layer and double layer potentials

$$(V^\omega\varphi)(x) = \int_{\partial S} D^\omega(x,y)\varphi(y)\,ds(y), \quad (W^\omega\varphi)(x) = \int_{\partial S} [T(\partial_y)D^\omega(y,x)]^{\mathrm{T}}\,\varphi(y)\,ds(y),$$

respectively, where φ is a (3×1)-vector function known as the density.

Some properties of the potentials are collected in the following assertion [2].

Theorem 2. (i) $V^\omega\varphi \in \mathcal{B}^\omega$ and $W^\omega\varphi \in \mathcal{B}^\omega$.

(ii) *If* $\varphi \in C(\partial S)$, *then* $V^\omega\varphi$ *and* $W^\omega\varphi$ *are analytic and satisfy* (2) *in* $S^+ \cup S^-$.

(iii) *If* $\varphi \in C^{0,\alpha}(\partial S)$, $\alpha \in (0,1)$, *then the direct values* $V_0^\omega\varphi$ *and* $W_0^\omega\varphi$ *of* $V^\omega\varphi$ *and* $W^\omega\varphi$ *on* ∂S *exist* (*the latter as principal value*), *the functions*

$$\mathcal{V}^{\omega+}(\varphi) = (V^\omega\varphi)|_{\bar{S}^+}, \quad \mathcal{V}^{\omega-}(\varphi) = (V^\omega\varphi)|_{\bar{S}^-}$$

are of class $C^\infty(S^+) \cap C^{1,\alpha}(\bar{S}^+)$ *and* $C^\infty(S^-) \cap C^{1,\alpha}(\bar{S}^-)$, *respectively, and*

$$T\mathcal{V}^{\omega+}(\varphi) = (W_0^{\omega*} + \tfrac{1}{2}I)\varphi, \quad T\mathcal{V}^{\omega-}(\varphi) = (W_0^{\omega*} - \tfrac{1}{2}I)\varphi$$

on ∂S, *where* $W_0^{\omega*}$ *is the adjoint of* W_0^ω *and* I *is the identity operator.*

(iv) *If* $\varphi \in C^{1,\alpha}(\partial S)$, $\alpha \in (0,1)$, *then the functions*

$$\mathcal{W}^{\omega+}(\varphi) = \begin{cases} (W^\omega\varphi)|_{S^+} & in\ S^+, \\ (W_0^\omega - \tfrac{1}{2}I)\varphi & on\ \partial S, \end{cases} \quad \mathcal{W}^{\omega-}(\varphi) = \begin{cases} (W^\omega\varphi)|_{S^-} & in\ S^-, \\ (W_0^\omega + \tfrac{1}{2}I)\varphi & on\ \partial S, \end{cases}$$

are of class $C^\infty(S^+) \cap C^{1,\alpha}(\bar{S}^+)$ *and* $C^\infty(S^-) \cap C^{1,\alpha}(\bar{S}^-)$, *respectively, and the equality* $T\mathcal{W}^{\omega+}(\varphi) = T\mathcal{W}^{\omega-}(\varphi)$ *holds on* ∂S.

We use Theorem 2 to derive boundary integral equations from which we can then construct the solutions of $(\mathrm{R}^{\omega+})$ and $(\mathrm{R}^{\omega-})$.

4. Indirect boundary integral equation method

Seeking the solutions of $(\mathrm{R}^{\omega+})$ and $(\mathrm{R}^{\omega-})$ in the form $\mathcal{V}^{\omega+}(\varphi)$ and $\mathcal{V}^{\omega-}(\varphi)$, respectively, and taking the boundary conditions (3) and (4) into account leads to the boundary integral equations

$$(W_0^{\omega*} + \sigma V_0^\omega + \tfrac{1}{2}I)\varphi = \mathcal{F}, \tag{5}$$

$$(W_0^{\omega*} + \sigma V_0^\omega - \tfrac{1}{2}I)\varphi = \mathcal{G} \tag{6}$$

for the unknown densities. This method is "indirect" in the sense that the unknown function φ has no physical meaning; it is merely a mathematical abstraction.

The solvability of (5) and (6) is discussed in [4].

Theorem 3. (i) *If condition* (C1) *holds and* ω *is not an eigenfrequency of* $(\mathrm{R}_0^{\omega+})$, *then* $(\mathrm{R}^{\omega+})$ *has a unique regular solution for any* $\mathcal{F} \in C^{0,\alpha}(\partial S)$, $\alpha \in (0,1)$, *which can be represented in the form* $u = \mathcal{V}^{\omega+}(\varphi)$, *with* $\varphi \in C^{0,\alpha}(\partial S)$.

(ii) *If condition* (C1) *holds and* ω *is an eigenfrequency of* $(\mathrm{R}_0^{\omega+})$, *then* $(\mathrm{R}^{\omega+})$ *has regular solutions if and only if*

$$\int\limits_{\partial S} \mathcal{F}^{\mathrm{T}}(y) u^{(k)}(y)\, ds(y) = 0, \quad k = 1, 2, \ldots, n,$$

where $\mathcal{F} \in C^{0,\alpha}(\partial S)$, $\alpha \in (0,1)$, *and* $\left\{u^{(k)}\right\}_{k=1}^{n}$ *is a complete set of linearly independent eigensolutions of* $(\mathrm{R}_0^{\omega+})$. *Each of these solutions can be represented in the form* $u = \mathcal{V}^{\omega+}(\varphi)$, *with* $\varphi \in C^{0,\alpha}(\partial S)$.

(iii) *If condition* (C2) *holds, then for any* $\mathcal{F} \in C^{0,\alpha}(\partial S)$, $\alpha \in (0,1)$, *and any* ω, $(\mathrm{R}^{\omega+})$ *has a unique regular solution, which can be represented in the form* $u = \mathcal{V}^{\omega+}(\varphi)$ *with* $\varphi \in C^{0,\alpha}(\partial S)$.

These results follow from the application of the Fredholm alternative to (5).

Theorem 4. *If condition* (C3) *holds, then for any* $\mathcal{G} \in C^{0,\alpha}(\partial S)$, $\alpha \in (0,1)$, *and any* ω, $(\mathrm{R}^{\omega-})$ *has a unique regular solution. If* ω *is not an eigenfrequency of* $(\mathrm{D}_0^{\omega+})$, *then the solution can be represented in the form* $u = \mathcal{V}^{\omega-}(\varphi)$ *with* $\varphi \in C^{0,\alpha}(\partial S)$. *If* ω *is an eigenfrequency of* $(\mathrm{D}_0^{\omega+})$, *then the solution can be represented in the form*

$$u = \mathcal{V}^{\omega-}(\varphi) + \sum_{k=1}^{n} c_k \left\{ \mathcal{W}^{\omega-}(\tilde{\psi}^{(k)}) + \mathcal{V}^{\omega-}(\sigma\tilde{\psi}^{(k)}) \right\}, \quad \text{where } c_k, \ k = 1, 2, \ldots, n, \text{ are}$$

constants, $\varphi \in C^{0,\alpha}(\partial S)$, *and* $\tilde{\psi}^{(k)} \in C^{1,\alpha}(\partial S)$, $k = 1, 2, \ldots, n$.

When ω is an eigenfrequency of $(\mathrm{D}_0^{\omega+})$, the solution of $(\mathrm{R}^{\omega-})$ given by Theorem 4 cannot be constructed from (6) because of solvability problems. The existence of a unique solution of $(\mathrm{R}^{\omega-})$ in this case is proved by considering a related integral equation that is solvable in theory but, unfortunately, cannot be dealt with in practice. For this reason we turn to an alternative integral equation formulation.

5. Direct boundary integral equation method

We consider only $(\mathrm{R}^{\omega-})$. This is the more interesting case because it is known to have a unique solution (Theorem 4), but the indirect method yields an integral equation which is not always uniquely solvable.

A representation formula for the regular solutions of (2) in S^- is derived in [11]. If $u \in C^2(S^-) \cap C^1(\bar{S}^-) \cap \mathcal{B}^\omega$ is a solution of (2) in S^-, then

$$-\mathcal{V}^{\omega-}\left(Tu|_{\partial S}\right) + \mathcal{W}^{\omega-}\left(u|_{\partial S}\right) = \begin{cases} 0 & \text{in } S^+, \\ u & \text{in } \bar{S}^-. \end{cases} \tag{7}$$

We introduce an operator $N_0^\omega : C^{1,\alpha}(\partial S) \to C^{0,\alpha}(\partial S)$ defined by

$$N_0^\omega \varphi = T\mathcal{W}^{\omega+}(\varphi) = T\mathcal{W}^{\omega-}(\varphi).$$

Substituting the boundary condition (6) into (7), we easily show that

$$u = \mathcal{W}^{\omega-}(\varphi) + \mathcal{V}^{\omega-}(\sigma\varphi) - \mathcal{V}^{\omega-}(\mathcal{G}) \tag{8}$$

is a solution of $(\mathrm{R}^{\omega-})$ provided that φ satisfies

$$(W_0^\omega + V_0^\omega \sigma - \tfrac{1}{2}I)\varphi = V_0^\omega \mathcal{G}, \tag{9}$$

$$(W_0^{\omega*} + \tfrac{1}{2}I)(\sigma\varphi) + N_0^\omega \varphi = (W_0^{\omega*} + \tfrac{1}{2}I)\mathcal{G}, \tag{10}$$

where $\varphi = u|_{\partial S}$. The solvability of this pair of equations is discussed in [5].

Theorem 5. *If condition* (C3) *holds and ω is not an eigenfrequency of* $(\mathrm{D}_0^{\omega+})$, *then* (9) *has a unique solution* $\varphi \in C^{1,\alpha}(\partial S)$ *for every* $\mathcal{G} \in C^{0,\alpha}(\partial S)$, $\alpha \in (0,1)$. *This solution also satisfies* (10).

Theorem 6. *If condition* (C3) *holds and ω is an eigenfrequency of* $(\mathrm{D}_0^{\omega+})$, *then* (10) *has a unique solution* $\varphi \in C^{1,\alpha}(\partial S)$ *for every* $\mathcal{G} \in C^{0,\alpha}(\partial S)$, $\alpha \in (0,1)$. *This solution also satisfies* (9).

The proofs of these results make use of the Fredholm alternative together with composition formulas for the various boundary integral operators.

From Theorems 5 and 6 we see that the pair of equations (9), (10) always has a unique solution φ. The unique solution of the boundary value problem $(\mathrm{R}^{\omega-})$ is then given by (8).

Instead of considering a pair of integral equations, it would be preferable to solve a single equation. For this reason, we propose a composite integral equation formed by taking a linear combination of (9) and (10).

Let Λ be a symmetric (3×3)-matrix, and consider the equation

$$(W_0^{\omega*} + \tfrac{1}{2}I)(\sigma\varphi) + N_0^\omega \varphi + \Lambda(W_0^\omega + V_0^\omega \sigma - \tfrac{1}{2}I)\varphi = (W_0^{\omega*} + \tfrac{1}{2}I)\mathcal{G} + \Lambda V_0^\omega \mathcal{G}. \tag{11}$$

The existence of at least one solution of (11) is guaranteed by Theorems 5 and 6. The question of uniqueness is answered by the following assertion (see [5]).

Theorem 7. *If condition* (C3) *holds and* $\mathrm{Im}\,\Lambda$ *is either positive definite or negative definite, then* (11) *has a unique solution* $\varphi \in C^{1,\alpha}(\partial S)$ *for every* $\mathcal{G} \in C^{1,\alpha}(\partial S)$, $\alpha \in (0,1)$.

Again, the unique solution of $(\mathrm{R}^{\omega-})$ is given by (8) with φ satisfying (11).

Some work has also been done on deriving single uniquely solvable second kind equations involving a modified matrix of fundamental solutions in a suitable way (see [12] and [13]).

References

1. G.R. Thomson and C. Constanda, On stationary oscillations in bending of plates, in *Integral methods in science and engineering*, vol. 1: *Analytic methods*, C. Constanda, J. Saranen and S. Seikkala, Eds., Addison Wesley Longman, Harlow-New York, 1997, 190–194.

2. G.R. Thomson and C. Constanda, Scattering of high frequency flexural waves in thin plates, *Math. Mech. Solids* **4** (1999), 461–479.

3. G.R. Thomson and C. Constanda, Uniqueness of solution for the Robin problems in plate oscillations, *Strathclyde Math. Research Report* **15** (1998).

4. G.R. Thomson and C. Constanda, The Robin problems for stationary oscillations in elastic plates, *Strathclyde Math. Research Report* **16** (1998).

5. G.R. Thomson and C. Constanda, Harmonic oscillations of elastic plates with Robin boundary conditions, *Strathclyde Math. Research Report* **17** (1998).

6. P. Schiavone and C. Constanda, Oscillation problems in thin plates with transverse shear deformation, *SIAM J. Appl. Math.* **53** (1993), 1253–1263.

7. G.R. Thomson and C. Constanda, The Newtonian potential in the theory of elastic plates, *Strathclyde Math. Research Report* **13** (1998).

8. G.R. Thomson and C. Constanda, The matrix of fundamental solutions in the theory of plate oscillations, *Strathclyde Math. Research Report* **7** (1998).

9. C. Constanda, Radiation conditions and uniqueness for stationary oscillations in elastic plates, *Proc. Amer. Math. Soc.* **126** (1998), 827–834.

10. G.R. Thomson and C. Constanda, The eigenfrequency spectra of the interior problems of oscillating plates, *Strathclyde Math. Research Report* **14** (1998).

11. G.R. Thomson and C. Constanda, Representation theorems for the solution of high frequency harmonic oscillations in elastic plates, *Appl. Math. Lett.* **11** (1998), 55–59.

12. G.R. Thomson and C. Constanda, Modified integral equations for the harmonic oscillations of thin plates, *Strathclyde Math. Research Report* **20** (1998).

13. G.R. Thomson and C. Constanda, A modified integral equation method for stationary plate oscillations, *Strathclyde Math. Research Report* **22** (1998).

Department of Mathematics, University of Strathclyde, Glasgow, UK

F. UNSACAR

Adapter and driver design for rotary encoders

1. Introduction

Rotary encoders are mounted directly to motors or shafts driven by motors. The traveled distance of driven table by the shaft is related to the rotational angle of the shaft. If the shaft stops after rotating n times plus β degrees, the angle is

$$L = 2\pi n + \beta. \tag{1}$$

Rotary encoders (RE) measure this angle. Two separate square shaped signals (A and B) or two separate sinusoidal signals (easily transformed to square) are produced when the RE shaft rotates. A', B', and C signal series are produced in most advanced REs. A' and B' are inverses of A and B, and C is only for counting the number of completed tours. It does not determin the rotational direction. A' and B' provide easy interface in some cases and are the main signal. C provides direct connection to the control board without interface when only the number of completed rotations.s required. When the RE's shaft rotates in the clockwise direction, B is 90 degrees behind A' and vice versa, B will be 90 degrees ahead of A if the direction is counterclockwise (see Fig. 1).

Angular velocity, rotational direction, swept angle, the position of table and virtual speed of table are calculated from these two signals. When the RE shaft completes one tour, the output of RE will be a number F of square shaped pulses. If we consider the tours as L degrees as in (1), then the output pulses K will be as follows:

$$K = F(L/360) = F((2\pi n + \beta)/360). \tag{2}$$

Since every complete tour produces F pulses, each pulse will be equivalent to $(360/F)$ degrees. The value of F is between 24 to 5000 for most REs, which can provide 18 to 0.072 degrees of accuracy [1]. These considerations are the same for linear encoders. Since the movements are forward and backward, the traveled distance, position, and average speed can easily be found.

2. Algorithms for finding the traveled distance and rotational direction

There are various approaches for finding direction, position and velocity by utilizing the two square-shaped signals that have a 90 degree phase difference between them [2,5].

The most popular method is to use output signals of RE as an input to microprocessor or microcontroller and calculate the parameters by a program. We have developed a different approach for this. One of the advantages of this approach is that it can catch even one single pulse by using only one signal series.

Pa and Pb are the pulses of A and B series, respectively, and S is the traveled distance. When there is no rotation, $PaPb = 0\ 0$ (see Fig. 2a and b). When there is a clockwise rotation, $PaPb = 0\ 0$ will be followed by $PaPb = 0\ 1$ code. For a counterclockwise rotation, $PaPb = 0\ 0$ will be followed by $PaPb = 0\ 1$ code.

The problem, however, could be easily solved if the case always started from $PaPb = 0\ 0$ condition. But as the rotational speed always varies, the possibility of starting condition as $PaPb = 0\ 0$ is not always matched in control processes, which is held in certain discrete times. One of the solutions in this case is to wait until the current pulse ends and then catch the next forthcoming couple of pulses and to find the rotational direction of shaft accordingly.

If the state at the beginning of the control process is $Pa = 0$ and $Pb = 0$, the condition is indefinite and whether the shaft is rotating or not cannot be defined (see Fig. 2a and b). If the state at the beginning of the control process is $Pa = 1$ and $Pb = 1$, when Pa becomes 0 it means that the rotation is in the clockwise direction (Fig. 2a). If the state is $Pb = 1$ and $Pa = 0$, when Pb becomes 0 the rotation is in clockwise direction (Fig. 2a). If the state is $Pb = 1$ and $Pa = 1$, when Pb becomes 0 the rotation is in counterclockwise direction (Fig. 2b). If the state is $Pa = 1$ and $Pb = 0$, when Pa becomes 0 the rotation is in counterclockwise direction (Fig. 2b).

As can be seen from the above algorithm, for each couple of pulses a program module (subprogram or interrupt) which consists of five Boole processor commands must be developed. Running this program takes at least 10 μs even with the fastest, widely used MCS51 family microcontrollers. When the speed is $Dmax = 5000$ rpm, the minimum period of the encoder shaft will be as follows:

$$Tmin = 60/5000 = 0.012 \text{ sec.} \tag{3}$$

The minimum period of pulse (for only one series considered) will be as follows:

$$Pmin = Tmin/F, \tag{4}$$

where F is the number of pulses in $Tmin$ of period. As the minimum period matches to the maximum frequency which we assume as $Fmax = 5000$, the minimum period of the output pulse can be found as follows:

$$Pmin = Tmin/5000 - 0.012/5000 = 2.4 \ \mu\text{s.} \tag{5}$$

We need 10 μs for doing calculations for one pulse, whereas we should check the input in less than 2.4 μs for not losing the next pulse. But in the general case, the maximum value of F and D can be less than 5000. And the value of $Pmin$ can be bigger than the required time to complete the pulse-related processes. The above algorithm can securely be used in these cases.

The disadvantage of the above algorithm is that the microprocessor has to wait until the shaft rerotates when there is logic 1 at one of the outputs of RE after it stopped. This disadvantage can be prevented by deciding that the shaft be stopped after waiting a certain amount of time. The easiest solution of this problem is to use interrupt (see Fig. 3).

If we produce C interrupt signal by using AND gate for A and B, and if we use C interrupt signal to invoke the below interrupt algorithm (see Fig. 4), then we can solve the problem.

3. The adapter for finding the traveled distance and rotational direction

We can use a special counter between the encoder and the microprocessor for the case when the required time of the microprocessor to process one couple of pulse is greater than the period of the signal. But there must be an adapter in this case for defining the counting mode (increase or decrease of content) of counter and finding the rotational direction. Adapter circuits used in industry [1,2,4,5] generally produce signals to define the rotational direction and pulse for the traveled distance between them. These adapter circuits have two disadvantages:

1. Because RC filters used for defining the rotational direction are very sensitive to pulse parameters, they can only work in a secure environment that employs delicate high-tech materials.

2. They are suitable for using asynchron-type counters in which the temporary situations may occur. They may perceive these temporary situations as real situations.

PLD16R8 adapters manufactured by Microchip Technology, Inc. for supporting PIC microcontrollers produce the Up and Down signals for synchron-type counters (5). However, as their working frequence is below 5 MHz, they can only be used when the following condition is met:

$$0 < F \cdot Dmax < 5000000.$$

However, there is a need for encoder adapters for supporting the synchron counters to work safely under the following conditions in the industrial environments:

$$0 < F \cdot Dmax < 25000000.$$

We describe below the encoder adapter consisting of only two D-type flip-flops which can work in every range of speed. If the shaft rotates only in one direction, one of the A or B series would be enough (see Fig. 1). But as the shaft rotates in both directions, the traveled distance increases in one direction and decreases when the direction is reversed. If the traveled distance is calculated by a synchoron counter, the pulses should be input into the Up end of the counter while the rotation is in the clockwise direction and the pulses should be input into the Down end of counter while the rotation is in the counterclockwise direction. The I (clockwise direction) signal should be produced form the signal of A to be applied into positive (Up) input of counter and the G (counterclockwise direction) signal should be produced form the signal of B to be applied into negative (Down) input of counter (see Fig. 5).

If we connect the A signal to the D input and the B signal to the C input of the D-type flip-flop, we can trigger the Bi pulse which follows Ai pulse with 90 degree phase difference (see Fig. 6a and 8). But in this case the flip-flop will respond only to the A signal, because the flip-flop was in $Q = 1$ state already as these pulses try to bring the flip-flop into $Q = 1$ state. So the flip-flop will not respond to these signals. Therefore, we should restore the effect of each pulse just after it is lost to provide the response of the flip-flop. This can be provided by connecting the A pulse into the R (Reset) input of the flip-flop (see Fig. 6b and 8). This flip-flop will respond to only the pulses produced while the clockwise rotation occurs. When the rotation is in the counterclockwise direction, the D input of flip-flop will always be 0, while the pulses coming into the C input were increasing.

As applying the second flip-flop in the same way described above for the opposite direction or rotation (see Fig. 7), we can control the signals produced while the rotation is in opposite direction. B signals go into the D and R inputs of flip-flop and the A signal goes into the C input of the flip-flop.

The connections of flip-flop providing the above diagrams can be done as in Fig. 8.

4. Conclusion

In this paper the disadvantages of the adapters used between encoders and microcontrollers are explained. The main advantage of the proposed design is its simplicity and reliability. It can be used for any speed range that may occur in an industrial environment. It can catch even one single pulse by using only one series of signals. The designed adapter is still under test in CNC lathes, and the system performed very well.

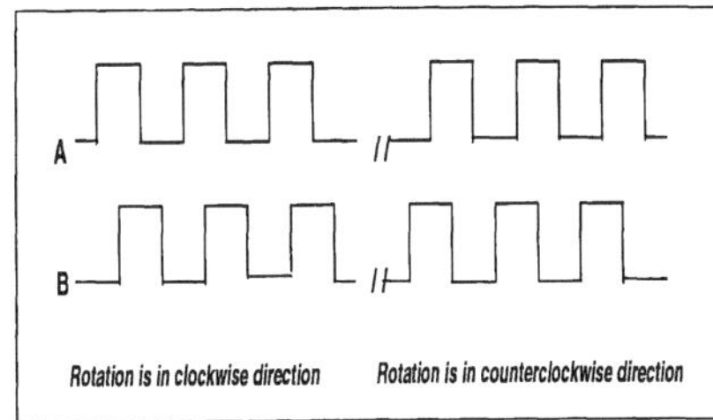

Fig. 1. Output signals of rotary encoder.

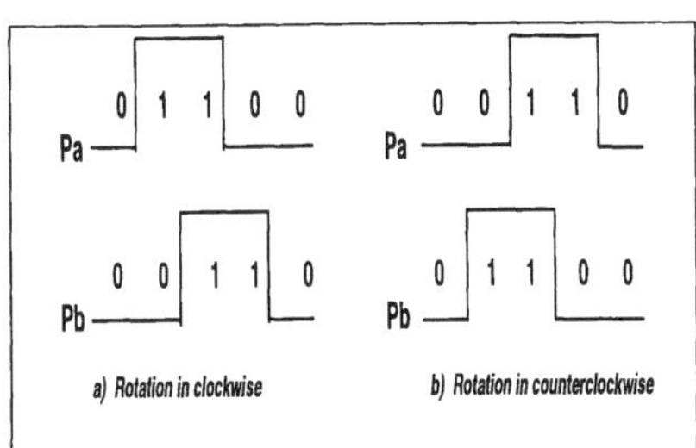

Fig. 2. Possible cases for defining rotational direction.

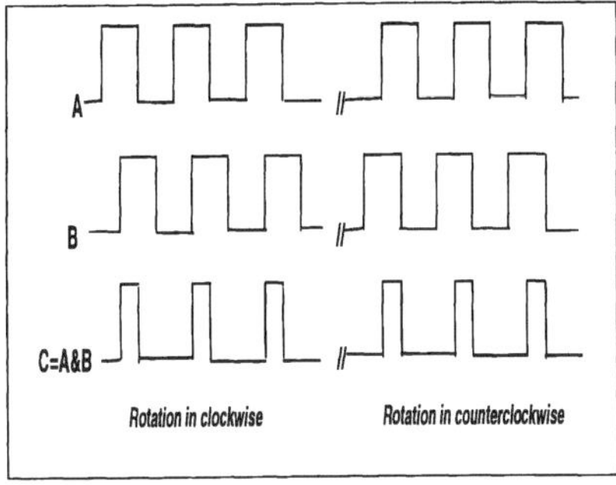

Fig. 3. Production of interrupt signals.

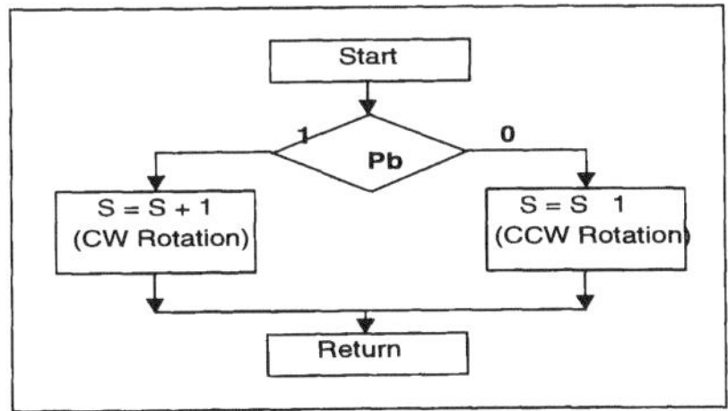

Fig. 4. Interrupt algorithm for identifying rotational direction.

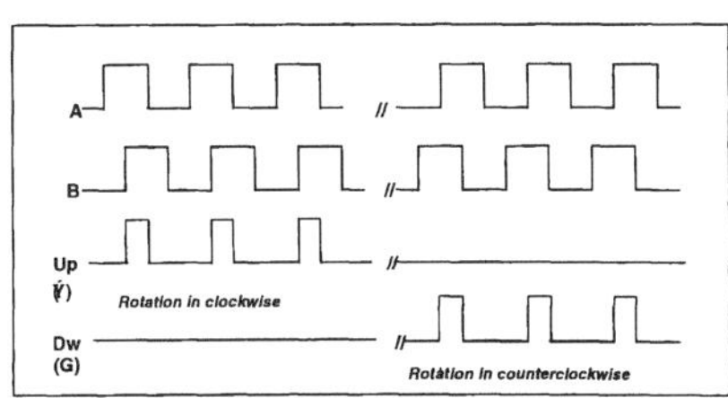

Fig. 5. Production of signals applied to the Up and Down counter inputs.

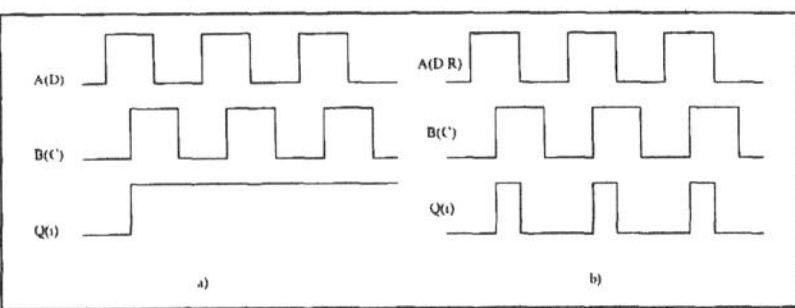

Fig. 6. Qi output states for clockwise rotation: (a) when the reset input is 1; (b) when the reset input is connected to the A signal.

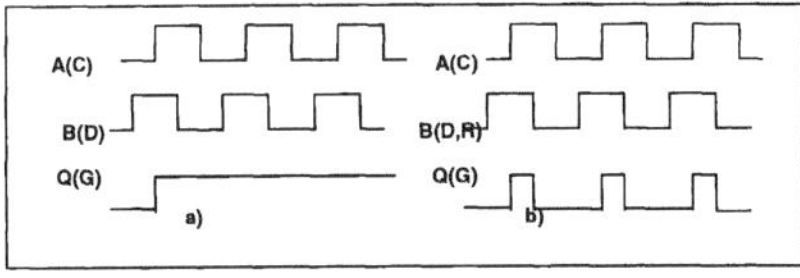

Fig. 7. Same situation as Fig. 6 for counterclockwise rotation.

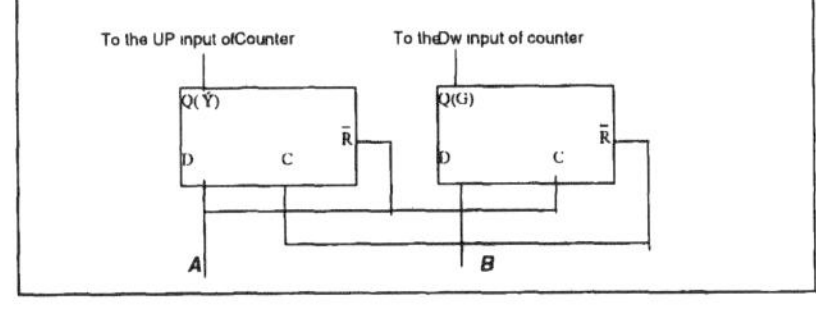

Fig. 8. The adapter connections between a rotary encoder and synchron counter.

References

1. J.T. Humperies and L.P. Sheets, *Industrial electronics*, Delmar Publishers, Albany, NY, 1993.

2. C.M. Gilmore, *Microprocessors; principles and applications*, International Ed., Singapore, 1989.

3. *Microchip data book*, Microchip Technology, 1994.

4. *Microcontroller handbook*, INTEL, 1986.

5. *PIC16/17 family cross-reference guide*, Microchip Technology, 1994.

Michigan Technological University, Houghton, MI 49931, USA

F. UNSACAR, S. HALKACI and A. MAJLESSI

Expert system design for fault diagnosing in CNC machine tools

1. Introduction

The structure of CNC machine tools is quite complex. The most difficult and time-consuming job is to find the source of a fault when it occurs. Correcting the fault is easy if the source is found. CNC service engineers know this from experience. In a CNC machine, there are hundreds of cables and connections; even one bad contact may put the machine into the alarm state. Finding out the cause of a problem may take 3 to 5 hours, depending on the experience of the service engineer. However it may take only a few minutes to correct the fault.

In this project the goal is to design an expert system to help CNC service engineers determine the cause of problems, reducing the extent of experience they would otherwise need.

2. The structure of CNC machine tools

It is necessary to know the structure of CNC machines to approach the subject. Any CNC machine is made up of the modules consisting of mechanic, hydraulic-pneumatic, electric, electronic, and computer. These are only simplified modules; the structure of CNC machines are more complex and detailed, with the computer having full control over the entire structure. Similar to the human body where the brain senses if the stomach is empty, the computer senses when the oil level is below a certain level.

The computer (widely called a "controller") in CNC machines, consists of hardware and software. Software itself consists of the system, ladder, and part program. System software (base software) and part program are outwith our discussion here. The PLC program which is called a ladder diagram [1] changes from machine to machine depending on the machine configuration. Service engineers must have a good knowledge of this, because when there is a fault the first thing service engineers do is to look for an explicit alarm message; when there is one then they go there, but there may not always be an explicit alarm message.

If there is a fault with no understandable alarm message, the engineer must check the diagnostic table (Fig. 1) consisting of binary numbers. This table is dynamically stored in the RAM of the CNC controller. The binary data in the diagnostic table shows the latest conditions of the real and logic contacts in the machine [2,3].

In this table each bit represents the state of a contact. If the contact is open, the number is 0, if closed 1. The normal condition of a contact is either open or closed. This can be seen from the ladder (Fig. 2) diagram. The address of each bit in the diagnostic table is defined by its row and column numbers. As an example the address of 20.3 show the address of a bit which is in 20th row and 4th column from right to left. 21.0 address represents the relay of oil sensor in Fig. 2.

The normal condition of 21.0 is 0. If the oil level decreases below the normal level, the relay of the oil sensor will close and bit in 21.0 will become 1.

Similarly the address of 21.4 belongs to the Emergency Stop button, this bit becomes 1 if the button is pressed. The bit in 48.1 will be 1 when the chuck is closed. The bit in 522.5 will be 1 when the coolant is on. When the above conditions reverse, all the above addresses will be 0.

	7 6 5 4 3 2 1 0
20	0 0 0 0 0 0 0 0
21	0 0 0 1 0 0 0 1
22	0 0 0 0 0 0 0 0
.....	....
48	0 0 0 0 0 0 1 0
49	0 0 0 0 0 0 0 0
.....	....
522	0 0 1 0 0 0 0 0

Fig. 1. Diagnostic table.

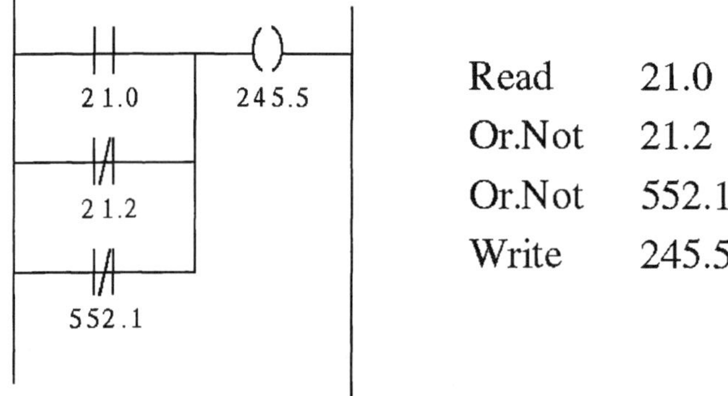

Fig. 2. Ladder diagram and program sample.

As shown, the latest conditions of the CNC are represented by bits in the diagnostic table. Real relays were given as examples above but the diagnostic table also contains logical relays.

These results in the diagnostic table are produced by the PLC (programable logic controller) program, which is widely known as a "ladder diagram" [1,4,5].

The program part in Fig. 4 will read the 21.0, 21.2, and 522.1 addresses and write the result into address 245.5. The symbol $-||-$ represents an open contact and $-|/|-$ represents a closed contact. In the above example, if 21.0 is 0, 21.2 is 1, and 522.1 is 1, the result in 245.5 will be 1.

3. Faults in CNC

There are many fault sources in CNCs, which can be in any one of the modules. Even with all the modules working properly, a fault can happen due to operator error, sending the machine into the alarm state.

328

Some faults have understandable alarm messages. OVERTRAVEL X, LUBRICA-
TION ALARM, SERVO ALARM, etc. But some faults have no alarm message. On
the other hand, alarm messages show only the area of the fault and not the fault itself
or its origin. As an example, the reason for a Lubrication Alarm can be the oil pump
itself or low oil level or low oil pressure, etc. Faults that require service engineers to
solve are generally more complex then the ones given here as examples.

4. Example of a fault finding method in a CNC

In this study, a knowledge-based expert system is introduced and a program developed
that will help service engineers with fault finding in CNCs. Let us take one example
to understand the fault finding procedure by following the ladder diagram.

If the bit in 245.5 is 1 the alarm lamp of the CNC will turn on (Fig. 3). In this
case the 21.0, 551.6, 552.0, 552.1 addresses need to be checked. Let us assume the bit
in 551.6 is 1 instead of 0. This address belongs to the clamping unit. We shall follow
the fault from this point. The related part of the ladder diagram is shown in Fig. 4.
Let us assume the bit in 521.7 is 0 instead of 1. This address belongs to clamping unit
ready signal. The fault must be followed from here. The related part of the ladder
diagram is shown in Fig. 5.

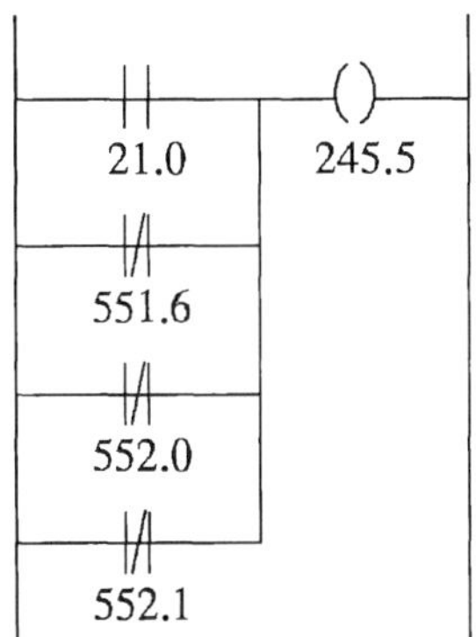

Fig. 3. A part of the ladder diagram.

Let us assume that the bit in 16.0 is 0 instead of 1. This address belongs to the
tailstock end forth switch, which is a real micro switch. The fault originated exactly
from here. The switch itself may be out of order, cables to it may be shorted, or a
contact may be loose. All of these possibilities must be examined to determine the
fault.

As can be seen in this example, finding the source of the fault is impossible without
following the ladder diagram correctly.

5. Development of expert systems for CNC fault finding

5.1. Introduction. Expert systems (ESs) were first introduced with the concept of
artificial intelligence and since 1970 have been applied successfully in medical diagnosis,
chemical analysis, and design of computer system configurations. For example, ES has
investigated 38 medical diagnosis systems [7]. Beside these, shells are developed and

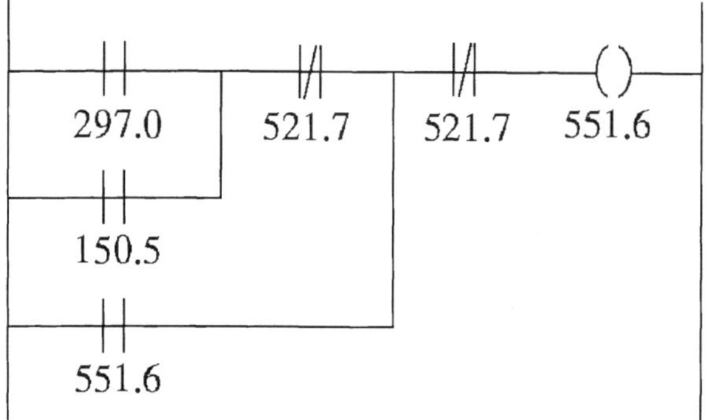

Fig. 4. The part of the ladder diagram belonging to clamping unit.

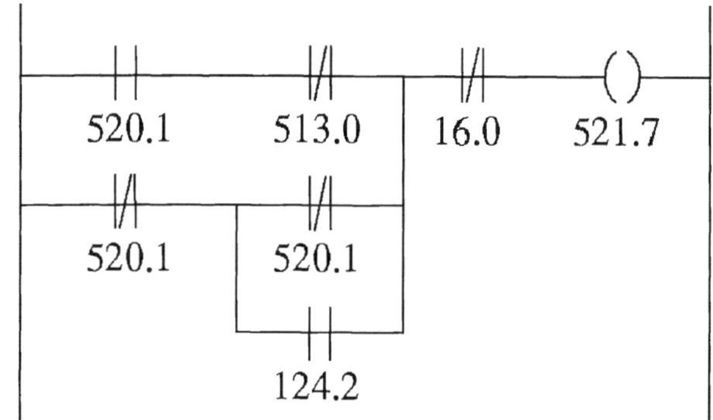

Fig. 5. The part of the ladder diagram belonging to tail stock.

used for speeding up trouble diagnosis, making training personnel easier and more efficient, and as a consulting assistant to experts. Generally the main parts of an ES are an inference engine mechanism, knowledge base, and data base (working storage).

5.2. Knowledge-base design for the expert system. It is clear that the knowledge base of ES trouble diagnosis may be applied in the form "IF...,THEN." Thus it has been decided to use the production rules method for the knowledge base of the present ES trouble diagnosis.

In reference to the above (Fig. 3–5) the rules will be as below:

```
Rule 10: IF 245.5 = 1 THEN CNC = FAULT
Rule 20: IF CNC = FAULT THEN 21.0 = 1
   OR   551.6 = 1
   OR   552.0 = 1
   OR   552.1 = 1
```

```
Rule 30: IF 551.6 = 1 THEN 297.0 = 1
  OR   150.5 = 1
  AND 521.7 = 1
  OR   551.6 = 1
  AND 521.7 = 0
Rule 40: IF 521.7 = 0 THEN 520.1 = 1
  AND 513.0 = 0
  OR   520.0 = 0
  AND 520.1 = 0
  AND 16.0  = 0
Rule 50: IF 16.0 = 0 THEN SWITCH = FAULT
Rule 60: IF SWITCH = FAULT THE CABLE = BAD CONTACT
  AND SWITCH = SHORT CUT
  AND SWITCH = DEFECTED.
```

All of the rules like these were developed using the ladder diagnostic table [2].

6. Realization of expert system and conclusion

After designing the rules they were converted into a computer program in the Prolog-2 programming language.

It is clear that the ES will work faster than an operator and will not require an expensive specialist. The developed system has all the advantages of a diagnostic ES for fault finding. It may help an inexperienced service engineer in environments with many CNC machine tools.

This study is an initial step towards developing Intelligent Manufacturing Systems for troubleshooting in CNC machine tools. Similar studies are necessary for other controllers as their ladder diagrams differ.

References

1. *Ladder diagram operator's handbook*, VEDA CNC-machine Tools Ltd., 1997.

2. *TS & M, Programmable Logic Controller programming manual*, Sofia, 1995.

3. *Fanuc System 01 operation manual*, Fanuc Inc., 1990.

4. T.A. Hughes, *Programmable controllers*, Instrument Society of America, North Carolina, 1989.

5. F.D. Petruzdla, *Programmable controllers*, McGraw-Hill, Ohio, 1991.

6. F. Unsacar, CNC Torna Tezghmda Calisma Emniyetinin PLC ile saglanmasi, Makine Market, Sayfa 84-88 Agustos 1996, Istanbul.

7. W. Regers et al., Computer aided medical diagnosis; literature review, in *Proc. 1st Conf. on Artificial Intelligence Applications*, IEEE Computer Society, 1984, 178–186.

8. A.I.A. Shobair, F. Unsacar, N. Allahverdu et al., Knowledge base of expert system for electric motor trouble diagnosis, in *Proc. 4th Symp. on Computer Communication*, 11-15 Dec. 1996, Bursa, 68–71.

9. N.M. Allahverdu, B.A. Huseynov and R.T. Mustafayev, Expert system in ex-Soviets, J-1 OTOMASYON, no. 43, 1996, Istanbul, 128–131.

Michigan Technological University, Houghton, MI 49931, USA

Selcuk University, Turkey

Michigan Technological University, Houghton, MI 49931, USA

D. URSUTIU, A. DUTA-CAPRA, D. NANU and P. COTFAS

Non-ideal liquid solutions modeling by means of integral methods

1. Introduction

The non-ideal behavior of liquid solutions influences all other properties and is important in vapor-liquid equilibrium calculations. Estimating the activity coefficient is the most common way to describe nonideality. It can be calculated using integral methods using an equation of state that allows the calculation of the pure component properties. Mixture properties are estimated using a set of mixing rules. This paper presents results obtained for binary hydrocarbon solutions, using the general equation of state (GEOS, 1986) and Van der Waals mixing rules. The dependence of the activity coefficient on temperature and compositions is found to be significant when the hydrocarbons are very different in size. One of the most important separation processes, distillation, is the result of vapor-liquid equilibrium, VLE. So, the design of the devices used in distillation installation is based on VLE calculations. This paper presents a LabVIEW program, which allows VLE prediction calculation, for binary systems at high pressures.

2. The vapor-liquid equilibria modeling

The general equilibrium condition is given—for a multicomponent system with NC components—by the equality of the chemical potential of each component in the phases that are in equilibrium. For real solutions this condition can be written as the equality of the fugacity of each component "i" in both phases, α, β considered to be in equilibrium:

$$T, p = \text{const}, \qquad f_i^{(\alpha)} = f_i^{(\beta)}, \qquad i = \overline{1, NC}. \tag{1}$$

In the particular case of vapor liquid equilibrium, we may design α to be the liquid (L) and β the vapor (V) phase.

An approach used for high pressure phase equilibrium allows the calculation of the fugacity of both phases using a fugacity coefficient Φ_i and the molar ratio of the "i" component (X_i in the liquid and Y_i in the vapor phase) [6–10],

$$\Phi_i^{(L)} X_i = \Phi_i^{(V)} Y_i. \tag{2}$$

Additional equilibrium conditions concern the composition of both phases:

$$\sum_{i=1}^{NC} X_i = 1, \qquad \sum_{i=1}^{NC} Y_i = 1. \tag{3}$$

An experimental point is characterized by the parameters: pressure (P), temperature (T), volume (V), and composition (n_i). These can be correlated with the fugacity coefficient using Lewis's equation

$$\ln \Phi_i = \frac{1}{RT} \int_V^\infty \left[\left(\frac{\partial P}{\partial n_i} \right)_{T,V,n_{i \neq j}} - \frac{RT}{V} \right] dV - \ln \frac{PV}{RT}, \tag{4}$$

333

where R is the gases constant (8.317 J/kmolK).

These parameters (P, T, V, n_i) describe the thermodynamic state and can be correlated using a mathematical relation called equation of state (EOS):

$$P = P(T, V, n_i),\tag{5}$$

so the derivative in (4) can be expressed by choosing a certain EOS.

This paper presents the results obtained when the general equation of state (GEOS) [3–5] is used. This is a cubic in volume EOS, with four parameters:

$$P = \frac{RT}{V - b} - \frac{a(T)}{(V - d)^2 + c},\tag{6}$$

where a, b, c, d are parameters depending on the components and can be calculated by using the critical constants (P_c, T_c). For pure components,

$$a_i = \frac{Z_{c,i}^2 R^2 T_{c,i}^2}{T_{r,i}^m P_{c,i}^2} \cdot A_i^1 \; ; \qquad\qquad b_i = \frac{Z_{c,i} RT_{c,i}}{P_{c,i}} \cdot B_i^1 ;\tag{7}$$

$$a_i = \frac{Z_{c,i}^2 R^2 T_{c,i}^2}{P_{c,i}^2} \cdot C_i^1 ; \qquad\qquad d_i = \frac{Z_{c,i} RT_{c,i}}{P_{c,i}} \cdot D_i^1 ,$$

where $A_i^1, B_i^1, C_i^1, D_i^1$ are parameters depending on the substance. They can be correlated with Riedel's criteria, α_c, the compressibility factor, $Z_{c,i} = P_{c,i} V_{c,i}/RT_{c,i}$, and the parameter m called the "isothermal contraction parameter":

$$A_i' = \frac{(\alpha_{c,i} - 1)^3}{Z_{c,i}^2 (\alpha_{c,i} + m_i)^3}; \qquad\qquad B_i^1 = 1 - \frac{1 + m_i}{Z_{c,i} (\alpha_{c,i} + m_i)},\tag{8}$$

$$C_i' = \frac{(\alpha_{c,i} - 1)^2 (3m_i + 4 - \alpha_{c,i})}{4Z_{c,i}^2 (\alpha_{c,i} + m_i)^3}; \qquad D_i^1 = 1 - \frac{(\alpha_{c,i} - 1)}{Z_{c,i} (\alpha_{c,i} + m_i)}.$$

For mixtures of two components, the parameters are calculated using the parameters of pure components, introduced in Van der Waals mixing rules [1]:

$$a = \sum_{i=1}^{NC} \sum_{j=1}^{NC} X_i X_j a_{ij}, \qquad a_{ij} = (a_i a_j)^{1/2} (1 - k_{ij}), \qquad i, j = \overline{1, 2},\tag{9}$$

$$b = \sum_{i=1}^{NC} b_i X_i,\tag{10}$$

where k_{ij} is called "interaction parameter" and is proportional with the non-ideal character of the system.

With the values for the parameters, calculated for each mixture composition (in liquid and in vapor phase) the value of the fugacity coefficient Φ_i^L and Φ_i^V are calculated. So, using GEOS, (3) has the particular form

$$\ln \Phi_i = \ln \frac{V}{V - b} + \frac{b + n\frac{\partial b}{\partial n_i}}{V - b} - \left(2a + n\frac{\partial a}{\partial n_i}\right) \frac{E}{RT} - \frac{a\left(d + n\frac{\partial d}{\partial n_i}\right)}{RT\left[(V - d)^2 + c\right]}\tag{11}$$

$$- \frac{a}{2RTc}\left(2c + n\frac{\partial c}{\partial n_i}\right) \left[\frac{V - d}{(V - d)^2 + c} - E\right] - \ln \frac{PV}{RT},\tag{12}$$

334

where the partial derivatives are the following:

$$n\frac{\partial a}{\partial n_i} = 2\left(\sum_{j=1}^{NC} X_j a_{ij} - a\right), \quad n\frac{\partial b}{\partial n_i} = b_i - b, \quad n\frac{\partial c}{\partial n_i} = 2\left(\sum_{j=1}^{NC} X_j c_{ij} - c\right), \quad n\frac{\partial d}{\partial n_i} = d_i - d$$

$$(13)$$

and

$$E = \frac{1}{\sqrt{c}}\operatorname{arctg}\frac{\sqrt{c}}{V+D} \text{ if } c > 0; \quad E = \frac{1}{2\sqrt{-c}}\ln\frac{V-D+\sqrt{-c}}{V-D-\sqrt{-c}} \text{ if } c < 0; \quad (14)$$

for $c = 0$:

$$\ln\Phi_i = \ln\frac{V}{V-b} + \frac{b + n\frac{\partial b}{\partial n_i}}{V-b} - \frac{\left(2a + n\frac{\partial a}{\partial n_i}\right)}{RT(V-d)} - \frac{a\left(d + n\frac{\partial d}{\partial n_i}\right)}{RT(V-d)^2} - \ln\frac{PV}{RT}. \quad (15)$$

The equilibrium conditions (2), (3) can be applied and (P, V, T, X_i, Y_i) sets can be obtained as "calculated" values.

Usually simplified calculations are required when a number of variables as considered as known (the experimental value is equal with those calculated) as is shown in Table 1.

Known Data	Calculated Data	VLE Calculation Type
P,X_1,X_2	T,Y_1,Y_2	Bubble T
T,X_1,X_2	P,Y_1,Y_2	Bubble P
P,Y_1,Y_2	T,X_1,X_2	Dew T
T,Y_1,Y_2	P,X_1,X_2	Dew P
P,T	X_1,X_2,Y_1,Y_2	Flash

Table 1. Types of VLE calculations in binary systems.

3. Results and discussion

Predictions were done at high pressures for binary systems of n-alkanes with low nonideality when the value of the interaction parameter k_{ij} was set null in Bubble P calculations. The calculation procedure is not new; it follows the basics given by Prauznitz [6–8] and has, as its main comparative source, the FORTRAN program created by D. Geana [3–5]. Our calculation can be described as a LabVIEW product and has the advantage of being a very compact and easy-to-handle program. LabVIEW allowed as an option a step-by-step run, so the program can be studied and understood also by nonchemists.

The results are plotted as dependencies of different parameters vs. compositions in liquid and vapor phase: $P = f(X_1, Y_1), V = f(X_1, Y_1), \rho = f(X_1, Y_1)$, etc. The quality of the results is judged in Bubble P calculations on the $P = f(X_1, Y_1)$ curve, therefore these plots also contain the experimental values for the systems we have chosen as examples. In Fig. 1 we present the VLE curves calculated with our program for the binary system propane-butane at five different temperatures.

In Fig. 2–4 we plot some other results obtained for a propane-pentane system, at 344.28 K. The same kind of plots can be calculated for any binary systems, [2]; If the non-ideality increases, the k_{ij} value must be obtained as the result of an optimization calculation, but this is not the goal of the present paper.

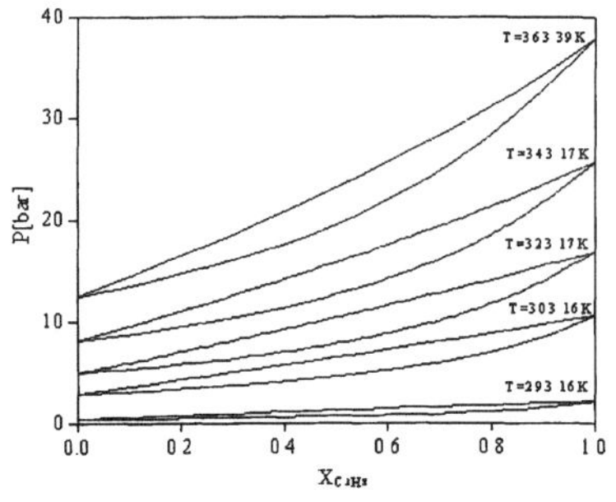

Fig. 1. VLE prediction for
$C_3H_8 - C_4H_{10}$.

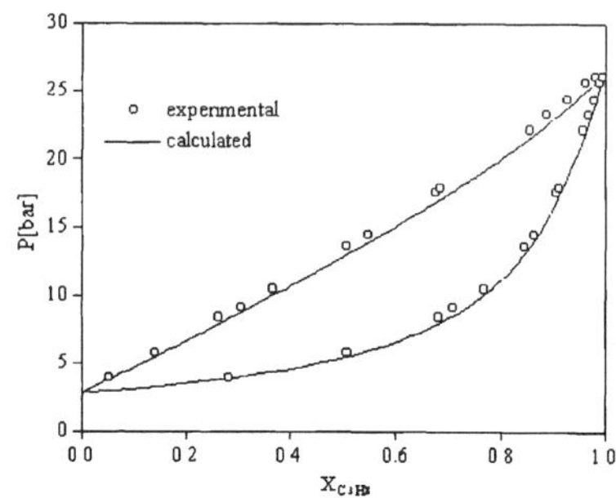

Fig. 2. VLE prediction for
$C_3H_8 - C_5H_{12}$,
$T = 344.28$ K [9].

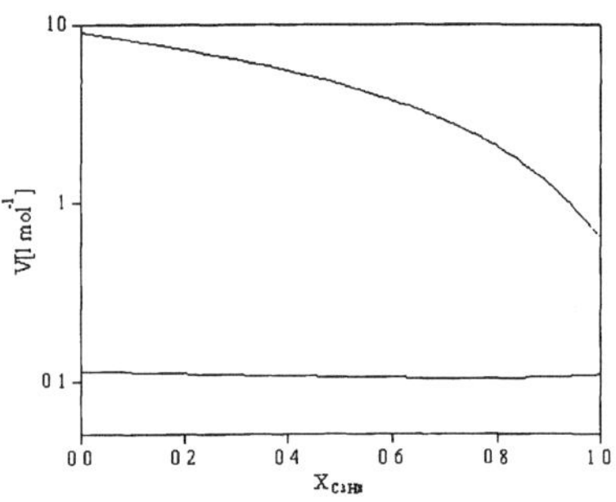

Fig. 3. Calculated volumes for the
liquid and vapor phases in
$C_3H_8 - C_5H_{12}$, $T = 344.28$ K.

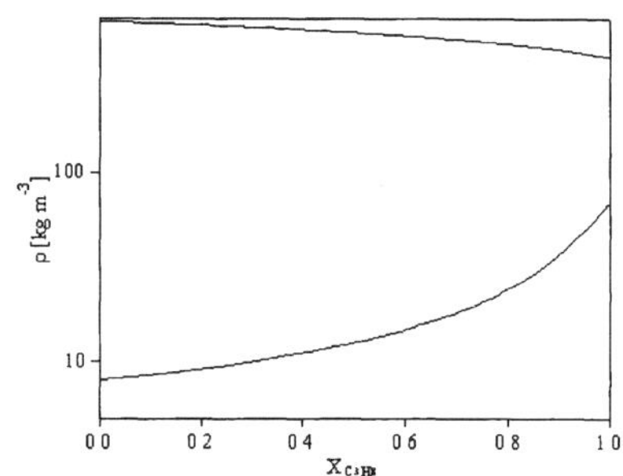

Fig. 4. Calculated densities for
liquid and vapor phases in
$C_3H_8 - C_5H_{12}$, $T = 344.28$ K.

4. Conclusions

The paper presents a new high-pressure VLE calculation program written using Lab-VIEW. As example, prediction calculations are done for some binary systems of n-alkanes at different temperatures. The program involves the latest results reported in high-pressure VLE calculations and LabVIEW proves to be a great tool, even for such scientific purposes, because the program is compact and easy to follow, and the results can be plotted in the most convenient and easily interpreted form.

References

1. P. Beranek and J. Wichterle, *Fluid Phase Equilibria* **279** (1981).

2. C.A. Duta and D. Geana, *J. Chimie* **332** (1996).

3. D. Geana, *J. Chimie* **303** (1986).

4. D. Geana, *J. Chimie* **951** (1986).

5. D. Geana, *J. Chimie* **975** (1987).

6. J.M. Prauznitz, C.A. Eckert, R.V. Orye and J.P. O'Conell, *Computer calculation for multicomponent vapor-liquid equilibra*, Prentice Hall, NJ, 1967.

7. J.M. Prauznitz and P.L. Chueh, *Computer calculation for high pressure vapor-liquid equilibra*, Prentice Hall, NJ, 1998.

8. J.M. Prauznitz, T.F. Anderson and E.A. Greus, *Computer calculation for multicomponent vapor-liquid and liquid-liquid equilibra*, Prentice Hall, N.J., 1980.

9. J. Vejrosta and I. Wichterle, *Coll. Czech. Chem. Comm.* **1246** (1974), 39.

10. J. Wichterle, *Modelierung von Phasengleichgewichte als Grundlage von Stofftrennprozessen*, Akademie-Verlag, Berlin, 1981.

Department of Physics, "Transilvania" University, Brasov, Romania;
e-mail: udoru@unitbv.ro, aduta@info.unitbv.ro

V. VASSILEV

Symmetry groups, conservation laws and group-invariant solutions of the Marguerre-von Kármán equations

1. Introduction

Marguerre's theory for large deflection of thin isotropic elastic shells [1] leads to the following system of two coupled nonlinear fourth-order partial differential equations,

$$
\begin{aligned}
D\Delta^2 w - \varepsilon^{\alpha\mu}\varepsilon^{\beta\nu} w_{;\alpha\beta}\, \Phi_{;\mu\nu} - \varepsilon^{\alpha\mu}\varepsilon^{\beta\nu} b_{\alpha\beta}\, \Phi_{;\mu\nu} &= p, \\
(1/Eh)\Delta^2 \Phi + (1/2)\varepsilon^{\alpha\mu}\varepsilon^{\beta\nu} w_{;\alpha\beta} w_{;\mu\nu} + \varepsilon^{\alpha\mu}\varepsilon^{\beta\nu} b_{\alpha\beta} w_{;\mu\nu} &= q,
\end{aligned}
\tag{1}
$$

in two independent variables—the coordinates on the shell middle-surface F, and two dependent variables—the transversal displacement function w, and Airy's stress function Φ, with right-hand sides appearing when the shell is subjected to an external transversal load and nonuniform heating. Here and throughout, $\varepsilon^{\alpha\beta}$ is the alternating tensor of F; $b_{\alpha\beta}$ is the curvature tensor of F; D, E, and h are the bending rigidity, Young's modulus, and thickness of the shell, respectively, which are supposed to be given constants; a semicolon is used for covariant differentiation with respect to the metric tensor $a_{\alpha\beta}$ of the surface F; Δ is the Laplace-Beltrami operator on F; Greek (Latin) indices range over 1, 2 (1, 2, 3), unless explicitly stated otherwise; the usual summation convention over repeated indices (one subscript and one superscript) is used.

This theory assumes that the intrinsic geometry of the shell middle-surface F should be Euclidean or approximately Euclidean in the following sense. Let (x^1, x^2, z) be a fixed right-handed rectangular Cartesian coordinate system in the three-dimensional Euclidean space in which the middle-surface F of a shell is embedded, and let this surface be given by the equation

$$
F : z = f(x^1, x^2), \ (x^1, x^2) \in \Omega \subset \mathbf{R}^2,
$$

where $f : \mathbf{R}^2 \to \mathbf{R}$ is a single-valued smooth function possessing as many derivatives as may be required on a certain domain of interest Ω. Let us take x^1, x^2 to serve as coordinates on the surface F. Then, relative to this coordinate system, the components of the fundamental tensors and the alternating tensor of F are given by the expressions

$$
a_{\alpha\beta} = \delta_{\alpha\beta} + f_{,\alpha} f_{,\beta}, \quad b_{\alpha\beta} = a^{-1/2} f_{,\alpha\beta}, \quad \varepsilon^{\alpha\beta} = a^{-1/2} e^{\alpha\beta},
\tag{2}
$$

where $a = \det(a_{\alpha\beta}) = 1 + (f_{,1})^2 + (f_{,2})^2$; $\delta_{\alpha\beta} = \delta^{\alpha\beta}$ is the Kronecker delta symbol; $e^{\alpha\beta}$ is the alternating symbol; subscripts after a comma at a certain function f denote its partial derivatives with respect to the coordinates on F. If the inequalities

$$
|f_{,\alpha}|\,|f_{,\beta}| \le \varepsilon^2 \ll 1, \quad \varepsilon = \text{const},
$$

hold for every point $(x^1, x^2) \in \Omega$ (such a shell is said to be shallow on the domain Ω), then the quadratic terms in the right-hand sides of expressions (2) are small compared

---

This work was partly supported by Contract No. MM 517/1995 with NSF, Bulgaria.

to unity, they may be neglected, and thus allowing for a relative error of order $O\left(\varepsilon^2\right)$ one may regard the intrinsic geometry of the shell middle-surface F as Euclidean and (x^1, x^2) may be thought of as an Euclidean coordinate system on F, in which

$$a_{\alpha\beta} = \delta_{\alpha\beta}, \ b_{\alpha\beta} = f_{,\alpha\beta}, \ \varepsilon^{\alpha\beta} = e^{\alpha\beta}, \tag{3}$$

and the mean curvature H of the surface F and its Gaussian curvature K read

$$H = (1/2)a^{\alpha\beta}b_{\alpha\beta} = (1/2)\delta^{\alpha\beta}f_{,\alpha\beta}, \ K = (1/2)\varepsilon^{\alpha\mu}\varepsilon^{\beta\nu}b_{\alpha\beta}b_{\mu\nu} = (1/2)e^{\alpha\mu}e^{\beta\nu}f_{,\alpha\beta}f_{,\mu\nu} \tag{4}$$

(note that the latter is not necessarily equal to zero within the allowed relative error).

Equations (1) are often referred to as Marguerre–von Kármán (MvK) equations to reflect the fact that they are an extension of the von Kármán equations for large bending of plates [2] (including the latter as a special case corresponding to $b_{\alpha\beta} = 0$) to the shallow shells. Actually (1) describe the state of equilibrium of the shell, but introducing, according to d'Alembert principle, the inertia force $-\rho w_{,33}$ in the right-hand side of the first MvK equation, $w_{,33}$ being the second derivative of the displacement field with respect to the time $t \equiv x^3$ and ρ the mass per unit area of the shell middle-surface, one can extend (1) to describe the dynamic behavior of shallow shells.

Applying the equivalence transformation $(x^1, x^2, w, \Phi) \mapsto (x^1, x^2, W, \Phi)$, $W = w + f$, to the time-independent MvK equations and $(x^1, x^2, x^3, w, \Phi) \mapsto (x^1, x^2, x^3, W, \Phi)$ to the time-dependent ones, one can map [3] the MvK equations to the von Kármán equations

$$\begin{aligned}
&D\Delta^2 W - \varepsilon^{\alpha\mu}\varepsilon^{\beta\nu}W_{;\alpha\beta}\,\Phi_{;\mu\nu} = P, \\
&(1/Eh)\,\Delta^2\Phi + (1/2)\,\varepsilon^{\alpha\mu}\varepsilon^{\beta\nu}W_{;\alpha\beta}W_{;\mu\nu} = Q,
\end{aligned} \tag{5}$$

and

$$\begin{aligned}
&D\Delta^2 W - \varepsilon^{\alpha\mu}\varepsilon^{\beta\nu}W_{;\alpha\beta}\,\Phi_{;\mu\nu} + \rho W_{,33} = P, \\
&(1/Eh)\,\Delta^2\Phi + (1/2)\,\varepsilon^{\alpha\mu}\varepsilon^{\beta\nu}W_{;\alpha\beta}W_{;\mu\nu} = Q,
\end{aligned} \tag{6}$$

respectively, where

$$P = 2Da^{\mu\nu}H_{;\mu\nu} + p, \quad Q = K + q. \tag{7}$$

Hereafter (5) and (6) will be referred to as the time-independent and time-dependent MvK equations, respectively. In both cases, the moment tensor $M^{\alpha\beta}$, membrane stress tensor $N^{\alpha\beta}$, and shear-force vector Q^α are given in terms of W and Φ by the expressions

$$\begin{aligned}
M^{\alpha\beta} &= D\left\{(1-\nu)a^{\alpha\mu}a^{\beta\nu} + \nu a^{\alpha\beta}a^{\mu\nu}\right\}\{W_{;\mu\nu} - f_{;\mu\nu}\}, \\
N^{\alpha\beta} &= \varepsilon^{\alpha\mu}\varepsilon^{\beta\nu}\Phi_{;\mu\nu}, \quad Q^\alpha = M^{\alpha\mu}_{;\mu} + N^{\alpha\mu}\{W_{;\mu} - f_{;\mu}\},
\end{aligned}$$

and the in-plane displacements v^α can be found solving the overdetermined system

$$v_{\alpha;\beta} + v_{\beta;\alpha} = (2/Eh)\left\{(1+\nu)\varepsilon^\mu_\alpha\varepsilon^\nu_\beta - \nu a_{\alpha\beta}a^{\mu\nu}\right\}\Phi_{;\mu\nu} - \{W_{;\alpha} - f_{;\alpha}\}\{W_{;\beta} - f_{;\beta}\}, \tag{8}$$

the second one of the MvK equations being its compatibility condition.

2. Symmetry groups [4]

The following is known [5] for the symmetry groups of the homogeneous MvK equations.

Proposition 1. *The homogeneous time-independent MvK equations (5) admit the group $G_{(S)}$ generated by the basic vector fields (operators)*

$$Y_1 = \frac{\partial}{\partial W}, \quad Y_2 = \frac{\partial}{\partial x^1}, \quad Y_3 = \frac{\partial}{\partial x^2}, \quad Y_4 = x^2 \frac{\partial}{\partial x^1} - x^1 \frac{\partial}{\partial x^2}, \quad Y_5 = x^1 \frac{\partial}{\partial \Phi},$$

$$Y_6 = x^2 \frac{\partial}{\partial \Phi}, \quad Y_7 = \frac{\partial}{\partial \Phi}, \quad Y_8 = x^1 \frac{\partial}{\partial W}, \quad Y_9 = x^2 \frac{\partial}{\partial W}, \quad Y_{10} = x^1 \frac{\partial}{\partial x^1} + x^2 \frac{\partial}{\partial x^2}.$$

Proposition 2. *The homogeneous time-dependent MvK equations (6) admit the group $G_{(D)}$ generated by the basic vector fields:*

$$X_1 = \frac{\partial}{\partial W}, \quad X_2 = \frac{\partial}{\partial x^1}, \quad X_3 = \frac{\partial}{\partial x^2}, \quad X_4 = \frac{\partial}{\partial x^3}, \quad X_5 = x^1 \frac{\partial}{\partial x^1} + x^2 \frac{\partial}{\partial x^2} + 2x^3 \frac{\partial}{\partial x^3},$$

$$X_6 = x^2 \frac{\partial}{\partial x^1} - x^1 \frac{\partial}{\partial x^2}, \quad X_7 = x^1 \frac{\partial}{\partial W}, \quad X_8 = x^2 \frac{\partial}{\partial W}, \quad X_9 = x^3 \frac{\partial}{\partial W}, \quad X_{10} = x^1 x^3 \frac{\partial}{\partial W},$$

$$X_{11} = x^2 x^3 \frac{\partial}{\partial W}, \quad X_{12} = x^1 f(x^3) \frac{\partial}{\partial \Phi}, \quad X_{13} = x^2 g(x^3) \frac{\partial}{\partial \Phi}, \quad X_{14} = h(x^3) \frac{\partial}{\partial \Phi},$$

where f, g, and h are arbitrary functions depending on the time only.

As for the symmetries of the nonhomogeneous MvK equations, we proved the following.

Proposition 3. *A nonhomogeneous time-independent MvK system is invariant under a vector field Y iff $Y = c^j Y_j$ $(j = 1, \ldots, 10)$, where c^j are real constants, and*

$$2P\xi^\mu_{,\mu} + \xi^\mu P_{,\mu} = 0, \quad 2Q\xi^\mu_{,\mu} + \xi^\mu Q_{,\mu} = 0, \tag{9}$$

for $\xi^\alpha = Y(x^\alpha)$, Y being regarded as an operator acting on the functions $\zeta : \Omega \to \mathbf{R}$, $\Omega \subset \mathbf{R}^2$.

Proposition 4. *A nonhomogeneous time-dependent MvK system is invariant under a vector field X iff $X = C^j X_j$ $(j = 1, \ldots, 14)$, where C^j are real constants, and*

$$P\xi^i_{,i} + \xi^i P_{,i} = 0, \quad Q\xi^i_{,i} + \xi^i Q_{,i} = 0, \tag{10}$$

for $\xi^i = X(x^i)$, X being regarded as an operator acting on the functions $\chi : \Omega \times T \to \mathbf{R}$, $\Omega \subset \mathbf{R}^2$, $T \subset \mathbf{R}$.

The above propositions imply the following group classification results.

Theorem 1. *The time-independent MvK equations (5) admit the one-parameter group G iff G is generated by a vector field $Y = c^j Y_j$ and the right-hand sides P and Q are invariants of G (when $c^{10} = 0$) or eigenfunctions (when $c^{10} \neq 0$) of its generator Y.*

Theorem 2. *The time-dependent MvK equations (6) admit the one-parameter group G iff G is generated by a vector field $X = C^j X_j$ and the right-hand sides P and Q are invariants of G (when $C^5 = 0$) or eigenfunctions (when $C^5 \neq 0$) of its generator X.*

3. Conservation laws

Both the time-independent and the time-dependent MvK equations constitute self-adjoint systems and are the Euler-Lagrange equations associated with the functionals

$$I^{(S)}[W, F] = \int \int \int L^{(S)} \, dx^1 dx^2, \ L^{(S)} = \Pi,$$

and

$$I^{(D)}[W, F] = \int \int \int L^{(D)} dx^1 dx^2 dx^3, \ L^{(D)} = (\mathrm{T} - \Pi),$$

respectively, where

$$\begin{aligned}
\Pi \ = \ & (D/2)\left\{(\Delta W)^2 - (1 - \nu)e^{\alpha\mu}e^{\beta\gamma}W_{,\alpha\beta}W_{,\mu\nu}\right\} \\
& - \ (1/2Eh)\left\{(\Delta\Phi)^2 - (1+\nu)e^{\alpha\mu}e^{\beta\nu}\Phi_{,\alpha\beta}\Phi_{,\mu\nu}\right\} + (1/2)\, e^{\alpha\mu}e^{\beta\nu}\Phi_{,\alpha\beta}W_{,\mu}W_{,\nu} \\
& - \ PW - Q\Phi
\end{aligned}$$

is the strain energy per unit area of the shell middle-surface and

$$\mathrm{T} = (\rho/2)\left(W_{,3}\right)^2$$

is the kinetic energy per unit area of the shell middle-surface.

In [6], the variational symmetries of the above functionals with $P = Q = 0$ are established and all Noether's conservation laws admitted by the smooth solutions of the homogeneous MvK equations are presented (see also Table 1 in [7] where the conservation laws associated with the time-dependent MvK equations are listed). The following statements hold for the nonhomogeneous MvK equations.

Proposition 5. *A conservation law of flux $A^\alpha_{(j)}$ and characteristic $\Lambda^\alpha_{(j)}$ ($j = 1, \ldots, 9$) admitted by the smooth solutions of the homogeneous time-independent MvK equations takes the form*

$$A^\mu_{(j),\mu} + S_{(j)} = 0, \ S_{(j)} = -\Lambda^1_{(j)}P - \Lambda^2_{(j)}Q \tag{11}$$

on the smooth solutions of the nonhomogeneous time-dependent MvK equations;

$$S_{(j)} = \widetilde{A}^\mu_{(j),\mu}$$

iff (9) hold, and then (11) can be written as a divergence-free expression (i.e., it becomes a proper conservation law in the sense appropriated in the group analysis of differential equations, see e.g. [4]), otherwise it has supply (production) $S_{(j)}$.

Proposition 6. *Each conservation law of density $\Psi_{(i)}$, flux $P^\alpha_{(i)}$ and characteristic $\Lambda^\alpha_{(i)}$ ($i = 1, \ldots, 14$) admitted by the smooth solutions of the homogeneous time-dependent MvK equations takes the form*

$$\Psi_{(i),3} + P^\mu_{(i),\mu} + S_{(i)} = 0, \ S_{(i)} = -\Lambda^1_{(i)}P - \Lambda^2_{(i)}Q, \tag{12}$$

on the smooth solutions of the nonhomogeneous time-dependent MvK equations;

$$S_{(i)} = \Psi_{(i),3} + \widetilde{P}^\mu_{(i),\mu},$$

iff (10) hold, and in this case (12) becomes a proper conservation, otherwise it has supply (production) $S_{(i)}$.

Note that the source therms in (11) and (12) appear due to the curvature of the shell.

4. Balance laws

Given a region Ω in the shell middle-surface with sufficiently smooth boundary Σ of outward unit normal n_α, a balance law

$$\int\limits_{\Sigma} A^{\alpha}_{(j)} n_\alpha d\Sigma + \int\limits_{\Omega} S_{(j)} dx^1 dx^2 = 0 \tag{13}$$

corresponds to each of the nine basic conservation laws of fluxes $A^{\alpha}_{(j)}$ characteristic $\Lambda^{\alpha}_{(j)}$ $(j = 1, \ldots, 9)$ admitted by the smooth solutions of the homogeneous time-independent MvK equations (these conservation laws are listed in Appendix B [6]).

The same holds true for the fourteen basic conservation laws of densities $\Psi_{(i)}$, fluxes $P^{\alpha}_{(i)}$ and characteristics $\Lambda^{\alpha}_{(i)}$ $(i = 1, \ldots, 14)$ admitted by the smooth solutions of the homogeneous time-dependent MvK equations (see [6, Appendix A] and [7, Table 1). Namely, to each of them there corresponds a balance low

$$\frac{d}{dt} \int\limits_{\Omega} \Psi_{(i)} dx^1 dx^2 + \int\limits_{\Sigma} P^{\alpha}_{(i)} n_\alpha d\Sigma + \int\limits_{\Omega} \int\limits_{T} S_{(i)} dx^1 dx^2 dx^3 = 0, \tag{14}$$

where T is a certain time interval.

Both (13) and (14) hold, just as the respective conservation laws, for every smooth solution of the nonhomogeneous MvK equations.

In the static case the balance laws (13) provide a set of path-independent integrals inherent to Marguerre's shell theory. Among them are the counterparts of the well-known and widely used in fracture mechanics J-, L-, and M-integrals. The applicability of the latter integrals in the analysis of cracked plates is discussed in [8] (see also the references therein). In the similar way, the path-independent integrals corresponding to the balance laws (13) can be used in the analysis of equilibrium and stability of shells undergoing stress concentrations near the tips of cracks and notches, since they allow computation of the stress intensity factors and energy release rates, the former characterizing the distribution of the stress field in a vicinity of a certain singular point, say the crack tip, and the latter characterizing the propagation of the crack through the shell.

In the dynamic case, the balance laws (14) provide the theoretical background for studying the propagation of waves of discontinuity in shallow shells since they are applicable in the domains where some important physical quantities suffer jump discontinuities at a certain curve. Using the balance laws (14), one can extend the "continuous" Marguerre shell theory in the same manner as it is done in [7] for the "continuous" von Kármán plate theory. Specifically, Definition 1 in [7] should be changed by supplying the integrals (4) with the right-hand sides

$$\int\limits_{\Omega} \int\limits_{T} P dx^1 dx^2 dx^3, \quad \int\limits_{\Omega} \int\limits_{T} Q dx^1 dx^2 dx^3.$$

Then Definition 1, Propositions 1–4, and the jump conditions listed in Table 2 in [7] remain the same under the assumption that P and Q are smooth functions.

5. Group-invariant solutions

In the cases when the MvK equations (5) or (6) admit a curtain subgroup of $G_{(S)}$ or $G_{(D)}$, respectively, it is worth looking for the corresponding group-invariant solutions. To obtain such solution one should follow the procedure described in detail in [4]. Here, we would like only to notice that the group-invariant solutions to the homogeneous time-dependent MvK equations (6) obtained in [6] and shown to determine acceleration waves in plates can be used for the same purpose in Marguerre's shell theory provided (according to Theorem 2) that P and Q are joined invariants of the group generated by the vector fields X_6 and eigenfunctions of X_5. In this case the reduced system reads

$$
\begin{aligned}
D(u'' - 4u' + 4u)'' + (De^{4s}/4 - \varphi')u'' + (De^{4s}/2 - \varphi'' + 2\varphi')u' &= P, \\
(\varphi'' - 4\varphi' + 4\varphi)'' + Eh(u'' - u')u' &= Q,
\end{aligned}
$$

where $s = (1/2)\ln(\sqrt{\rho/D}\,r^2/t)$, $r^2 = (x^1)^2 + (x^2)^2$, $u(s)$ and $\varphi(s)$ are the new dependent variables, and the prime denotes differentiation with respect to the argument s.

As for the group-invariant solutions describing traveling waves in plates discussed in [7, Section 5], now, according to Theorem 1, P and Q are to be joint invariants of the group generated by X_3 and $X_2 + (1/c)X_4$.

References

1. K. Marguerre, Zur Theorie der gekrümmten Platte großer Formänderung, in *Proc. Fifth Internat. Congr. Appl. Mech.*, Cambridge, Massachusetts, 1938, 93.

2. Th. von Karman, Festigkeitesprobleme im Maschinebau, in *Encyklopädie der Mathematischen Wissenschaften*, vol. IV, Taubner, Leipzig, 1910, 311.

3. V. Vassilev, Symmetry groups and equivalence transformations in the nonlinear Donnell-Mushtari-Vlasov theory for shallow shells, *J. Theor. Appl. Mech.* **27** (1997), 43–51.

4. P.J. Olver, *Applications of Lie groups to differential equations*, 2nd ed., Graduate Texts in Mathematics **107**, Springer-Verlag, New York, 1993.

5. F. Shwarz, Lie symmetries of the von Kármán equations, *Comput. Phys. Comm.* **31** (1984), 113–114.

6. P. Djondjorov and V. Vassilev, Conservation laws and group-invariant solutions of the von Kármán equations, *Internat. J. Non-Linear Mech.* **31** (1996), 73–87.

7. P. Djondjorov and V. Vassilev, Acceleration waves in the von Kármán plate theory, in *Integral methods in science and engineering*, Pitman Res. Notes Math. Ser., Chapman & Hall/CRC, Boca Raton, FL (this volume).

8. H. Sosa, P. Rafalski and G. Herrmann, Conservation laws in plate theories, *Ingenieur-Archiv* **58** (1988), 305–320.

Institute of Mechanics, Bulgarian Academy of Sciences, Acad. G. Bontchev St., Bl. 4, 1113 Sofia, Bulgaria; e-mail: vassil@bgcict.acad.bg

A. VERMA, P.J. HARRIS and R. CHAKRABARTI

Modeling the motion of an underwater explosion bubble

1. Introduction

In recent times there has been a significant interest in the problem of determining the motion of an explosion bubble, especially when there is a submerged structure close to the point of explosion. The phenomenon of jet formation during the collapse of an underwater explosion bubble in the neighborhood of a submerged structure is postulated as a possible mechanism for causing damage to the structure. The direction of this jet is dependent on the geometry of the fluid region in which the bubble exists.

In this paper, we show how the boundary integral method, with an axisymmetric formulation, can be used to model the motion of a bubble close to a fixed rigid structure, such as a sphere, cylinder or a plate, immersed in an infinite fluid. Section 2 introduces a simple mathematical model which can be used to study the motion of a bubble in a fluid. In Section 3 we reformulate the problem as an integral equation over the surfaces of the bubble and rigid structure, and present a numerical scheme for obtaining its solution. Section 4 presents the results for the motion of a bubble close to a number of different rigid structures under the influence of gravity.

2. Mathematical model

This section presents a suitable model for determining the motion of a bubble close to a fixed rigid structure by making the standard assumptions [5] that the fluid is inviscid, incompressible, and irrotational. Therefore, the flow field can be described by a velocity potential ϕ, which is the solution of the Laplace's equation [7]

$$\nabla^2 \phi = 0 \tag{1}$$

The contents of the bubble (if any) are assumed to be ideal and the thermodynamic processes are assumed to be adiabatic with a constant γ. Hence the pressure P_b of the gas inside the bubble is given by

$$P_b = P_0 \left(\frac{V_0}{V(t)} \right)^\gamma, \tag{2}$$

where V_0 is the initial volume of the fluid, $V(t)$ is the volume of the bubble at some later time t, and P_0 is the initial pressure inside the bubble. If we let P_∞ denote the far-field pressure in the plane $z = 0$, it is possible to write Bernoulli's equation for any point in the fluid as [7]

$$\frac{\partial \phi}{\partial t} + \frac{1}{2} |\nabla \phi|^2 + \frac{P}{\rho} + gz = \frac{P_\infty}{\rho}, \tag{3}$$

where ρ is the density of the fluid, g is the acceleration due to gravity assumed to be directed parallel to the negative z-axis, and P denotes the pressure at the point in the fluid. Since the fluid pressure at the surface of the bubble must be equal to the pressure inside the bubble, Bernoulli's equation yields

$$\frac{\partial \phi}{\partial t} + \frac{1}{2} |\nabla \phi|^2 + \frac{P_0}{\rho} \left(\frac{V_0}{V(t)} \right)^\gamma + gz = \frac{P_\infty}{\rho} \tag{4}$$

344

at all points on the surface of the bubble. There cannot be a fluid flow perpendicular to the surface of a fixed rigid structure, therefore $\frac{\partial \phi}{\partial n} = 0$ on all such surfaces. The initial conditions of the system are that the initial potential on the surface of the bubble and the internal pressure of the bubble are known. For a cavitation bubble the initial potential is taken as the Rayleigh solution [9] and the internal pressure is assumed to be zero, while for an explosion bubble the initial potential is zero but there is a large excess pressure inside the bubble.

3. Numerical analysis

In this formulation the fluid domain (Ω) is assumed to be unbounded and so there are obvious problems with the use of domain-based numerical methods, such as the finite element method, to solve the underlying differential equation problem. For this reason, the boundary integral method has proved popular for solving problems such as this, as the three-dimensional infinite domain differential equation problem is transformed into a two-dimensional integral equation defined on a finite region, namely the surfaces of the bubble and the rigid structure. It can be shown that the velocity potential ϕ and the normal derivative of the potential $\frac{\partial \phi}{\partial n}$ must satisfy Green's second theorem [6]

$$\int_S \left\{ \phi(p)\frac{\partial G}{\partial n}(p,q) - G(p,q)\frac{\partial \phi}{\partial n}(q) \right\} dS_q = \left\{ \begin{array}{ll} \phi(p) & \text{if } p \in \Omega \\ \frac{1}{2}\phi(p) & \text{if } p \in S \\ 0 & \text{otherwise} \end{array} \right. . \tag{5}$$

where S denotes the union of the surface of the bubble and the surface of the rigid structure and $G(p,q)$ is the free space Green's function

$$G(p,q) = \frac{1}{4\pi|p-q|}. \tag{6}$$

For $p \in S$, (5) yields a first kind Fredholm integral equation for $\frac{\partial \phi}{\partial n}$ if ϕ is known and a second kind Fredholm integral equation for ϕ if $\frac{\partial \phi}{\partial n}$ is known. If $p \in S_b$ or $p \in S_r$, where b denotes the quantities on the bubble and r denotes the quantities on the rigid structure, the integrals appearing in (5) can be split to yield

$$\begin{aligned}
\frac{1}{2}\phi_b(p) &= \int_{S_b} \left\{ \phi_b(p)\frac{\partial G}{\partial n}(p,q) - G(p,q)\frac{\partial \phi_r}{\partial n}(q) \right\} dS_q \\
&\quad + \int_{S_r} \left\{ \phi_r(p)\frac{\partial G}{\partial n}(p,q) - G(p,q)\frac{\partial \phi_r}{\partial n}(q) \right\} dS_q, \quad p \in S_b, \tag{7} \\
\frac{1}{2}\phi_r(p) &= \int_{S_b} \left\{ \phi_b(p)\frac{\partial G}{\partial n}(p,q) - G(p,q)\frac{\partial \phi_r}{\partial n}(q) \right\} dS_q \\
&\quad + \int_{S_r} \left\{ \phi_r(p)\frac{\partial G}{\partial n}(p,q) - G(p,q)\frac{\partial \phi_r}{\partial n}(q) \right\} dS_q, \quad p \in S_r. \tag{8}
\end{aligned}$$

This analysis can be generalized to more than two surfaces. The effects of an infinite rigid plane on the bubble's motion can be included by using the modified Green's function as used by Blake et al. [5]. To further simplify the problem we assume that

345

fluid domain and the flow are axisymmetric about the z-axis. The surface of the bubble
and the structure are generated by rotating some appropriate curve about the z-axis.
For the bubble's surface S_b the parametric functions $r(s)$ and $z(s)$ are interpolated
by using clamped cubic splines [3]. Similarly as the value of ϕ_b is known at all node
points, it can be interpolated by a clamped cubic spline with $\frac{\partial \phi}{\partial s}$ at both ends. The
unknown normal derivative of the potential, ψ, is approximated by using a piecewise
linear interpolation scheme. For the surface of the rigid structure S_r the generating
curves $r(s)$ and $z(s)$ and the potential are interpolated using a simple piecewise linear
interpolation scheme, as is the normal derivative of the potential ψ_r. Further details
of these approximations can be found in Amini et al. [1]. Using the approximate
surfaces, potentials and the normal derivative of the potential described above, it is
possible to discretize (7) and (8) using the collocation method [2] to obtain a block
matrix equation of the form

$$\begin{pmatrix} L_{bb} & -M_{br} \\ L_{rb} & -M_{rr} \end{pmatrix} \begin{pmatrix} \Psi_b \\ \Phi_r \end{pmatrix} = \begin{pmatrix} M_{bb} & -L_{br} \\ M_{rb} & L_{rr} \end{pmatrix} \begin{pmatrix} \Phi_b \\ \Psi_r \end{pmatrix}, \tag{9}$$

where Φ_b is the vector of the nodal values of ϕ_b and Φ_r, Ψ_b and Ψ_r are similarly defined.
The block matrix equation can be solved for Ψ_b and Φ_r. Once both the potential and
the normal derivative of the potential are known on the surface of the bubble, it is
possible to compute the components of fluid velocity $\frac{\partial \phi}{\partial r}$ and $\frac{\partial \phi}{\partial z}$ at each node by solving

$$\frac{\partial \phi}{\partial n} = \frac{\partial \phi}{\partial r} n_r + \frac{\partial \phi}{\partial z} n_z, \tag{10}$$

$$\frac{\partial \phi}{\partial s} = \frac{\partial \phi}{\partial r} \frac{\partial r}{\partial s} + \frac{\partial \phi}{\partial z} \frac{\partial z}{\partial s}, \tag{11}$$

where $\frac{\partial \phi}{\partial s}$, $\frac{\partial r}{\partial s}$, and $\frac{\partial z}{\partial s}$ are obtained by differentiating the appropriate interpolating cubic
splines and n_r and n_z denote the r and z components of the unit normal, respectively.
The location of the surface of the bubble and the surface potential can now be updated
by a simple Euler's scheme as described in [8].

4. Computational results and conclusions

All calculations are made in terms of the nondimensional variables [4]. The validity
of the model has been verified [8] by showing that our results are consistent with the
previous work [5].

Fig. 1 shows the growth and collapse of a bubble which is 1.0 unit above a fixed
rigid cylinder. The radius of the cylinder is 1.0 unit and its height is 2.5 units. The
entire configuration is in the buoyancy field with a buoyancy parameter, $\delta = 0.1$. The
figure illustrates that if the bubble is close to the rigid structure then it jets downward
toward the rigid structure, whereas it would jet upwards due to the buoyancy field if
the structure were not there.

Fig. 2 shows the displacement of the centroid of the bubble at different initial
distances above a fixed rigid sphere, in a mild buoyancy field (with the buoyancy
parameter $\delta = 0.1$). The radius of the sphere is 3.5 units and the bubble is placed
1.0, 2.0, 3.0, 4.0 units, respectively, above the rigid sphere. The figure shows that the
motion of the bubble, i.e., toward or away from the structure, is dependent on the
initial distance of the bubble from the structure. The effect is similar to that observed
by Blake et al. [5] for a bubble near an infinite rigid boundary.

346

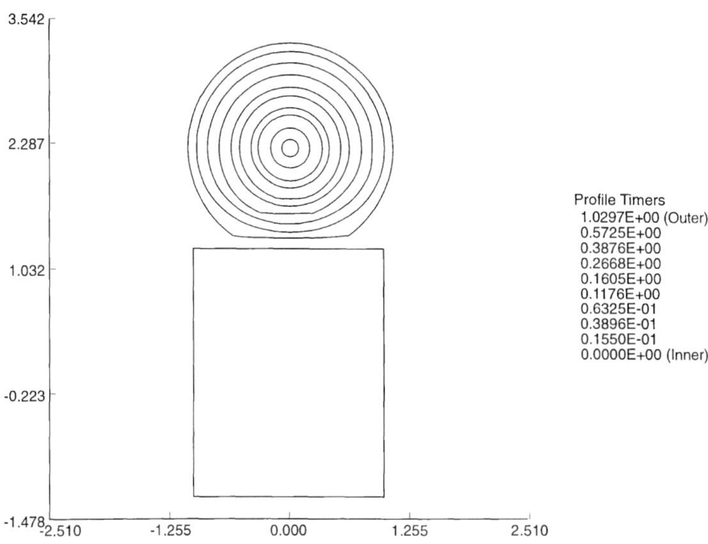

Fig. 1a. Growth and collapse of an explosion bubble above a fixed rigid cylinder.

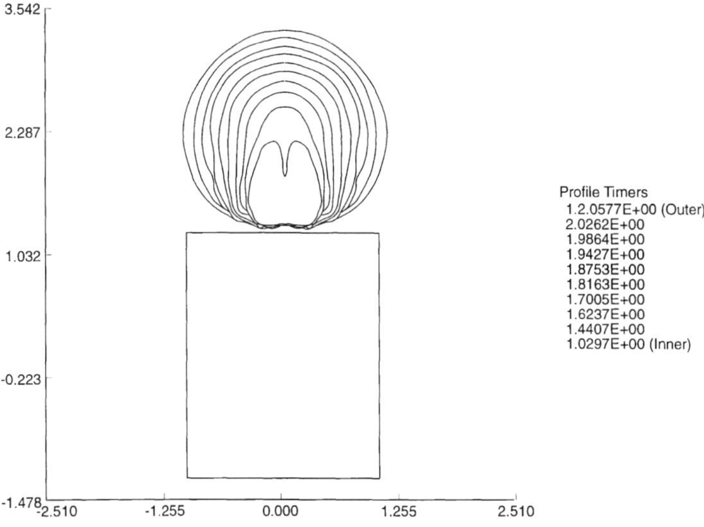

Fig. 1b. Growth and collapse of an explosion bubble above a fixed rigid cylinder.

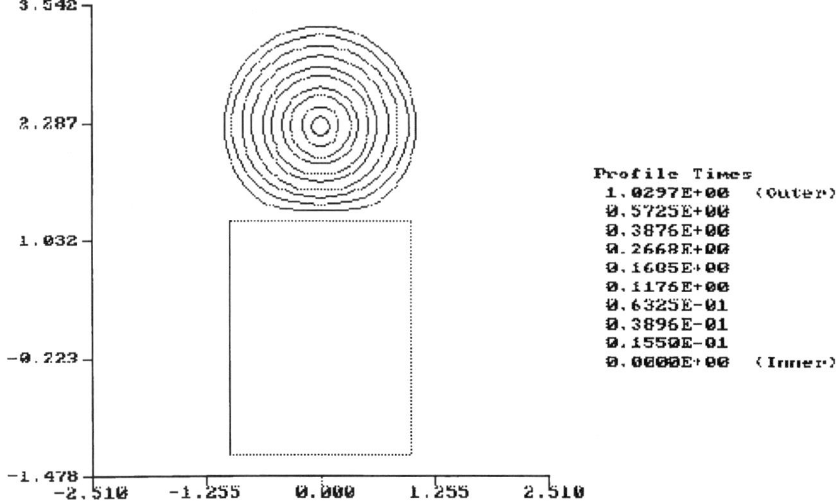

Fig. 2. Displacement of the centroid of the explosion bubble at various initial distances above a fixed rigid sphere.

The boundary element method has shown to be a useful tool in predicting the motion of the bubble in an infinite and semi-infinite potential flow situation. This paper has shown how the basic method can be modified to include finite rigid structures immersed in the fluid close to the bubble. For relatively large structures, the bubble essentially behaves like a bubble close to an infinite rigid boundary, whereas for structures that are approximately the same size as that of the bubble, the interaction is more complicated.

References

1. S. Amini, P.J. Harris and D.T. Wilton, *Coupled boundary and finite element methods for the solution of the dynamic fluid-interaction problem*, Springer-Verlag, 1992.

2. K.E. Atkinson, *A survey of numerical methods for the solution of Fredholm integral equations of the second kind*, SIAM, Philadelphia, 1976.

3. K.E. Atkinson, *An introduction to numerical analysis*, Wiley, 1978.

4. J.P. Best, *The dynamics of the underwater explosions*, Ph.D. thesis, The University of Wollongong, New South Wales, 1991.

5. J.R. Blake, B.B. Taib and G. Doherty, Transient cavities near boundaries. Part 1. Rigid boundaries, *J. Fluid Mech.* **170** (1986), 476–497.

6. M.A. Jaswon and G.T. Symm, *Integral equation methods in potential theory and elastostatics*, Academic Press, London, 1977.

7. J. Lighthill, *An informal introduction to theoretical fluid mechanics*, IMA Monograph Series, Oxford University Press, 1986.

8. P.J. Harris, A. Verma and R. Chakrabarti, Interaction of an explosion bubble with a fixed rigid structure, *Internat. J. Numer. Methods Fluids* (to appear).

9. Lord Rayleigh, On the pressure developed in a liquid during the collapse of a spherical void, *Phil. Mag.* **34** (1917), 96–98.

School of Computing and Mathematical Sciences, University of Brighton,
Watts Building, Lewes Road, Brighton, UK

N. VIRCHENKO

On some new systems of N-ary integral equations

1. Introduction

Systems of dual (triple, N-ary) integral equations are found often in the solving of a wide class of mixed boundary value problems in mathematical physics, elasticity, etc. [1]. Such systems of integral equations have been investigated in detail. At the same time systems of dual integral equations with more complicated functions have not been investigated and therefore are interesting for study.

In this paper, systems of dual (triple) integral equations with generalized Legendre functions, Fox's H-functions, generalized Watson functions, and with Whittaker's functions are solved. These systems are solved by different methods. In particular, the system of dual integral equations with Whittaker functions $W_{k,i\tau}$ using Weyl fractional integral and generalized Kontorovich-Lebedev transform is reduced to the system of linear equation for the unknown functions.

A system of N-ary integral equations with generalized Legendre functions $P_{-1/2+i\tau}^{m,n}(\cosh \alpha)$ in the kernel is solved and investigated by using the generalized integral Mehler–Fock transform. The case $N = 3$ is considered in detail.

2. The general N-ary system

As is known [1], the system of N-ary integral equations is the set of integral equations of the following form:

$$\int_{D_i^j} \sum_{k=1}^{l} a_{ik}^j \phi_k(\tau) K_i^j(\tau, x) d\tau = f_i^j(x), \quad (i = \overline{1, N};\ j = \overline{1, l};\ x \in I_i^j), \tag{1}$$

where D_i^j and I_i^j are intervals of the real axis, $\phi_k(\tau)$ are unknown functions designated on the set $\bigcup_{i=1}^{N} \bigcup_{j=1}^{l} D_i^j$; $K_i^j(\tau, x)$ is the i kernel of j equation; f_i^j are the given functions at I_i^j; and a_{ik}^j are the weight functions.

3. A system of triple integral equations

Let us consider the system of triple integral equations:

$$\int_0^\infty \omega(\tau) \sum_{k=1}^{3} a_{lk}(\tau) \phi_k(\tau) P_{-1/2+i\tau}^{m,n}(\cosh \alpha) d\tau = f_l(\alpha), \quad (0 < \alpha < a),$$

$$\int_0^\infty \omega(\tau) \sum_{k=1}^{3} a_{lk}(\tau) \phi_k(\tau) P_{-1/2+i\tau}^{m,n}(\cosh \alpha) d\tau = g_l(\alpha), \quad (a < \alpha < b), \tag{2}$$

$$\int_0^\infty \omega(\tau) \sum_{k=1}^{3} a_{lk}(\tau) \phi_k(\tau) P_{-1/2+i\tau}^{m,n}(\cosh \alpha) d\tau = h_l(\alpha), \quad (\alpha > b),$$

350

where $\phi_k(\tau)$ are the unknown functions; $a_{lk}(\tau)$ are given functions for $\tau \in [0, \infty)$; $f_l(\alpha), g_l(\alpha), h_l(\alpha)$ are given functions for $\alpha \in [0, a), \alpha \in (a, b), \alpha \in (b, \infty)$, respectively,

$$
\begin{aligned}
\omega(\tau) &= 2^{m-n-2}\Gamma\left(\frac{1-m-n}{2}+i\tau\right)\Gamma\left(\frac{1-m+n}{2}+i\tau\right) \\
&\times \Gamma\left(\frac{1-m-n}{2}+i\tau\right)\Gamma\left(\frac{1-m-n}{2}-i\tau\right)\pi^{-1}\Gamma^{-1}(2i\tau)\Gamma(-2i\tau),
\end{aligned}
$$

and $P^{m,n}_{-1/2+i\tau}(\cosh \alpha)$ is the generalized associated Legendre function [2].

Let us suppose that

$$
\int_0^\infty \omega(\tau)\sum_{k=1}^3 a_{lk}(\tau)\phi_k(\tau)P^{m,n}_{-1/2+i\tau}(\cosh \alpha)d\tau = r_l(\alpha), \quad (a < \alpha < b), \tag{3}
$$

Using formulas of the generalized integral Mehler-Fock transform [3], from (1) we find that

$$
g_l(\alpha) = \int_0^a f_l(t)\sinh t F(t,\alpha)dt + \int_a^b r_l(t)\sinh t F(t,\alpha)dt + \int_b^\infty h_l(t)\sinh t F(t,\alpha)dt; \tag{4}
$$

here

$$
\begin{aligned}
F(t,\alpha) &= \frac{2^{n-m}\sinh^m\alpha\,\sinh^m t}{\Gamma^2(1/2-m)}\int_0^{\min(\alpha,t)}\frac{(\cosh\alpha-\cosh s)^{-m-1/2}}{(\cosh t-\cosh s)^{m+1/2}} \\
&\times {}_2F_1\left(\frac{n-m}{2},-\frac{m+n}{2};\frac{1}{2}-m;\frac{\cosh\alpha-\cosh s}{1+\cosh\alpha}\right) \\
&\times {}_2F_1\left(\frac{n-m}{2},-\frac{m+n}{2};\frac{1}{2}-m;\frac{\cosh t-\cosh s}{1+\cosh t}\right)ds.
\end{aligned}
$$

After transformations, (3) is brought to the form

$$
\begin{aligned}
N_1(\alpha) &= \int_a^\alpha M_l(s)(\cosh\alpha-\cosh s)^{-m-1/2} \\
&\times {}_2F_1\left(\frac{m-n}{2},-\frac{m+n}{2};\frac{1}{2}-m;\frac{\cosh\alpha-\cosh s}{1+\cosh\alpha}\right)ds, \tag{5}
\end{aligned}
$$

where

$$
\begin{aligned}
M_l(s) &= \int_s^b r_l(t)\sinh^{m+1}(t)(\cosh t-\cosh s)^{-m-1/2} \\
&\times {}_2F_1\left(\frac{n-m}{2},-\frac{m+n}{2};\frac{1}{2}-m;\frac{\cosh t-\cosh s}{1+\cosh t}\right)dt,
\end{aligned}
$$

$$N_l(\alpha) = \frac{2^{m-n}\sinh^{-m}\alpha}{\Gamma^{-2}(1/2-m)}G_l(\alpha)$$

$$-\int_0^a (\cosh\alpha - \cosh\tau)_2^{-m-1/2}\, {}_2F_1\left(\frac{n-m}{2}, -\frac{m+n}{2}; \frac{1}{2}-m; \frac{\cosh\alpha - \cosh\tau}{1+\cosh\alpha}\right)d\tau$$

$$\times\int_a^b r_l(t)\sinh^{m+1}(\cosh t - \cosh\tau)^{-m+1/2}$$

$$\times\, {}_2F_1\left(\frac{n-m}{2}, -\frac{m+n}{2}; \frac{1}{2}-m; \frac{\cosh t - \cosh\tau}{1+\cosh t}\right)dt.$$

Using the solution of the integral equation [4]:

$$\int_a^x \phi(t)[r(x) - \psi(t)]^{-k}\, {}_2F_1\left(p, q; s; \frac{r(x) - \psi(t)}{r(x) + d}\right)dt = \Psi(x), \tag{6}$$

with $k = m + \frac{1}{2}$, $p = \frac{n-m}{2}$, $q = -\frac{m+n}{2}$, $s = \frac{1}{2} - m$, $a = 0$, $d = 1$, $r = \cosh x$, and $\psi(t) = \cosh t$, we bring (4) to the form

$$M_l(s) = \frac{\Gamma^{-1}(1/2-m)}{\Gamma(1/2+m)}\frac{d}{ds}\left[(\cosh s + 1)^{(n-m)/2}\int_a^s (\cosh s - \cosh\alpha)^{m-1/2}\right. \tag{7}$$

$$\times(\cosh\alpha + 1)^{(m-n)/2}\, {}_2F_1\left(\frac{m-n}{2}, \frac{1+m-n}{2}; \frac{1}{2}+m; \frac{\cosh s - \cosh\alpha}{1+\cosh s}\right)$$

$$\left.\times N_l(\alpha)\sinh\alpha\, d\alpha\right]. \tag{8}$$

Thus, we define $r_l(t)$ by

$$r_l(t) = \frac{\Gamma^{-1}(1/2-m)\sinh^{-m-1}t}{\Gamma(1/2+m)}$$

$$\times\frac{d}{dt}\left[(\cosh t + 1)^{(n-m)/2}\int_t^b (\cosh s - \cosh t)^{m-1/2}(\cosh s + 1)^{(m-n)/2}\right.$$

$$\left.\times\, {}_2F_1\left(\frac{m-n}{2}, \frac{1+m-n}{2}; \frac{1}{2}+m; \frac{\cosh s - \cosh\alpha}{1+\cosh t}\right)M_l(s)\sinh s\, ds\right].$$

On the other hand, $r_l(t)$ can be written in the form

$$r_l(t) = -\Gamma^{-1}\left(\frac{1}{2}-m\right)\Gamma^{-1}\left(\frac{1}{2}+m\right)\sinh^{-m-1}t\,\frac{dH_l}{dt}, \tag{9}$$

where

$$H_l(t) = (\cosh t + 1)^{(n-m)/2}\int_t^b (\cosh r - \cosh t)^{m-1/2}(\cosh r + 1)^{(m-n)/2}$$

$$\times\, {}_2F_1\left(\frac{m-n}{2}, \frac{1+m-n}{2}; \frac{1}{2}+m; \frac{\cosh r - \cosh t}{1+\cosh t}\right)M_l(r)\sinh r\, dr. \tag{10}$$

After some unwieldy transformations, from (5)–(8) we obtain a Fredholm integral equation of the second kind with respect to $M_l(s)$:

$$M_l(s) = \frac{\Gamma(1/2 - m)}{\Gamma(1/2 + m)} L_l(s) - \frac{\Gamma^{-2}(1/2 - m)}{\Gamma^2(1/2 + m)} \int_a^b M_l(r) K(s, r) dr, \qquad (11)$$

where

$$
\begin{aligned}
K(s, r) &= (\cosh r + 1)^{(m-n)/2} \sinh r \\
&\quad \times \int_0^a S(\tau, r) \frac{d}{ds}\left[(\cosh s + 1)^{(n-m)/2} \int_a^s (\cosh s - \cosh \alpha)^{m-1/2} \right. \\
&\quad \times {}_2F_1\left(\frac{n-m}{2}, -\frac{m+n}{2}; \frac{1}{2} - m; \frac{\cosh \alpha - \cosh \tau}{1 + \cosh \tau} \right) \\
&\quad \left. \times (\cosh \alpha + 1)^{(m-n)/2} (\cosh \alpha - \cosh \tau)^{-m-1/2} \sinh \alpha \, d\alpha \right] d\tau. \quad (12)
\end{aligned}
$$

From (6) we now find $r_l(\alpha)$; then, applying the inverse formula of the Mehler-Fock integral transform, we immediately obtain

$$\sum_{k=1}^{3} a_{lk}(\tau)\phi_k(\tau) = \int_a^b r_l(t) \sinh t P^{m,n}_{-1/2+i\tau}(\cosh t) dt. \qquad (13)$$

We remark that (11) is a solvable Fredholm integral equation if the functions $f_l(\alpha), g_l(\alpha), h_l(\alpha)$ satisfy the conditions

$$f_l(\alpha) \in L_2(0, a), \begin{cases} f_l(\alpha) = O(\alpha^{-1/2-m+\beta_1}), & \alpha \to 0, \\ f_l(\alpha) = O((a-\alpha)^{-1/2-m+\beta_2}), & \alpha \to a, \end{cases} \quad \beta_1 > 0, \beta_2 > 0;$$

$$g_l(\alpha) \in L_2(a, b), \begin{cases} g_l(\alpha) = O((\alpha-a)^{-1/2-m+\gamma_1}), & \alpha \to a, \\ g_l(\alpha) = O((b-\alpha)^{-1/2-m+\gamma_2}), & \alpha \to b, \end{cases} \quad \gamma_1 > 0, \gamma_2 > 0;$$

$$h_l(\alpha) \in L_2(b, \infty), h_l(\alpha) = O\left((\alpha-b)^{-1/2-m+\delta_1}\right), \quad \alpha \to b, \delta_1 > 0.$$

4. A dual equation example

Let us consider the system of dual integral equations with Whittaker's function $W_{k,i\tau}$:

$$
\begin{aligned}
\int_0^\infty \omega(k, \tau) W_{k,i\tau}(x) A(\tau) \Phi(\tau) \, d\tau &= F(x), & 0 < x < a, \\
\int_0^\infty \omega(k, \tau) W_{k+\mu,i\tau}(x) A(\tau) \Phi(\tau) \, d\tau &= G(x), & x > a.
\end{aligned}
\qquad (14)
$$

The solution of the system (14) has the following closed form:

$$
\begin{aligned}
\Phi(\tau) = \pi^2 A^{-1}(\tau) \left\{ \int_0^a \frac{1}{x^2} W_{k,i\tau}(x) F(x) dx \right. \\
\left. + \int_a^\infty e^{x/2} x^{-1-k-\mu} W_{k,i\tau}(x) I_x^\mu [y^{k-1} e^{-y/2} G(y)] dx \right\}, \qquad (15)
\end{aligned}
$$

where $I_x^\mu[f]$ is the Weyl fractional integral operator.

353

5. A triple equation example

Let us write the solution of the system of triple integral equations with Watson's functions $\omega_{\nu,\lambda}(x) = 2\sqrt{x} \int_0^\infty J_\nu(t) J_\lambda\left(\frac{x}{t}\right) t^{-1} dt$ in the kernel:

$$
\begin{aligned}
S^\alpha_{\nu-\mu,\lambda-\mu}[A\Phi(x)] &= F_1(x), \ x \in (0, a), \\
S_{\nu,\lambda}[B\Phi(x)] &= K_2(x), \ x \in (a, b), \\
S^\beta_{\nu+\gamma,\lambda+\gamma}[C\Phi(x)] &= G_3(x), \ x \in (b, \infty),
\end{aligned}
\tag{16}
$$

where $S^\alpha_{\nu-\mu,\lambda-\mu}, S^\alpha_{\nu,\lambda}, S^\beta_{\nu+\gamma,\lambda+\gamma}$ are generalized Hankel integral operators:

$$
\begin{aligned}
S^\alpha_{\nu,\lambda}[f(x)] &= x^\alpha S_{\nu,\lambda}[x^\alpha f(x)], \\
S_{\nu,\lambda}[f(x)] &= \int_0^\infty \omega_{\nu,\lambda}(x\tau) f(\tau) d\tau.
\end{aligned}
$$

The solution of (16) can be reduced to a system of Fredholm equations of the second kind with the restrictions

$$
\nu > -\frac{1}{2} + \max\left\{\mu, -\gamma, \frac{\mu-\alpha}{2}, 2\mu - 2\alpha - 1\right\}, \ \lambda > -\frac{1}{2} + \max\left\{\mu, -\gamma, \mu - \alpha\right\},
$$

$$
|\mu + \alpha| < 2, |\gamma + \beta| < 2, 0 < \gamma - \beta < 1, 0 < \mu - \alpha < 1.
$$

References

1. N. Virchenko, *Dual (triple) integral equations*, Vyshcha Shkola, Kyiv, 1989.

2. L. Kuipers and B. Meulenbeld, On a generalization of Legendre's associated differential equation, *Proc. Koninkl. Nederl. Akad. Wet.* **60**, 436–443.

3. F. Götze, Verallgemeinerung einer Integraltransformation von Mehler–Fock durch den von Kuipers und Meulenbeld eingefuhrten Kern $P_k^{m,n}(z)$, *Proc. Koninkl. Nederl. Acad. Wet.* **68**, 396–404.

4. N. Virchenko, Integral equation with the hypergeometric function in the kernel, *Dokl. Acad. Sci. Ukraine*, 1984, no. 9, 3–5.

Department of Physics and Mathematics, National Technical University of Ukraine, Kyiv, Ukraine

G. YU and A. NARAIN

Computational simulations and flow domain classification for laminar/laminar annular/stratified condensing flows

1. Introduction

The proposed two-dimensional numerical solution scheme for internal and external flows solves the full Navier-Stokes equations in each phase, and locates the phase change interface while satisfying all the conditions at the interface, inlet, outlet, and the walls. It is used to solve classical smooth-interface external film condensation problems and gives results [8] in excellent agreement with classical solutions [12–15]. The research results show that conditions at the exit of an internal condensing duct flow, unlike single phase or other noncondensing gas-liquid incompressible flows, often significantly influence the flow upstream and are responsible for determining the nature of wavy-interface quasi-steady flows.

Fig. 1 shows a typical configuration of condensing flows in an inclined channel. Film condensation occurs at the lower plate of the channel. A dry, slightly superheated condition is assumed for the upper plate of the channel. We assume the inlet vapor is at saturation temperature (or slightly superheated), and both the vapor and the condensate flows are laminar. The flow rates are assumed to be small enough to allow the assumption of a smooth or nearly smooth interface [1].

2. Governing equations and nondimensionalization

We denote the liquid and vapor phases in the flow (see Fig. 1) by a subscript I: $I = 1$ for liquid and $I = 2$ for vapor. The fluid properties (density ρ, viscosity μ, specific heat C_p, and thermal conductivity k, each with subscript I) are assumed to take their representative constant values for each phase ($I = 1$ or 2). Let $\mathcal{T}_I$ be the mean temperature fields, p_I be the mean pressure fields, $\mathcal{T}_s(p)$ be the saturation temperature of the vapor as a function of the pressure p, Δ be the mean film thickness, $\dot{\mathcal{M}}$ be the local interfacial mass transfer rate per unit area, $\mathcal{T}_w(x)(< \mathcal{T}_s(p))$ be the known temperature variation of the bottom plate, $\mathbf{v}_I = \mathcal{U}_I\mathbf{i} + \mathcal{V}_I\mathbf{j}$ be the mean steady velocity fields. (If $\mathcal{T}_w(x)$ is not given, the correct value can be iteratively computed, using appropriate information about the coolant flow, by solving the *conjugate* problem dealing with conduction in the bottom plate and convection to the coolant flow underneath.) The distinction between mean and actual flow variables vanishes for smooth interface steady laminar flows. Furthermore, let h be a characteristic length (the channel height in this case) for the flow, g_x and g_y be the components of gravity along x and y axes, p_0 be the inlet pressure, $\Delta\mathcal{T} \equiv \mathcal{T}_s(p_0) - \mathcal{T}_w(0)$ be a representative controlling temperature difference between the vapor and the bottom plate, $h_{fg}^0 \equiv h_g - h_f$ be the latent heat of vaporization at the inlet temperature $\mathcal{T}_s(p_0)$, and U be the average inlet vapor speed determined by the inlet mass flux. For the results given in this paper, all the above-mentioned property values for relevant fluids (R-113, etc.) were obtained from the ASHRAE Handbook, SI Edition [2]. With $(\mathcal{X}, \mathcal{Y})$ representing physical distances of a point with respect to the axes in Fig. 1 ($\mathcal{X} = 0$ is at the inlet and $\mathcal{Y} = 0$ is at the condensing surface), we nondimensionalize the variables as

$$\begin{aligned}
\{\mathcal{X}, \mathcal{Y}, \Delta, \mathcal{U}_1, \dot{\mathcal{M}}\} &\equiv \{hx, hy, h\delta, Uu_1, U\dot{m}\}, \\
\{\mathcal{V}_I, \mathcal{T}_I, p_I\} &\equiv \{Uv_I, (\Delta\mathcal{T})\theta_I, p_0 + \rho_I U^2 \pi_I\}.
\end{aligned} \tag{1}$$

The nondimensional differential forms of mass, momentum (x and y components), and energy equations for steady flow in the interior of either of the phases are

$$\frac{\partial u_I}{\partial x} + \frac{\partial v_I}{\partial y} = 0, \qquad u_I \frac{\partial \theta_I}{\partial x} + v_I \frac{\partial \theta_I}{\partial y} \approx \frac{1}{\mathrm{Re}_I \mathrm{Pr}_I}\left(\frac{\partial^2 \theta_I}{\partial x^2} + \frac{\partial \theta_I}{\partial y^2}\right),$$

$$u_I \frac{\partial u_I}{\partial x} + v_I \frac{\partial u_I}{\partial y} = -\left(\frac{\partial \pi_I}{\partial x}\right) + (Fr_x)^{-1} + \frac{1}{\mathrm{Re}_I}\left(\frac{\partial^2 u_I}{\partial x^2} + \frac{\partial^2 u_I}{\partial y^2}\right), \qquad (2)$$

$$u_I \frac{\partial v_I}{\partial x} + v_I \frac{\partial v_I}{\partial y} = -\left(\frac{\partial \pi_I}{\partial y}\right) + (Fr_y)^{-1} + \frac{1}{\mathrm{Re}_I}\left(\frac{\partial^2 v_I}{\partial x^2} + \frac{\partial^2 v_I}{\partial y^2}\right),$$

where $\mathrm{Re}_I \equiv \rho_I U h / \mu_I$, $\mathrm{Pr}_I \equiv \mu_I C p_I / k_I$, $(Fr_x)^{-1} \equiv g_x h / U^2$, and $(Fr_x)^{-1} \equiv g_y h / U^2$.

After scaling approximations [3], the more exact interface conditions ([4,5]) for condensing flows reduce to the simpler form [6] appropriate for the flows considered here. The interface conditions [7–9] that apply at $y = \delta(x)$, under gentle slope and other approximations ($\delta'(x)^2 \ll 1$, etc.) for locations away from $x \sim 0$, are

$$u_1(x, \delta(X)) = u_2(x, \delta(x)) \equiv u_f(x), \qquad \pi_1(x, \delta(x)) \approx \tfrac{\rho_2}{\rho_1} \pi_2(x, \delta(x)),$$

$$\frac{\partial u_1}{\partial y}(x, \delta(x)) = \frac{\mu_2}{\mu_1}\frac{\partial u_2}{\partial y}(x, \delta(x)) \equiv \tau_v^i,$$

$$\dot{m} \equiv \frac{\rho_2}{\rho_1}\left[u_2(x, \delta(x))\frac{d\delta}{dx} - v_2(x, \delta(x))\right] = \left[u_1(x, \delta(x))\frac{d\delta}{dx} - v_1(x, \delta)\right], \qquad (3)$$

$$\theta_1(x, \delta(x)) = \theta_2(x, \delta(x)) = \tfrac{1}{\Delta T}\tau_S(p^i) \equiv \theta_S(\pi(x)),$$

$$\dot{m}(x) \approx \frac{Ja}{\mathrm{Re}_1 \mathrm{Pr}_1}\left\{\frac{\partial \theta_1}{\partial y}(x, \delta(x)) - \frac{k_2}{k_1}\frac{\partial \theta_2}{\partial y}(x, \delta(x))\right\},$$

where $\tau_v^i \equiv (1/(\mu_1 U/h))(T_v^i)_{xy}$ represents the *nondimensional* value of interfacial shear $(T_v^i)_{xy}$, and $Ja \equiv C_{p_1}\Delta T / h_{fg}^0$.

The wall conditions, inlet conditions (e.g., $u_2(0, y) = U$, $v_2(0, y) = 0$, etc.), and appropriate exit conditions are specified for each flow situation in accord with established practice and each of these conditions are computationally handled in standard ways.

All of the condensing flow problems solved in this paper assume that the vapor temperatures at the inlet, outlet, and dry walls (if present) are close to the saturation temperature $\mathcal{T}_s(p_0)$. This makes vapor temperature nearly a constant ($\approx \mathcal{T}_s(p_0)$) at all points, and therefore the energy equation for the interior of the vapor is automatically satisfied.

An inspection of all the nondimensional governing equations, interface conditions, and boundary conditions reveal the fact that the flows considered here are affected by the following set of nondimensional parameters:

$$\left\{\mathrm{Re}_{\mathrm{in}}, \mathrm{Pr}_1, Ja, (Fr_x)^{-1}, Z_e; (Fr_y)^{-1}, \frac{\rho_2}{\rho_1}, \frac{\mu_2}{\mu_1}\right\} \quad \text{where } Z_e \equiv \int_{\delta(x_e)}^{1} u_2(x_e, y)dy; \qquad (4)$$

Z_e is the ratio of exit vapor mass flow rate to total mass flow rate, $\mathrm{Re}_{\mathrm{in}} \equiv \rho_2 U h / \mu_2 \equiv \mathrm{Re}_2$, and the definitions of other parameters are same as in (2) and (3). The normal gravity parameter $(Fr_y)^{-1}$ can be ignored for smooth interface condensing flows but it is important for wavy-interface situations where it—along with a surface-tension parameter—plays a more significant role of providing restoring forces for the transverse interface oscillations [10].

356

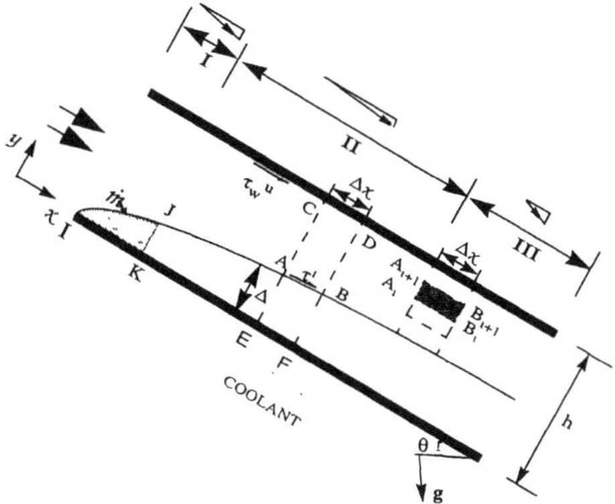

Fig. 1 Configuration of condensing flows in an inclined channel ($0^\circ \leq \theta \leq 90^\circ$)

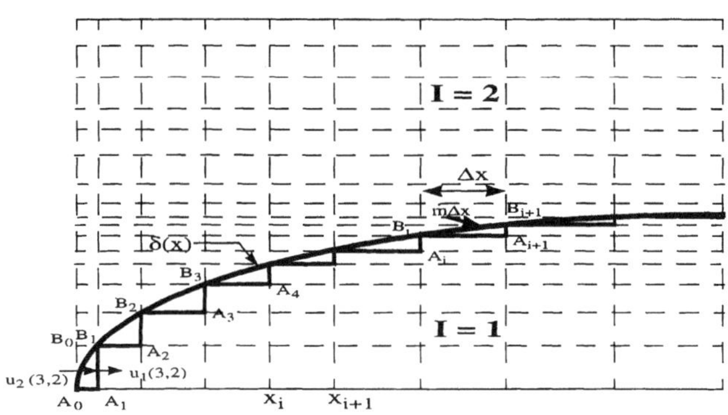

Fig. 2 Sketch depicting the grid and the notations used.

3. Numerical approaches

At each iteration, the computational domains for each phase ($I = 1$ or 2) is defined and grid is generated in the manner shown in Fig. 2. The grid is nonuniform. The smooth interface can be replaced by a stair-step type surface. The grid generation technique allows us to choose a denser grid near the flat part of the interface. A control volume method [11] has been chosen to develop the desired code. Detailed computational strategy are given in Yu's thesis [8] and a forthcoming paper [9].

357

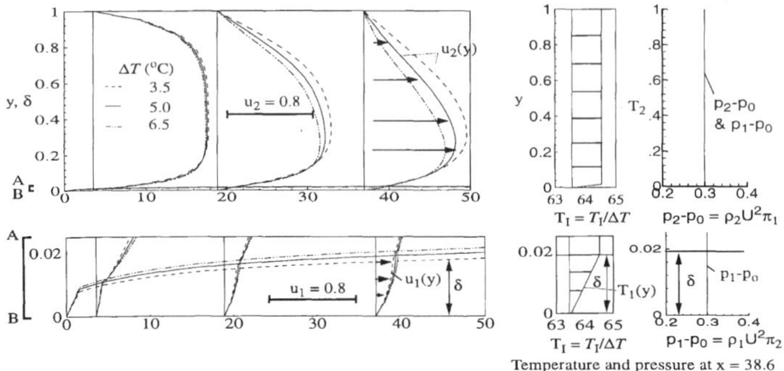

Fig. 3: Effect of changes in ΔT. The above predictions are for a vertical ($\alpha = 90^\circ$) flow of R-113 specified by: $T_s(p_0) = 322.6$ K, $U = 0.41$ m/s, $x_e = 46.685$, and h = 0.004 m. The exit quality for $\Delta T = 3.5$ °C, 5 °C, and 6.5 °C are respectively $Z_e = 0.5824$, 0.4587, and 0.3450.

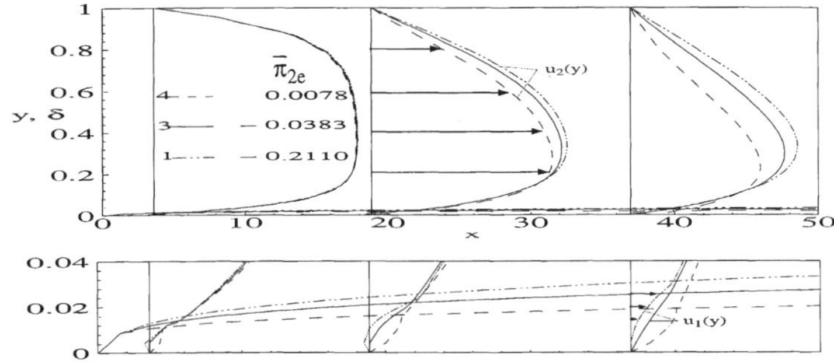

Fig. 4: Effect of different exit prescriptions of vapor quality Z_e. The flows are same as in Fig. 4 and $\bar{\pi}_{2e}$ values respectively correspond to Z_e values of 0.6645, 0.4388, and 0.3673. We expect some of these solutions (particularly curves 1 and 3 for the liquid velocity) are likely to be unstable in the sense that smooth-interface solutions will be replaced by quasi-steady wavy-interface situations.

4. Simulation results and discussion

Some computation results for uniform wall temperature $\mathcal{T}_w$ cases are shown in Fig. 3 and 4. Simulation results show that a typical flow could consist of three zones with regard to the impact of downstream conditions on the values of flow variables at an upstream location. Starting from the inlet, these zones (I, II, and III in Fig. 1) are respectively termed *singularly elliptic, parabolic,* and *elliptic.* However, in Fig. 3 and 4, parabolic zone II is absent because inlet speeds U are small.

358

5. Conclusions

- We proposed and successfully implemented a new two-dimensional computational simulation capability for condensing flows.

- Within the theoretical/computational framework giving Fig. 3 and 4, we showed, for the first time, the role and significance of exit conditions for internal condensing flows.

- This simulation capability has been used to propose a nearly exactly interfacial shear model which provides the framework for subsequent modifications applicable to wavy-interface flows [8].

References

1. Q. Lu, An experimental study of condensing heat transfer with film condensation in a horizontal rectangular duct, Ph.D. thesis, ME-EM Department, Michigan Technological University, 1992.

2. *ASHRAE Handbook*, Fundamentals SI Edition, American Society of Heating, Refrigeration and Air-Conditioning Engineers, Atlanta, GA, 1985.

3. I.Y. Chen and G. Kocamustafaogullari, *Internat. J. Heat Mass Transfer* **30** (1987), 1133–1148.

4. J.M. Delhaye, *Internat. J. Multiphase Flow* **1** (1974), 395–409.

5. A. Narain and Y. Kizilyalli, *Internat. J. Non-Linear Mech.* **26**, 501–520.

6. V.P. Carey, *Liquid-vapor phase-change phenomena*, Hemisphere, Washington, DC, 1992.

7. G. Yu and A. Narain, Simulation and interfacial shear for annular/stratified condensing flows, in *FED Conference Volume*, Internat. Mech. Engrg. Congress and Explosion, Dallas, TX, 1997.

8. G. Yu, Development of a CFD code for computational simulations and flow physics of annular/stratified film condensation flows, Ph.D. thesis, ME-EM Department, Michigan Technological University, 1999.

9. G. Yu and A. Narain, Computational simulation and flow physics for stratified/annular condensing flows: smooth-interface laminar/laminar considerations, *J. Fluid Mech.* (to appear).

10. Q. Liu, Computational simulation and interfacial shear models for wavy annular condensing downward flows in a vertical pipe: turbulent vapor and laminar condensate, Ph.D. thesis, ME-EM Department, Michigan Technological University, 1999.

11. S.V. Patankar, *Numerical heat transfer and fluid flow*, Hemisphere, Washington, DC, 1980.

12. T. Fujii and H. Uehara, *Internat. J. Heat Mass Transfer* **15** (1972), 217–233.

13. J.C.Y. Koh, E.M. Sparrow and J.P. Hartnett, *Internat. J. Heat Mass Transfer*
2 (1961), 69–82.

14. A. Narain, G. Yu and Q. Liu, *Inernat. J. Heat Mass Transfer* **40** (1997), 3559–
3575.

15. W. Nusselt, Die Oberflächenkondesation des Wasserdampfes, *Z. Ver. Dt. Ing.*
60 (1916), 541–546.

Department of Mechanical Engineering–Engineering Mechanics,
Michigan Technological University, Houghton, MI 49931, USA